Daniel Schoerling • Alexander V. Zlobin
Editors

Nb_3Sn Accelerator Magnets

Designs, Technologies and Performance

Editors
Daniel Schoerling
CERN (European Organization for Nuclear Research)
Meyrin, Genève, Switzerland

Alexander V. Zlobin
Fermi National Accelerator Laboratory (FNAL)
Batavia, IL, USA

https://doi.org/10.1007/978-3-030-16118-7

Foreword

Colliders of highly energetic particle beams are a crucial tool for fundamental research in high-energy physics (HEP), allowing for the investigation of highest mass particles and smallest length scales. Accelerator magnets are essential for steering and focusing such particle beams. The development and the practical implementation of superconducting (SC) accelerator magnets, in particular dipoles and quadrupoles, for the Fermilab Tevatron in the 1970s and 1980s enabled a breakthrough jump in technology and allowed for hitherto unprecedented particle-beam energies and collision rates. The Large Hadron Collider (LHC, in operation since 2008) at the European Organization for Nuclear Research (CERN) represents the current state of the art of large SC colliders. At present, CERN is preparing the high-luminosity LHC (HL-LHC) upgrade to increase the collision rate even further and to fully exploit the LHC potential.

For the post-LHC era, various colliders are under study, including linear lepton colliders (Compact Linear Collider (CLIC) and International Linear Collider (ILC)), and circular colliders (for electron–positron and proton–proton collisions). At CERN, the long-term goal of the Future Circular Collider (FCC) Study is to push the energy frontier much beyond other proposed accelerators, so as to increase the discovery reach, in energy, by an order of magnitude with respect to LHC in an affordable and energy-efficient manner. Two FCC options are currently under study, and, depending on the available time span, they can possibly be housed, successively, in the same tunnel, as it had been the case for the Large Electron–Positron Collider (LEP) and LHC at CERN. The FCC hadron collider as second stage would provide a unique opportunity to probe the nature at the smallest distance scales ever explored by mankind; to discover, if existent, new particles with exceedingly tiny Compton wavelengths; to thoroughly examine the dynamics of electroweak symmetry breaking; and to test the fundamental principles that have guided progress for decades.

To enable highest energy hadron colliders, new, reliable, and cost-effective magnet technologies are indispensable. Currently, only Nb_3Sn SC seem to be technically and commercially mature enough to be considered as candidate material for the magnets of such a future collider, to be constructed in the coming decades. Although work on Nb_3Sn magnets started already in the 1960s, a significant effort is still required to optimize both the SC material and the magnet designs to prepare for mass production. A major milestone will be the first-time implementation of Nb_3Sn dipole and quadrupole accelerator magnets in the HL-LHC. In parallel, a worldwide conductor and magnet R&D program has been launched, toward the challenging design goals for the FCC. This global effort is strongly supported by the FCC Study, the EuroCirCol Design Study co-funded by the European Commission, and the U.S. Magnet Development Program (MDP).

This book provides a critical review of the existing worldwide experience in the area of Nb_3Sn dipole magnets and will play a vital role in supporting this truly global effort toward the next generation of SC high-field accelerator magnets.

Genève, Switzerland

Michael Benedikt
FCC Study Leader
Michael.Benedikt@cern.ch

Preface

The goal of this book is to summarize and review the vast experience with Nb_3Sn accelerator dipole magnets accumulated in the United States, Europe, and Asia since the discovery and production of Nb_3Sn composite conductors. Interest in Nb_3Sn accelerator magnets is soaring, and their further development is rapidly gaining momentum worldwide, thanks to the growing maturity of this technology and its great potential for particle accelerators used in high-energy physics. This book is intended to contribute to the transfer of the accumulated experience with the design, technology, and performance of such magnets in view of the challenging requirements set by the needs for ever-higher collision energies in future colliders. Engineers and physicists working in the field of particle accelerators, as well as students studying courses in particle accelerator physics and technologies, may find it an indispensable source of information on Nb_3Sn accelerator magnets. Readers with a general interest in the history of science and technology may also find useful information that was obtained over a long period of time, from the late 1960s to the present day.

The book contains 16 chapters, structured within 5 sections.

The first section includes three introductory chapters. It starts with a brief description of the general problems of accelerator magnet design and operation (Chap. 1), followed by historical overviews of the research and development (R&D) of Nb_3Sn wires and cables for accelerator magnets (Chap. 2), and the early period of Nb_3Sn magnet R&D—a time during which this technology was competing with Nb-Ti magnets in the same field range (Chap. 3). It took almost 25 years (1965–1990) to advance the performance of Nb_3Sn accelerator magnets to fields above 10 T—a field range beyond the limits of Nb-Ti accelerator magnets.

The next three sections describe the period from the early 1990s to the present day. This period is characterized by the appearance of powerful numerical computer programs for the electromagnetic, mechanical, and thermal analysis of superconducting magnets, advanced superconducting and structural materials and fabrication techniques, and significant progress in magnet instrumentation and test methods. The great progress in these areas allowed significant advances in the magnet design process, improving the magnets' operating parameters and deepening the understanding of their performance. A key result of this progress is that the maximum field in Nb_3Sn accelerator magnets has approached at the present time 15 T. In this period, three main dipole designs (cos-theta, block type, and common coil) were thoroughly explored. Their design features, technologies, and performances are described in detail in Chaps. 4, 5, 6, 7, 8, and 9 (cos-theta), Chaps. 10, 11, and 12 (block type), and Chaps. 13, 14, and 15 (common coil). In each of the three sections, the chapters follow chronological order to demonstrate the progress made within each design approach. The structure of the material presented in the chapters follows the main theme of the book: magnet design, technology, and performance. This approach is used to ease the finding of appropriate information inside each chapter and to simplify the comparison of similar data presented in the various chapters.

The last section of the book outlines the future needs and the target parameters of the next generation of Nb_3Sn accelerator magnets and briefly summarizes the main open issues for their design and performance (Chap. 16). One of the main challenges in the next decade will be increasing the nominal operation field in accelerator magnets towards 16 T. To provide sufficient operation margin, it will require raising the magnet maximum field above 18 T and approaching the limit of the Nb_3Sn accelerator magnet technology. New cost-effective materials and technology affordable for the next generation of particle accelerators have to be also developed. The discussion presented in this session could be considered also as an invitation to the reader to take part in this new, exciting R&D phase of Nb_3Sn accelerator magnet technologies.

Genève, Switzerland Daniel Schoerling
Batavia, IL, USA Alexander V. Zlobin

Acknowledgments

The editors and authors thank the forefront experts from the research and development (R&D) projects, programs, and fields treated here for openly sharing with us their work and their enthusiastic engagement in the preparation of this book. We also thank the many other colleagues who have helped us in finding material spread over the archives of the laboratories involved in this field over the last six decades and for providing their valuable insights and comments on the book's content, in particular Daniel Dietderich (LBNL), Michael Fields (B-OST), René Flükiger (University of Genève and CERN), Eugeny Yu. Klimenko (Kurchatov Institute), David C. Larbalestier (ASC-FSU), Peter Lee (ASC-FSU), Clément Lorin (CEA-Saclay), Alfred D. McInturff (LBNL), Jean-Michel Rifflet (CEA-Saclay), Tiina-Maria Salmi (UoT), William B. Sampson (BNL), Ronald M. Scanlan (LBNL), Manfred Thoener (B-EAS), Peter Wanderer (BNL), Akira Yamamoto (KEK), and Franz Zerobin (ELIN-UNION). We would also like to acknowledge the technical staff of BNL, CEA-Saclay, CERN, FNAL, KEK, LBNL, TAMU, and the University of Twente for their contributions to magnet design, fabrication, and testing. Most names are indicated in the corresponding references.

Our thanks are also due to the copy editors from Sunrise Setting for editing and proofreading the text, Simon-Niklas Scheuring (Dreamlead Pictures) for image processing and coloring, and Pierre-Jean François and his team (Intitek) for drawing and sketch preparation. We thank Jens Vigen (Head of the CERN library) for his efforts toward publishing this book as an open access publication and Salomé Rohr (CERN library) for her great support in finding and archiving the references. Special thanks also go to Springer Nature and its editorial staff, in particular Hisako Niko, who supported this project from the beginning, and their valuable help in the publication process. Without the large effort and patience of all these people, this book would have not been possible.

Contents

Glossary of Terms

Accelerator magnet Accelerator magnets are a component of particle accelerators used to act on the beam properties. They typically have to meet stringent requirements in terms of design, technologies, and performance to allow a reliable operation of the accelerator

Arc Portion of a ring accelerator occupied by a regular structure of dipole, quadrupole, sextupole and octupole magnets

ARL Accelerator Research Laboratory (ARL) at the Texas Agricultural and Mechanical (A&M) University

ASC-FSU Applied Superconductivity Center at the joint college of engineering of Florida A&M University (FAMU) and Florida State University (FSU)

Beam pipe Ultrahigh vacuum chamber in which the beam is being transported

Bi-2212 Bismuth strontium calcium copper oxide ($Bi_2Sr_2CaCu_2O_8$), a high-temperature superconductor

BICC Boundary-induced coupling currents

Block type Dipole magnet type based on racetrack coils with flared ends

BNL Brookhaven National Laboratory in Upton, Brookhaven, NY

BSCCO Bismuth strontium calcium copper oxide, a family of high-temperature superconductors of which Bi-2212 is one variant

CCT Canted-cosine-theta, a magnet type based on pairs of conductor layers wound and powered such that their transverse field components sum and axial (solenoidal) field components cancel. For dipoles, the single layers resemble tilted solenoids.

CEA Saclay Commissariat à l'énergie atomique et aux énergies alternatives (CEA) de Saclay (English: French Alternative Energies and Atomic Energy Commission at Saclay)

CERN European Organization for Nuclear Research

CLIC The Compact Linear Collider (CLIC) study is an international collaboration working on a concept for a machine to collide electrons and positrons (antielectrons) head on at energies up to several teraelectronvolts (TeV).

CLIQ Coupling Loss Induced Quench (CLIQ), a system allowing to bring a superconducting magnet rapidly to the normal-conducting state

Coldmass Assembly of superconducting magnet coils, a mechanical structure and a helium vessel

Collider Particle accelerator for acceleration of charged particles which are brought to collision

Common-Coil Dipole magnet type based primarily on flat racetrack coils which are common to both apertures (twin-aperture magnets only)

Copper-to-non-copper ratio Area ratio of the copper stabilizer to the non-copper in superconducting strands

Cos-theta A magnet type with a winding scheme following a cosine current distribution: for a current distribution following cos θ, where θ is the angle around the aperture, a dipolar field is generated; for a current distribution following cos 2θ, a quadrupolar field is generated and so on.

Critical surface Graph of the critical current density J_c as a function of the modulus of the magnetic flux density B and the operation temperature T

Curing Process during coil production which is performed after winding to glue together the windings of a coil

D20 Cos-theta Nb_3Sn dipole magnet designed, manufactured, and tested at LBNL

DESY Deutsches Elektronen-Synchrotron (English: German Electron Synchrotron)

ELIN ELIN-UNION at Weiz, Austria, was an Austrian electric company

FCC The Future Circular Collider (FCC) study develops options for potential high-energy frontier circular colliders at CERN for the post-LHC era

FEM The finite element method (FEM) is a numerical method for approximating the solution of differential equations describing problems of engineering and mathematical physics.

FNAL Fermi National Accelerator Laboratory

FRESCA2 Upgrade of the Facility for the Reception of Superconducting Cables (FRESCA2)

G-10 Grade G-10 is constructed from a continuous filament woven glass fabric with an epoxy resin binder. The epoxy resin is made from an epichlorohydrin/bisphenol A epoxy resin and contains no other halogenated compounds, except residuals from the manufacture of the base resin. This grade is not manufactured from a brominated epoxy resin and is not flame-retardant (NEMA Standards Publication LI 1-1998 (R2011), Specification Sheet – 21, NEMA Grade G-10)

HD Helmholtz dipole series built at LBNL

Heat treatment Process in which the precursors of Nb_3Sn are reacted and the Nb_3Sn phase forms

HEP High Energy Physics

HERA Hadron-Elektron-Ring-Anlage (HERA) (English: Hadron Electron Ring Facility) was a particle accelerator colliding leptons and protons at DESY, Germany. It was operated from 3 July 1983 until 29 September 2011

HFDA Series of cos-theta dipole magnets which were fabricated and tested at FNAL

HL-LHC The High-Luminosity LHC (HL-LHC) is an upgrade of the LHC to achieve instantaneous luminosities, a factor of five larger than the LHC nominal value

Hot spot temperature Hottest spot after a quench in a superconducting coil

IGC Intermagnetics General Corporation (IGC), a US company

ILC The International Linear Collider is an international endeavor aiming at building a machine to collide electrons and positrons (antielectrons) head on at energies of up to 500 gigaelectronvolts (GeV)

ISCC Interstrand Coupling Currents

ITER ITER ("the way" in Latin) is a project aiming to produce energy with fusion.

KEK Kō-enerugī kasokuki kenkyū kikō (English: The High Energy Accelerator Research Organization)

LBNL Lawrence Berkeley National Laboratory

LEP The Large Electron-Positron Collider was operated from 14 July 1989 until 2 November 2000 at CERN

LHC The Large Hadron Collider (LHC) is housed in the former LEP tunnel at CERN. It first started up on 10 September 2008

LHe Liquid helium

Magnet training Typical process in which superconducting magnets reach at an initial powering campaign after each quench a slightly higher current and magnetic field

Magnetic aperture Magnetic aperture of the magnet, contrary to the mechanical aperture, which is the minimum aperture of the storage ring

MBH Abbreviation nominating the 11 T dipole magnets used for replacing a regular Nb-Ti bending magnet in a dispersion suppressor region of LHC

MDP US Magnet Development Program, a US program to develop high-field magnets for future circular colliders

Mica tape Inorganic electric insulation material based on sheets of silicate minerals

MIIT Heat load of the normal zone of a quenching magnet. MIIT $\equiv 10^6$ A^2 s

Mirror coil Coil tested in a structure of iron which "mirrors" the missing coils to resemble the field distribution in the magnet

MJR Modified jelly-roll (MJR) process, a fabrication process of Nb_3Sn multifilamentary wires

MSUT Model Single of the University of Twente (MSUT) dipole

n-Value of a superconductor Exponent obtained in a specific range of electric field or resistivity when the voltage/current curve is approximated by $U = I^n$

OST Oxford Superconductor Technologies, a former US company which is now part of Bruker Corporation

Persistent currents Induced eddy currents in the superconductors which are persistent, due to the fact that there is no resistivity

PIT Powder-in-tube process, a fabrication process of Nb_3Sn multifilamentary wires

PMF Pressure Measurement Film

Quench Transition from the superconducting to the normal-conducting state

Quench heaters Heaters which are fired to bring a superconducting magnet rapidly to the normal-conducting state

RD Series of racetrack dipoles (RD) built at LBNL

REBCO $REBa_2Cu_3O_7$ (REBCO), where RE stands for rare earth element, a group of high-temperature superconductors

RHIC The Relativistic Heavy Ion Collider (RHIC) is a heavy-ion collider at BNL, USA. It first started up in 2000

RRR Residual resistivity ratio: the ratio of the electrical resistivity at 273 K to that at 4.2 K

Rutherford cable Multistrand flat or slightly keystoned (trapezoidal) two-layer cable being identical to the Roebel bar

S2 glass S2 glass is a special glass used for insulation consisting out of 65 wt% SiO_2, 25 wt% Al_2O_3, and 10 wt% MgO

Short sample limit The short sample limit is the theoretical current and field limit a superconducting magnet can reach, calculated based on test results in solenoidal background fields of short samples wound around normalized barrels which are heat treated together with the superconducting coils

SSC Superconducting Super Collider was a particle accelerator complex under construction in the vicinity of Waxahachie, Texas, aiming at reaching a collision energy of 40 TeV. The project was cancelled in 1993

Strand Composite wire containing superconducting filaments dispersed in a matrix with suitably small electrical resistivity properties

Synchrotron A synchrotron is a type of particle accelerators in which the magnetic field is synchronized to the beam energy, so that the particles travel on the same path while being accelerated

TAMU Series of block type magnets being developed at the Accelerator Research Laboratory (ARL) at the Texas Agricultural and Mechanical (A&M) University (TAMU)

Tevatron The Tevatron is a particle accelerator colliding protons and antiprotons at FNAL. It was operated from 3 July 1983 until 29 September 2011

Thermal cycle Cool down from room temperature (293 K) to cryogenic temperature (4.2 K or 1.9 K), heat back to room temperature (293 K), and cool down again to cryogenic temperature (4.2 K or 1.9 K) of a superconducting magnet

Transfer function Current-to-field correspondence in accelerator magnets

TWCA Teledyne Wah Chang Albany, a US company

Twin-aperture magnet A magnet housing two apertures in the same yoke

UNK Uskoritel'no Nakopitel'nyj Kompleks (UNK) (English: Accelerator and Storage Complex) was a particle accelerator complex under construction in Protvino, near Moscow, Russia, at the Institute for High Energy Physics, aiming at reaching a collision energy of 3 TeV. The project was cancelled

VLHC Very Large Hadron Collider (VLHC) study

Part I
Introduction

Chapter 1
Superconducting Magnets for Accelerators

Alexander V. Zlobin and Daniel Schoerling

Abstract Superconducting magnets have enabled great progress and multiple fundamental discoveries in the field of high-energy physics. This chapter reviews the use of superconducting magnets in particle accelerators, introduces Nb_3Sn superconducting accelerator magnets, and describes their main challenges.

1.1 Circular Accelerators and Superconducting Magnets

Circular accelerators are the most important tool of modern high-energy physics (HEP) for investigating the largest mass and the smallest space scales. A key element of a circular accelerator is its magnet system (Wolski 2014). The magnet system is composed of large number of various magnets, mainly dipoles and quadrupoles, to guide and steer the particle beams. The main function of the majority of the magnets (the so-called arc magnets, which are periodically placed along a ring) is to keep the beam on a quasi-circular orbit and confine them in a relatively small and well-defined volume inside a vacuum pipe. Magnets are also used to transfer beams between accelerator rings in so-called transfer lines, to match beam parameters from the transfer line into the injection insertions or into extraction lines and beam dumps, to direct or separate beams for the accelerating radio frequency cavities, and to focus beams for collision at the interaction points where the experiments reside.

One of the most important parameters of colliders is the beam energy, as it determines the physics discovery potential. The energy E in GeV of relativistic particles with a charge q in units of the electron charge in a circular accelerator is limited by the strength of the bending dipole magnets B in Tesla and the machine radius r in meters

---

A. V. Zlobin (✉)
Fermi National Accelerator Laboratory (FNAL), Batavia, IL, USA
e-mail: zlobin@fnal.gov

D. Schoerling
CERN (European Organization for Nuclear Research), Meyrin, Genève, Switzerland
e-mail: Daniel.Schoerling@cern.ch

D. Schoerling, A. V. Zlobin (eds.), *Nb_3Sn Accelerator Magnets*, Particle Acceleration and Detection, https://doi.org/10.1007/978-3-030-16118-7_1

$$E \approx 0.3qBr.$$

Thus, high magnetic fields are an efficient way towards higher energy machines for hadron and ion collisions.

The value of the magnetic field in a circular accelerator needs to be synchronized with the beam energy. It is achieved by using electromagnets that allow the field strength to be varied by changing the electric current in the coil. The maximum field of traditional electromagnets with copper or aluminum coils is limited, however, by Joule heating, which limits the current density in a magnet coil typically to ~10 A/mm^2.

In 1911, the Dutch physicist H. Kamerlingh-Onnes discovered the phenomenon of superconductivity—the vanishing of electrical resistance in some metals at very low (<10 K) temperatures (Wilson 2012). This discovery inspired him 2 years later to propose a 100,000 Gauss (10 T) solenoid based on a superconducting coil cooled with liquid helium. He believed that superconductivity would allow the current in a coil to be increased and, thus, a larger magnetic field to be generated. Yet, it took more than 50 years of hard work to realize this dream in practice.

The design and construction of superconducting magnets have become possible only after the discovery and development in the early 1960s of technical superconductors. Technical superconductors are defined as a class of superconducting materials that provide high current densities in the presence of high magnetic fields.

Earlier attempts to use technical superconductors in superconducting magnets failed due to premature magnet transitions to the normal state, called quenches. These quenches were caused by the abrupt movement of magnetic flux inside a superconductor, the so-called flux jump effect. The analysis of flux jumps led to the development of stability criteria for technical superconductors (Wilson, 1983; Rogalla and Kes 2012). Stability of the superconducting state with respect to small field or temperature perturbations can be guaranteed only if the superconductor transverse size does not exceed a maximum value proportional to the material's specific heat, and inversely proportional to its critical current density. For example, for a Nb-Ti superconductor at 5 T and 4.2 K the maximum filament size has to be smaller than 50 μm.

The other main concern was protection of the superconductor in the case of a quench. All superconductors in the normal state have high resistivity. In the case of a quench they are likely to be damaged by Joule heating due to the high current density they carry, if no adequate measures for protection are taken. To minimize heating after a quench, a superconductor should therefore be surrounded by a normal conductor with a low resistance.

The two abovementioned conditions (stability and protection) have led to the concept of composite superconductors, in which small superconducting filaments are embedded in a normal conducting matrix with low resistance and large thermal diffusivity. The matrix decreases the Joule heating when the superconductor becomes normal, conducts the heat away from the surface of the superconducting filaments thanks to its high thermal conductivity, and absorbs a substantial fraction of heat due to its high specific heat. To provide stability of a composite superconductor to flux jumps and reduce the eddy currents induced by varying external magnetic fields, the superconducting filaments are twisted along the conductor axis. Flux jumps not only limit the size of the filaments, but also the size of the multifilament composite wire due to self-field instabilities related to the non-uniform distribution of transport current inside the wire. For example, for a Nb-Ti composite wire at 5 T and 4.2 K, the maximum diameter is limited to ~2 mm.

A composite superconductor placed in a varying magnetic field becomes magnetized with two components: one is related to persistent currents in superconducting filaments, and the second is caused by coupling eddy currents between filaments. Both components are diamagnetic in an increasing field and paramagnetic in a decreasing field. The hysteretic behavior of wire magnetization leads to energy dissipation, also called alternating current (AC) losses. Likewise the magnetization, the AC loss power in a composite superconductor has two main components: one is related to persistent currents in superconducting filaments and the other one to coupling eddy currents in composite wires. The magnetization of composite wires plays an important role in superconducting accelerator magnets, which have demanding requirements on field uniformity. The AC losses are important for cryogenic cooling of superconducting coils during magnet operation and quench, and contribute to the heat load on a magnet's cooling system.

The critical current density J_c is a key parameter, which controls the current carrying capability, stability, magnetization, and AC losses of technical superconductors and, thus, the performance of superconducting magnets. As the resistive transition from superconducting to the normal conducting state in a composite superconductor is smooth, the definition of J_c is not straightforward (Warnes and Larbalestier 1987). The most commonly used criterion at the present time, for superconducting accelerator magnets in particular, defines J_c at the resistivity of $10^{-14}\ \Omega\,\mathrm{m}$.

Important features of practical materials for superconducting magnets include not only the appropriate combination of critical parameters, but also their reproducibility in long lengths, compliance with mass production, and affordable cost.

Large accelerator magnets use large, high-current cables to reduce the magnet's inductance, which is an important parameter for magnet protection during a quench. To achieve the required high current level, several strands are connected in parallel, and twisted or transposed in the axial direction (Wilson 1983). Multi-strand cables allow a reduction of the piece length requirement for wire manufacturing, the number of turns in a magnet coil, and also allow for current redistribution between strands in the case of a localized defect of some strands or quench. The Rutherford cable is the most widely used cable type in accelerator magnets (Gallagher-Daggit 1973). To adapt the cable design to the magnet design, it is produced with either a rectangular or a slightly keystoned cross-section. A Rutherford cable is composed of fully transposed twisted composite strands. Strand transposition reduces coupling losses and ensures uniform current distribution, also aided by the electrical contact between the strands. The cable critical current I_c is normally the sum of the strand's critical currents, which depend on wire degradation during cabling and current distribution in the cable cross-section. Due to strand coupling inside the cable, its magnetization and AC losses have additional eddy current components controlled by the interstrand resistance and the cable twist pitch.

1.2 Accelerator Magnet Design and Operation

1.2.1 Magnetic Design

The desired magnetic field in superconducting magnets is produced by a current I in a coil and is calculated using the Biot–Savart law

$$\vec{B} = \frac{\mu_0}{4\Pi} \int_C \frac{I \mathrm{d}\vec{l} \times \vec{r}}{r^3},$$

where Idl is a current element and r is the radius vector from the current element to the field point. The total field in a given point of a magnet is obtained by integrating the current elements over the coil volume.

A perfect dipole field can be generated by two infinite slabs, by two intersecting ellipses (cylinders) with uniform and equal currents of opposite direction, or by a cylinder with a cos-theta current density distribution (Fig. 1.1) (Brechna 1973). The field strength B in the dipole aperture is defined as

$$B \sim J(B) w,$$

where w is the slab or cylinder thickness.

To mimic the ideal current configurations, they are approximated with a number of cables arranged in blocks and separated by wedges (Russenschuck 2010). The block positions and cross-sections are optimized to approach the ideal coil cross-section and achieve the required field quality. In practice, in line with the three different types of dipolar configurations in Fig. 1.1, three coil types are used: block, shell, and cos-theta.

Coils of accelerator magnets are usually surrounded by an iron yoke, which serves as a return path for the magnetic flux and contributes to the magnet bore field. Three different classes of magnets can be identified based on the way in which the wanted field is achieved: (a) iron dominated or superferric magnets, in which the shape of iron poles determines the field pattern; (b) iron-free magnets, in which the field configuration is dominated by the coil shape; and (c) magnets, in which the field configuration is provided by both the coils and the iron yoke.

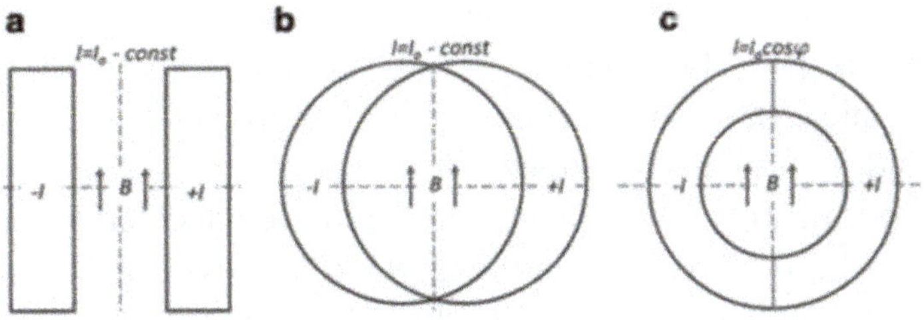

Fig. 1.1 Pure dipole configurations: (**a**) two infinite slabs; (**b**) intersecting ellipses; and (**c**) cylinder with cos-theta azimuthal current density distribution

1.2.2 Field Quality

Since real coils only approximate the ideal dipole cross-sections, the magnetic fields in accelerator magnets are not perfect. Excluding the coil ends, the magnetic field in the bore of long and slender accelerator magnets is two-dimensional and can be represented by the power series (Mess et al. 1996; Russenschuck 2010)

$$B_y + iB_x = B_1 \sum_{n=1}^{\infty} (b_n + ia_n) \left(\frac{x + iy}{R_{\text{ref}}} \right)^{n-1},$$

where B_x and B_y are the horizontal and vertical transverse field components, B_1 is the dipole field component, and b_n and a_n are the "normal" and "skew" n-pole coefficients, also called the field harmonics, at a reference radius R_{ref}. The reference radius is usually chosen at two-thirds of the magnet free aperture.

Due to the symmetry of the magnet's cross-section, only normal multipole coefficients allowed by symmetry are expected to be non-zero (Brechna 1973; Mess et al. 1996; Russenschuck 2010). These so-called allowed multipole coefficients can be minimized by iterating on the coil cross-section parameters. In addition, the magnetic properties of coil and structural materials, and geometrical errors produce non-allowed field harmonics. For instance, a top/bottom asymmetry in a dipole magnet produces a skew quadrupole a_2, while a left/right asymmetry produces a normal quadrupole b_2. These unwanted harmonics can be minimized by selecting the appropriate materials and improving the precision of coil and structural components, tooling, and assembly procedures.

At the magnet design phase, two classes of field errors are distinguished: systematic and random errors. The systematic errors include geometrical errors, which originate from the imperfections of the coil and the iron yoke cross-sections, as well as by iron yoke saturation and coil magnetization. The random errors are mainly due to random variations of the coil geometrical parameters (inner and outer coil radii, coil pole angles, wedge geometry and position, etc.).

To achieve accelerator field quality, the coil and the yoke cross-sections and their relative position typically must have an accuracy better than ~0.1 mm including coil deformations under the Lorentz forces. The errors from Lorentz forces vary during magnet excitation, and their level depends on the rigidity of the coil and magnet mechanical structure. The minimization of these errors is one of the key parts of mechanical structure optimization.

The random errors are typically estimated by means of Monte Carlo simulations imposing a random displacement of the coil blocks. Since these displacements do not respect the magnet's symmetry, they produce a whole spectrum of field harmonics. Based on the random error analysis, fabrication tolerances for the main coil and structural components, tooling, and assembly process are formulated, which must be provided to achieve the required accelerator field quality.

The iron yoke is saturated when the level of field in the iron exceeds 2 T. The relative magnetic permeability of saturated iron is dramatically reduced depending on the field level in each point of the yoke. As a result, the iron contribution is a non-linear function of the transport current. In a single-aperture magnet with a symmetrical iron yoke, the saturation effect is mainly observed in the magnet's main field B_1 (or magnet transfer function B_1/I) and in the normal sextupole b_3. The variation of sextupole field due to iron saturation is reduced by optimizing the iron inner and outer dimensions, and introducing special holes at appropriate places in the yoke. In a twin-aperture dipole, the central part of the yoke saturates before the outer parts, resulting in left/right asymmetries in the yoke contributions affecting the normal quadrupole b_2. The saturation effects in b_2 are of opposite sign in the two apertures. This effect is controlled by asymmetrical holes in the iron yoke and by choosing an inter-beam distance large enough to minimize these effects.

Three main components, which add to the field quality distortion from coil magnetization, are the persistent currents in the superconducting filaments; the eddy current between filaments within the strands; and the eddy currents between the strands of a cable, the so-called interstrand coupling currents. The field distortions produced by coil magnetization are most significant at low fields and become negligible at high fields. The field errors from coil magnetization change sign during current ramp up or down and, as a result, the main field and the allowed lower-order harmonics show ample hysteresis as a function of transport current and magnet excitation history.

The persistent currents in the filaments can be reduced by reducing their filament size. The eddy current component within the strands is controlled by the wire twist pitch, and the interstrand coupling currents depend on the interstrand resistance in the cable. The interstrand resistances should not be too large to allow current sharing among cable strands. If the cable's magnetic properties are uniform, only the main field and low-order allowed field harmonics (mainly b_3 and b_5) are disturbed. In the case of non-uniform magnetic properties, all lower-order harmonics will be affected.

1.2.3 Mechanical Design

The coil turns carrying a transport current in the magnetic field are exposed to electromagnetic (Lorentz) forces. The Lorentz force per unit length F/l of the conductor with current I in magnetic field B is

$$\vec{F}/_l = \vec{I} \times \vec{B}.$$

The force is directed perpendicularly to the current and field vectors. The value and distribution of forces inside the magnet coils, and the associated mechanical stress and deformations, depend on the magnet size and configuration, the value of the magnetic field, and the mechanical properties of the coil and the magnet structure. The analysis of mechanical forces, stresses, and deformations is a complex task, which is usually performed using finite element codes.

The Lorentz forces in superconducting accelerator magnets are very large (Mess et al. 1996; Ašner 1999). To stabilize the magnetic field characteristics in the operating field range and to reduce the probability of spontaneous quenches, it is necessary to ensure the mechanical stability of turns in the coil. Mechanical stability is achieved by applying a preload to the coil during magnet assembly and by supporting the compressed coil during operation with a rigid support structure. The required minimum preload value is determined by the magnet design, the level of the operating field, and the thermal contraction and mechanical rigidity of the structural materials. The allowed coil pre-stress is limited by the maximum stress that the coil can sustain before the superconductor starts degrading or before insulation damage. The preload applied to the magnet coils at room temperature has to be sufficient to compensate for the preload decrease due to coil creep after magnet assembly, differences in thermal contraction of the coil and structure, and coil deformations under Lorentz forces during magnet excitation.

The horizontal component of the Lorentz force bends the coil horizontally with a maximum displacement at the magnet mid-plane. This bending leads to additional coil stress and to a coil deformation that generates field distortions. The coil support against the horizontal component of the Lorentz force is provided by a special stiff support structure placed between the coil and the iron yoke. This structure is optimized to limit the horizontal coil deflections. To increase the field enhancement from the iron yoke, however, it is placed rather close to the coil, which limits the thickness and, thus, rigidity of the support structure. To compensate for this, the yoke and strong metallic shell outside the yoke or the helium vessel shell are also used as part of the coil support system.

The axial component of the Lorentz force stretches the coil axially, increasing stresses and turn displacements in the coil ends. To minimize these effects, it is essential to provide a stiff support (and even some initial axial preload to compensate for different thermal contraction of the magnet coils and mechanical structure) against the axial component of the Lorentz force using thick stainless-steel end plates welded to the shell or connected by thick axial rods.

1.2.4 Operation Temperature, Fields and Margins

Superconducting magnets are operated at temperatures well below the superconductor's critical temperature. Liquid helium, which has a boiling temperature of around 4.22 K at atmospheric pressure (Weisend 1998), is usually used for this purpose. At temperatures below 2.17 K, called the lambda point, liquid helium turns into a superfluid due to a phase transition. The superfluid phase has extremely high thermal conductivity and extremely low viscosity. This combination of properties is beneficial for the cooling of superconducting magnets as it allows the superfluid helium to penetrate into a porous coil and magnet structure, and to transfer heat from the magnet to a heat sink in a stagnant liquid superfluid helium bath.

The superconductor critical current density is a function of its temperature and the applied magnetic field. Figure 1.2 shows the dependence of the conductor critical

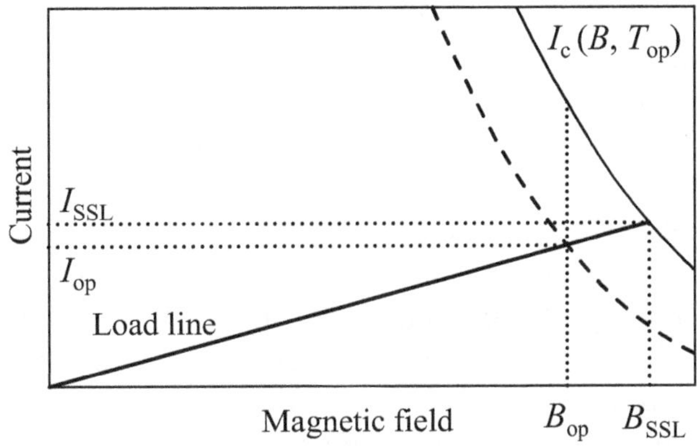

Fig. 1.2 Conductor I_c vs. magnetic field B and magnet load line

current, defined as the maximum transport current carried through the superconductor area vs. the magnetic field for a given operating temperature T_{op}. The magnet load line describing the dependence of the magnetic field in the coil and the magnet current is also shown. The intersection of these two functions defines the magnet's ultimate parameters such as the maximum field B_{SSL} and the maximum current I_{SSL}, also known as the magnet short sample limits (SSL).

To ensure stable performance, superconducting magnets operate at currents below the conductor critical current with some margin. Important magnet operation margins include the margin along the load line, defined as the ratio of the operation current to the magnet short sample limit, $i = I_{op}/I_{SSL}$, and the critical current margin, defined as the ratio of the operation current to the conductor critical current at B_{op} and T_{op}

$$i = I_{op}/I_c(B_{op}, T_{op}).$$

The magnet temperature margin ΔT, associated with the critical current margin, is defined as the difference between the operation temperature T_{op} and the current sharing temperature T_{cs} at which the operation current is equal to the critical current. Note that the critical current margin and the temperature margin are different for each turn, due to the field and related critical current variations inside the coil.

1.2.5 Magnet Thermal Stabilization

Superconducting magnet coils are exposed to a variety of disturbances such as conductor displacements during excitation, epoxy cracking, AC losses, splice heating, beam-induced heat deposition, fluctuations of the helium temperature in the cryogenic system, etc. These disturbances may raise the coil temperature above the superconductor critical temperature T_c, as the transport current is expelled into the low-resistivity matrix. The main goal of coil thermal stabilization is ensuring its ability to stay superconducting, or to recover superconductivity after a disturbance.

To achieve complete magnet thermal stability with respect to any perturbations, the cross-section of the normal stabilizer has to be rather large and well-cooled (Brechna 1973; Weisend 1998). Considerations related to cost and performance optimization, however, require reducing the stabilizer cross-section in superconducting accelerator magnets to a minimum, as dictated by the internal stability of the composite wires to flux jumps and by quench protection. In this case, complete magnet stability is not provided. Some improvement of magnet stability can be achieved by limiting the conductor motion in a coil by applying coil preload, by coil impregnation (e.g., with epoxy or other suitable agents), and by reducing friction on various conductor surfaces inside a magnet.

The perturbation spectrum is usually not known for superconducting magnets. In this case, magnet pre-conditioning plays an important role for releasing large perturbations accumulated during magnet fabrication. In early experiments with superconducting magnets it was found that the magnet maximum current gradually increases in consequent quenches. This process is known as magnet training. After a limited number of quenches a magnet often reaches a stable plateau. If the magnet quench current at the plateau is lower than the magnet short sample limit, it indicates a magnet current degradation. Magnets usually remember their training after warming-up and cooling-down cycles, or show only a very short retraining. This memory is extremely important for accelerator magnets, which are installed and connected in series in large-scale accelerators.

Despite the disadvantages associated with possible instabilities and training, partially stabilized accelerator magnets provide the highest engineering current densities in the coils and, thus, enable the most compact magnets and the highest fields at the lowest cost.

In the case of AC losses, radiation-induced heat deposition, splice heating, and other steady or pulsed heat losses, the perturbation power is known and the coil peak temperature T_{coil} is defined by

$$T_{\text{coil}} - T_{\text{op}} = \frac{q_{\text{v}}A}{hP}$$

for a steady heat deposition, or

$$W = \int_{T_{\text{op}}}^{T_{\text{coil}}} C_{\text{p}}(T)\text{d}T$$

for a pulsed heat deposition, where q_{v} is the average volumetric heat load, W is the heat pulse energy, A is the cable cross-sectional area, h is the heat transfer coefficient, P is the conductor cooling perimeter, and T_{op} is the operating temperature. To prevent magnet quenches, the temperature everywhere in the magnet coil should be kept below T_{c}. Magnet stability is ensured by choosing an appropriate temperature margin and coil cooling conditions.

1.2.6 Quench Protection

The stored energy in superconducting accelerator magnets is large. In the case of a quench, a large portion of this stored energy is typically dissipated in the relatively small coil volume and, thus, the coil is heated to quite high temperatures, which may damage or even destroy the coil. Hence, the magnet and conductor design and the protection system parameters have to be optimized to provide for reliable magnet protection during a quench and to limit the coil temperature, thermal stress, and voltages within acceptable values (Wilson 1983; Mess et al. 1996; Iwasa 2009).

The peak quench temperature T_{max} is usually estimated using an adiabatic approach by equating the Joule heat in the coil during the quench to the increase of the coil enthalpy

$$\int_0^{\infty} I(t)^2 \mathrm{d}t = \lambda S^2 \int_{T_q}^{T_{max}} \frac{C(T)}{\rho(B,T)} \mathrm{d}T$$

where $I(t)$ is the current decay after a quench, T_q is the conductor quench temperature, S is the total conductor cross-section, λ is the stabilizer fraction in the conductor cross-section, $C(T)$ is the average conductor specific heat, and $\rho\,(B,T)$ is the conductor normal resistivity. For a given value of the so-called quench integral usually measured in million ampere squared seconds (MIITS, 10^6 A^2 s), T_{max} can be minimized by reducing the matrix resistivity, or by increasing the conductor cross-section and the average specific heat. The value of the total quench integral is controlled by the quench detection and validation time, the protection circuit response, and the current decay time.

If the quench propagates rapidly inside a magnet, the stored energy can be practically uniformly dissipated in the magnet coil. The magnet will be self-protected if the enthalpy of the quenched part of the coil is sufficient to keep the magnet's maximum temperature below the acceptable level, usually 300–400 K. Quench protection of large accelerator magnets requires a special protection system based either on traditional resistive quench heaters or on the recently developed Coupling Loss Induced Quench (CLIQ) system (Ravaioli 2015). Both approaches provide a rapid quench of a large coil volume, sufficient to absorb the magnet storage energy without coil overheating.

Protection heaters are usually made of thin stainless-steel strips, which are placed on the coil surface. Since the heaters are electrically insulated from the coil, the heater electrical insulation introduces a thermal barrier. As a result, there is a delay between heater excitation and the coil quench. The heaters and their integration into the coil structure are optimized to reduce this delay. In the CLIQ system, capacitor banks are discharged over parts of the coil, inducing eddy currents that quench the coil quasi-instantaneously.

The natural quench propagation in the coil, as well as the heat transfer between the quenched coil and the magnet support structure, increases the effective magnet volume involved in the energy dissipation and allows the dissipation of some fraction of the stored energy outside the coil. It reduces the hot-spot temperature and the temperature under the quench heaters. In some cases, the magnet current decay is fast enough that eddy currents induced in the conductor may also quench superconducting parts of the coil. This phenomenon, known as the "quench-back" effect, evidently helps to reduce the peak coil quench temperature. Using a protective heating system also helps to provide a more balanced temperature and voltage distributions, and reduce the thermal and electrical stresses in the coil.

1.3 Nb-Ti Accelerator Magnets and Technologies

The work on Nb-Ti superconducting magnets for accelerators started in the late 1960s (Prodell 1968) and culminated from the 1980s to the 2000s in the construction of a series of large superconducting accelerators in the USA and in Europe (Edwards 1985; Meinke 1991; Anerella et al. 2003; Evans and Bryant 2008). These powerful superconducting accelerators have enabled several fundamental discoveries, including the charm and top quarks, the tau lepton, the gluon, the *Z* and *W* bosons and, most recently, the Higgs boson, which have shaped and confirmed our understanding of the particle physics Standard Model.

1.3.1 Nb-Ti Composite Wire

The superconducting Nb-Ti alloy, discovered in the 1960s, is the most successful practical superconductor. Since the 1970s, it has been the workhorse for superconducting accelerator magnets, thanks to its ductility, high current carrying capability, and the successful industrialization of the composite wires based on this material. Its critical temperature at zero magnetic field, T_{c0}, is 9.8 K; and the upper critical magnetic field at zero temperature, B_{c20}, is 14.5 T.

The Nb-Ti composite wires are manufactured by a co-drawing process with intermediate heat treatments to achieve an optimal critical current density. The final annealing provides a high residual resistivity ratio (RRR) for the copper stabilizer. A thin Nb barrier separates the Nb-Ti alloy from the copper to avoid the formation of a brittle CuTi intermetallic composite during the high-temperature extrusion and the annealing heat treatments.

Diameters of practical wires are in the range 0.1–3 mm, and piece lengths are of a few kilometers. The number of filaments in a wire may reach ~10^5, and the filament diameter is typically 5–20 μm (the practical low limit is ~1–2 μm). The filament size is determined mainly by the stability criteria and by hysteresis magnetization or AC loss requirements. The critical current density in a superconductor at 5 T and 4.2 K in Nb-Ti composite wires used in accelerator magnets increased from ~1.5 kA/mm^2 to more than 3 kA/mm^2 today. The typical RRR values of the copper matrix are in the range 50–200.

1.3.2 Nb-Ti Accelerator Magnet Designs and Technologies

The first large superconducting accelerator based on Nb-Ti magnets was the Tevatron (1983–2011) at the Fermi National Accelerator Laboratory (FNAL, USA). It was followed by the Hadron Elektron Ringanlage (HERA, 1991–2007) at the Deutsches Elektronen-Synchroton (DESY, Germany), the Relativistic Heavy Ion Collider (RHIC, since 2000) at Brookhaven National Laboratory (BNL, USA), and the Large Hadron Collider (LHC, since 2008) at the European Organization for Nuclear Research (CERN).

Cross-sections of the Nb-Ti dipole coils used in arc dipoles in these projects are shown in Fig. 1.3. Coil and cable parameters are summarized in Table 1.1.

Figure 1.4 shows cross-sections of the dipole magnets in their cryostats, so-called cryo-magnets. The main arc dipole parameters are shown in Table 1.2.

These figures and the data in Tables 1.1 and 1.2 show the main tendencies in the evolution of the coil and magnet designs.

The Tevatron was the highest-energy hadron collider until its shutdown in 2011. It had a circumference of ~6.9 km and consisted of 774 dipoles and 240 quadrupoles, as well as more than 200 corrector spool pieces. A key element for the success of the Tevatron was the use of superconducting dipoles wound from Rutherford cables as two-layer shell-type coils. The mechanical structure was based on precise stainless-steel collars to apply coil azimuthal pre-stress and radial support. To ensure protection, quench heaters were used to accelerate the normal zone propagation in the coil during a quench. Tevatron magnets also featured a compact cryostat design with a warm iron yoke. This approach led to some problems with the centering and alignment of the collared coils inside the warm iron yoke, and to quite large heat

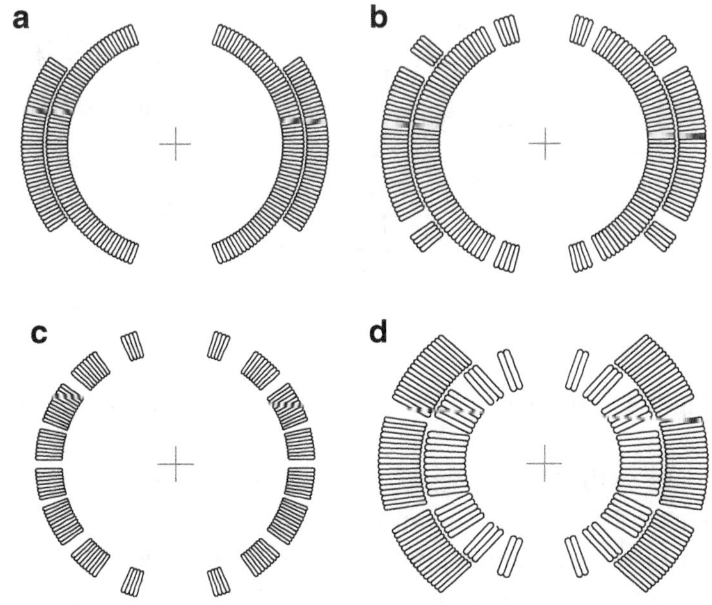

Fig. 1.3 Magnet coil cross-sections: (**a**) Tevatron; (**b**) HERA; (**c**) RHIC; (**d**) LHC

Table 1.1 Coil and cable parameters

	Tevatron	HERA	RHIC	LHC
Coil aperture (mm)	76	75	80	56
Number of turns	35/21[a]	32/20[a]	31	15/25[a]
Strand diameter (mm)	0.68	0.84	0.65	1.065/0.825[a]
Superconducting filament diameter (μm)	9	14	6	7/6[a]
Cu:Superconductor ratio	1.8	1.8	2.25	1.65/1.95[a]
Number of strands in cable	23	24	30	28/36[a]
Cable thickness (mm)	1.26	1.48	1.16	1.9/1.48[a]
Cable width (mm)	7.77	10.0	9.7	15.1/15.1[a]

[a] Inner/outer layers

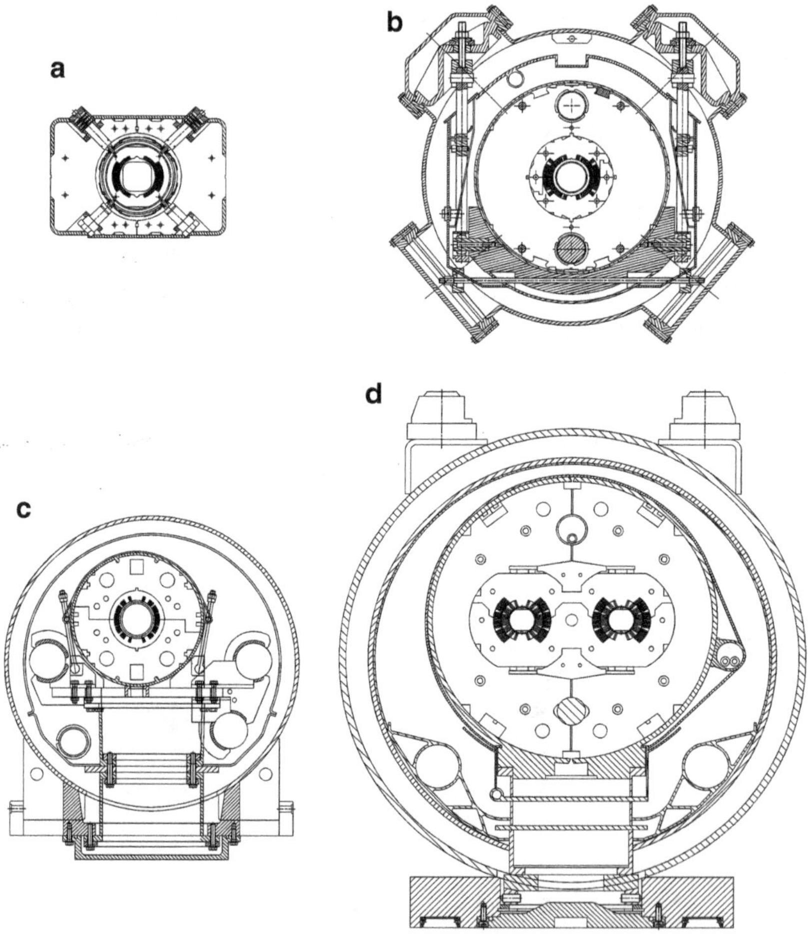

Fig. 1.4 Accelerator dipole cross-sections: (**a**) Tevatron; (**b**) HERA; (**c**) RHIC; (**d**) LHC

leaks to the helium vessel through the short plastic supports. The alignment problem was solved by the invention of spring-loaded smart bolts. The high level of heat leaks through the support system did not, however, allow the achievement of the design operation temperature and, thus, the target beam energy of 1 TeV. Despite these small hick-ups, the design and production of the Tevatron superconducting magnets opened the era of large superconducting accelerators.

Table 1.2 Arc dipole parameters

	Tevatron	HERA	RHIC	LHC
Aperture (mm)	76	75	80	56
Magnetic length (m)	6.1	8.8	9.45	14.3
Nominal bore field (T)	4.3	5.3	3.5	8.3
Nominal current (kA)	4.3	5.7	5.1	11.9
Stored energy at I_{nom} (MJ)	0.30	0.94	0.35	6.93
Operation temperature (K)	4.6	4.5	4.3–4.6	1.9

The next large superconducting accelerator, HERA, comprised a 30 GeV electron storage ring (SR) with conventional electromagnets and a 820 GeV superconducting proton SR. The 820 GeV ring had a circumference of ~6.3 km and consisted of 422 main dipoles and ~225 main quadrupoles, along with approximately the same number of superconducting correcting elements. The dipole magnets of the proton ring were developed at DESY, while the quadrupole magnets were developed at the Commissariat à l'Energie Atomique at Saclay (CEA/Saclay) near Paris, France. HERA dipoles, designed to produce 4.7 T at 4.6 K, subsequently operated at 5.5 T by reducing the magnet operation temperature below 4 K. The HERA dipoles used thick, free-standing aluminum collars and a cold iron yoke. The cold iron yoke design selected, greatly simplified the collared coil alignment inside the yoke and reduced heat leaks to the magnet cold mass by using long plastic suspensions and better thermal insulation at the expense of a much larger cold mass. The HERA arc dipoles were manufactured in industry, which required a significant improvement of magnet technology, technology transfer to industry, and the provision of reliable quality control throughout magnet production. This brave and cost-saving step was adopted later by the next large-accelerator projects.

The RHIC is designed to collide beams of nuclei as heavy as gold, accelerated in two identical SRs to energies between 7 GeV and 100 GeV per beam and per unit of atomic mass. RHIC beams are guided by 3.5 T dipole magnets. Each of the two separate superconductor SRs is ~3.8 km in circumference and they intersect at six points. Each ring consists of ~1740 superconducting magnets, including 264 arc dipoles and 276 arc quadrupoles. An important goal of the RHIC magnet design was to minimize magnet cost, which was achieved by using a single-layer shell-type coil design, thin plastic coil–yoke spacers instead of expensive metallic collars, and a cold iron yoke close to the coil, which adds ~30% to the field in the bore. The coils, surrounded by glass-reinforced phenolic spacers, are preloaded and supported by a horizontally split iron yoke and a stainless-steel skin. The magnet cold mass is installed inside a vacuum vessel on Superconducting Super Collider (SSC)-type plastic support posts, that were used later in the LHC magnets. Several improvements in the dipole design included the optimization of the field quality in operation field range with a saturated iron yoke, high-J_c superconducting strand with 5–6 μm filaments, and 10 mm wide Rutherford cable. As with HERA, the RHIC dipoles were produced by industry, including systematic warm magnetic measurements for quality control.

The LHC is the largest superconducting proton–proton collider in the world, with an SR of ~27 km in circumference and the highest operation dipole field of 8.33 T. It is located in an underground tunnel previously constructed and used for the Large Electron–Positron Collider (LEP). The ring is filled with 1276 superconducting dipoles and ~425 quadrupoles. The dipole magnets were developed at CERN, while the quadrupole magnets were developed at CEA/Saclay. The dipole design is based on two-layer shell-type graded coils with two 15 mm wide Rutherford cables in the inner and outer layers. The coils are preloaded with thick stainless-steel collars and supported by a vertically split cold iron yoke. The LHC dipoles adopted for the first time a two-in-one design concept in which two apertures with opposite field directions are placed inside a common collar and iron yoke. The LHC magnets use high-performance composite wires made of high homogeneity Nb-Ti alloy and are cooled by superfluid helium at 1.9 K, which further enhances the performance of the Nb-Ti conductor and, thus, increases the magnet operating field. As with the HERA and RHIC magnets, all of the LHC dipoles were produced in industry. The production of the LHC dipoles was shared by three European companies.

In parallel with the development and construction of superconducting magnets for HERA, from the 1980s until the early 1990s, Nb-Ti magnets were developed for the 3 TeV Uskoritel'no Nakopitel'nyj Kompleks (UNK) (Accelerator and Storage Complex) in Protvino, near Moscow, Russia, at the Institute for High Energy Physics (Ageyev et al. 1980), and for the 40 TeV SSC near Waxahachie, Texas, (Jackson 1986). Both dipole designs were similar to the HERA dipole. The UNK dipoles had an aperture of 80 mm and were designed to operate at a nominal field of 5 T at 4.5 K. The magnets used circular stainless-steel collars. The magnet cold masses were suspended inside the vacuum vessel using long Ti-alloy rods. The SSC dipoles had an aperture of 50 mm (the initial design had an aperture of 40 mm) and a nominal operating field of 6.6 T at 4.3 K. Special plastic posts were developed for the SSC magnets to support the cold masses inside the vacuum vessel. Although a series of short dipole models and full-scale prototypes have been produced and successfully tested in the framework of both projects, these two projects were terminated in the middle of the 1990s.

1.4 Next Generation of Superconducting Accelerator Magnets

A field of ~10 T is considered to be the practical limit for Nb-Ti accelerator magnets. As shown above, using magnets with a higher field is the most efficient way to achieve higher collision energies in those accelerators. To produce higher fields, new superconducting materials with better properties are needed. The only practical material that is readily available on an industrial scale is an intermetallic Nb_3Sn compound, which belongs to the A15 crystallographic family.

1.4.1 Nb_3Sn Composite Wires

The superconducting properties of the Nb_3Sn compound, as for the Nb-Ti alloy, were also discovered in the 1960s. The work on Nb_3Sn accelerator magnets started practically immediately in the late 1960s, motivated by the better critical parameters of Nb_3Sn. The critical temperature T_{c0} of Nb_3Sn is 18 K and the upper critical magnetic field B_{c20} is 28 T.

Nb_3Sn composite wires are currently produced using three main methods: bronze, internal tin (IT), and powder-in-tube (PIT) (Rogalla and Kes 2012). The bronze process is based on a large number of Nb filaments dispersed in a Sn-rich bronze matrix. The bronze route provides the smallest filament size (~2–3 μm), but has a relatively low critical current density due to the limited Sn content in bronze available for reaction to Nb_3Sn. The IT process is based on assembling a large number of Nb filaments and pure Sn or Sn-alloy rods in a copper matrix. Restacking of assemblies allows the reduction of the final sub-element size. Due to the optimal amount of Sn, this process provides the highest critical current density, but limits the minimum sub-element size currently achievable in the final wire to ~40–50 μm. The PIT process is based on stacking thick-wall Nb tubes, filled with fine $NbSn_2$ powder in a high-purity Cu matrix. This method provides a reasonable combination of small filament size (<50 μm) and high critical current density comparable with the IT process. The cost of PIT wire is, however, still considerably higher than the cost of IT wire. In all of these methods, the Nb_3Sn phase is produced during a final multi-stage high-temperature heat treatment with a maximum temperature ~650–700 °C for 50–100 h.

As the accelerator construction costs are driven by the ring circumference, superconducting magnets for particle accelerators rely on the highest possible current density to minimize the coil volume and magnet cost. Thus, IT and PIT composite wires are the most appropriate for use in high field accelerator magnets. The highest value of the critical current density at 12 T and 4.2 K in state-of-the-art Nb_3Sn composite wires developed for use in accelerator magnets exceeds 3 kA/mm^2.

Nb_3Sn composite wires are compatible with the Rutherford cable design and technologies, making the Rutherford cable a primary option for use in Nb_3Sn accelerator magnets. Due to the more delicate structure and mechanical properties of the Nb-Sn precursor, however, forming the compact keystoned cable cross-section without degrading the strand critical current is challenging. This problem, as well as strand fusing in cables during high-temperature heat treatment of coils under pressure, requires special attention and optimization.

1.4.2 Design Issues of Nb_3Sn Accelerator Magnets

All the abovementioned accelerators with Nb-Ti magnets have been built based on the shell-type (also known as cos-theta) designs. This design is also suitable for use with the Nb_3Sn superconductor. In this design, keystoned Rutherford cables are wound in blocks, which are spaced with wedges, around a circular bore. The number of cables in each block and the dimensions of the wedges are selected such that field errors and the conductor peak field to bore field ratio are minimized. Thanks to the Roman arch principle, no internal support structure is required in the cos-theta structure. Consequently, no valuable real estate inside the aperture is lost.

An alternative, the block coil design, is based on racetrack coils with a vertical turn position and flared ends. A special support structure is usually required in high field dipoles to support the pole blocks and flared ends. In the late 1990s, a common-coil approach, first proposed in the early 1980s for low field Nb-Ti magnets, was considered for high field dipole magnets. The common-coil design is a twin-aperture version of the block-type dipole with vertically positioned apertures. In this design flat racetrack coils are shared between the two apertures. This coil arrangement and simple coil geometry provide some technological advantages. In particular, the coil radii in this design are defined by the distance between apertures rather than by aperture size, making it suitable for the winding of brittle conductors such as Nb_3Sn.

In the late 1990s, it was postulated that the block-type coil design is better suited for high field Nb_3Sn magnets than the cos-theta design (Sutter and Strauss 2000). This statement started to be intensively disputed within the accelerator magnet community. The main debate points are the coil winding issues (racetrack coils with flared ends vs. coils wound around a cylinder); the relative position of high field and high stress areas (cos-theta magnets have the highest stress in the coil mid-plane where the field level is also large whereas block coils have the largest stress level in the low field parts); the possibility of using wider cables in block coils than is feasible in the cos-theta coils, allowing for smaller inductance and larger operating current; and the amount of conductor required to achieve the same field level.

All three different coil types based on Nb_3Sn Rutherford cables have been designed, manufactured, and tested. In the following chapters, they are thoroughly described and discussed. The question of which of these options is best suited for the hadron collider continues to be passionately debated within the scientific community.

1.4.3 Nb_3Sn Magnet Technology

The fabrication of Nb_3Sn magnets is a complicated process due to the delicate processing, brittleness, and stress/strain sensitivity of Nb_3Sn composite wires and cables. It imposes serious limitations upon the use of reacted Nb_3Sn conductors for coil winding (react-and-wind technology). To overcome the limitations of working with brittle conductors, the so-called wind-and-react technology was introduced. In this approach, a coil is wound with a ductile conductor made of the Nb-Sn precursors. Afterwards, the coil is reacted at temperatures of up to 650–700 °C for several days to form the brittle superconducting Nb_3Sn phase. The high temperature during the coil reaction excludes the use of organic materials for cable and coil insulation as well as for various coil components.

The phase transition during the formation of the Nb_3Sn leads to a radial and azimuthal coil expansion after reaction. In the axial direction, this effect is compensated for by the coil contraction due to a release of stress accumulated in the strand matrix during cable fabrication. In practice, the exact dimensional changes of the coil depend on the cable compaction and on the insulation used. No empirical fitting-law for the exact prediction of the dimensional change of the conductor during reaction has yet been established. It is determined for each cable and coil design empirically in trial coil cross-sections. The empirical data for each specific project need to be established and considered during coil assembly and preloading.

The Nb_3Sn conductor, if exposed to stress/strain, shows a degradation of the critical current density. Depending on the stress/strain value the degradation is reversible or irreversible. Stress/strain analysis and optimization of the magnet assembly process are important parts of magnet design and technology development. For example, applying excessive pre-stress during magnet assembly, to compensate for pre-stress lost while the magnet cools down to the operating temperature, could cause irreversible degradation of the Nb_3Sn coils. The engineering of mechanical structures using materials with different thermal contraction coefficients to avoid pre-stress loss, or even to achieve pre-stress increase, is important for exploring the potential of Nb_3Sn accelerator magnets.

Good field quality, as discussed above, is mandatory in accelerator magnets. All the major components—geometric, iron saturation, and coil magnetization effects—require special attention in Nb_3Sn magnets. Geometric field errors, controlled by the position of the coil turns, are impacted by the turn expansion and possible displacements during coil reaction as well as by turn motion under large Lorentz forces. The iron saturation effects are larger due to the larger field level and larger field variations in the yoke. Persistent current field errors are larger due to the large superconductor filament sizes and higher critical current density in Nb_3Sn filaments. In addition, the field errors related to the interstrand eddy currents in cable can also be very large due to strand fusing during coil reaction.

Lastly, due to the large scale of future circular hadron colliders, the cost-optimization of Nb_3Sn magnets becomes of high importance.

1.4.4 Magnet Performance Test

Sophisticated analysis tools, available at the present time to magnet designers, allow for accurate magnet analysis (magnetic, mechanical, thermal) and the exact definition of the magnet target parameters based on the properties of materials and the operating conditions. Many uncertainties in material properties as well as variations of technological procedures inevitably, however, affect magnet performance. As a result, the difference between the real magnet performance and the magnet target parameters can be significant. Therefore, detailed material characterization, quality control during all production steps, and final comprehensive magnet tests are important components of a successful magnet production process development.

Before performing the final magnet tests at the nominal operating conditions, tests at ambient temperature are typically conducted. The final magnet tests are performed at the nominal operation conditions. Both magnet production and final cold tests not only provide the final performance parameters but also their reproducibility from magnet to magnet in large series. During the research and development phase, these tests provide very important feedback for the optimization of magnet design, technology, and performance.

The main objectives of the magnet cold tests are the examination of all magnet design parameters including the quench performance study, magnetic measurements, quench protection studies, and measurements of magnet mechanical properties. The list of tests, their sequence, and conditions are defined in a magnet test plan, which is prepared for each magnet. This plan consists of the standard test procedures common for all magnets of a given series, as well as some specific tests for each magnet.

To measure the magnet performance parameters, each magnet is equipped with appropriate instrumentation. Some elements of the instrumentation (voltage taps, strain and temperature gauges, heaters, etc.) are implemented during magnet assembly, whereas the others are placed inside the magnet bore (quench antenna coils, and Hall probes or rotating coils for magnetic measurements) or on the magnet skin and end plates (strain gauges, bullet gauges, thermometers).

The following examples illustrate the importance of magnet final tests and the important feedback that they provide. Quenches below the magnet short sample limit point to possible energy depositions in the magnet coil due to turn frictional motions under the Lorentz force, heat dissipation from losses in the cable, or non-uniform current distribution among cable strands. The non-uniform distribution of the transport current in the cable is also visible in longitudinal oscillations of the field harmonics. The current imbalances may be caused by non-uniform properties of strands in the cable, non-uniform solder joints connecting the coils to the current leads, or large eddy current loops formed by small local interstrand resistance.

Other effects that are observed during cold tests of superconducting accelerator magnets and the way they are being addressed during the magnet development phase will be presented and discussed in the next chapters of this book.

References

Ageyev AI, Balbekov VI, Dmitrevsky YP et al (1980) The IHEP accelerating and storage complex (UNK) status report. In: Newman WS (ed) 11th international conference on high-energy accelerators, Genè, 7–11 July 1980. Springer Basel AG, Basel, p 60

Anerella M, Cottingham J, Cozzolino J et al (2003) The RHIC magnet system. Nucl Instrum Meth A 499:280–315

Ašner FM (1999) High field superconducting magnets. Oxford University Press, Oxford

Brechna H (1973) Superconducting magnet systems. Springer, Berlin/Heidelberg

Edwards HT (1985) The Tevatron energy doubler: a superconducting accelerator. Ann Rev Nucl Part Sci 35(1):605–660. https://doi.org/10.1146/annurev.nucl.35.1.605

Evans L, Bryant P (2008) LHC machine. J Instrum 3(8):S08001. https://doi.org/10.1088/1748-0221/3/08/s08001
Gallagher-Daggit G (1973) Superconductor cables for pulsed dipole magnets. Technical report, Rutherford High Energy Laboratory Memorandum, No. RHEL/M/A25, Chilton, Didcot
Iwasa Y (2009) Case studies in superconducting magnets: Design and operational issues. Springer, New York
Jackson JD (1986) Conceptual design of the superconducting super collider. SSC-SR-2020, SSC Central Design Group, Berkeley
Meinke R (1991) Superconducting magnet system for HERA. IEEE Trans Magn 27(2):1728–1734. https://doi.org/10.1109/20.133525
Mess K-H, Wolff S, Schmüser P (1996) Superconducting accelerator magnets. World Scientific, Singapore
Prodell AG (ed) (1968) BNL summer study. Brookhaven National Laboratory, Upton. http://www.bnl.gov/magnets/Staff/Gupta/Summer1968/contents.html
Ravaioli E (2015) CLIQ. A new quench protection technology for superconducting magnets. PhD Thesis, University of Twente
Rogalla H, Kes PH (eds) (2012) 100 years of superconductivity. CRC Press, Boca Raton
Russenschuck S (2010) Field computation for accelerator magnets. Wiley-VCH, Weinheim
Sutter DF, Strauss BP (2000) Next generation high energy physics colliders: technical challenges and prospects. IEEE Trans Appl Supercond 10(1):33–43. https://doi.org/10.1109/77.828171
Warnes WH, Larbalestier DC (1987) Determination of the average critical current from measurements of the extended resistive transition. IEEE Trans Magn 23(2):1183–1187. https://doi.org/10.1109/tmag.1987.1065081
Weisend JG II (ed) (1998) Handbook of cryogenic engineering. Taylor & Francis, Reading
Wilson MN (1983) Superconducting magnets. Oxford University Press, New York
Wilson MN (2012) 100 years of superconductivity and 50 years of superconducting magnets. IEEE Trans Appl Supercond 22(3):3800212. https://doi.org/10.1109/tasc.2011.2174628
Wolski A (2014) Beam dynamics in high energy particle accelerators. Imperial College Press, London

Chapter 2
Nb_3Sn Wires and Cables for High-Field Accelerator Magnets

Emanuela Barzi and Alexander V. Zlobin

Abstract Since the discovery of superconducting Nb_3Sn in 1954, continuous developments have led to a strong improvement of its properties. This chapter describes the evolution of Nb_3Sn fabrication techniques since the 1960s and the main properties of modern Nb_3Sn composite wires and Rutherford cables.

2.1 Introduction

The intermetallic compound Nb_3Sn, discovered by Matthias et al. (1954), is a type II superconductor having a stoichiometric composition ranging from 18–25 atomic percent (at. %) Sn. Its crystal structure is classified as A15 (also known as β-tungsten). The critical temperature T_{c0} is ~18 K, and the upper critical magnetic field B_{c20} can reach 30 T. As a comparison, the ductile alloy Nb-Ti has a T_{c0} of 9.8 K and a B_{c20} of 14.5 T, which makes Nb-Ti adequate only for operational magnetic fields of up to 8–9 T. Thanks to its larger critical current density J_c, Nb_3Sn enables operating fields in accelerator magnets above 10 T. This field is larger than any achieved with present Nb-Ti particle accelerator magnets. Some of the challenges with Nb_3Sn are that it requires high-temperature processing to create the superconducting Nb_3Sn phase. It is also a brittle material, which makes it strain-sensitive, i.e., high strain on the sample may reduce or totally destroy its superconductivity (Foner and Schwartz 1981).

The first laboratory attempt to produce Nb_3Sn wires was in 1961 at Bell Laboratories (Kunzler et al. 1961) by filling Nb tubes with crushed powders of Nb and Sn and drawing them into long wires. This primitive powder-in-tube (PIT) technique required reaction at very high temperatures, in the range of 1000–1400 °C, to form the superconducting Nb_3Sn phase. Nevertheless, that same year it was used to fabricate the first 6 T solenoid. An initial alternative to PIT wires and the first commercial production of Nb_3Sn were tape conductors fabricated by either chemical vapor deposition (1963) or by surface (or liquid solute) diffusion of Sn into a Nb

E. Barzi · A. V. Zlobin (✉)
Fermi National Accelerator Laboratory (FNAL), Batavia, IL, USA
e-mail: barzi@fnal.gov; zlobin@fnal.gov

D. Schoerling, A. V. Zlobin (eds.), *Nb_3Sn Accelerator Magnets*, Particle Acceleration and Detection, https://doi.org/10.1007/978-3-030-16118-7_2

substrate (1964). Although successful in demonstrating the use of Nb_3Sn in high-field magnets, none of these techniques were very practical. The large filaments in the case of the PIT wire, and the inherently large aspect ratio of the tape, invariably resulted in large trapped magnetization and flux jump instabilities.

A major step for practical Nb_3Sn production occurred with the discovery of the solid-state diffusion process. It was demonstrated in 1970 that the superconducting Nb_3Sn phase can be produced at the interface of Nb and a Cu-Sn alloy at lower temperatures compared to binary Nb and Sn couples. This principle allowed composite conductors to be formed in fine multifilamentary configurations by cold working, and before heat treatment. The bronze route (Howlett 1970; Kaufmann and Pickett 1971; Tachikawa 1971) was the first application of the above principles. To overcome the J_c limits of the bronze method the internal tin (IT) process was introduced in 1974 (Hashimoto and Yoshizaki 1974). The PIT, bronze, and IT methods were further developed into multifilamentary composites, and practically all commercially available wires are now produced by one of these methods.

In the 1980s and early 1990s, the development of Nb_3Sn conductor was mainly steered by fusion magnet programs. The Nb_3Sn toroidal magnetic system of the Tokamak T-15 was built and tested in the USSR at the end of 1988. The construction of the T-15 motivated the use of Nb_3Sn wires for the International Thermonuclear Experimental Reactor (ITER). The conductor-development programs for accelerator magnets were focused on Nb-Ti composite wires and were driven by the needs of accelerators such as the Tevatron, the Superconducting Super Collider (SSC), and the Large Hadron Collider (LHC). Since the late 1990s the high-energy physics (HEP) community has taken leadership in the development of Nb_3Sn wires for post-LHC accelerators to be used in high-field accelerator magnet research and development (R&D).

In this chapter we describe the development of commercial Nb_3Sn composite wires and tapes from the 1960s to the 1990s. We then illustrate the most important wire properties and their progress, as achieved in HEP by 2010. Finally, Rutherford cables are touched upon. These developments on wires and cables were crucial for the first reproducible accelerator-quality Nb_3Sn dipole and quadrupole magnets and their scale-up in the 2000s to the 2010s. Definitions of key wire and cable parameters are presented in Appendix 1. Properties and fabrication methods for Nb_3Sn composite wires and Rutherford cables have also been reviewed elsewhere (Dietderich and Godeke 2008; Bottura and Godeke 2012; Barzi and Zlobin 2016).

2.2 Development of Nb_3Sn from the 1960s to the 1990s

2.2.1 Nb_3Sn Development from the 1960s to the 1970s

The very first commercial Nb_3Sn composite conductor, which was used for accelerator magnet models in the period from the 1960s to the 1970s, was in the form of a flexible tape (also called a ribbon). A schematic of a Nb_3Sn ribbon is shown in Fig. 2.1.

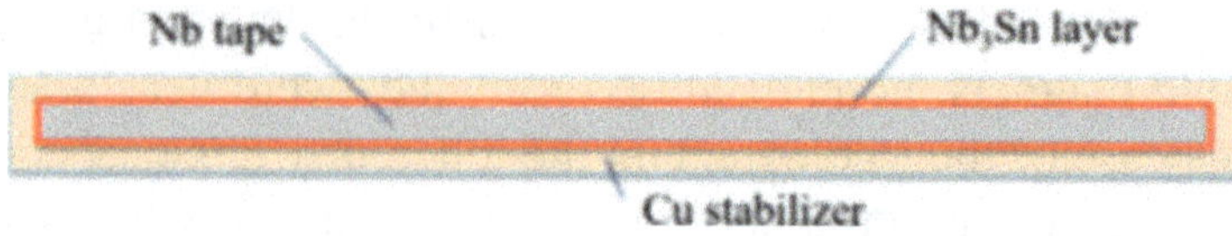

Fig. 2.1 Schematic of a cross-section of a Nb_3Sn ribbon

Tape conductors had been fabricated by chemical vapor deposition since 1963, and by surface diffusion of Sn into a Nb substrate since 1964. The latter process, also called Liquid Solute Diffusion, consisted of the introduction of a thin Nb tape into a molten Sn bath. The Sn-coated tape was then heated to ~1000 °C in a vacuum or inert atmosphere to form Nb_3Sn on both sides of the substrate. The tape was then Cu-cladded or Cu-plated for electrical stabilization. This technique was readily developed for long length production of ~500 m, and was first available in 1964 from the National Research Corporation in Massachusetts. Next, in 1965 the Compagnie Générale de Télégraphie sans Fil in France commercially produced a tape that was double the width and half the thickness, to allow for coil bending around smaller radii. In the same year, General Electric (GE) at Schenectady, NY, produced wide tapes with currents of 150 A, 300 A, and 600 A at 10 T. This was possible by fabricating tapes with a double substrate. GE also provided stainless-steel cladding, in addition to the Cu cladding option. And finally, in 1968 the Plessey Company in England introduced a diffusion-type Nb_3Sn narrow ribbon, which was Cu-cladded on one or both sides.

In parallel, chemical vapor deposition was developed, which is accomplished by hydrogen reduction of a mixture of chlorides of the constituent elements. By passing chloride over the metallic elements at ~1000 °C, gaseous chlorides such as $SnCl_2$ and $NbCl_4$ are produced. The chlorides are reduced by hydrogen and helium gasses, and the Sn and the Nb are deposited in the proper proportions onto substrates such as stainless steel. The A15 layer thickness was controlled by the rate at which the substrate was passed through the reaction chamber. Following deposition of Nb_3Sn, the ribbon was electroplated with a thin layer of nickel and then clad with silver or copper. The sole commercial producer of this material since 1963 had been the Radio Corporation of America (RCA), at Harrison, New Jersey, offering wide tape at current ratings of 300 A, 600 A, 900 A, and 1200 A at 10 T. It is to be noted that the vapor-deposited ribbon had a critical temperature of only 15 K, compared to 18 K for the diffusion process material. Whereas the commercial diffusion-type tapes were offered either bare or insulated with varnish, the vapor-deposited ribbon was just offered bare.

Conductor testing at that time included critical current and resistive critical temperature measurements of short samples, as well as continuous testing of the ribbon magnetization when passed through a 5 T field and measured with pick-up coils. This latter method produced the diamagnetic strength of the superconducting layer and allowed the detection of defects both in the superconductor and at the interface between superconductor and conducting metal. Unfortunately, information on the performance of these ribbons is limited. The best engineering current density of 500 A/mm^2 at 4.2 K and 12 T was obtained in a small-bore solenoid coil wound with RCA tape. The lowest current density of 270 A/mm^2 at 4.2 K and 2 T was seen in a 0.6 m long quadrupole with a large aperture.

It is interesting that, at that time, instabilities had already been recognized as a challenge. Ribbons had anisotropic properties and suffered from flux jumps in magnetic fields perpendicular to the tapes' wide face. In solenoidal coils, flux jump instabilities already occurred in magnetic fields below 8–10 T. Flux jumping was observed as voltage spikes much larger than expected by magnet inductance when ramping the current up and down. An experiment showed that by doubling the purity of the normal Cu conductor the current limits increased, but were still not at the short sample limit (SSL). When coils were instead used as inserts into a magnet and operated in background fields, the short sample limit was more likely to be achieved.

Requirements for superconductor stability with respect to magnetic flux jumps and superconductor protection in case of transition to the normal state led to the concept of composite superconducting wire, in which thin superconducting filaments are distributed in a normal low-resistance matrix. This matrix provides several important functions. It conducts heat away from the surface of the superconducting filaments because of high thermal conductivity, absorbs a substantial fraction of heat due to its high specific heat, and decreases Joule heating when the superconductor becomes normally conducting. To reduce eddy currents induced by varying external fields and improve the stability of composite wires to flux jumps, these filaments are twisted along the wire axis.

Multifilamentary Nb_3Sn composite wires were possible through the discovery of the solid-state diffusion process and by finding that Nb_3Sn can be produced at the interface of Nb and a Cu-Sn alloy at lower temperatures than for binary Nb-Sn couples. The formation of the brittle compound could be therefore postponed until the desired geometrical configurations for conductors (such as a multifilamentary wire) were achieved.

In the binary Nb-Sn system, single-phase Nb_3Sn is formed by solid-state diffusion above ~930 °C, where the only stable phase is Nb_3Sn. At temperatures below 845 °C, the two non-superconducting phases $NbSn_2$ and Nb_6Sn_5 are also stable, and all three phases will grow at the interface, with $NbSn_2$ most rapidly formed and Nb_3Sn being the slowest. In the ternary system (Nb-Cu-Sn), however, the only relevant stable phase is Nb_3Sn, even at lower temperatures. The diffusion path from the Cu-Sn solid solution to the Nb-Sn solid solution passes through the A15 phase field alone, preventing formation of the non-superconductive phases. In short, the addition of Cu strongly lowers the A15 formation temperature from well above 930 °C to practical values in industrial wires of ~650 °C, thereby also limiting grain growth and retaining a higher grain boundary density, as required for flux pinning.

This principle was used to fabricate multifilamentary Nb_3Sn composite wires by the so-called bronze route, which is still today one of the leading techniques for manufacturing Nb_3Sn wires for various applications. The bronze process is based on a large number of thin Nb filaments dispersed in a Sn-rich bronze matrix (Fig. 2.2).

The initial billet is made of hundreds of Nb rods, and it is drawn into a hexagonal element of intermediate size. The rods are then cut and assembled in a second billet, which is extruded, annealed, and drawn to the final wire size. The bronze core is surrounded by a high-purity Cu matrix that is separated by a thin Nb or Ta diffusion

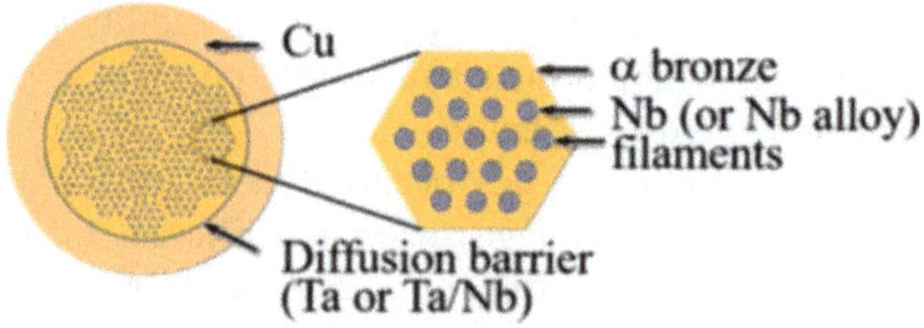

Fig. 2.2 Schematic of Nb_3Sn composite wire based on the bronze route. (Courtesy Peter Lee, FSU)

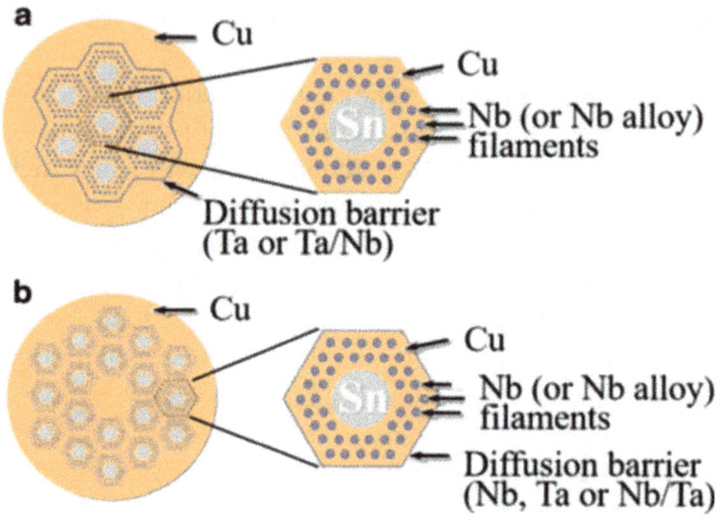

Fig. 2.3 Schematics of Nb_3Sn composite wire based on the IT process with: (**a**) single diffusion barrier; and (**b**) distributed diffusion barrier. (Courtesy Peter Lee, FSU)

barrier. The composite is twisted to increase electromagnetic stability and reduce AC losses. The bronze route provides the smallest filament size (~2–3 μm), but has a relatively low critical current density J_c due to the 13 wt% limited content of Sn in bronze.

The external diffusion process, a modification of the bronze method, was introduced in 1972. This process consists of extruding a Nb-Cu composite and drawing it to a final size. The wire is then coated with a layer of Sn before heating it to form a Cu-Sn alloy matrix and then Nb_3Sn filaments. The main benefit of this variant was that the Sn in the matrix was not limited as in the case of the bronze process. Its primary difficulty was delamination of the outer layer of the wire for plated Sn layers thicker than 5 μm. The maximum wire size was therefore limited to 0.25 mm.

The IT diffusion process, another modification of the bronze method, was first reported in 1974. This process consists of assembling Cu-clad Nb bars and a Cu-clad Sn-Cu alloy rod in a Cu tubing such that the Sn-Cu rod is at the center (Fig. 2.3). The assembly is then cold-drawn to final size. By restacking a number of assemblies and redrawing the billet down, a larger number of filaments and reduced bundle sizes are obtained. The advantage over the external diffusion process was that the Cu stabilizer, and a Ta barrier to protect the Cu from Sn diffusion, could be placed either outside of a wire in a single barrier configuration (Fig. 2.3a), or around each filament bundle in a distributed barrier configuration (Fig. 2.3b).

Another interesting method that was proposed in 1973 was the in situ process. It consisted of casting a Nb-Cu melt to form a dendritic network of Nb in Cu, and of drawing it to wire. Sn could be added in the initial melt or afterward before heat treatment. The most interesting aspect of this method was the great enhancement of the wire mechanical properties, due to the close proximity of the finely divided filaments.

Several companies were already active in producing commercial Nb_3Sn wires by that time. US companies included Airco, Inc., Intermagnetics General Corporation (IGC), Supercon, Inc., and Teledyne Wah Chang Albany (TWCA).

Airco, Inc. had started to manufacture superconductors in the mid-1960s at their Central Research and Development Laboratories in Murray Hill, NJ. They moved their manufacturing plant to Carteret, NJ in 1979. Their standard bronze wire was based on a modular approach, which meant combining one of three first extrusions containing 19, 55, and 187 rods, respectively, into second extrusion billets. In the 1970s Airco supplied 14 kg of 0.3 mm diameter 30% Nb in situ wire, Sn-plated on a commercial plating line in Airco's plant, to Brookhaven National Laboratory (BNL) for Nb_3Sn dipoles. In parallel, in cooperation with BNL, Airco developed a technique to wrap a multifilamentary Cu matrix conductor around a high-Sn Cu alloy core. The cable was then sheathed in Cu on a continuous tube mill to produce long lengths of finished wire. This was called the "distributed tin" process and reduced some of the problems encountered by external tinning. Airco also developed and brought to commercial maturity a high-rate magnetron sputtering process for producing Nb_3Sn tape for transmission lines by a variation of the bronze process.

In Europe, Siemens/Vacuumschmelze, Hanau, Germany, was producing bronze route wires on an industrial scale. Europa Metalli, LMI, Fornaci di Barga, Italy, and GEC Alstom Intermagnetics, Belfort, France, were producing IT wires.

In the USSR, the collaboration between the Kurchatov and Bochvar institutes and several plants of the nuclear industry focused on the development and production of Nb_3Sn wires for new thermonuclear facilities and for research magnets with high fields (Chernoplekov 1978). Studies of Nb_3Sn tapes were also conducted from 1962 to 1972. Equipment for the industrial production of tape by tinning was manufactured and installed in industry, but the work was stopped after the discovery of the solid-phase diffusion method. In 1972, intensive testing of multifilamentary samples based on Nb_3Sn composite wires began. The number of tests in subsequent years reached 3000 samples per year. These studies led to the Tokamak T-15 project with Nb_3Sn toroidal coils (Chernoplekov 1981). For this project, in the 1980s the USSR industry produced 30 tons of 1.5 mm diameter Nb_3Sn bronze wire with a Sn content of 10.5%.

Several Japanese companies also developed and produced Nb_3Sn composite wires. Sumitomo Electric Industries developed Nb_3Sn conductors in which the Nb_3Sn layer is formed by solid–liquid reaction between Nb tubes and the Cu-Sn alloy core (Yamasaki and Kimura 1982). This simpler method was later called the "internal-tin tube" or "tube-type" method. Mitsubishi Electric Co. was working on the internal Sn diffusion process, and Toshiba R&D Center and Showa Electric Wire and Cable Co. developed a Nb tube method. Furukawa Electric Company was working on bronze Nb_3Sn conductors with a large current capacity. Hitachi Ltd.

and Hitachi Cable Co. focused on large-scale multifilamentary Nb_3Sn composite wires.

By this time, critical current measurements of wires were performed within the bore of a solenoid on either straight samples a few centimeters long in a transverse field, or on a few turns wound around a 3.5–4 cm diameter barrel. The longer sample length allowed the use of a critical current criterion of 10^{-14} Ω m (i.e., smaller resulting voltages) as opposed to an electrical field criterion of 1 μV/cm. Inductive critical temperature measurements of short samples were sometimes added to conductor characterization as a means to identify lower critical temperature T_c regions in the material. The degradation of the critical current I_c due to the bending of the wire was measured as function of the bending diameter. The critical current density at 4.2 K and 12 T in the non-Cu samples ranged from 500 to 700 A/mm^2 in round wires.

2.2.2 Nb_3Sn Development from the 1980s to the 1990s

In 1980, TWCA proposed another innovative process called "jellyroll." In this process Nb foil was slit on a conventional slitter used to produce expanded mesh and rolled in a "jellyroll" together with bronze sheet. After wrapping with a diffusion barrier and a Cu stabilizer, the billet was compacted and then extruded and drawn in a conventional manner.

By the end of the 1990s, Oxford Superconducting Technology (OST), previously Airco, was fabricating Nb_3Sn multifilamentary wires with a modified jellyroll (MJR) process. In this version of the IT process, thin Nb sheets (with a thickness from 50 to 500 μm) were slit and expanded with controlled interconnection distances of 5–150 mm. The expanded Nb sheet (45–85% open) was then rolled up with the desired matrix material (i.e., bronze, Cu, Al, etc.) in the fashion of a jellyroll, inserted into a Cu container, sealed, extruded, and processed as conventional wire. A diffusion barrier was inserted if needed.

To overcome some of the constraints of the MJR process, OST also worked in the early 2000s on an internal Sn strand made with a restacking process called hot extruded rod (HER) process (Parrell et al. 2002). In this method, small Sn rods were inserted into the washed-out salt holes after extrusion. This technique, which was successfully demonstrated for a 37 restack, resulted in better bonding and had the potential for economic production of large quantities of conductor, as is the case for Nb-Ti. It was later abandoned, however, due to the relatively large effective filament diameter, or D_{eff}.

In September 1980, the Energy Research Foundation (ECN) from the Netherlands presented an alternative process that used both conventional and powder technologies. The core of a Nb tube was filled with fine grain size $NbSn_2$ powders inserted into a Cu tube and the resulting wire reacted at temperatures of 650–700 °C to form a continuous Nb_3Sn layer. The so-called PIT process was further optimized by the Shape Metal Innovation Company (SMI) (Lindenhovius et al. 1999) (Fig. 2.4).

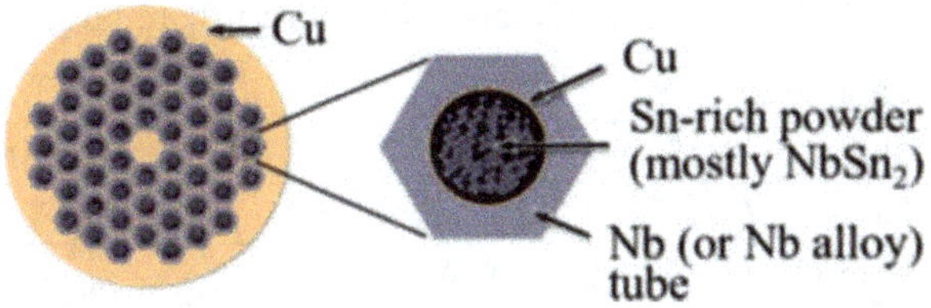

Fig. 2.4 Schematics of Nb_3Sn composite wire based on the powder-in-tube process. (Courtesy Peter Lee, FSU)

By the end of the 1990s, the strand samples used for I_c measurements were typically wound on grooved cylindrical Ti-alloy (Ti-6Al-4V) barrels, and held in place by two removable end rings. This method is still current today. After heat treatment, the Ti-alloy end rings are replaced by Cu rings, and voltage–current (*V–I*) characteristics are measured in boiling He at 4.2 K, in a transverse magnetic field provided by a solenoid. The voltage is measured along the sample by means of voltage taps placed 50 cm apart. The I_c is determined from the *V–I* curve by either the 10^{-14} Ω m resistivity or the 0.1 μV/cm electrical field criterion. The relative directions of the external magnetic field and the transport current are such as to generate an inward Lorentz force. Due to the latter and the differential thermal contraction between sample and barrel, the specimen is subject to a tensile strain of up to 0.05% at 12 T and 4.2 K. This strain leads to an overestimate of I_c between 3% and 5%. The *n*-values are determined in the $V(I_c)$-$10V(I_c)$ range by fitting the *V–I* curve with the power law $V \sim I^n$. Using this procedure, typical I_c uncertainties can be as low as ±1% at 4.2 K and 12 T, and about ±5% for the *n*-values. Magnetization measurements are performed with either balanced coil or vibrating-sample magnetometers.

Another new type of characterization included measuring the I_c of strands extracted from Rutherford-type cables. Plastic deformation during cabling affects the strand integrity. This phenomenon depends on the specific strand design and structure, and I_c degradation after cabling can be an indicator of internal damage.

2.3 Modern High-J_c Nb_3Sn Composite Wires

2.3.1 Target Parameters of Nb_3Sn Wires for HEP

IT and PIT are the two Nb_3Sn composite wire technologies, with sufficiently high J_c and availability in large quantities from industry, that were used in magnets for HEP applications in the new millennium. Commercial Nb_3Sn wires based on these two processes were further developed to meet the needs of accelerator magnets.

State-of-the-art PIT and IT wires benefited from at least two HEP-related conductor programs. The US Conductor Development Program (CDP) (Scanlan 2001) was started in 1999 by the US Department of Energy (DOE) as a collaborative effort of US industry, national laboratories, and universities with the goal of increasing the critical current density of Nb_3Sn IT wires for HEP applications including high-field accelerator magnets. Parallel R&D started in the early 2000s in the European Union as part of the Next European Dipole (NED) program (Devred et al. 2004). The target

Table 2.1 CDP, NED, and FCC Nb_3Sn wire target parameters

Parameter	CDP	NED	FCC
Non-Cu J_c (12 T, 4.2 K) (A/mm^2)	3000	–	–
Non-Cu J_c (15 T, 4.2 K) (A/mm^2)	–	1500	–
Non-Cu J_c (16 T, 4.2 K) (A/mm^2)	–	–	1500
Wire diameter (mm)	0.7–1.0	<1.25	1.0
Effective filament diameter D_{eff} (μm)	<40	<50	<50
Residual resistivity ratio RRR	–	>200	>150
Heat treatment time (h)	<200	–	–
Unit length (km)	>10	–	>5

parameters of Nb_3Sn wires for the superconductor R&D efforts of CDP and NED are summarized in Table 2.1.

Both programs had similar performance requirements in terms of J_c and D_{eff}, but NED was focused on the development of PIT composite Nb_3Sn wires of larger diameter (up to 1.25 mm) for high-field magnet applications. The effort was later continued by CERN with Bruker-EAS for the High Luminosity LHC (HL-LHC) upgrades.

Development of Nb_3Sn wires for high-field accelerator magnets continues. At present, properties are driven by the design studies of a Future Circular Collider (FCC) (Ballarino and Bottura 2015). The FCC Nb_3Sn wire specifications are also shown in Table 2.1. They are focused on reaching higher J_c at 16 T and on a significant reduction of conductor cost. The push for J_c increase is a stimulus for improving flux pinning using artificial pinning centers (APC) and increasing the upper critical field B_{c2} of Nb_3Sn. CERN is presently co-funding the R&D of several superconductor companies in Europe, Asia, and Russia to achieve the target parameters in commercial Nb_3Sn wires and to involve a sufficient number of companies in the longer-term production of large conductor volumes.

2.3.2 PIT Nb_3Sn Wires

The latest PIT process is based on stacking thick-wall round Nb tubes filled with fine $NbSn_2$ powder in a high-purity Cu matrix. The stacked assembly is drawn or extruded to the final wire size. This method allows an optimal combination of small filament size (<50 μm) and high J_c comparable with the IT process. The current cost of PIT wire is, however, a factor of two to three higher than the cost of IT wire.

The PIT process, first developed by the Netherlands Energy Research Foundation (ECN), was further optimized by the Shape Metal Innovation Company (SMI). In 2006 Bruker-EAS in Germany purchased the "know-how" for the PIT technology from SMI to industrialize this type of conductor. Cross-sections of some PIT composite wires produced by SMI and later by Bruker-EAS are shown in Fig. 2.5.

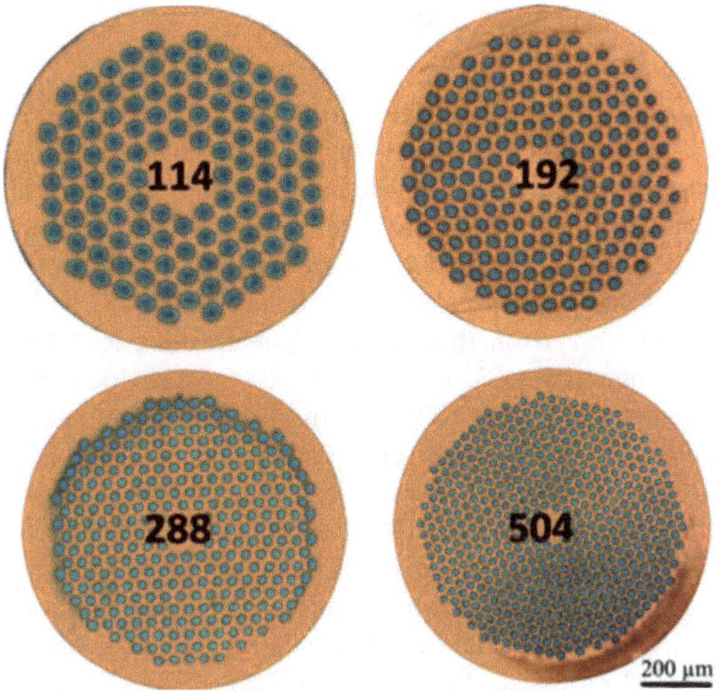

Fig. 2.5 Cross-sections of PIT wires of different designs (Courtesy of SMI and Bruker-EAS). The numbers represent the total number of superconducting tubes

The development of this technique has allowed the production of kilometer-long wires with 192 filaments. Shorter laboratory-scale wire samples with 504 and even 1332 filaments were also obtained. Wires were manufactured at Bruker-EAS in about 50 kg net production units. In 2008 the commercial PIT wire design started using round filaments instead of hexagonal ones to keep both the J_c and residual resistivity ratio (RRR) high during heat treatment and to reduce cabling degradation. These positive effects had been first demonstrated by SMI and Fermi National Accelerator Laboratory (FNAL) in 2004.

In PIT wires the Nb_3Sn A15 phase is formed in a solid-state diffusion reaction, typically after a few days at ~675 °C. The Sn diffusion and Nb_3Sn phase formation processes in the PIT route are described in detail elsewhere (Godeke et al. 2008). The $NbSn_2$ powder turns first into Nb_6Sn_5 and then into the Nb_3Sn phase. After 16 h, the initial Nb_6Sn_5 phase is converted into large grains. The void fraction in these regions is attributed to the reduced volume of Nb in Nb_3Sn, relative to the Nb_6Sn_5 phase. The Nb_3Sn phase formation ends after about 64 h at 675 °C, due to Sn depletion of the core–A15 interface region. Thus, a longer reaction does not increase the Nb_3Sn fraction. The outer boundary of the Nb_3Sn area is controlled to prevent Sn diffusion into the high-purity Cu matrix, and the resulting decrease in RRR. The heat treatment of commercial PIT composite wires without a diffusion barrier was optimized with respect to the area of reacted Nb to provide good RRR values. Most recently, PIT binary and ternary, mono and multi-filamentary wire configurations were successfully used at Hypertech Research Inc. to implement ZrO_2 precipitates as artificial pinning centers (APC) in the Nb_3Sn by means of internal oxidation of Nb-Zr.

2.3.3 IT Nb_3Sn Wires

As mentioned above, the IT process includes several versions, which differ, for example, in the design of the Nb filaments, the diffusion barrier position, the Sn distribution in the composite cross-section, sub-element and billet processing. Their potentials and limitations in terms of performance and large-scale production vary widely. IT composite wires were produced by several companies. In the USA, first IGC (Outokumpu since 2000, Luvata since 2005), and later OST (Bruker-OST since 2016) worked on this process. The Restack Rod Process® (RRP®), which was developed by OST in the early 2000s, provided the best results for accelerator magnet applications until February 2019, when short samples of APC PIT wires achieved the FCC J_c specifications shown in Table 2.1.

The IT RRP process is based on assembling a large number of Nb filaments and pure Sn or Sn-alloy rods in a Cu matrix (Parrell et al. 2004). The assembly is surrounded by a Nb barrier to prevent Sn diffusion into the high-purity Cu matrix, and it is then cold-drawn down to final size. Restacking of assemblies allows further reduction of the final sub-element size. Due to the fact that the diffusion barrier also reacts to form Nb_3Sn (and contributes to the J_c), there is a balance between sub-element size, metal ratios, heat treatment parameters, J_c, and Cu purity, which must be considered and optimized for every strand design.

During heat treatment of IT wires, several Cu-Sn phases are created and eliminated in the course of the Cu-Sn diffusion and Nb_3Sn formation processes. The presence of liquid phases in IT wires may cause motion of Nb filaments, allowing contact with adjacent ones; and the presence of voids may hinder the diffusion process. In addition, wire bursts can damage the wires. These problems were solved by using a three-step heat treatment cycle.

In the first step, temperature dwells below 227 °C allow the formation of a thin layer of a higher melting point Cu-Sn phase (ε-phase) that works as a container for the liquid Sn above 227 °C. A 1–3 day 210 °C dwell followed by 1–2 days at 400 °C provide appropriate Sn diffusion inside each sub-element, but also prevent Sn leaks through the Cu matrix, which is typically strongly deformed in the Rutherford cabling process. The superconducting Nb_3Sn phase is formed during the third step of the heat treatment cycle between 620 °C and 750 °C. During this stage the optimal phase microstructure, critical for flux pinning, is also formed. The Nb_3Sn microstructure is controlled by the temperature and the duration of this stage. Reactions at higher temperatures usually take the shortest time, but produce the largest grains. The choice of temperature and duration of the third stage is a compromise between an optimal pinning structure leading to high J_c, and Sn diffusion through the barrier leading to Cu pollution and to an increase of the matrix electrical and thermal resistivity.

Since 2002, when the RRP process was commercially introduced for HEP applications, OST has produced tons of Nb_3Sn wires. Cross-sections of some wire

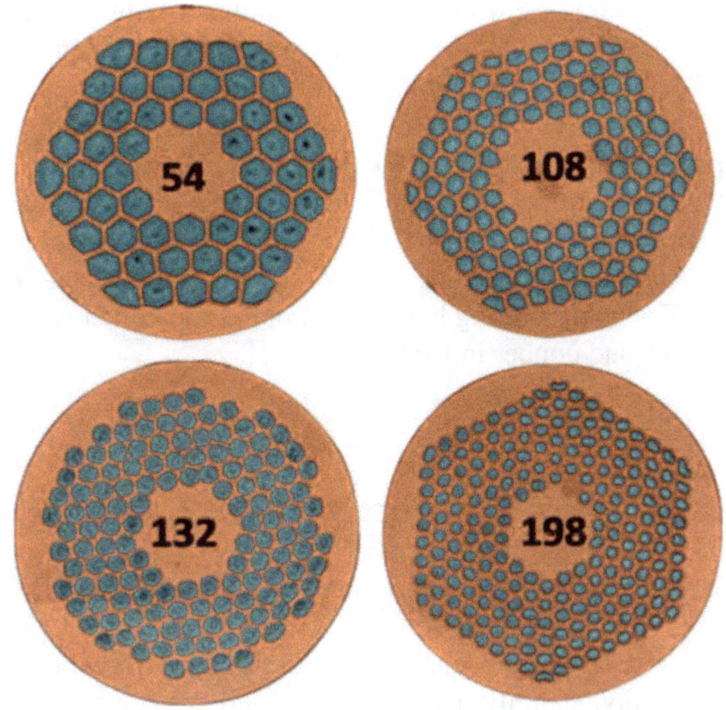

Fig. 2.6 Cross-sections of some commercial RRP wires (Courtesy of OST). The numbers represent the total number of superconducting sub-elements

designs manufactured by OST for accelerator magnets are shown in Fig. 2.6. In a configuration with 54 superconducting elements in a 61-restack wire (also represented as 54/61) of 0.8 mm size, the sub-element diameter is ~80 μm. Because of the impact of D_{eff} on conductor and magnet stability, OST focused on increasing the stack count in a billet to decrease the sub-element diameter.

As we will see below, in the Rutherford cables used in magnets, the Nb_3Sn filaments coalesced with each other. This "merging" randomly produced locally some superconducting areas several times larger than the size of the individual undeformed filaments (i.e., as in the round wire). These enlarged superconducting areas, which in cables may be up to five to six times the size of the undeformed filaments, clearly increase magnetic instabilities in cables. In 2006, OST produced for FNAL a 60/61 RRP billet with extra Cu spacing between sub-elements to reduce sub-element merging during cabling. Studies at FNAL showed that the mechanism by which the extra Cu thickness in the new OST design is effective is by providing a barrier to merging during reaction (i.e., not as much during the deformation process itself). Shortly afterward a 108/127 stack billet was produced with the same concept, and in 2007 OST adopted the spaced-filaments design as standard for HEP wires.

The 127-stack design entered production in 2008, with several tons utilized in HEP at 0.7–0.8 mm diameter and a D_{eff} of 45–52 μm. A third-generation wire with a 169-stack design followed in 2011. This wire has a D_{eff} of 40–58 μm for sizes of 0.7–1 mm. Integrated volume production of 169-stack RRP billets at OST is approaching that of the 127-stack billets. The development of 217-stack billets still continues.

Another improvement that OST obtained in RRP wires was that of interspersing Ti rods among the Nb filaments inside the sub-elements instead of using Ta-doped Nb filaments. This kind of doping allowed lowering the wire optimal reaction temperatures from ~695 °C for Ta-doped wire to ~665 °C for a Ti-doped wire, thereby increasing the wire J_c at high fields, better preserving RRR, and also improving the irreversible strain limit (Cheggour et al. 2010).

In 2016, OST was purchased by Bruker, which recently stopped standard PIT production, making the RRP conductor a workhorse for Nb_3Sn accelerator magnets. Bruker continues to use the PIT technology in R&D as a possible platform for future high-J_c wires with APCs. Also, based on promising results in increasing the heat capacity of Nb_3Sn composite wires for improved stability (Xu et al. 2019), Bruker-OST is implementing hexagonal copper rods filled with mixed powder of rare earth oxides and copper in RRP re-stacks.

2.3.4 Main Properties of Nb_3Sn Composite Wires

2.3.4.1 Critical Current Density J_c and Critical Current I_c

The critical current density J_c is a key parameter, which controls the current-carrying capability, stability, magnetization, and AC losses of a superconducting wire, and thus the design field and performance of superconducting magnets. Improvement of the critical current density at 12 T and 4.2 K in commercial IT and PIT composite wires since 1980, when the best J_c for Nb_3Sn wires reached 0.5–0.7 kA/mm^2, is shown in Fig. 2.7.

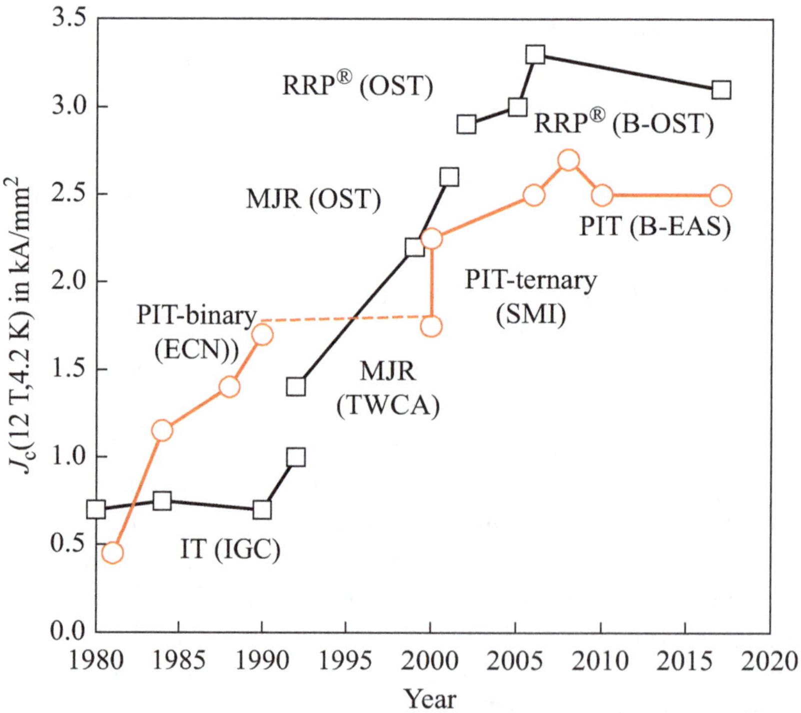

Fig. 2.7 J_c(12 T, 4.2 K) progress of IT and PIT composite wires

The J_c of PIT Nb_3Sn is affected by design parameters such as filament size, number of Nb tubes, use of binary Nb_3Sn or ternary $(NbTa)_3Sn$, and the quality and size of the $NbSn_2$ powder. During the 1980s the J_c of PIT wires was increased by ECN to 1.7 kA/mm^2 at 4.2 K and 12 T. The next step to ~2.3 kA/mm^2 was made by SMI in 2000. The maximum non-Cu J_c at 12 T and 4.2 K in PIT reached ~2.7 kA/mm^2 in 1.25 mm wires with 288 filaments of 50 μm, developed by SMI for the NED program. In the most recent composite wire, produced by Hypertech Research Inc. in 2019 using PIT technology and APC, the J_c in short samples approached to 1.5 kA/mm^2 at 16 T field, which corresponds to J_c of ~3.6 kA/mm^2 at 12 T. For commercial PIT wire production at Bruker-EAS, the J_c at 12 T and 4.2 K was ~2.5 kA/mm^2.

Optimization of the IT wire design and processing at OST from 1999 to 2006 produced fast progress in J_c at 12 T and 4.2 K, from ~1.5 kA/mm^2 to more than 3 kA/mm^2. The record J_c(12 T, 4.2 K) of 3.3 kA/mm^2 was achieved first in a RRP wire of 54/61 design and a diameter of 0.7 mm. The peak value of J_c(12 T, 4.2 K) in RRP wire production has been essentially stable since 2005. A J_c(12 T, 4.2 K) at or above 2.5 kA/mm^2 is routinely achieved in commercial wires.

The critical current density of Nb_3Sn is a function of magnetic field B, temperature T, and strain ε. Several empirical scaling laws were proposed to parameterize the J_c data (see Appendix 1). One of the practical purposes of parameterization is that of calculating the expected performance of a magnet from I_c measurements of strand samples used as witnesses during coil reaction. The intersection of the critical surface of each coil at the various magnet test temperatures with the B_{peak} load line of the magnet produces the expected coil SSL at that temperature. It is to be noted that SSL values for accelerator magnets typically have very little sensitivity (<0.3%) to the scaling law that is used.

Figure 2.8 shows an example of the critical current for a Nb_3Sn wire vs. magnetic field measured at 4.2 K. Measurements were performed by either current ramping in a fixed magnetic field (V–I method) or field ramping at a fixed transport current (V–H method). Closed symbols represent I_c data measured in a smooth

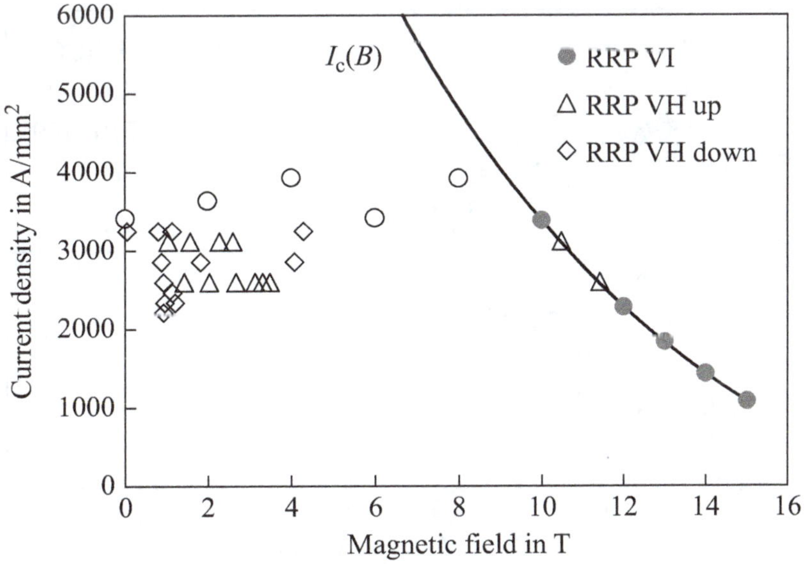

Fig. 2.8 Critical current of a Nb_3Sn wire vs. magnetic field measured at 4.2 K

voltage–current transition, whereas open symbols denote the maximum current I_q reached before a premature quench due to instabilities. An $I_c(B,T)$ parameterization curve is shown by the solid line.

2.3.4.2 Flux Jump Instabilities

Flux jumps in Nb_3Sn composite wires, predicted by stability criteria at fields below certain levels (Wilson 1983), are observed in critical current and magnetization measurements. In critical current measurements the flux jumps are recorded as large voltage spikes and premature quenches below the superconductor critical surface $I_c(B,T)$ during either *V–I* or *V–H* measurements. An example of flux jump instabilities in critical current measurements of RRP wires is shown in Fig. 2.8. In magnetization measurements they are seen as a sawtooth pattern at the low fields (Fig. 2.9). Flux jump instabilities may cause magnet premature quenches and also complicate quench detection in magnets.

2.3.4.3 Nb_3Sn Wire Magnetization

Magnetization of a composite wire plays an important role in superconducting accelerator magnets, which have demanding requirements on field uniformity. Magnetization loops measured at low field ramp rates ($dB/dt < 0.02$ T/s) for IT (IGC), RRP, and PIT wires are shown in Fig. 2.9 per non-Cu volume.

The magnetization of a composite superconducting wire has eddy current and hysteretic components (see Appendix 1). The eddy current component of magnetization in Nb_3Sn composite wires is suppressed by using a small wire twist pitch l_p. For $l_p < 15$ mm and a rather low matrix transverse resistivity $\rho \sim 10^{-10}\ \Omega$ m, the eddy current

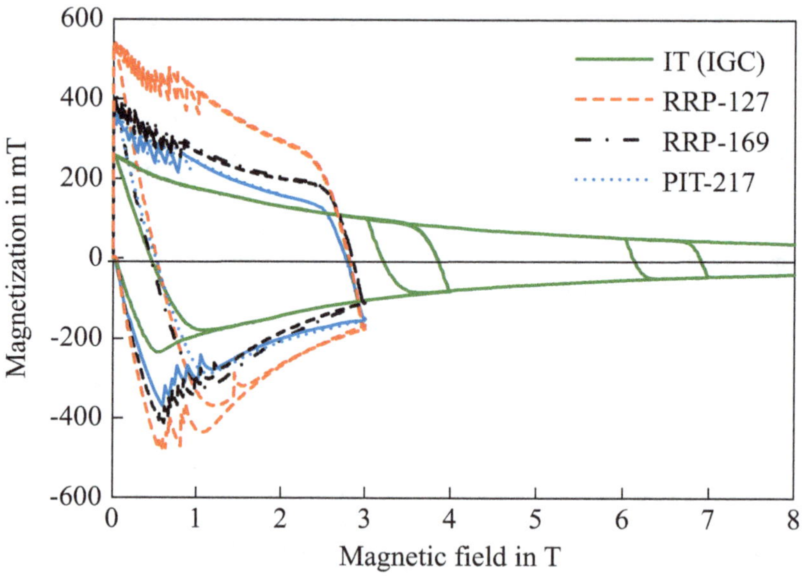

Fig. 2.9 Nb_3Sn wire magnetization per non-Cu volume

magnetization component is less than 1% of the hysteretic component at $dB/dt < 0.1$ T/s, which are typical maximum field variation rates in accelerator magnets.

As expected for the hysteretic component, the magnetization loop width is larger for wires with higher J_c and larger D_{eff}. Due to the larger J_c and D_{eff}, the level of wire magnetization as well as the range of wire re-magnetization when dB/dt changes sign are more than an order of magnitude larger than for Nb-Ti wires used in accelerator magnets. A large magnetization leads to field quality deterioration. Flux jumps also produce field fluctuations from cycle to cycle in accelerator magnets at low fields.

2.3.4.4 Cu Matrix RRR

RRR is a measure of Cu matrix purity, which is important for strand stabilization and magnet quench protection. It is defined as the ratio of the Cu matrix resistivity at room temperature R_{300} to its residual resistivity $R_{4.2}$ at $T = 4.2$ K. When measuring the RRR of the Nb_3Sn wire matrix, $T = 19$ K. A low RRR indicates damage of the internal structure of the strand and Sn leakage into the Cu matrix.

RRR depends on the type of Cu used, on the amount of Sn in the billet, on the diffusion barrier thickness, and on the heat treatment cycle. For instance, when using the same heat treatment cycle for IT wires, RRR values ranging from about 20, to 60, to 160 were obtained for barrier thicknesses of 3 μm, 4.2 μm, and 6 μm, respectively.

Typical RRR values for the latest PIT and RRP round wires are about 200 to 300. Bruker-OST uses Cu 101 with a RRR rating of 300 for a majority of their billets. To maintain a good RRR, OST had also been working on optimizing the Sn fraction in the billet, as well as the diffusion barrier thickness. It is to be noted, however, that RRR, using as numerator the Cu matrix resistivity value at room temperature and zero magnetic field, reduces because of the magnetoresistivity effect, i.e., its value strongly decreases with increasing magnetic field. The same can be expressed in a so-called reduced Kohler plot, where the normalized magnetoresistance increase is shown against the magnetic field multiplied by the resistance ratio at 294 K over 19 K.

The magnetoresistivity effect is stronger for a higher purity Cu matrix, thereby reducing the importance of high RRR at larger fields.

2.4 Rutherford Cables Based on PIT and RRP Wires

2.4.1 Cable Design and Fabrication

Accelerator magnets use high-current, multi-strand superconducting cables to reduce the number of turns in the coils, and thus magnet inductance. The Rutherford cable, developed at the Rutherford Appleton Laboratory (RAL) in the early 1970s (Gallagher-Daggit 1973), has played a crucial role in establishing Nb-Ti accelerator magnet technology due to its excellent mechanical, electrical and thermal properties.

Fig. 2.10 Nb_3Sn Rutherford cables: (**a**) large face view of cable with stainless-steel (SS) core; (**b**) cross-sections of 40-strand cable; and (**c**) 28-strand cable with SS core

It is also widely used in Nb_3Sn magnets. Examples of Nb_3Sn Rutherford cables are shown in Fig. 2.10.

Rutherford cables are produced using special cabling machines, with single-pass and multi-pass approaches. In the former case, the cable is formed into its final cross-section in one single step, i.e., through the largest plastic deformation. In the multi-pass case, a cable with intermediate cross-sections with respect to its final geometry is produced first, which imparts plastic deformation more gradually. This method allows the use of an annealing heat treatment between cabling steps to release tension in the cable.

During cabling, quality monitoring includes checking for strand crossovers, when two adjacent strands alternate their position in the cable by crossing over each other during cable fabrication, and for sharp edges. Crossovers are deleterious to cable performance, and sharp edges are dangerous for the insulation, as shorts may be created. Cable width and thickness are measured periodically or continuously to keep their values within the required tolerances, which are usually of ± 6 μm for thickness and of ± 25 μm for width.

The Rutherford cable geometry is characterized by a cable aspect ratio, a geometrical cross-section area, a pitch angle, a packing factor (*PF*) and, if applicable, a keystone angle. Commonly used definitions of these parameters are summarized in Appendix 1. Large strand-plastic deformations, which were acceptable for a ductile superconductor like Nb-Ti, are not suitable for the more delicate Nb_3Sn strand structure. Examples of RRP strand cross-sections, as deformed after cabling, are shown in Fig. 2.11. Local sub-element deformations due to barrier breakage and merging are shown.

It was found that RRP and PIT strands behave differently under plastic deformation. Despite relative filament deformation being similar for the two, at increasing deformations RRP wires manifest some breaking, but also increasing merging between sub-elements, whereas PIT tubes show only some breaking under shearing.

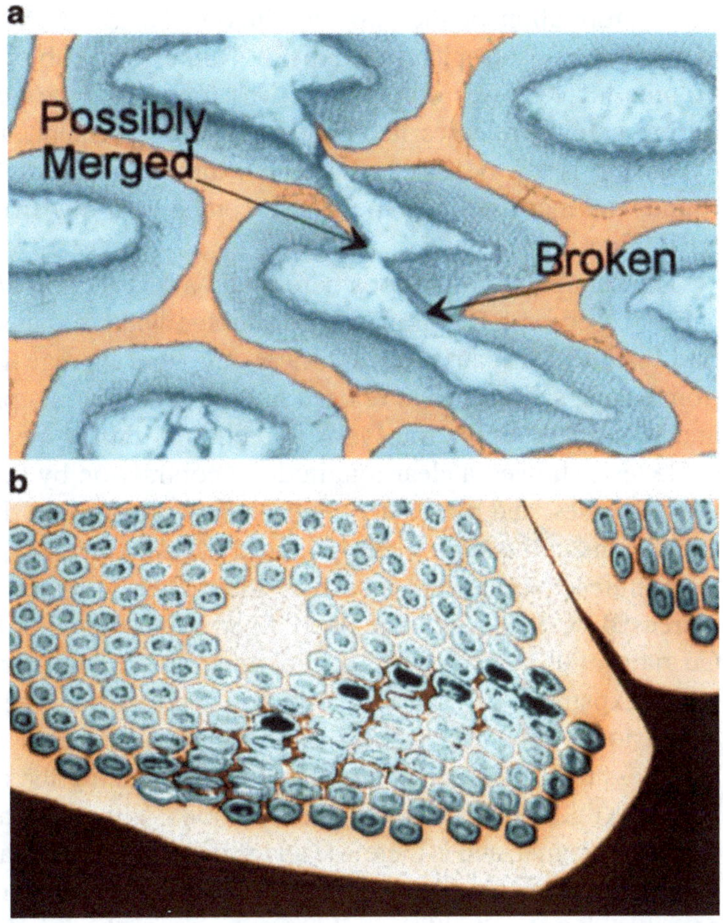

Fig. 2.11 Examples of: (**a**) a deformed RRP strand in a cable with local sub-element damage and merging; and (**b**) tube breakage in PIT wires

In both keystoned and rectangular cables the largest deformation values are found in the strands at the cable edges. The average strand deformation is lower in a less-compacted cable. Deformation of strand cross-sections in cables also causes deformation of the sub-elements, which depends on the cable packing factor *PF*.

The small edge deformation in Nb_3Sn keystoned cables should be larger than 0.85, and the width deformation should be slightly larger than 1.0, typically between 1.0 and 1.03, to avoid excessive strand deformation at the cable's thin edge. The limits on small edge deformation and cable width define a value for the optimal keystone angle of the cable cross-section.

The nominal cable *PF* for Nb_3Sn cables is within 84–87%. This *PF* range allows keeping the critical current degradation of Nb_3Sn Rutherford cables sufficiently low and at the same time provides adequate cable compaction to achieve the mechanical stability that is needed for coil winding.

2.4.2 Cable Size Changes after Reaction

A precise cable cross-section is important for achieving the required field quality in accelerator magnets. It is known that Nb-Sn composite strands expand after reaction due to formation of the superconducting Nb_3Sn A15 phase. In round wires this expansion is isotropic. For Nb_3Sn Rutherford cables, however, an anisotropic volume increase was observed. It was found that the plastic deformation imparted during cabling releases itself through heat treatment, and therefore the thickness expansion is always larger than the width expansion. The volume expansion of a round wire increases with the Nb-Sn content, varying from 2 to 3% for both PIT and RRP wires.

The change in dimensions before and after reaction was measured for keystoned cables based on RRP strands. The average width expansion was 2.6%, the average mid-thickness expansion was 3.9%, and the average length decrease was 0.3%. Some cable samples were reacted under two different conditions, i.e., "unconfined" and "confined" in the transverse direction within a special fixture. The unconfined cable tests showed a clear longitudinal contraction by 0.1–0.2% for the cable made in two passes, and by 0.2–0.3% for the one-pass cable. The thickness and the width expansions in this case were 4% and 2%, respectively. When transversally confined to the unreacted cable cross-section, the cables elongated by ~0.4% and the thickness increased by only ~2%. The unconfined cables made of PIT strands with a Cu/non-Cu ratio of 1.2–1.3 showed a thickness expansion after reaction of 3–3.6% and a width expansion of ~1.5%. It was also reported that the cable width and thickness expansions were affected by the insulation. Braided insulation increased cable thickness expansion while simultaneously reducing the cable width expansion.

Cable thickness expansion after reaction is an important parameter for cables to be used in a magnet, and is therefore an important part of a magnet development plan. For magnetic design optimization, it is the reacted thickness and width values that are used as cable dimensions. The coil dimensions in the winding and curing tooling are determined by the unreacted cable cross-section, whereas the coil dimensions in the reaction and impregnation tooling are based on the reacted cross-section.

2.4.3 *Cable Performance Parameters*

The most important parameters that define the performance of a Rutherford cable in a magnet include the critical current I_c and average critical current density J_a, the Cu/non-Cu ratio, cable RRR, and interstrand resistances R_c and R_a, where R_c is the resistance between strands from the two cable layers and R_a is the resistance between adjacent strands within a layer. The parameters of the heat treatment cycle affect I_c, RRR, and contact resistances R_c and R_a.

2.4.3.1 Cable Critical Current I_c and Flux Jump Instabilities

The critical current I_c evaluation of Rutherford cables is performed by either testing short cable samples or individually extracted strands. The good correlation usually found between cable and extracted strand test results at high fields, as shown for instance in Fig. 2.12, confirms the validity of both approaches. As in Fig. 2.8, closed

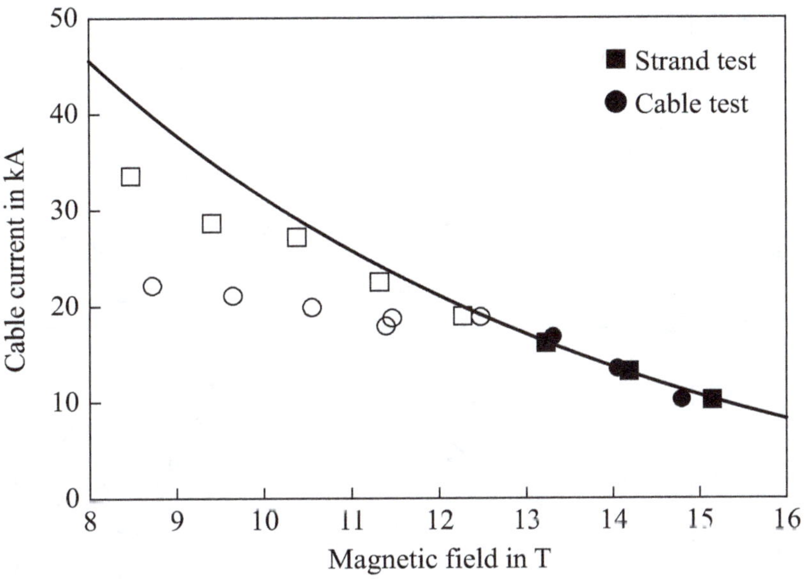

Fig. 2.12 Maximum current vs. magnetic field of a 40-strand Nb_3Sn cable with RRP strands. The solid line represents an $I_c(B)$ parameterization

symbols represent I_c data measured in a smooth voltage–current transition, whereas open symbols denote the maximum current I_q as reached before an abrupt quench due to instabilities. The solid line represents an $I_c(B)$ parameterization. A good correlation between extracted strand and cable test results also reveals a small variation of strand properties within the billets used in the cable, and is consistent with a uniform transport current distribution.

The effect of flux jump instabilities observed at low fields in Nb_3Sn wires was also seen in tests of short cable samples (see Fig. 2.12). Analysis and comparison of flux jump instabilities in Rutherford cables and their corresponding round wires show that these instabilities are larger in cables, due to sub-element deformation and possible sub-element merging (Fig. 2.11a), which in both cases lead to an increase of the local D_{eff}. The reduction of local RRR after cabling also increases flux jump instabilities in cables with respect to virgin wires.

2.4.3.2 Effect of Strand Plastic Deformation in Cables

Cable degradation studies and optimization are an important part of the magnet development process. The effect of cable plastic deformation on the critical current I_c, minimal stability current I_S and matrix RRR was studied using strands extracted from cables with different *PF*s. The results of I_c measurements made on extracted strands were compared with those made on the round wire from billets used in the cables. The cable I_c and critical current density J_a per cable cross-section area at 4.2 K and 12 T normalized to the I_{c0} and J_{a0} of a cable made of undeformed round strands ($PF = 78.5\%$), respectively, are plotted in Fig. 2.13 vs. cable *PF*. Whereas I_c

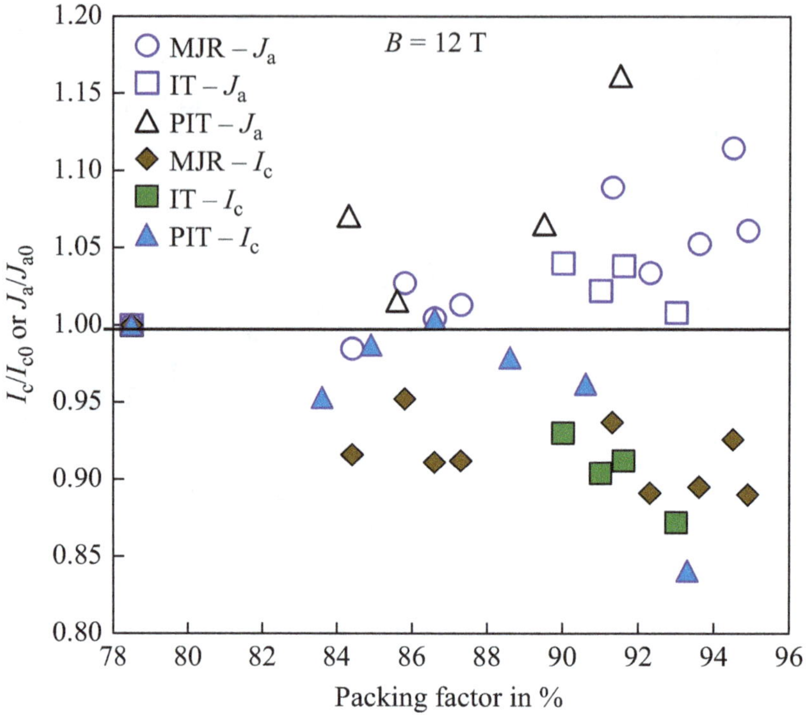

Fig. 2.13 Normalized cable I_c (filled markers) and J_a (unfilled markers) at 4.2 K and 12 T vs. *PF* for cables made with IT, MJR, and PIT Nb_3Sn strands

decreases with increasing *PF*, the normalized critical current density J_a in compacted cables is larger than in the undeformed cable, since the I_c degradation is less than the reduction of cable geometrical cross-section.

Some early IT wires demonstrated a relative I_c degradation of up to 80% at *PF*s above 84%. A large I_c degradation of up to 60% was also observed in early PIT strands. In 2004, however, SMI and FNAL found that by using round filaments in PIT wires, the I_c degradation could be reduced to 15% or less at *PF*s up to 94%. At a *PF* between 84% and 87%, which is typical for Nb_3Sn Rutherford cables, the I_c degradation in well-optimized cables is usually ~5% or less. Similar measurements performed at CERN on cables made with modern RRP and PIT strands are consistent with these data.

The effect of strand plastic deformation due to cabling on the stability current I_S and on the RRR is much stronger than on the I_c due to sub-element damage, and merging and Cu matrix contamination. It was also found that RRP strands with extra spacing between sub-elements were able to maintain a higher I_S in the higher *PF* range (above 90%).

Based on the results of I_c degradation in Nb_3Sn Rutherford cables, high *PF* values of 92–95% provide the highest J_a. Large I_S and RRR degradation values due to large deformations and possible damage and merging of the delicate sub-elements, however, impose an optimal *PF* within 84–87%.

2.4.3.3 Cable RRR

Due to the larger strand deformation at the cable edges, it was expected that RRR would vary along a strand. The first resistivity measurements made on extracted strands from high-edge compacted cables showed significant RRR degradation from an RRR $\approx$ 120 measured on strand segments on the cable faces. On the edges the results were an order of magnitude smaller, i.e., RRR $\approx$ 13, consistent with local Sn leakages through the diffusion barriers caused by the plastic deformation at the cable edges. Since this reduction is local, the average RRR value was still large, around 81. Such a large RRR degradation at the edges is often found even in cables with low packing factors. Recent measurements were performed at CERN on optimized cables made out of RRP132/169 wires with high RRR (between 100 and 250) for the 11 T dipole. The RRR reduction on the small edge was between 40% and 50%, and that on the large edge between 70% and 75% of the maximum values obtained on the cable flat parts.

2.4.3.4 Effect of Transverse Pressure

Transverse stress is the largest stress component in accelerator magnets, and can therefore damage brittle Nb_3Sn coils. To determine I_c sensitivity to transverse pressure, electro-mechanical tests are typically performed on either cables or encased wires. Transverse pressure studies are made by applying pressure to impregnated cable or wire samples and testing their transport current at several magnetic fields. Strain sensitivity, as previously indicated, increases with magnetic field. There are two components of the critical current degradation: a reversible component, which is fully recovered when removing the load, and an irreversible component. The latter is permanent. The irreversible limit is defined as the pressure leading to a 95% recovery of the initial I_c, or I_{c0}, after unloading the sample.

Institutions where transverse pressure measurements are performed include FNAL, the University of Twente, the National High Magnetic Field Laboratory, CERN, and the University of Genève. The I_c degradation strongly depends on conductor technology, but also on sample preparation and setup design. The former has an impact on possible stress concentration; the latter determines the sample's actual stress–strain state.

Figure 2.14 shows the total I_c degradation at 4.2 K and 12 T (unless otherwise indicated in figure caption), at transverse pressures up to ~200 MPa, of epoxy-impregnated Rutherford cables and encased wires made of PIT and RRP. Most of the load in this plot was applied cold. Older data (not shown) indicate that cables made of low-J_c strands are less sensitive to transverse pressure than those made with state-of-the-art, high-J_c strands. In some cases a stainless-steel core inside the cable also reduced pressure sensitivity.

In reference to magnet assembly and operation, it is important to distinguish the areas with maximum stress from those with maximum field, as well as the actual temperature conditions of the load application. Recent experiments performed at

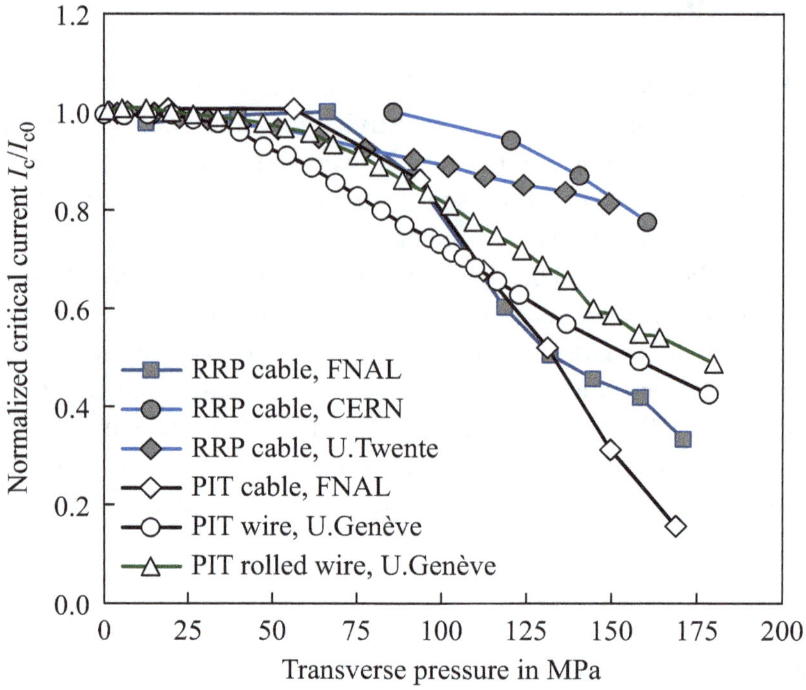

Fig. 2.14 Normalized I_c at 4.2 K vs. transverse pressure on Rutherford cable face or on encased wire for epoxy impregnated samples. All the Ic data correspond to 12 T, except for the PIT round and rolled wires which were measured by U. Genève at 19 T

FRESCA to measure the irreversible component with the load applied warm showed that the irreversible component is negligible at least up to 150 MPa. It is important that these studies continue.

2.4.3.5 Interstrand Resistance

Contact resistances between adjacent strands inside each layer, R_a, and between strands from different layers, R_c, control eddy current magnetization and AC losses, as well as current and heat sharing between strands in a Rutherford cable. To reduce eddy current effects without compromising cable electrical stability, one needs to increase R_c while keeping R_a low. Calculations show that for field ramp rates up to 0.1 T/s, R_c has to be larger than ~10 μΩ, and R_a larger than ~0.1 μΩ. In this case, the eddy current effect will be negligible with respect to the persistent current effect.

Direct measurements of R_c and R_a contact resistances performed under transverse pressures of 10–100 MPa gave $R_c = 1.1$–1.4 μΩ and $R_a = 8$–16 μΩ for Nb_3Sn cables without a core, and $R_c = 150$–275 μΩ and $R_a = 1.5$–1.9 μΩ for cables with a 25 μm stainless-steel core. Very low R_c values of ~0.1–0.4 μΩ were measured in Nb_3Sn Rutherford cables without a core reacted in coils under pressure. In cables with a full-width stainless-steel core, a high R_c of 246 μΩ was found based on AC loss measurements. For comparison, in the LHC Nb-Ti cables, typical R_c values under pressure are ~20 μΩ, i.e., an order of magnitude larger than in a Nb_3Sn cable without a resistive core, and an order of magnitude lower than in a Nb_3Sn cable with resistive core. Typical R_a values for Nb-Ti cables are ~170 μΩ.

Since the R_a limit is much smaller than the actual R_a data, their large variations will not affect magnet field quality or AC losses, whereas low values and large variations of R_c data in cables without a core will produce a large impact on magnet field quality and quench performance. Studies of interstrand contact resistances in Nb_3Sn Rutherford cables have shown that using a stainless-steel core is very efficient in reducing the level of eddy current effects (magnetization, AC loss) in cables. This approach is simpler than using a resistive coating, such as chromium, on individual strands. Resistive coating increases both R_c and R_a to 100 μΩ and higher, and may negatively impact the cable's electrical stability.

2.5 Conclusion

High-performance composite wires and Rutherford cables are key components of superconducting accelerator magnets. Whereas Nb-Ti has been the workhorse for HEP applications for almost 50 years, in the past 25 years Nb_3Sn wires and cables have made steady progress in wire fabrication techniques, understanding of their properties, and large-scale commercial production. They have now approached the necessary maturity to be used in accelerator magnets. Coupled with innovative development in the Nb_3Sn magnet technology itself, conductor progress has made it possible for accelerator magnets to achieve magnetic fields above 10 T, which is beyond the reach of Nb-Ti technology. In 2010, Nb_3Sn was adopted as the baseline conductor for the HL-LHC at CERN. Still, the potential of Nb_3Sn wires and cables is not fully apprehended. Work on Nb_3Sn composite wires for accelerator magnets continues, focusing on a better understanding and optimization of their technical performance and of their market parameters.

Appendix 1—Wire and Cable Parameterizations

The Rutherford cable geometry is characterized by a cable aspect ratio α and a cross-sectional area S_{cbl}, keystone and pitch angles, and cable packing factor *PF*.

Cable Aspect Ratio

The cable aspect ratio is the ratio of the cable width w to its mean thickness t

$$\alpha = w/t.$$

Cable Cross-Section

The cable cross-section is defined as the cable cross-section envelope

$$S_{\text{cbl}} = w \cdot t.$$

Pitch or Transposition Angle θ

The cable pitch angle affects the cable's mechanical stability and the critical current degradation. Typical values of pitch angle in Nb-Ti cables used in accelerator magnets were within 13° to 17°. A study of the possible pitch angle range for Nb_3Sn Rutherford cables was performed using 1 mm and 0.7 mm non-annealed A101Cu strands. For 1 mm strands, below 12° the cable showed mechanical instability, and for angles larger than 16°, popped strands, sharp edges, and cross-overs occur. For 0.7 mm strands, the stable range of transposition angles was between 9° and 16°.

Cable Packing Factor PF

The cable packing factor, *PF*, is defined as the ratio of the total cross-section of the strands to the cable cross-section envelope $S_{\text{cbl}} = w \cdot t$

$$PF \cong \frac{\pi N \cdot D^2}{4(w \cdot t - A_{\text{core}}) \cdot \cos\theta},$$

where *N* is the number of strands in the cable, *D* is the strand diameter, *w* and *t* are the average cable width and thickness, *θ* is the cable transposition angle, and A_{core} is the cross-sectional area of the core.

The minimal *PF* for a Rutherford cable, i.e., one having a non-deformed cross-section, has a value of ~π/4 = 0.785. To provide cable mechanical stability and precise width and thickness (parameters that are important for accelerator magnet coils), Rutherford cables are usually compacted by compressing their cross-section in both transverse directions. For an I_c degradation limited to 5–10%, increasing the cable *PF* also allows raising the cable average current density J_a, which is defined as

$$J_a = I_c / S_{\text{cbl}}.$$

Cable Edge and Width Deformations R_e, R_w

The critical current degradation is determined mainly by the amount of cable cross-section deformation. The deformations of cable edge R_e and width R_w are defined as

$$R_e = \frac{t}{2D},$$
$$R_w \cong \frac{2w \cdot \cos\theta}{N \cdot D},$$

where D is the strand diameter, N is the number of strands in the cable ($N = N + 1$ in the case of odd N), and θ is the cable transposition angle. They can also be expressed as edge c_t and width c_w compactions (Bottura and Godeke 2012), which are related to R_e and R_w as

$$c_t = R_e - 1,$$
$$c_w = R_w.$$

Critical Current Density

Critical current density is a key parameter, which controls the current carrying capability, stability, magnetization, and AC losses of a superconducting wire, and thus the performance of superconducting magnets. Parameterizations of the critical current density as a function of magnetic field B, temperature T and strain ε, can be generally expressed as

$$J_c(B, T, \varepsilon) = \frac{C(t, \varepsilon)}{B}(1 - t^m)^n b^p (1 - b)^q,$$

where

$$b = B/B^*_{c2}(T, \varepsilon),$$
$$t = T/T_{c0}(\varepsilon),$$
$$B^*_{c2}(T, \varepsilon) = B^*_{c20}(0, \varepsilon)(1 - t^\nu)k(t).$$

Parameters m, n, p, q, and ν as well as functions $C(t,\varepsilon)$, $T_{c0}(\varepsilon)$, $k(t)$, and $B^*_{c20}(0,\varepsilon)$ are usually determined by fitting experimental data of Nb_3Sn wires. For the practical strain range of $-1 < \varepsilon < 0.5$, the experimental data are well fitted with $m = n = q = 2$, $p = 0.5$, $\nu = 1.7$–2, and $C(t,\varepsilon) = C(\varepsilon)$.

Engineering Current Density

The engineering current density J_E is defined as the critical current density per total conductor cross-section. It depends on the superconductor J_c and superconductor fraction in the composite cross-section

$$\lambda = 1/(1 + r),$$

where r is the Cu/non-Cu ratio.

Magnetization and AC Losses

A composite superconductor placed in a varying magnetic field becomes magnetized (Wilson 1983) with a magnetization described by

$$M(B, \dot{B}) = -\mu_0 \left[\frac{2}{3\pi} \lambda J_c(B) D_{\text{eff}} + \frac{l_p^2 \dot{B}}{4\pi^2 \rho(B)} \right],$$

where D_{eff} is the effective filament diameter, l_p is the filament twist pitch, $\rho(B)$ is the effective transverse resistivity of the matrix, and $J_c(B)$ is the critical current density in the superconductor. The first term represents the component related to persistent currents in the superconducting filaments, and the second term represents the component associated with coupling eddy currents between filaments. Both components are diamagnetic in an increasing field and paramagnetic in a decreasing field. Composite wire magnetization plays an important role in superconducting accelerator magnets, which have demanding requirements on field uniformity.

Magnetic hysteresis leads to energy dissipation in superconducting composites (Wilson 1983). Similarly to magnetization, the power of AC losses P in a composite superconductor has two main components related to persistent and coupling eddy currents. The AC loss power per unit volume of composite wire after full flux penetration in the superconducting filaments can be represented as

$$P(B, \dot{B}) \cong \frac{2}{3\pi} \mu_0 \lambda J_c(B) D_{\text{eff}} \dot{B} + \frac{\mu_0 l_p^2 \dot{B}^2}{4\pi^2 \rho(B)}.$$

AC losses in composite superconductors play an important role in the thermal stabilization of superconducting coils during magnet operation and quench, and contribute to the heat load on a magnet cooling system.

Due to electromagnetic coupling between strands, the Rutherford cable magnetization and AC loss power include additional eddy current contributions, which depend on the cable geometry and interstrand contact resistances as

$$\vec{M}_c \cong -\mu_0 \left(\frac{8\alpha^2 L^2}{15\rho_c} \frac{d\vec{B}_\perp}{dt} + \frac{L^2}{3\rho_a} \frac{d\vec{B}_\perp}{dt} + \frac{L^2}{4\alpha^2 \rho_a} \frac{d\vec{B}_\parallel}{dt} \right),$$

$$P_c \cong \frac{8\alpha^2 L^2 \dot{B}_\perp^2}{15\rho_c} + \frac{L^2 \dot{B}_\perp^2}{3\rho_a} + \frac{L^2 \dot{B}_\parallel^2}{4\alpha^2 \rho_a},$$

where *4L* is the cable transposition pitch, α is the cable aspect ratio, $B_\perp$ and $B_\parallel$ are the perpendicular and parallel components of the magnetic field to the cable wide surface, and ρ_c and ρ_a are the effective cable resistivity between cable layers and within a layer, respectively. The first term in both formulae provides the main contribution due to the large value of α. The parameters ρ_c and ρ_a are related to the measurable contact resistances R_c and R_a as

$$\rho_c = \frac{4\alpha L}{N^2} R_c \text{ and } \rho_a \cong \left(\frac{2L}{\cos\theta} \right)^2 R_a,$$

where R_c is the contact resistance of two crossing strands and R_a is the resistance per unit length of two adjacent strands in a cable.

References

Ballarino A, Bottura L (2015) Targets for R&D on Nb_3Sn conductor for high energy physics. IEEE Trans Appl Supercond 25(3):1–6. https://doi.org/10.1109/tasc.2015.2390149

Barzi E, Zlobin AV (2016) Research and development of Nb_3Sn wires and cables for high-field accelerator magnets. IEEE Trans Nucl Sci 63(2):783–803. https://doi.org/10.1109/tns.2015.2500440

Bottura L, Godeke A (2012) Superconducting materials and conductors, fabrication, and limiting parameters. In: Chao AW, Chou W (eds) Reviews of accelerator science and technology, vol 5. World Scientific, Singapore, pp 25–50

Cheggour N, Goodrich LF, Stauffer TC et al (2010) Influence of Ti and Ta doping on the irreversible strain limit of ternary Nb_3Sn superconducting wires made by the restacked-rod process. Supercond Sci Technol 23:052002. https://doi.org/10.1088/0953-2048/23/5/052002

Chernoplekov NA (1978) Superconducting magnet systems for plasma physics research in the USSR. In: Polak M et al (eds) Sixth international conference on magnet technology (MT-6), 29 Aug–2 Sep 1977. ALFA, Bratislava, pp 3–12

Chernoplekov NA (1981) Status and trends of SC magnets for thermonuclear research in the USSR. IEEE Trans Magn 17(5):2158–2167. https://doi.org/10.1109/tmag.1981.1061408

Devred A, Baynham DE, Bottura L et al (2004) High field accelerator magnet R&D in Europe. IEEE Trans Appl Supercond 14(2):339–344. https://doi.org/10.1109/tasc.2004.829121

Dietderich DR, Godeke A (2008) Nb_3Sn research and development in the USA—Wires and cables. Cryogenics 48(7–8):331–340. https://doi.org/10.1016/j.cryogenics.2008.05.004

Foner S, Schwartz BB (1981) Superconductor material science: metallurgy, fabrication, and applications. Plenum Press, New York

Gallagher-Daggit G (1973) Superconductor cables for pulsed dipole magnets. Technical report, Rutherford High Energy Laboratory Memorandum, No. RHEL/M/A25, Chilton, Didcot

Godeke A, den Ouden A, Nijhuis A et al (2008) State of the art powder-in-tube niobium–tin superconductors. Cryogenics 48(7–8):308–316. https://doi.org/10.1016/j.cryogenics.2008.04.003

Hashimoto Y, Yoshizaki K, Tanaka M (1974) Processing and properties of superconducting Nb_3Sn filamentary wires. In: Mendelssohn K (ed) Proceedings of the 5th international cryogenics engineering conference, 7–10 May 1974. Cryogenic Association of Japan/IPC Science and Technology Press, Guildford/Kyoto, pp 332–335

Howlett EW (1970) US Patent 3, 728, 165, 19 Oct 1970; UK Patent 52, 623/69, 27 Oct 1969

Kaufmann AR, Pickett JJ (1971) Multifilament Nb_3Sn superconducting wire. Bull Am Phys Soc 15:838. https://doi.org/10.1063/1.1659651

Kunzler JE, Buehler E, Hsu FSL et al (1961) Superconductivity in Nb_3Sn at high current density in a magnetic field of 88 kgauss. Phys Rev Lett 6(3):89. https://doi.org/10.1103/PhysRevLett.6.89

Lindenhovius JH, Hornsveld EM, den Ouden A et al (1999) Progress in the development of Nb_3Sn conductors based on the "powder in tube" method with finer filaments. IEEE Trans Appl Supercond 9(2):1451–1454. https://doi.org/10.1109/77.784664

Matthias BT, Geballe TH, Geller S et al (1954) Superconductivity of Nb_3Sn. Phys Rev 95(6):1435. https://doi.org/10.1103/physrev.95.1435

Parrell J, Zhang Y, Hentges RW et al (2002) Nb_3Sn strand development at Oxford Superconducting Technology. In: Balachandran B, Gubser D, Hartwig TK (eds) Advances in cryogenic engineering: proceedings of the international cryogenic materials conference, Madison, Wisconsin, 16–20 July 2001, Melville, vol 48. American Institute of Physics, New York, pp 968–977

Parrell JA, Field MB, Zhang Y et al (2004) Nb_3Sn conductor development for fusion and particle accelerator applications. Adv Cryo Eng (Materials) 50B:369–375. https://doi.org/10.1063/1.1774590854

Scanlan RM (2001) Conductor development for high energy physics-plans and status of the US program. IEEE Trans Appl Supercond 11(1):2150–2155. https://doi.org/10.1109/77.920283

Tachikawa K (1971) Studies on superconducting V_3Ga tapes. In: International cryogenics engineering conference Berlin, 25–27 May 1970. Iliffe Science and Technology Publications, Guildford, p 339

Wilson MN (1983) Superconducting magnets. Oxford University Press, New York

Xu X, Zlobin AV, Peng X, Li P (2019) Development and study of Nb_3Sn wires with high specific heat. IEEE Trans Appl Supercond 29(5):6000404

Yamasaki H, Kimura Y (1982) Fabrication of Nb_3Sn superconductors by the solid–liquid diffusion method using Sn rich CuSn alloy. Cryogenics 22(2):89–93. https://doi.org/10.1016/0011-2275(82)90100-x

Chapter 3
Nb_3Sn Accelerator Magnets: The Early Days (1960s–1980s)

Lucio Rossi and Alexander V. Zlobin

Abstract Since Nb_3Sn became available in the form of relatively long tapes in the early 1960s, it was considered for high-field magnets thanks to its high upper critical magnetic field and critical temperature. This chapter is a review of the effort accomplished by the community in the first 25 years of development of Nb_3Sn accelerator magnets, and an attempt to understand why it took so long before this technology became successful.

3.1 Introduction

Discovery and major advancements in the understanding and development of practical type II superconductors in the 1950s and 1960s made it possible to employ them in superconducting magnets, particularly, in accelerator magnets used for steering charged particle beams. The superconducting properties of Nb_3Sn were discovered in the early 1960s, and soon it was available in form of tapes (or ribbons) produced by the surface diffusion process. Shortly afterwards an alternative concept based on solid-state diffusion was introduced. This principle has been used since the 1970s to fabricate Nb_3Sn composite wires by the so-called bronze route. In 1974 the internal tin (IT) process was developed, which allowed overcoming the limitation of the bronze method to reach a high critical current density J_c due to the limited content of tin in bronze. The availability of IT composite wires eventually proved to be a key asset for high-field Nb_3Sn magnet research and development (R&D) for accelerators. The year-by-year history of Nb_3Sn wire development can be found in Foner and Schwartz (1981).

L. Rossi (✉)
CERN (European Organization for Nuclear Research), Meyrin, Genève, Switzerland

University of Milan, Physics Department, Milano, Italy
e-mail: Lucio.Rossi@cern.ch

A. V. Zlobin
Fermi National Accelerator Laboratory (FNAL), Batavia, IL, USA
e-mail: zlobin@fnal.gov

D. Schoerling, A. V. Zlobin (eds.), *Nb_3Sn Accelerator Magnets*, Particle Acceleration and Detection, https://doi.org/10.1007/978-3-030-16118-7_3

From the 1960s through the 1980s various teams set up programs in the USA, Europe, and Japan to develop Nb_3Sn magnets as an alternative to Nb-Ti magnets in the 6–10 T field range. In those days, a field of 5–6 T was considered to be the maximum field attainable at 4.2 K in Nb-Ti magnets. It was known that the use of superfluid helium considerably boosts Nb-Ti performance. Complexity, reliability, and cost of superfluid helium cryogenic systems in the early 1980s suggested, however, that Nb_3Sn magnets operated at 4.2 K could be a serious alternative to Nb-Ti magnets operated at 1.9 K for the 8–10 T field range. For this reason, various high-energy physics (HEP) laboratories around the world started exploring the feasibility of Nb_3Sn magnets in the 8–10 T range for future colliders beyond the Tevatron (Edwards 1985), which was under construction at the Fermi National Accelerator Laboratory (FNAL, also known as Fermilab) from the late 1970s to the early 1980s. A more detailed historical overview of Nb_3Sn conductor R&D for HEP, which found the limitations and provided breakthroughs for Nb_3Sn accelerator magnet R&D programs, is presented in Chap. 2 of this book.

The effort in the USA was mainly concentrated at Brookhaven National Laboratory (BNL), along with Nb-Ti magnet R&D for the Intersecting Storage Accelerator (ISABELLE), later renamed the Colliding Beam Accelerator (CBA) project (cancelled in 1983) (Dahl et al. 1973), and later at Lawrence Berkeley National Laboratory (LBNL). In Europe, both the European Organization for Nuclear Research (CERN), Switzerland, which developed and operated the first low-beta superconducting quadrupoles in the Intersecting Storage Rings (ISR) (Billan et al. 1979), and the Commissariat à l'Energie Atomique (CEA), France, launched R&D programs for a possible hadron collider for the post-Large Electron–Positron (LEP) collider era. At the same time, the High Energy Accelerator Research Organization (KEK), Japan, also started a Nb_3Sn-based program for accelerator magnets. Since it is beyond the scope of this chapter to account for all the efforts at those laboratories, we have selected a few endeavors to illustrate progress in an almost chronological order (our sincere apologies go to the projects not reported here). The main parameters of the projects described below are summarized in the table at the end of this chapter.

3.2 Nb_3Sn Accelerator Magnets in the 1960s

3.2.1 *First Nb_3Sn Quadrupoles at BNL*

The first Nb_3Sn accelerator magnets were built at BNL in the middle of the 1960s, only a few years after the first Nb-Ti accelerator magnets. The magnets had a quadrupole configuration and were built using a Nb_3Sn ribbon (in modern terminology, a tape) and the react-and-wind (R&W) method. The first Panofsky-type quadrupole model with square current blocks is shown in Fig. 3.1. It had a 30 mm bore and produced a field gradient of 100 T/m (Britton 1968). Despite the reasonable performance, this coil design was abandoned due to several deficiencies, e.g., the large field strength concentration in the coil corners, rather large coil ends and large

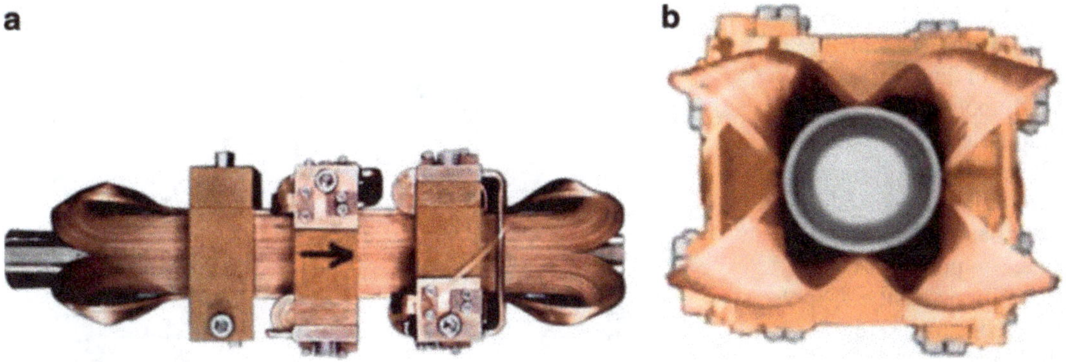

Fig. 3.1 Side (**a**) and end (**b**) views of the first BNL Nb_3Sn quadrupole model with square current blocks. (Modified from Britton 1968)

Fig. 3.2 The shell-type Nb_3Sn quadrupole. (Modified from Britton and Sampson 1967)

end field errors, and the complicated coil winding technology. Even so, several quadrupoles of this type were fabricated.

The next BNL magnets were developed using a theoretical model based on current arrays or blocks placed around a circular or elliptical bore. The first quadrupole using this type of coil (also known as the shell-type coil) is shown in Fig. 3.2. It had a 76 mm bore and a 200 mm long straight section (Sampson 1967; Britton and Sampson 1967). Unfortunately, the details of magnet fabrication are not known at the present time. The magnet was tested in January 1966 and produced a maximum field gradient of 85 T/m at 630 A current, with a peak field on the walls of 3.2 T. The

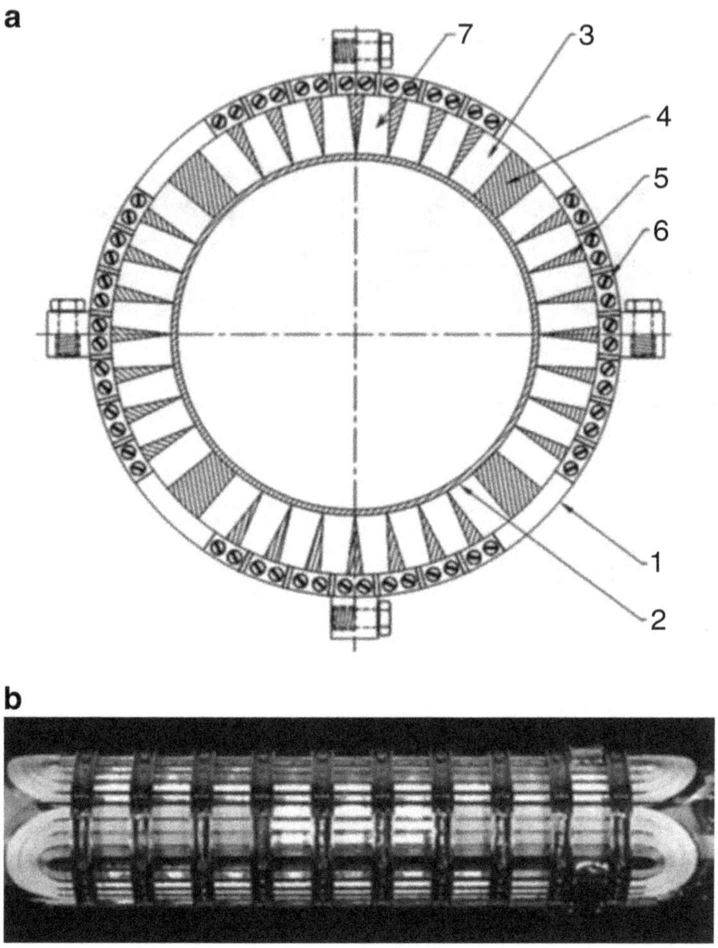

Fig. 3.3 (**a**) The cross-section; and (**b**) a picture of the beam line Nb_3Sn quadrupole coil (Modified from Britton and Sampson, 1967; Britton, 1968): *1* – Lorentz force constraining outer bands; *2* – mounting tube; *3* – pole block with lowest turn density; *4* – pole; *5* – wedge; *6* – screw to align wedge; 7 – mid-plane block with highest turn density

field was nearly two times higher compared to similar iron-dominated quadrupoles with water-cooled copper coils.

All the BNL models, except those with square blocks, used a 12.7 mm wide by 0.125 mm thick copper-plated ribbon (the thickness of the Nb_3Sn layer was at least 0.075 mm). The available ribbon length varied between 300 and 950 m, which was sufficient to avoid internal splices. Typically, this ribbon carried 600 A at 4 T, which corresponded to a current density of 400 A/mm^2.

The insulation of the Nb_3Sn ribbons used in all the BNL beam magnets was either a thin stainless-steel tape for magnets with a stored energy above 10 kJ or, for smaller magnets, a varnish coating applied on the ribbon by the vendor.

Longer quadrupoles with wider apertures were also planned for a beam line doublet. The cross-section of a 100 mm aperture, 0.6 m long quadrupole is shown in Fig. 3.3. These magnets had to be wound with a better conductor to generate a coil peak field as high as 6 T. To our knowledge, however, the status and results of this work were never reported.

a
b
c

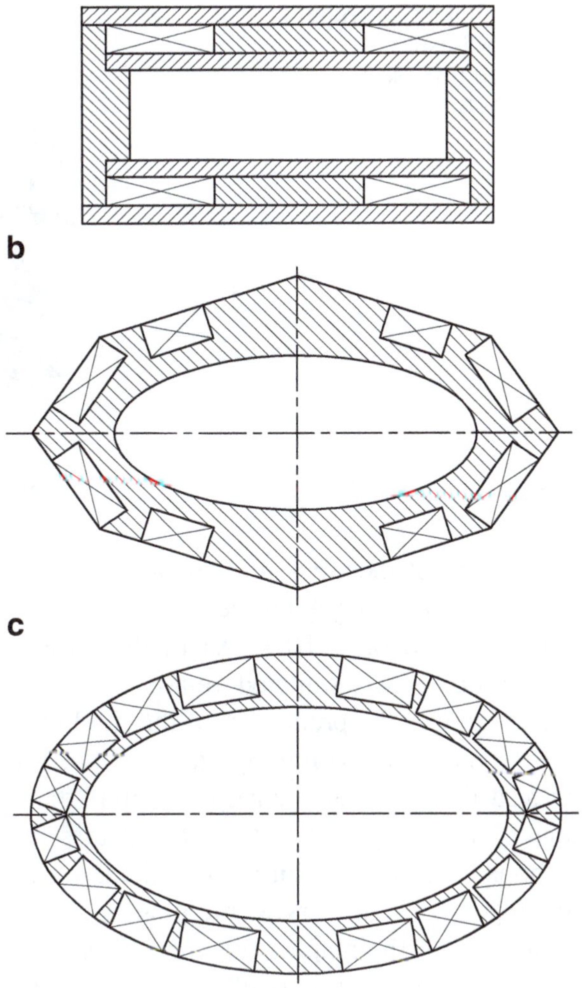

Fig. 3.4 Cross-sections of: (**a**) Nb_3Sn racetrack dipole coil; and (**b**, **c**) Nb_3Sn shell-type dipole coil. (Modified from Britton 1968)

3.2.2 *First Nb_3Sn Dipoles for Synchrotron at BNL*

In parallel with the quadrupoles, several dipole magnets were developed and constructed at BNL to study magnet quench performance and alternating current (AC) losses (Sampson et al. 1967). A 350 mm long dipole model with two flat racetrack coils separated by a 50 mm gap, and a dipole model with elliptical aperture were built and tested. Their cross-sections are shown in Fig. 3.4a, b. Figure 3.4c shows a cross-section of a real accelerator dipole with good field quality and field direction designed to be parallel to the ribbon face in the magnet straight section to take into consideration the J_c anisotropy in the ribbons. The magnets were made of 12.7 mm wide Nb_3Sn ribbon, which provided the highest current density available at that time.

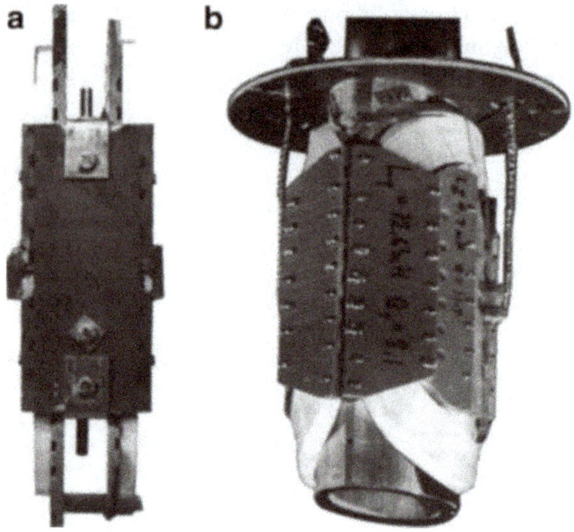

Fig. 3.5 (**a**) BNL racetrack and (**b**) elliptical-aperture shell-type dipoles. (Modified from Sampson et al. 1967)

Pictures of racetrack and shell-type dipole models are shown in Fig. 3.5. These magnets were designed to be used in a synchrotron where the field had to be cycled with a period of 1–10 s. As in the case of the quadrupoles, the proper current distribution was simulated relatively easily using the current block model. The challenge was the prediction of the field level and the level of the AC losses during the expected fast cycling operation. Since multi-filamentary Nb_3Sn composite wires with sufficient current density and small filament size were not yet available, the field quality and, especially, the power losses were a concern, when working with ribbons. The latter required the efficient removal of heat from the coil and increased the heat load on the cryogenic plant.

Results of the magnet quench performance tests for the racetrack and shell-type dipole models are not available. The AC loss measurements demonstrated that the losses were too high for operation in a synchrotron. Even for a relatively slow 10 s cycle, a refrigeration power of at least 20–100 W/m would have been required to remove the heat from the coils.

3.2.3 Nb_3Sn Magnet Design Based on Braided Cable

Experimental and theoretical studies showed that to reduce the AC losses in superconducting accelerator magnets, high-current cables based on twisted multi-filament composite wires were needed. To address this issue, BNL developed a conductor based on braided cable, which became a distinctive characteristic of that laboratory for many years. The braid provided a complete transposition of many parallel strands, which, combined with the twisting of each single strand, drastically reduced the losses.

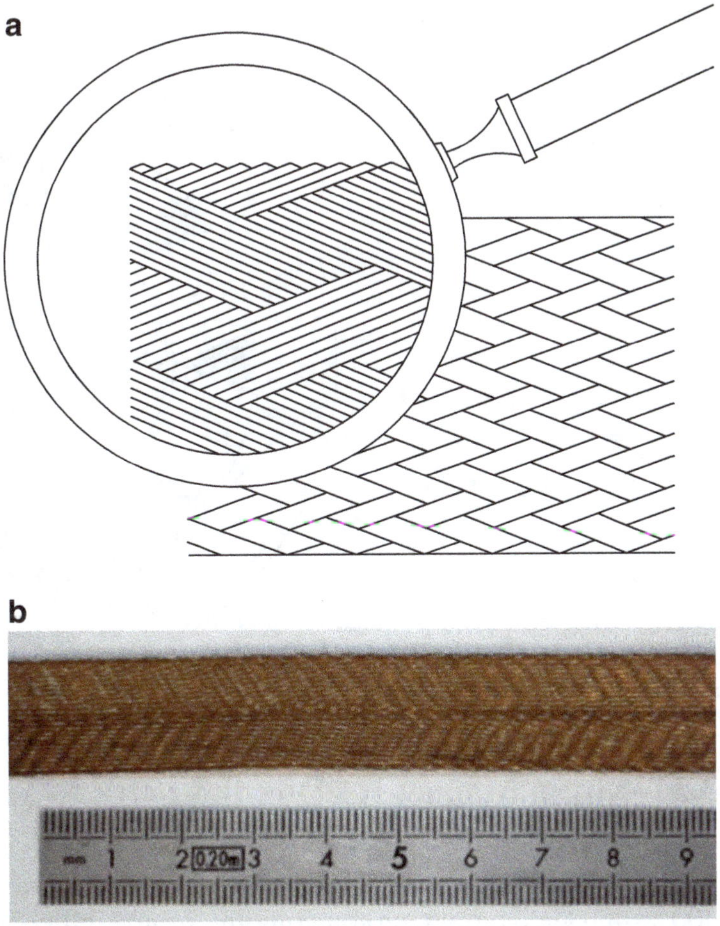

Fig. 3.6 (**a**) The first BNL braided cable concept from 1969 (modified from Sampson et al. 1969); (**b**) the 16.9 mm wide, 1.04 mm thick 97 strands braid used in the RHIC bus (courtesy of P. Wanderer, BNL)

The first braid design was quite complex, see Fig. 3.6a. It had insulated fine Nb_3Sn strands and smaller steel strands implemented for mechanical stability. It was expected that by weaving a braid with superconducting filaments of around 10 μm in diameter, a heat dissipation of less than 10 W/m would be achievable. Another version of the BNL braided cable that was later used in the Relativistic Heavy Ion Collider (RHIC) at BNL bus is shown in Fig. 3.6b.

The cross-section of the dipole model designed at BNL to test the new braided cable is shown in Fig. 3.7 (Sampson et al. 1969). It had four current blocks in each quadrant and was similar to the previous BNL shell-type magnets. The magnet was 300 mm long and had a circular aperture 50 mm in diameter.

Initially, it was planned to wind the magnet with a high-current Nb_3Sn ribbon to check field calculations and measure the end field distribution. Then, as soon as sufficient low-loss braided cable became available, it would be used in place of the conventional ribbon for the AC loss studies. It is not known if this dipole model was ever fabricated and tested (with ribbon or with braided cable).

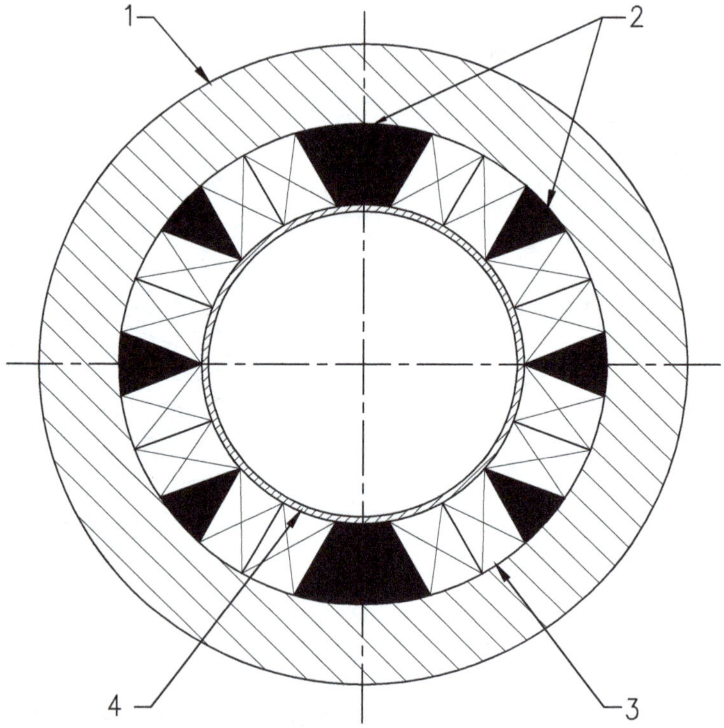

Fig. 3.7 Dipole design with braided cable (Modified from (Sampson et al. 1969): *1* – phenolic support rings; *2* – phenolic spacers; *3* – windings; *4* – non-magnetic beam tube

3.3 Nb_3Sn Magnet R&D in the 1970s

In the 1970s, BNL was working on 5 T Nb-Ti magnets for the ISABELLE 400 × 400 GeV proton–proton colliding beam facility. The first ISABELLE dipole design was based on a single-layer cos-theta coil with 130 mm aperture and a cold iron yoke of 457 mm outer diameter. The coil was made of a high aspect ratio superconducting braid with 97 composite Nb-Ti wires of 0.3 mm diameter filled (soldered) with a Sn-3wt%Ag alloy to provide mechanical rigidity. The cable was insulated with an epoxy-impregnated 0.05 mm thick fiberglass tape. The ISABELLE Nb-Ti dipole cross-section is shown in Fig. 3.8.

In parallel, it was decided to build and test similar dipole models, made of Nb_3Sn braided cable using the R&W technique. It was expected that the thin, high aspect ratio Nb_3Sn braid would ensure an acceptable degradation at the bends around the pole blocks. On the other hand, the R&W technique allowed the use of a standard cable insulation and a coil winding technology similar to those employed for Nb-Ti braid. Therefore, the idea was to substitute the Nb-Ti coil with a Nb_3Sn one with virtually no change to the magnet design and assembly procedure. The use of Nb_3Sn superconductor would be advantageous for ISABELLE, whose magnet cooling scheme was based on helium gas. To verify this idea, three 1 m long dipole magnets were constructed using reacted braided Nb_3Sn cable (Sampson et al. 1977, 1979).

The braid was formed from 95 wires, each of 0.3 mm diameter and containing 1045 filaments of niobium in a Cu-10%Sn matrix surrounded by a pure copper layer. The latter was separated from the core by a tantalum diffusion barrier. After braiding and compacting, the cable was heat-treated in an inert atmosphere for 100 h at

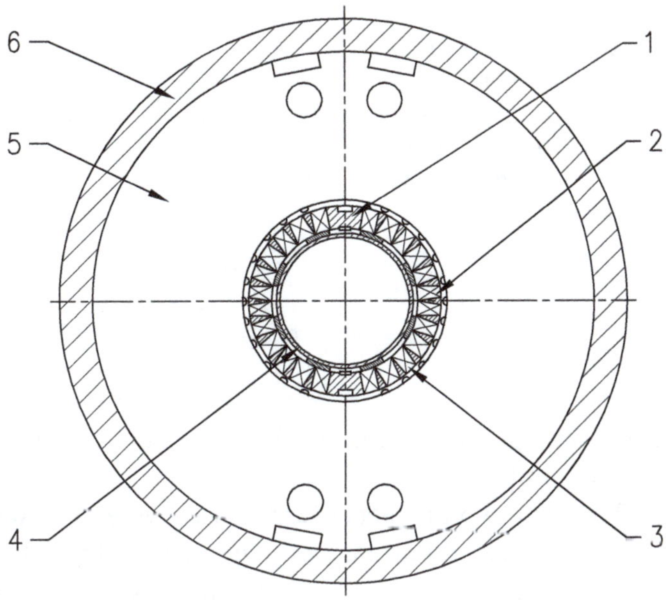

Fig. 3.8 ISABELLE Nb-Ti dipole made of a braided cable: *1* – pole block; *2* – coil-yoke spacer; *3* – coil block; *4* – inner bore; *5* – iron yoke; *6* – outer shell

700 °C to form Nb_3Sn filaments. This heat treatment converted about 80% of the Nb to Nb_3Sn, leaving a small niobium core. The braid was then filled with soft In+7%Pb alloy and insulated with glass–epoxy tape. During processing the braid was never subjected to bends smaller than 160 mm in radius.

The layout of the ISABELLE-type dipole coil is shown in Fig. 3.9. To reach the proper current distribution, a copper braid of identical size was wound as a spacer in parallel with the Nb_3Sn braid. The number of such spacer braids varied from three in two blocks near the pole to none in the rest of the current blocks. The sharp transition bend at the coil pole was replaced with a shaped copper piece co-laminated with Nb_3Sn tapes. This connection between the magnet halves was the weak point in the first magnet. The coil connection problem was solved in the next magnets by using short preformed pieces of braid soldered to the coils during magnet assembly. The resistance of this joint was about 1 nΩ.

In the pole turn, where the bending radius was minimal, a maximum strain of 4.6% was reached, which was almost an order of magnitude larger than typically accepted for Nb_3Sn. The critical current degradation due to braid bending was compensated for by using parallel superconducting braids in the pole blocks instead of copper braids as in the first Nb_3Sn coils. Thus, the current density in the pole blocks with the largest bending deformation was reduced by a factor of 3; and by a factor of 2 in the next current blocks, where the bending radius was somewhat larger.

For the remainder, the technology was very similar to that used for the Nb-Ti ISABELLE dipoles. The Nb_3Sn coils were assembled with a split iron yoke and tested in a liquid helium bath. It was found that the braid degradation due to bending was consistent with expectations. The first magnet was, however, severely limited by heating in the resistive coil connection. At 2.5 kA, it dissipated about 10 W of Joule heating, which was beyond the value at which the heat starts propagating into the

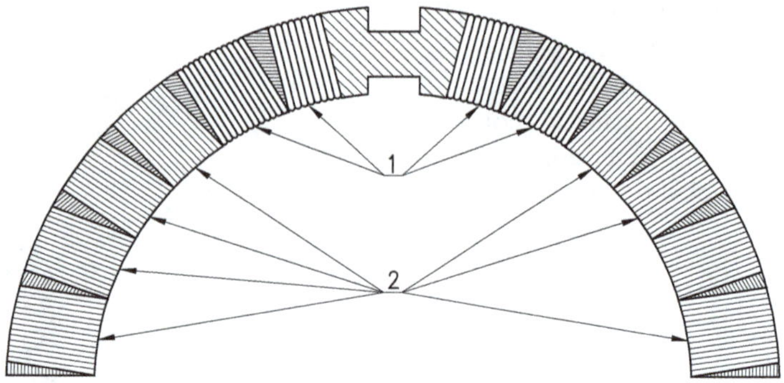

Fig. 3.9 Cross-section of the Nb_3Sn ISABELLE-type dipole coil (Modified from Sampson et al. 1977, 1979): *1* – coil blocks with spaced turns; *2* – coil blocks with single turns

coil. The maximum coil current was 2.65 kA, or about 50% of the magnet short sample limit.

The next two magnets of this series were built with slight variations in the cable size and in the coil arrangement geometry. The latter dipole model was tested with coil cooling by high-pressure helium gas over a temperature range 4.2–15 K. At 4.2 K, the magnet reached a maximum bore field of 4.8 T, which was very close to the short sample limit. At higher temperatures, the magnet maximum field decreased approximately linearly with temperature. It is remarkable that at 8 K the field achieved in the magnet was still almost 4 T. This result shows that the magnet has realized its full potential. The successful test of this dipole magnet demonstrated further possibilities and important conditions for using the R&W technique for high-field Nb_3Sn accelerator magnets.

3.4 Nb_3Sn R&D in the 1980s

In the 1980s, various HEP laboratories started "conquering" the 8–10 T field region, to double the nominal magnetic field of the Tevatron and Hadron-Elektron-Ring-Anlage (HERA) projects. Nb_3Sn multi-filamentary wires based on the bronze process were commercially available and produced in relatively large amounts at that time. The critical current density at 8 T and 4.2 K of bronze-route Nb_3Sn wires was already slightly higher than that of Nb-Ti wires, and above 9 T Nb_3Sn wires were clearly superior. In addition to a superior critical current density at 10 T and 4.2 K, the Nb_3Sn bronze-route wires were very appealing thanks to their intrinsic small filament diameter of 2–3 μm, reducing the persistent current field errors for colliders and power losses for cycled synchrotrons with respect to Nb-Ti wires. A much higher critical temperature, together with superior intrinsic stability thanks to a small filament size, was also seen as a way to make more stable magnets less prone to quench.

An alternative approach was using traditional Nb-Ti magnets cooled by superfluid helium at 1.9 K. The critical current density curves of Nb-Ti wires at 1.9 K are shifted towards a higher field by about 3 T making feasible the use of Nb-Ti magnets for fields of 8–10 T at operating temperatures around 1.9 K. The complexity and the cost of superfluid helium cryogenics for operation at 1.9 K were such, however, at that time that Nb_3Sn magnets operated at 4.2 K with all their technological difficulties and unsolved problems were still an attractive option to reach a target field range of 8–10 T.

The use of multi-strand Rutherford cables based on multi-filamentary composite wires, which after the Tevatron became the recognized standard for accelerator magnets, was an important element of the R&D toward 10 T dipoles and quadrupoles operating at 4.2 K. In addition, new types of Nb_3Sn composite wires, based on the IT process, were explored to overcome the J_c limitation above 10 T. This eventually proved to be the key technology to enable high-field Nb_3Sn dipole magnets well beyond the 10 T Nb-Ti uncontested reign.

3.4.1 The Nb_3Sn Dipole at CEA-Saclay, 1983

At the early 1980s, the CEA laboratory in Saclay (France), which was strongly engaged in the development and construction of superconducting magnets for various projects such as the Russian Ускорительно-Накопительный Комплекс (UNK) (Accelerator and Storage Complex) and the German HERA, started an R&D program on Nb_3Sn synchrotron magnets. A short dipole magnet was designed and manufactured using Nb_3Sn Rutherford cable and the wind-and-react (W&R) technique (Perot 1983). To reduce costs and save time, the magnet design had to make use of the tooling and laminated collars that were available from the UNK Nb-Ti dipole models. Therefore, the coil geometry was predetermined and the bore field was limited to 6 T, based on the critical current density available at that time in Nb_3Sn bronze composite wires. The aim of this project was not a record field, but rather exploring the Nb_3Sn technology.

3.4.1.1 Conductor

The magnet coil used a Rutherford cable with 23 strands, each 0.7 mm in diameter. The cable was 7.8 mm wide and had a keystoned cross-section with a small edge of 1.2 mm and a large edge of 1.4 mm. The Nb_3Sn cable geometry was identical to the cable of the previous Nb-Ti UNK dipole.

The strand was based on the copper-stabilized bronze-route Nb_3Sn wire produced by the Magnetic Corporation of America (MCA). Each strand consisted of 4675 niobium filaments, each 3 μm in diameter, embedded in a bronze matrix. The bronze matrix was surrounded by a 0.017 mm thick tantalum barrier and a 0.070 mm thick outer ring of pure copper. The copper area was only 35% of the total strand area, thus the Cu/non-Cu ratio was only 0.54. More copper would be highly desirable for magnet quench protection, but it would decrease the wire's average current density, losing any advantage with respect to Nb-Ti wires.

The cable was insulated by a 0.13 mm thick fiberglass tape wrapped with overlap. Cable reaction was kept at a relatively low temperature, compensated for by a rather long time (660 °C during 200 h in argon). The short sample critical current at 4.2 K was 7500 A at 6 T and 5700 A at 8 T.

3.4.1.2 Magnet Design and Manufacturing

The dipole featured two-layer coils, clamped by stainless-steel collars. The iron yoke was not a part of the magnet's mechanical structure. The magnet had a 90 mm bore and was 0.64 m long. The relatively low current density of the Nb_3Sn bronze wire and the predetermined size of the cable and the coil led to a moderate design field of 6 T for this Nb_3Sn magnet. At the bore field of 6 T the magnet current was 6300 A and the stored energy was 100 kJ.

The two coil layers were wound separately on corresponding stainless-steel mandrels. Each layer was fitted in its own mold with some tolerance and then reacted. G10 end saddles and end pole pieces were installed during coil impregnation. Special care was paid to the cable insulation, which become very fragile after the sizing agent was removed, to the complex Nb_3Sn/Nb-Ti joints manufactured within the impregnated coil, and to the connections among the coils by means of flexible Nb-Ti cables.

After impregnation, the coil layers were assembled together to form a pole. The upper and lower poles were then surrounded by ground insulation and precision collars. Figure 3.10 shows two impregnated coils ready for collar installation. The coil collaring procedure was the same as for Nb-Ti dipoles. This procedure provided precise conductor placement, coil pre-stress, and support.

The key issue of the described design was magnet protection during a quench. With a copper matrix of only 35%, the maximum current density in the copper during a quench was about 2000 A/mm^2, which was considered an excessive value for a standard quench protection system based on quench heaters and/or an external dump resistor. To provide fast energy extraction, a secondary coil made of copper Rutherford cable of the same size as the superconducting one was wound and

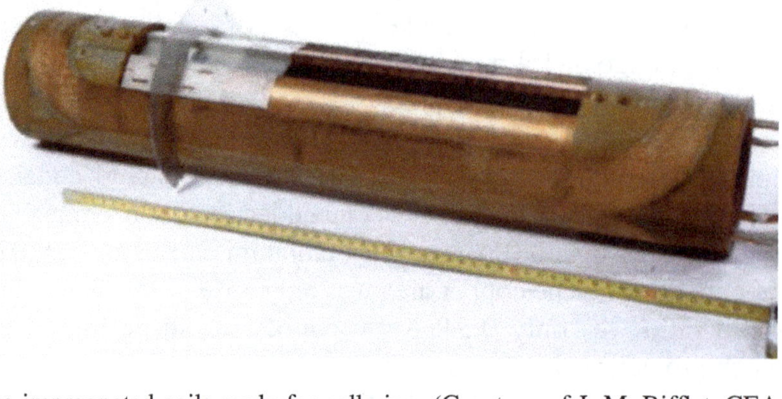

Fig. 3.10 Two impregnated coils ready for collaring. (Courtesy of J.-M. Rifflet, CEA-Saclay)

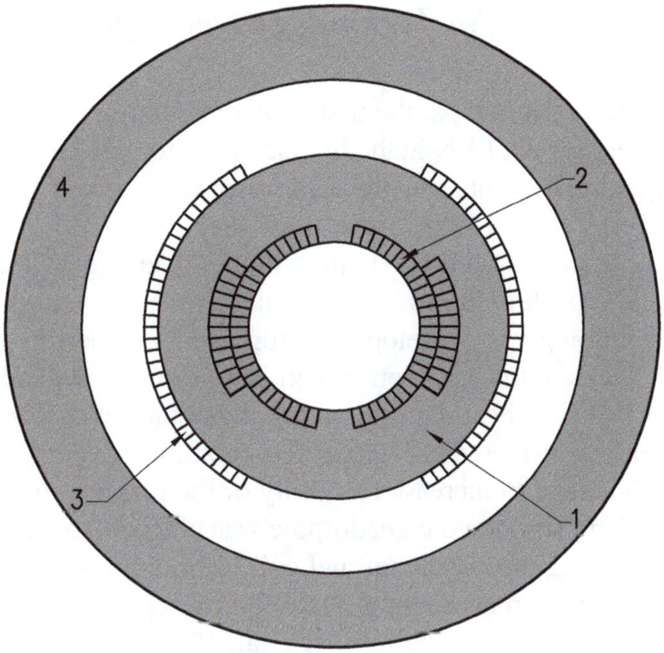

Fig. 3.11 Cross-section of the CEA Nb_3Sn dipole (Modified from Perot 1983): *1* – collar; *2* – Nb_3Sn dipole coils; *3* – secondary copper coil; *4* – warm iron yoke (around collared coil)

installed between the collars and the iron yoke (see Fig. 3.11). The coupling was not perfect since the secondary coil was rather far from the main coil. Nevertheless, this approach was sufficient to extract up to 30% of the stored energy in the magnet.

3.4.1.3 Test Results

A very conservative 100 K "hot-spot" temperature was targeted for this magnet. Limiting the maximum voltage to 650 V with the given magnet inductance, the time constant to satisfy the hot-spot temperature limit was only 50 ms. It was much shorter than the opening time of the mechanical breakers available at that time. Therefore, it was decided to equip the magnet with a superconducting switch, activated by a capacitor bank for a few milliseconds. The switch was tested before using it in the protection circuit. It quenched at a steady-state current of 5.50 kA, thus setting the limit for the magnet power system.

The Nb_3Sn dipole quenched three times, at currents of 5.15, 5.16, and 5.35 kA. Then it went up to 5.55 kA, a current at which the switch quenched. The maximum bore field was 5.3 T while the coil peak field reached 6 T, or 85% of the magnet short sample limit. Despite the apparent success, the achieved field was not impressive since it was similar to the field of the Nb-Ti dipole. Today we remain with the question of why the dipole was not simply pushed up without the switch, accepting a hot-spot temperature of 200–300 K, as we do today (but, after 30 years of experience!) in modern Nb_3Sn magnets. Such an experiment would be extremely interesting.

3.4.2 Nb_3Sn Technology Development at CERN

In the early 1980s, the first superconducting accelerator magnets were successfully operated at CERN in the Intersecting Storage Rings (ISR). The eight Nb-Ti low-beta quadrupoles pushed the ISR to proton–proton collisions with record luminosity (Perin et al. 1979). Since they were only used in a steady-state regime, the coils were wound with Nb-Ti monolithic conductor, with a warm bore of about 170 mm and a peak field of about 5 T. It was then a natural choice for CERN to use Nb_3Sn technology to develop quadrupoles of superior gradient and operating margin. Thanks to the experience with Nb-Ti low-beta quadrupoles, which were essentially exposed to heat depositions by radiation debris coming from collisions, the CERN team considered the larger temperature margin of Nb_3Sn coils as a considerable advantage to increase reliability of the magnets in accelerator operation.

The work on the quadrupole was preceded by a technology-learning phase aimed at testing the conductor and coil technologies in solenoidal configuration, which is simpler than accelerator quadrupole magnets; meanwhile, the design of the quadrupole was worked out (Asner et al. 1981, 1983). During this work, valuable experience was gained in handling the delicate brittle superconductor and solving problems of insulation, reaction, and coil splicing.

3.4.2.1 Conductor

The cable cross-section is shown in Fig. 3.12. The cable was 1.1 mm thick and 2.2 mm wide. It was made by Vacuumschmelze in Germany. The superconducting strands were based on bronze-route multi-filamentary Nb_3Sn composite wire, similar to that used in the CEA dipole. Differently from the CEA conductor, however, the CERN Nb_3Sn wire had a smaller diameter of 0.4 mm vs. the 0.7 mm of the CEA wire. Supplementary stabilizing copper was added to the cable by three elements: two 0.4 mm diameter copper strands, and a central copper core 0.4 mm thick and 1.5 mm wide. To prevent contamination of the three stabilizing elements with tin during heat treatment, they were surrounded by a tantalum barrier and copper-clad. Both the copper and the superconducting strands were tinned on the surface to avoid

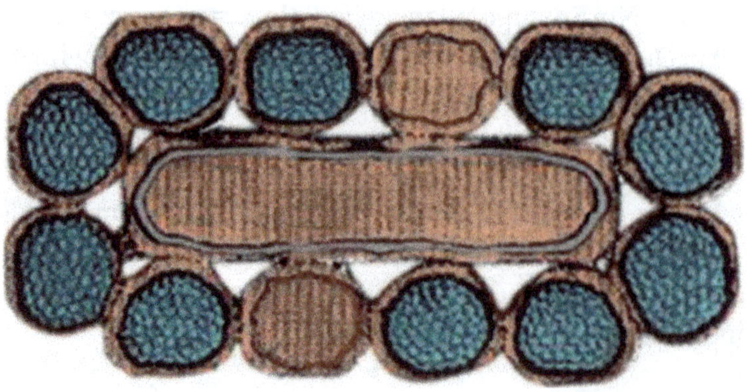

Fig. 3.12 The Nb_3Sn cable used in CERN Nb_3Sn magnets. (Modified from Asner et al. 1983)

tin depletion from the Nb_3Sn strands. Thus, after reaction the copper outside the tantalum barrier became low conductivity bronze. To keep the cable current density high, the pure copper content was only 27.2% of the total metal cross-section (a Cu/non-Cu ratio of only 0.37). Each superconducting strand had 3721 twisted niobium filaments, each ~3 μm in diameter.

3.4.2.2 Coil Fabrication Technology

Small radii of curvature in the quadrupole magnet made the W&R approach the preferred choice for coil fabrication. Extensive cable insulation R&D ended with the choice of quartz glass with a residual content of the carbon sizing agent of less than 0.03% after heat treatment. Despite various tests, however, the quartz glass proved to be fragile and easily deteriorated during winding and coil handling. The solution was to wet the quartz glass with paraffin to reinforce the insulation during handling. Then, a special heat treatment cycle was used to remove the paraffin and, thus, reduce the carbon remains in the coil. The coil reaction cycle used at CERN is presented in Table 3.1.

The coil lead joints were made by placing the Nb_3Sn cables between two Nb-Ti cables, yielding a joint resistance of less than 10 nΩ. Finally, after reaction, the coils were vacuum-impregnated with epoxy resin.

To test the whole coil technology a 0.5 m long solenoid with 23 mm bore and 90 mm outer diameter was built with the cable, insulation, and coil technology described above. The solenoid, tested at 4.2 K, went straightforwardly and repeatedly to its maximum current of 1370 A, corresponding to 8.75 T of peak field in the coil. Thanks to a positive strain effect, the solenoid maximum current was slightly above the expected short sample limit. This was an important demonstration of the developed coil technology in a real accelerator magnet.

3.4.2.3 Nb_3Sn Quadrupole Design and Fabrication

The Nb_3Sn quadrupole design inherited CERN's 10-year experience with Nb-Ti quadrupoles and dipoles for beam lines and low-beta triplets. The cross-section of the CERN Nb_3Sn quadrupole is shown in Fig. 3.13. It consisted of an aluminum

Table 3.1 Heat treatment cycle at CERN

Step	Temperature (°C)	Time (h)	Comments
1	75	10	Outgassing, paraffin melting
2	200	24	Argon, tin diffusion
3	270	10	Vacuum, paraffin gas extraction
4	400	24	Argon, bronze stress relief
5	700	72	Nb_3Sn phase formation

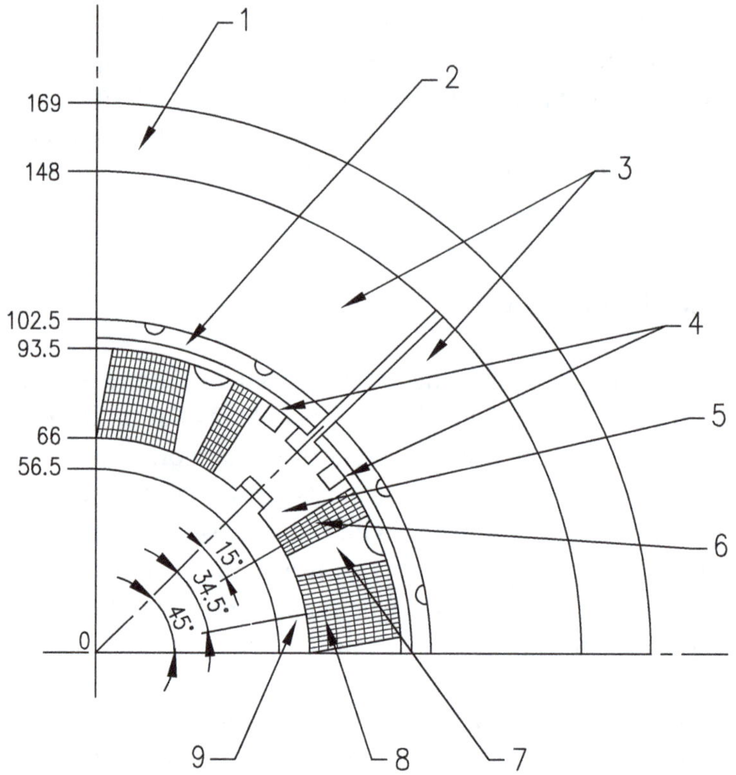

Fig. 3.13 Cross-section of the CERN Nb_3Sn quadrupole (Modified from Asner et al. 1983): *1* – aluminum shrinking cylinder; *2* – spacer with helium channel; *3* – iron yoke; *4* – ground insulation; *5* – copper pole block; *6* – inner coil block; *7* – copper wedge; *8* – outer coil block; *9* – cold support tube

shrinking cylinder, spacer with helium channel, iron yoke, ground insulation, coil with copper pole block and two copper wedges, and cold support tube.

The quadrupole length was 0.9 m and the aperture was 90 mm. To save cost and time, the structure and tooling were borrowed from the Nb-Ti quadrupole, at that time already operating in a secondary beam line of the Super Proton Synchrotron (SPS) at CERN (also known as the "Castor" quadrupole).

The coils were divided in two blocks, the one near the pole containing 60 turns and the one near the mid-plane containing 160 turns. Each coil was wound on a thick stainless-steel cylinder. The pole blocks and wedges, made of copper and insulated with Al_2O_3, were fixed to the winding mandrel. The pole blocks were split axially to allow for different thermal expansion and contraction between the coil and tooling during reaction and impregnation.

The copper end blocks were installed in place after winding. The lateral stainless-steel bars and grids were added with clearance to confine the coil before reaction. After reaction, each Nb_3Sn lead was sandwiched between two Nb-Ti conductors and soldered using indium solder. The coil, surrounded with additional fiberglass and polyimide ground insulation, was impregnated in a precise mold to achieve an accurate geometry.

Four impregnated coils were assembled onto the 10 mm thick 316LN stainless-steel tube acting as the radial inner support and inner helium enclosure. The coils were then surrounded with radial spacers with helium channels. Two iron half-rings with a cut in the pole region formed the flux return yoke. At the coil ends the iron half-rings were replaced by stainless-steel ones to reduce the magnetic field in the coil ends. A cold iron yoke, introduced at CERN in 1973, became a trademark of CERN, providing field enhancement and compactness.

Coil azimuthal and radial preloads were delivered by an external clamping cylinder made of an aluminum alloy. To apply the pre-stress, the cylinder had a slightly smaller inner diameter and was heated and shrink-fitted around the iron halves. The mechanical structure described provided coil target compression to about 65 MPa at 4.2 K. With the applied electromagnetic forces the coil compression in the pole regions reduced to 20 MPa. The maximum azimuthal and radial stresses in the coil were selected to be acceptable for brittle Nb_3Sn coils.

In the axial direction, the coil preload was applied by means of spring-loaded screws attached to the inner stainless-steel tube. This approach allowed for compensation of the differential thermal contraction between the stainless-steel tube and the coil, and partially counterbalanced the electromagnetic forces in the coil ends.

3.4.2.4 Quadrupole Test Results

The quadrupole was tested in 1981 at 4.2 K without the protection resistor, since the stored energy was 100 kJ, which is relatively small. The measured resistances of the Nb_3Sn/Nb-Ti connections were comfortably low, between 20 and 50 nΩ. The first quench occurred at a nominal current of 1.0 kA, and in four quenches the current reached its maximum value of 1.1 kA, which was within 93–99% of the magnet short sample limit based on witness sample tests. The corresponding maximum field gradient was 70.5 T/m (compared to 55 T/m of the Nb-Ti Castor quadrupole) and the maximum field in the coil was 7.8 T.

The quadrupole was tested again in 1982 after shunting of each coil with a 50 mΩ protection resistor. The magnet went straight up to a current of 1.1 kA, confirming that the maximum current was effectively reached. Magnetic measurements showed a reasonable field quality. The deviations from ideal field harmonics were in the range 20–70 units.

This program successfully validated the Nb_3Sn wire and cable, and the W&R technology developed at CERN for accelerator magnets. This success (as well as the results of CEA) provided a strong base for CERN's next project to build a 10 T Nb_3Sn accelerator dipole. This magnet was successfully developed and tested at the end of the 1980s (Asner et al. 1990).

3.4.3 Nb_3Sn Dipole at LBNL

LBNL undertook the development of a 10 T dipole based on a Nb_3Sn conductor (Taylor and Meuser 1981; Taylor et al. 1983, 1985). This development was carried out in parallel with the Nb-Ti magnet R&D for the Superconducting Super Collider (SSC).

3.4.3.1 Main Design Concept

To produce a cost-effective dipole design, a small bore of 50 mm and a high field of 10 T were chosen as initial design parameters. Analysis of magnets with these parameters showed that a high critical current density of about 1200 A/mm^2 at 4.2 K and 10 T was required.

These parameters had a strong influence on the conductor choice: (a) the bronze route for Nb_3Sn wires was abandoned since it was at least 20–30% short of the critical current density needed. Instead, the development and scale-up of the IT approach for Nb_3Sn high critical current density wires became an important part of this program; (b) the conductor had to carry a large current to reduce inductance and magnet discharge time; (c) due to the small coil diameter, the W&R technique was considered for coil fabrication.

A cross-section of the LBNL 10 T dipole is shown in Fig. 3.14. The coil consisted of four racetrack windings per pole, each layer having different winding diameters

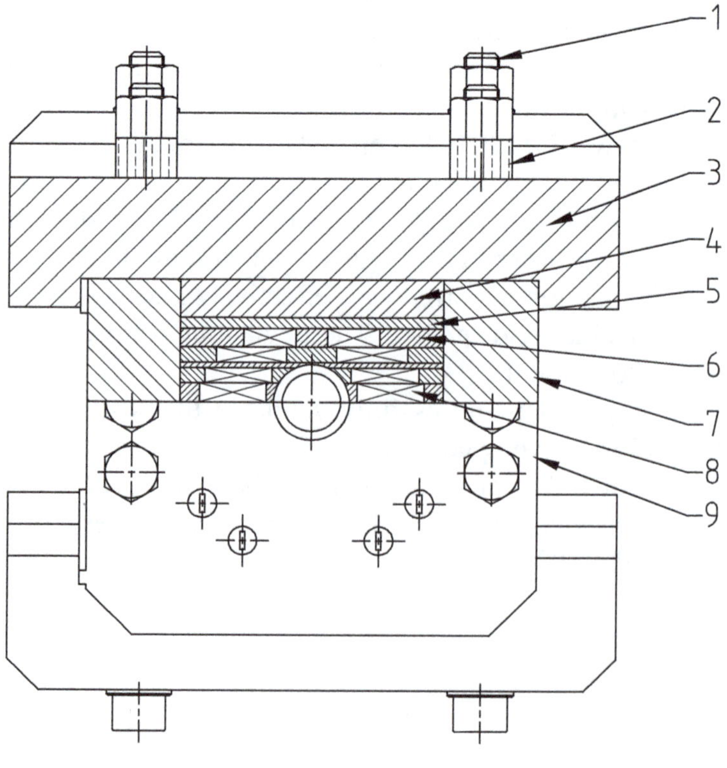

Fig. 3.14 Cross-section of the LBNL Nb_3Sn dipole (Modified from Taylor et al. 1983): *1* – stainless-steel stud; *2* – brass spacer; *3* – aluminum clamp; *4* – stainless-steel bearing plate; *5* – stainless-steel winding form; *6* – fiberglass epoxy; *7* – stainless-steel side plate; *8* – coil; *9* – stainless-steel end plate

and thicknesses, to mimic a shell cos-theta configuration. The coils were not graded and used the same Rutherford cable with a rectangular cross-section. The coils featured flared ends to clear the bore. The inner tube, needed for structural reasons, reduced the useful bore to 40 mm from the 60 mm coil aperture. The magnet was assembled without an iron yoke. The coil was clamped to a high pre-stress to avoid separation from the structure under the Lorentz forces. An advantage of this configuration is that the maximum stress at high fields occurs in the low-field regions and, therefore, it is less prone to degradation due to strain in the Nb_3Sn filaments. Calculations of the stress at 10 T in this dipole predicted an average compressive stress of 83 MPa along the mid-plane, and local compressive stresses of over 138 MPa.

3.4.3.2 Conductor

The main novelty of this program was the use of IT Nb_3Sn composite wire, produced using a new process, in those years still in its early stage compared to the much more mature bronze route. A Nb_3Sn composite wire with the following specifications was ordered at Intermagnetics General Corporation (IGC): strand diameter 1.7 mm, 50% copper, and a target J_c(10 T) of 1200 A/mm^2. This Nb_3Sn wire was used to produce 300 m of an 11-strand Rutherford cable.

The cable cross-section is shown in Fig. 3.15. Dimensions of the uninsulated cable were 11.0 mm by 3.0 mm. The cable was compacted to around 80% and the minimum J_c(10 T) after cabling was 975 A/mm^2. After fabrication, the cable was annealed at 200 °C for 3 h to soften the copper and facilitate winding. Presumably, the main causes of the large J_c degradation allowed due to cabling were the large strand diameter (more than double what was used at that time for other accelerator magnets) and the higher sensitivity of the IT wires to stress.

3.4.3.3 Magnet Construction

The magnet was composed of four double-pancake coils, and each coil was wound from a single piece of cable. The cable was insulated on-line during coil winding with an S-glass tape. The tape was 11.3 mm wide and 1.625 mm thick, wrapped onto the cable with 50% overlap.

A winding base with a straight section and ends sloped downward at 10° was mounted on a winding table, and the stainless-steel pole pieces were attached to this

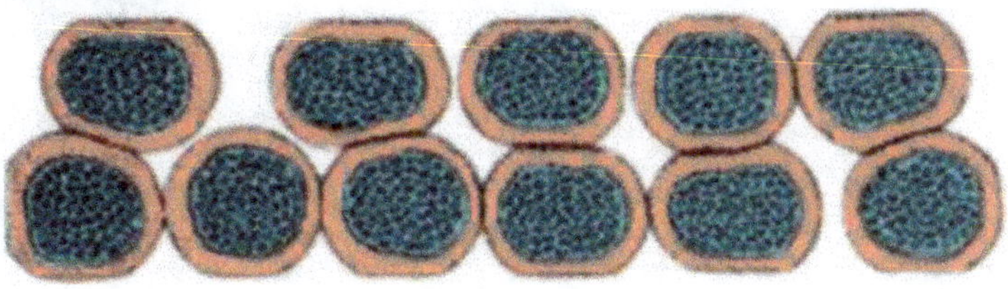

Fig. 3.15 Rutherford cable for LBNL Nb_3Sn dipole. (Modified from Taylor et al. 1985)

base. Three layers had 14 turns while the fourth layer had 11 turns. After winding, the remaining sides and ends of the stainless-steel coil forms were assembled around the coils. The coils were compressed an average of 1.2% from the slightly clamped dimensions.

The reaction cycle consisted of four steps: 200 h at 200 °C, 24 h at 375 °C, 240 h at 500 °C, and 100 h at 700 °C. After reaction, some change of the insulation color was seen, but the turn-to-turn insulation remained adequate and the insulation showed no signs of deterioration. No tin leakage was found at the ends of the strands.

Prior to potting, Nb-Ti cable leads were spliced onto the Nb_3Sn cable leads. The splice joint was made over the 500 mm long straight section using 62 wt% Sn, 2 wt% Ag, and 0.35% Sb balance Pb solder alloy. Before potting, aluminum forms sealed with rubber rings were installed and the coil was pumped in a vacuum chamber. After baking for 12 h, epoxy was introduced into the coil and cured at 50 °C for 5 h, 60 °C for 8 h, and 80 °C for 8 h. The epoxy was a mixture of 50 parts Epon 826 and 50 parts DER 736, catalyzed with 25 parts Tonex. The picture of a D10 dipole coil is shown in Fig. 3.16.

The potted layers were installed on each side of a stainless-steel bore tube insulated with a 0.25 mm thick layer of bi-axially oriented polyethylene terephthalate (Mylar). Coil compression was provided with stainless-steel side plates and top and bottom aluminum bars. The assembly was clamped to achieve a moderate horizontal compression of 34 MPa, chosen for initial testing to avoid possible damage to the Nb_3Sn coils. After the assembly, clamps were removed and the coil compression decreased to about 17 MPa. Note that these numbers are significantly smaller than those produced by the Lorentz forces.

Because of the long lead time to develop and procure the IT Nb_3Sn wire, a Nb-Ti magnet, called D10B, was built to validate the process and the technology, in particular the design chosen for the flare ends (Gilbert et al. 1983). The magnet was designed to produce a 10 T dipole field in a 40 mm diameter bore. The Nb-Ti cable was half as thick as the Nb_3Sn cable, and thus the Nb-Ti dipole operated at half the operating current of the Nb_3Sn dipole. The Nb-Ti model, tested at 1.8 K, allowed direct comparison between the two conductor technologies.

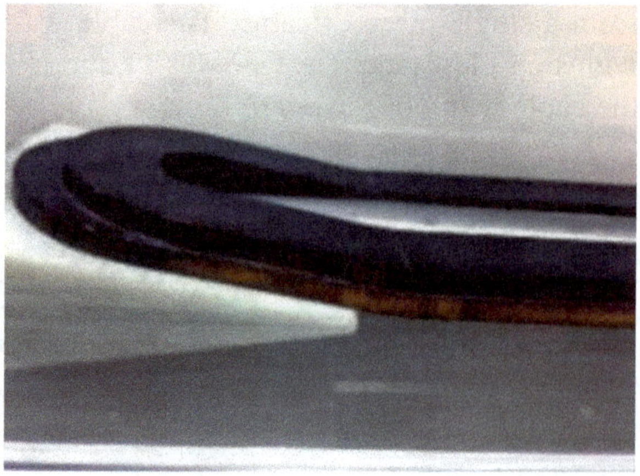

Fig. 3.16 The D10 dipole coil with flared ends. (Courtesy of S. Caspi, LBNL)

3.4.3.4 Test Results

The Nb-Ti D10B dipole was tested at 4.2 K and then at 1.8 K. At 4.2 K, magnet training started at 80% of the short sample limit. The magnet short sample limit was reached after about 25 quenches, producing a central bore field of 7 T. The magnet was then cooled to 1.8 K, and it reached its short sample limit (within a few percent) in a dozen quenches, providing a central field of 9.1 T (almost 10 T coil peak field).

The Nb_3Sn magnet was tested at 4.2 K at a current ramp rate of about 60 A/s, equivalent to 0.04 T/s in the magnet bore. Magnet training started at 7 T and proceeded at a reasonable rate to 8 T after 15 quenches. The quenches were distributed over all layers and did not start at the inner turns exposed to the highest field. At the 16th quench at 12 kA (bore field of 8 T), the extraction system failed and all the stored energy was dumped into the magnet. Afterwards, the maximum quench current was around 300 A lower. Quenches in superfluid helium at 1.8 K were at the same current, indicating that the magnet did not reach its short sample limit. The observed limitations were associated with conductor motion or heating in the Nb_3Sn/Nb-Ti splices. The available instrumentation was not sufficient to disentangle the issue.

In principle, D10 was a successful magnet. The coil technology based on the W&R approach and impregnation with epoxy as well as the strong mechanical structure was proved to work. The IT approach was capable of producing long lengths of high-quality Nb_3Sn composite wire. The direct comparison with the Nb-Ti dipole of similar design, which reached a higher field, clearly indicated that the Nb_3Sn technology was still far from being mature or even usable. This dipole was, however, among the most innovative Nb_3Sn accelerator magnets concerning the conductor and the mechanical structure. In particular, the development of IT Nb_3Sn wire was a fertile seed for future magnet programs.

3.4.4 *Nb_3Sn Magnet R&D at KEK, Japan*

At the end of the 1970s and the start of the 1980s, development and studies of accelerator magnets based on Nb_3Sn superconductor were also performed in Japan. It is remarkable that the Japanese magnet R&D program, in parallel with the bronze Nb_3Sn wires, started using new IT-type composite wires developed by Japanese industry to overcome the intrinsic J_c limitations of the bronze-route wires.

3.4.4.1 KEK-Toshiba Small-Scale Nb_3Sn Dipole (1980)

A small 0.35 m long dipole with a free bore of 62 mm was designed, manufactured, and tested by a collaboration between KEK and the Toshiba Research and

Fig. 3.17 Cross-section of the IT Nb_3Sn conductor for the KEK-Toshiba small scale dipole. (Modified from Ishibashi et al. 1981)

Development Center (Ishibashi et al. 1981). To our knowledge, it was perhaps the first direct involvement of industry in Nb_3Sn accelerator magnet R&D.

The conductor, produced by the Showa Electric Wire and Cable Company, was a rectangular monolithic wire, 2 mm wide and 1 mm thick, with 258 niobium tubes filled with tin and distributed in a copper matrix (Murase et al. 1979). The conductor cross-section is shown in Fig. 3.17. The geometrical diameter of the niobium tubes was 60 μm. At the end of the reaction, the pure copper fraction was 0.635, which corresponds to a fairly high Cu/non-Cu ratio of more than 1.7. The conductor had a relatively low critical current density, a non-Cu J_c of 1600 A/mm^2 at 4.2 K and 5 T, which would yield just above 600 A/mm^2 at 10 T. This test was intended to prove the conductor and magnet fabrication technology, rather than to achieve a record field.

The coil construction technology was based on the W&R approach. The conductor was insulated with E-glass insulation, whose sizing agent was removed with an intermediate heat treatment. The tin diffusion process was rather slow, requiring a heat treatment of 790 h at 700 °C.

The four coils were arranged in two pairs of shell layers with constant current density. The cos-theta current distribution was approximated by optimizing the azimuthal extension of the coil layers. Each shell layer was composed of four layers of conductor winding. Due to a small conductor cross-section, this magnet had 408 turns and a large inductance of 24 mH. Each coil shell was reacted and epoxy impregnated. Afterwards the coils were assembled on a stainless-steel mandrel with an inner diameter of 62 mm, and covered with an insulation layer. An aluminum cylinder was heated and shrink-fitted around the coil. A cross-section of the magnet assembly is shown in Fig. 3.18.

The magnet was tested at 4.2 K, and started quenching at very low field, reaching a plateau after seven quenches at 3.5 T, which corresponded to 88% of the magnet short sample limit. Based on investigations, the rather small radius of the inner layer pole and heating in the joints were ruled out as a possible cause of the critical current degradation. Therefore, only mechanical degradations remained as a possible explanation, without any other precise indication.

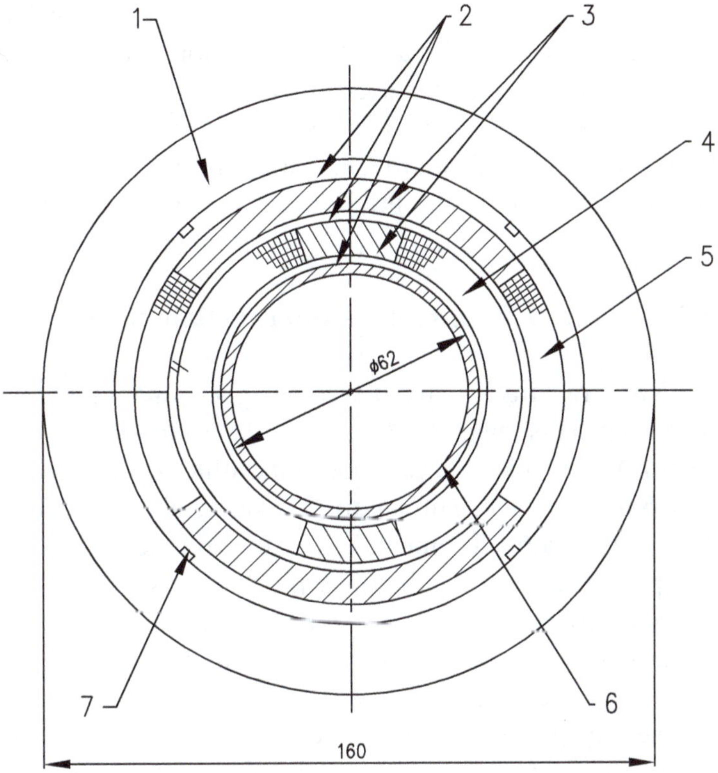

Fig. 3.18 Cross-section of the small-scale KEK-Toshiba Nb_3Sn dipole (Modified from Murase et al. 1979): *1* – aluminum pipe; *2* – epoxy impregnated glass tape; *3* – SUS 316 L stainless-steel pole blocks; *4* – coil inner layer; *5* – coil outer layer; *6* – SUS 304 L stainless-steel support tube; *7* – cooling channel

3.4.4.2 The KEK Effort Toward a 10 T Dipole

To explore the 10 T dipole field region, two development paths were explored at KEK in the first part of the 1980s (Hirabayashi 1983), similar to what occurred in the same years at LBNL and at CERN. One path was the investigation of the performance of Nb-Ti magnets cooled by pressurized superfluid helium. The other path was devoted to developing the design and manufacturing technology for an accelerator-quality dipole using Nb_3Sn bronze conductor.

A 10 T Nb_3Sn multi-shell dipole was designed with a layout of three pairs of double-shell coils, see Fig. 3.19. The angle and thickness of each pair was adjusted to generate an accurate dipole field in the magnet aperture. Each layer was graded to maximize the current density using conductors with optimal keystone angles.

The monolithic multi-filamentary Nb_3Sn conductors were manufactured using the bronze route. Each conductor had a pure copper content of 30–40%. The design value of the central field was 10 T at a magnet current of 5.16 kA, and the magnet maximum stored energy was 959 kJ/m. The detailed parameters of the various layers and their conductors are reported in Yan et al. (1987).

To develop and test the technology for Nb_3Sn accelerator magnets, a single-layer racetrack magnet and a double-shell dipole model (coils #3 and #4) were built and tested. Based on the experience gained from those two magnets, a second double-shell dipole (coils #5 and #6) was also constructed and tested. The program was based on the W&R approach. All of the magnets were manufactured with bronze conductors produced by the Furukawa Electric Company.

3.4.4.3 R&D Racetrack as Intermediate Step

An R&D racetrack magnet was designed, manufactured, and tested as an intermediate step. This magnet (Mito et al. 1984) was almost 1 m long (800 mm long straight section) and had 30 turns. The monolithic rectangular conductor was 2.3 mm thick and 6.0 mm wide, similar to that designed for layer #2. Its critical current was 11.7 kA at 4.65 T and 4.2 K. The conductor was insulated with a mica-glass composite, a new insulation type that was later used in some Nb_3Sn magnets. After reaction, the coil was not impregnated with epoxy to allow direct helium cooling as a mean of coil thermal stabilization.

The magnet was tested in a vertical cryostat. After a first quench at 5.3 kA, it reached the power supply limit of 7.0 kA, which was about 80% of the magnet short sample limit. A second training campaign after improvement of the end support was, however, limited by quenching at 6.82 kA. The reason was attributed to conductor movement in the coil ends due to insufficient support.

3.4.4.4 Double-Layer Dipole Test

A double-layer dipole was assembled, as described above, with coils #3 and #4 (see Fig. 3.19). The coil pre-stress was applied by compression of the iron yoke. The magnet had a 132 mm diameter bore and a 400 mm long straight section. The total magnet length was 870 mm. The iron yoke had a 184 mm inner diameter and a 348 mm outer diameter. The cross-section of the KEK double-layer dipole is shown in Fig. 3.20, and a picture of the magnet is shown in Fig. 3.21.

The double-layer dipole with coils #3 and #4 was tested twice. During the first test eight quenches occurred, and during the second one 25 quenches occurred. A relatively long training and a large degradation were observed during magnet tests. The maximum quench currents ranged from 4.2 to 5.6 kA, whereas the magnet's short sample limit was about 10 kA. Thus, the magnet reached only 3.8 T (the maximum field in the coil was about 4.5 T), which was about 56% of the expected short sample limit. It was concluded that conductor motion and the joints were not acceptable, indicating that further improvements were required.

Comparison of the test results of the racetrack magnet and the double-layer dipole with coils #3 and #4 suggested that conductor motion was the main cause for the training and degradation of both these magnets, although the racetrack magnet showed better performance than the double-shell dipole.

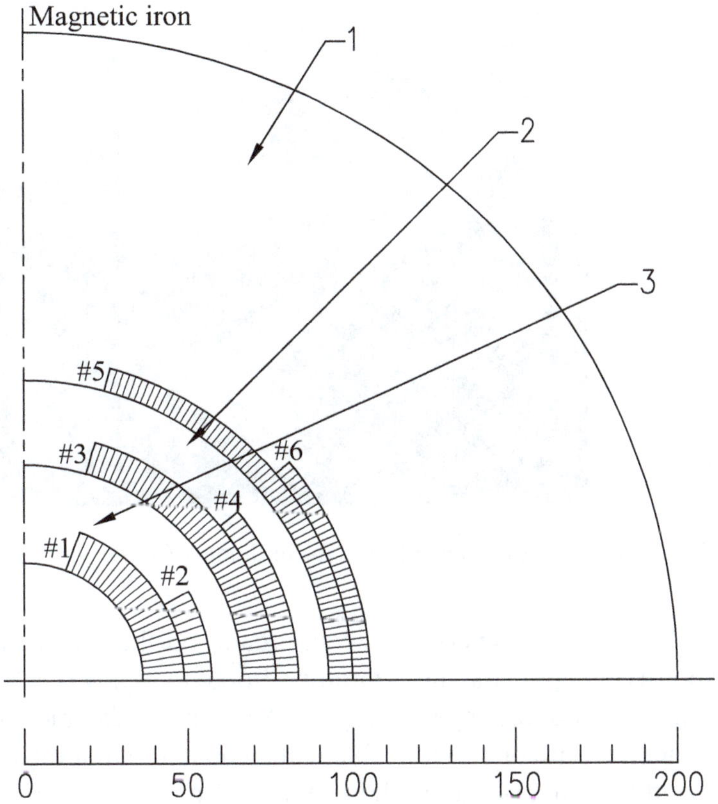

Fig. 3.19 Cross-section of the KEK six-layer Nb_3Sn dipole (Modified from Hirabayashi 1983): *#1–6* – coils; *1–3* – stainless-steel collars

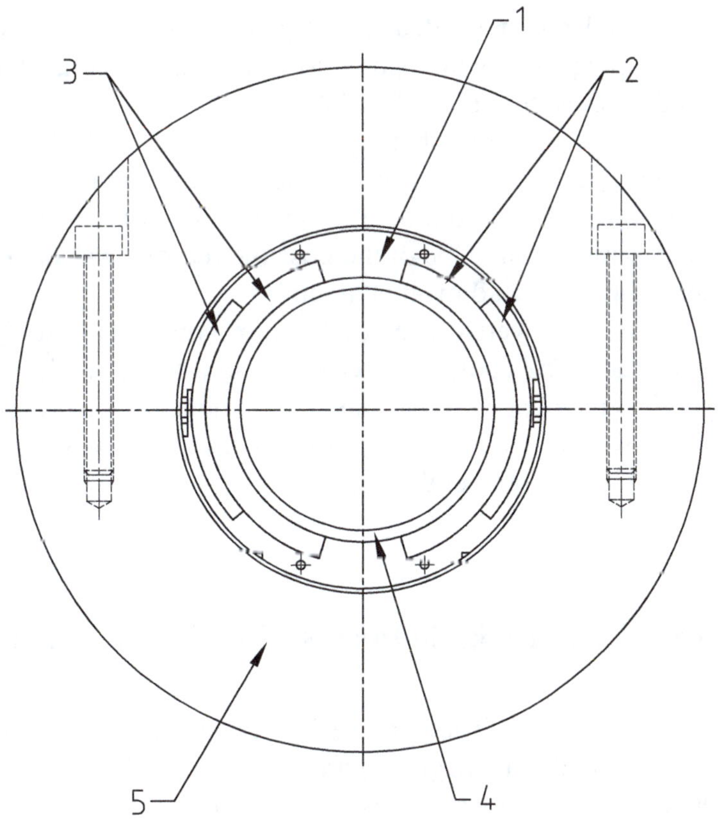

Fig. 3.20 Cross-section of the KEK double-layer Nb_3Sn dipole (Modified from Hirabayashi and Tsuchiya 1986): *1* – stainless-steel collar; *2* – G10 insulation; *3* – dipole coils; *4* – G10 bore tube; *5* – iron yoke

Fig. 3.21 The KEK double-layer Nb_3Sn dipole. (Courtesy of A. Yamamoto, KEK)

With all the experience accumulated on the racetrack and the first double-layer shell-type dipole magnets, another dipole model was assembled with coils #5 and #6 (the outer double-layer in Fig. 3.19) in a similar mechanical structure, but with a larger aperture of 186 mm. The coils had a 400 m long straight section and were first wound in a flat racetrack shape and then pressed to form the shell-type coils desired. Two layers of 0.13 mm thick mica-glass tape provided the conductor insulation in the magnet. After heat treatment at a temperature of 700 °C for 48 h, the coils were assembled, compressed, and clamped by stainless-steel collars. Coils #5 and #6 were not impregnated with epoxy to allow direct contact of liquid helium with the conductor, as for coils #3 and #4.

The test of the double-layer magnet with coils #5 and #6 showed even poorer quench performance. The magnet maximum current was about 3 kA, which corresponded to a bore field of about 2.9 T, or less than half of the 7.3 kA magnet short sample limit. One of the possible reasons was damage to the conductor during coil fabrication and magnet assembly.

Many tests were carried out on those magnets to study their stability and quench properties. The cooling effect was found to be small, with measurements being in good agreement with the adiabatic models. The results above led to the main conclusion that coil impregnation with epoxy was mandatory to avoid conductor motion and to protect the coils.

3.5 Nb_3Sn Dipole for a Cable Test Facility at BNL (1991)

After various R&D activities, BNL made another attempt to use the R&W approach in a Nb_3Sn dipole magnet approximately 1 m long with an 80 mm bore (McClusky et al. 1993). The magnet was intended to provide the background field for a cable test

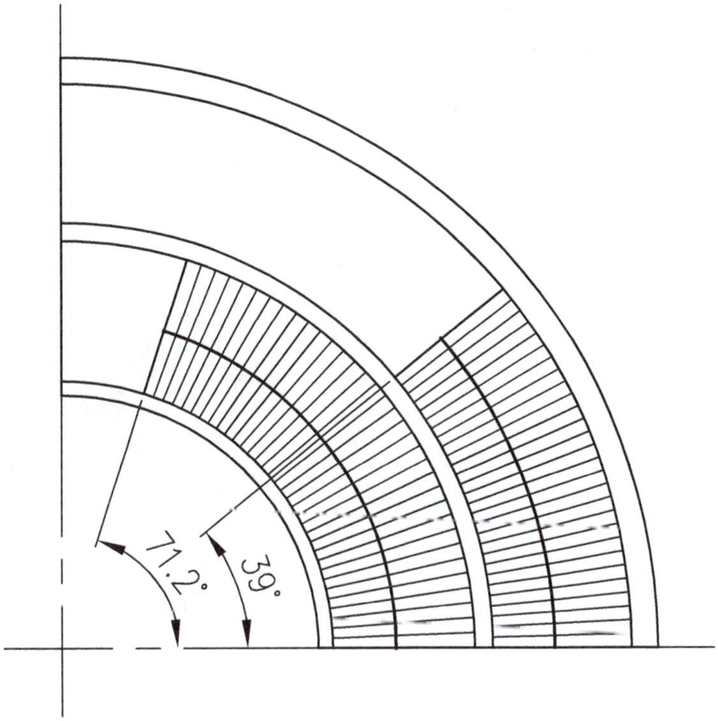

Fig. 3.22 Cross-section of the BNL Nb_3Sn dipole coil. (Modified from McInturff et al. 1997)

facility for the SSC and RHIC projects. To test the magnet construction techniques, Nb-Ti coils of identical dimensions were also built with the SSC cable.

The Nb_3Sn magnet used a Rutherford cable made from Nb_3Sn multi-filamentary composite wires. The cross-section of the coil is shown in Fig. 3.22. The coil featured four layers: each set of layers was separately constrained with aramid fiber (Kevlar) reinforced epoxy bands, and the complete coil assembly was pre-stressed by a horizontally split laminated iron yoke, 508 mm in diameter. The cross-section of the magnet is shown in Fig. 3.23. The yoke was fabricated from the standard ISABELLE/CBA blocks that were formed from thin low carbon steel laminations glued together under pressure.

3.5.1 *Conductor*

The outer cable was made of a low-magnetization modified jelly roll (MJR) Nb_3Sn composite wire (a special IT diffusion process) that was developed in the previous years for Nb_3Sn solenoids. To our knowledge, this was the first use of MJR wires in accelerator magnets. Thanks to the high critical current density and relatively low cost, this type of Nb_3Sn wire dominated the scene in the 1990s until the first years of the new century.

The inner cable was also made of a MJR-type wire with higher Nb_3Sn content to allow for a very high overall current.

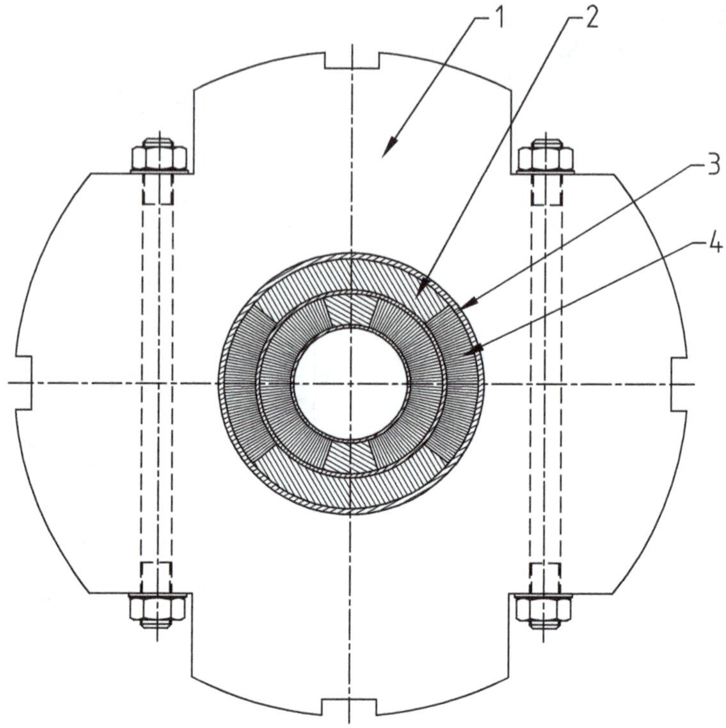

Fig. 3.23 Cross-section of the BNL Nb_3Sn dipole (Modified from McClusky et al. 1993): *1* – iron yoke; *2* – pole block; *3* – spacer; 4 – coil

3.5.2 Coil Fabrication

The four-layer coil was fabricated in pairs using the double-pancake technique to eliminate internal joints. The two inner layers were manufactured using the W&R technique to avoid conductor degradation during winding around narrow poles. The two outer layers were manufactured using the R&W technique, which was considered to be safe due to the wide poles.

The Nb_3Sn cable for the outer layers was treated with Mobil-1® synthetic oil before reaction to avoid strand sintering, and heat-treated on a 1 m diameter stainless-steel drum with alumina–paper separators. The outer layer cable was insulated with polyimide and B-stage epoxy–fiberglass insulation, and the outer coils were molded in a high-pressure fixture.

The inner layer cable was insulated by a 10 mm wide and 0.1 mm thick S-glass tape wrapped with 50% overlap, providing a total insulation thickness of 0.2 mm. The inner coils were reacted as wound in a special fixture, which maintained their shape during reaction. The inner coils were given the same heat treatment as the cable used in the outer coils. A low-pressure argon atmosphere was maintained in the furnace during the four-step reaction cycle: 200 h at 220 °C, 48 h at 340 °C, 24 h at 580 °C, and 150 h at 650 °C.

Both inner and outer coils were not impregnated with epoxy to get the maximum benefit from helium cooling, similar to the Japanese R&D program described above.

3.5.3 Magnet Assembly

After assembly, the coils were clamped onto an inner stainless-steel tube with 25 mm thick aluminum rings, spaced by 50 mm. Kevlar–epoxy composite wire was wound between the aluminum blocks to apply pre-stress to the coils. Once the Kevlar–epoxy composite was cured, the aluminum rings were removed and their space filled again with the same Kevlar–epoxy composite to apply pre-stress to these sections. The outer radius of the Kevlar–epoxy banding was then machined to fit exactly the inner diameter of the yoke. The yoke pre-stressed the coils by means of bolts and supported the Lorentz forces during magnet excitation. The coils were connected in series by pre-shaped Nb-Ti cables, located at the end of the magnet in a low field region.

3.5.4 Test Results

The short sample bore field of the Nb-Ti dipole was 7.8 T at 3.5 kA, while the short sample limit of the Nb_3Sn dipole was 9.8 T at 5.15 kA. The first quench in the Nb-Ti dipole occurred at 3.1 kA (7 T bore field, 7.3 T peak field). After 15 quenches, the magnet reached its maximum current of 3.5 kA, corresponding to a bore field of 7.8 T and a peak field in the coil of 8.27 T.

The first quench of the Nb_3Sn dipole occurred at approximately 4 kA (the bore field 7.6 T, the coil peak field 8.0 T), i.e., at nearly 80% of the expected short sample limit. During subsequent training at the fourth quench (the first one occurred in the inner coils) at 4.35 kA something happened, and from there on the magnet was limited to less than 4 kA. By measuring the decay rate while in the persistent mode, a significant field-dependent resistance was found in the windings. Disassembly of the coils could not reveal any damage. It was observed, however, that the inner coil was detached from the pole by more than 2 mm!

It is interesting that in the original paper the magnet's performance was considered disappointing. The reason for the disappointment was the resistance developed in the coil at relatively low currents. Today it is clear that this was due to the choice of using non-impregnated coils. The large 2 mm displacement of the coil pole turns from the poles also indicated that the coil pre-stress was less than adequate. Perhaps this displacement would have been smaller and would have had less impact on magnet performance if the coils had been impregnated. While the helium was of little help for stabilization, as also concluded by the KEK program, the lack of good thermal conduction may have had a detrimental effect during quench, constraining transverse heat propagation.

3.6 Conclusion

This review covered the first 25 years of Nb_3Sn accelerator magnet R&D, which started in the USA and was then quickly taken up in Europe and Japan. The key design and performance parameters of the magnets described above are summarized in Table 3.2.

The development of Nb_3Sn accelerator magnet technology proceeded in parallel with Nb-Ti accelerator magnets, which made steady progress and quickly filled all the field space up to their limit of 10 T, thanks to the development of large superfluid helium refrigerators and cooling technologies that were affordable and reliable. During the same period of time Nb-Ti accelerator magnet technology made quick progress from short model magnets, throughout technology scale-up, to mass production and use in large accelerators and colliders such as the Tevatron, HERA, and RHIC.

On the contrary, the progress of Nb_3Sn accelerator magnet technology was very slow, being restricted by a limited number of short technological models and demonstrating very volatile results. Nevertheless, these initial R&D efforts on Nb_3Sn accelerator magnets were not useless. Indeed, various types of Nb_3Sn wires and cables, shell-type and block-type coils, R&W and W&R techniques, various support structures, and coil preload concepts were tested, and the first knowledge of Nb_3Sn magnet performance was obtained. Finally, at the beginning of the 1990s, the first Nb_3Sn dipole magnets to break the 10 T wall were built and tested (Asner et al.

Table 3.2 Main characteristics of the Nb_3Sn magnets described

Year	Laboratory	Magnet type	Magnet bore (mm)	Max. gradient (Q) or bore field (D)
1967	BNL	Quadrupole	76	85 T/m (3.2 T[a])
1979	BNL	Dipole	130	4.8 T[a]
1980	KEK-Toshiba	Dipole	62	3.5 T[a]
1983	CEA-Saclay	Dipole	90	5.3 T[a]
1983	CERN	Quadrupole	90	71 T/m (7.8 T[b])
1985	LBNL	Dipole	50	8 T[a]
1986	KEK	Dipole	132	3.8 T[a]/4.5 T[b]
1991	BNL	Dipole	80	7.6 T[a]/8.0 T[b]

[a]Maximum field in aperture; [b]maximum field in coil

1990; den Ouden et al. 1997; McInturff et al. 1997). These exciting results paved the way towards new, previously unattainable, fields for accelerator magnets. The results of these works are presented and discussed in the next chapters of this book.

References

Asner A, Becquet C, Hagedorn D et al (1981) Development and testing of high field, high current density solenoids and magnets, wound with stabilized filamentary Nb_3Sn cable and reacted after winding. IEEE Trans Magn 17(1):416–419. https://doi.org/10.1109/tmag.1981.1061064

Asner A, Becquet C, Rieder H et al (1983) Development and successful testing of the first Nb_3Sn wound, in situ-reacted, high-field superconducting quadrupole of CERN. IEEE Trans Magn 19 (3):1410–1416. https://doi.org/10.1109/tmag.1983.1062338

Asner A, Perin R, Wenger S et al (1990) First Nb_3Sn, 1m long superconducting dipole model magnets for LHC break the 10 Tesla field threshold. In: Sekiguchi T, Shimamoto S (eds) 11th international conference on magnet technology (MT-11), vol 1, Tsukuba, 28 Aug–1 Sep 1989. Springer, Dordrecht, pp 36–41

Billan J, Henrichsen KN, Laeger H et al (1979) A superconducting high-luminosity insertion in the intersecting storage rings (ISR). IEEE Trans Nucl Sci 26(3):3179–3181. https://doi.org/10.1109/tns.1979.4329976

Britton RB (1968) Brookhaven superconducting DC beam magnets. In: Prodell AG (ed) 1968 BNL school on superconductivity June 10–July 1968, BNL. Upton, New York, pp 893–907

Britton RB, Sampson WB (1967) Superconducting beam handling equipment. IEEE Trans Nucl Sci 14(3):389–392

Dahl PF, Damm R, Jacobus DD et al (1973) Superconducting magnet models for Isabelle. IEEE Trans Nucl Sci 20(3):688–692. https://doi.org/10.1109/TNS.1973.4327216

Den Ouden A, Wessel S, Krooshoop E et al (1997) Application of Nb_3Sn superconductors in high-field accelerator magnets. IEEE Trans Appl Supercond 7(2):733–738. https://doi.org/10.1109/77.614608

Edwards HT (1985) The Tevatron energy doubler: a superconducting accelerator. Annu Rev Nucl Part Sci 35(1):605–660. https://doi.org/10.1146/annurev.ns.35.120185.003133

Foner S, Schwartz BB (1981) Superconductor material science. Plenum Press, New York. (chapter 4, p 201)

Gilbert W, Caspi S, Hassenzahl W et al (1983) 9.1 T iron-free Nb-Ti dipole magnet with pancake windings. IEEE Trans Nucl Sci 30(4):2578–2580. https://doi.org/10.1109/tns.1983.4332888

Hirabayashi H (1983) Dipole magnet development in Japan. IEEE Trans Magn 19(3):198–203. https://doi.org/10.1109/tmag.1983.1062355

Hirabayashi H, Tsuchiya K (1986) Superconducting dipole and quadrupole magnets in KEK. KEK Preprint 86–25, June 1986

Ishibashi K, Koizumi M, Hosoyama K et al (1981) Nb_3Sn dipole magnet by wind and react process. IEEE Trans Magn 17(1):428–431. https://doi.org/10.1109/tmag.1981.1061080

McClusky R, Robins KE, Sampson WB (1993) A Nb_3Sn high field dipole. IEEE Trans Magn 27 (2):1993–1995. https://doi.org/10.1109/20.133596

McInturff AD, Benjegerdes R, Bish P et al (1997) Test results for a high field (13 T) Nb_3Sn dipole. In: Comyn M, Craddock MK, Reiser M et al (eds) Proceedings of the 1997 particle accelerator conference (Cat. No.97CH36167), Vancouver, 12–16 May 1997, IEEE, Piscataway, vol 3, pp 3212–3214

Mito T, Tsuchiya K, Hosoyama K et al (1984) A single-layer race-track magnet using a niobium-tin conductor. Adv Cryo Eng 29:79–86. https://doi.org/10.1007/978-1-4613-9865-3_9

Murase S, Koizumi M, Horigami O et al (1979) Multifilament niobium-tin conductors. IEEE Trans Magn 15(1):83–86. https://doi.org/10.1109/tmag.1979.1060198

Perin R, Tortschanoff T, Wolff R (1979) Magnetic design of the superconducting quadrupole magnets for the ISR high luminosity insertion. Internal report CERN ISR-BOM 79–02, CERN, Geneva

Perot J (1983) Construction and test of a synchrotron dipole model using Nb_3Sn cable. IEEE Trans Magn 19(3):1378–1380. https://doi.org/10.1109/tmag.1983.1062270

Sampson WB (1967) Superconducting magnets for beam handling and accelerators. In: Hadley H (ed) Proceedings of the second international conference on magnet technology (MT-2), 11–13 July 1967, The Rutherford Laboratory, Oxford, Chilton, pp 574–578

Sampson WB, Britton RB, Morgan GH et al (1967) Superconducting synchrotron magnets. In: Mack RA (ed) Sixth international conference on high energy accelerators, 11–15 Sep 1967. Cambridge Electron Accelerator, Cambridge, pp 393–396

Sampson WB, Britton RB, Morgan GH et al (1969) Superconducting synchrotron magnets. IEEE Trans Nucl Sci 16(3):720–722. https://doi.org/10.1109/tns.1969.4325342

Sampson WB, Suenaga M, Kiss S (1977) A multifilamentary Nb_3Sn dipole magnet. IEEE Trans Magn 13(1):287–289. https://doi.org/10.1109/tmag.1977.1059452

Sampson WB, Kiss S, Robins K et al (1979) Nb_3Sn dipole magnets. IEEE Trans Magn 15 (1):117–118. https://doi.org/10.1109/tmag.1979.1060162

Taylor CE, Meuser RB (1981) Prospects for 10 T accelerator dipole magnets. IEEE Trans Nucl Sci 28(3):3200–3204. https://doi.org/10.1109/tns.1981.4332052

Taylor C, Meuser R, Caspi S et al (1983) Design of a 10 T superconducting dipole magnet using niobium-tin conductor. IEEE Trans Magn 19(3):1398–1400. https://doi.org/10.1109/tmag.1983.1062261

Taylor C, Scanlan R, Peters C et al (1985) A Nb_3Sn dipole magnet reacted after winding. IEEE Tran Magn 21(2):967–970. https://doi.org/10.1109/tmag.1985.1063680

Yan L, Tsuchiya K, Hirabayashi H (1987) Some results from the first test of the double-shell Nb_3Sn dipole. KEK Report 87–14, Sep 1987, National Laboratory for High Energy Physics, Oho-machi, Tsukuba-gun

Part II
Cos-Theta Dipole Magnets

Chapter 4
CERN–ELIN Nb_3Sn Dipole Model

Romeo Perin

Abstract This chapter reports on the European Organization for Nuclear Research (CERN)–ELIN Nb_3Sn dipole program. In this program a 1 m long model magnet was built, using Nb_3Sn superconductor. The Nb_3Sn magnet reached field levels of 9.5 T in the full dipole and 10.2 T in the magnetic mirror configuration in a 50 mm bore.

4.1 Introduction

In the second half of the 1980s, in the initial phase of the development of the magnets for the Large Hadron Collider (LHC), a crucial issue was the choice of the superconductor. The only industrially available materials at that time were Nb-Ti and Nb_3Sn. The former had already been successfully used in large solenoids such as the OMEGA (Morpurgo 1970) and the Big European Bubble Chamber (BEBC) magnets (Haebel and Wittgenstein 1971), in the Intersecting Storage Rings (ISR) superconducting quadrupoles (Billan et al. 1976) at the European Organization for Nuclear Research (CERN), and then massively in the Tevatron (Wilson 1978) and Hadron-Elektron-Ring-Anlage (HERA) (Wolff 1988) particle accelerators. Nevertheless, the field level required for the LHC, between 8 and 10 T, could be reached only by lowering the operating temperature to about 1.9 K, requiring a new cryogenic system, still to be developed.

At that time, Nb_3Sn had been used only in a few short model magnets, with mixed success. Its brittleness forbids winding reacted conductor around small curvature radii (about 5 mm for the LHC dipole coils) and, thus obliges that coils are wound with conductors containing precursors of the superconducting compound. The wound coil then has to be submitted to a heat treatment at about 700 °C to form the superconducting phase. This heat treatment step at about 700 °C required the development and use of insulation systems able to withstand this treatment, the construction of a complex tooling to maintain the coil strongly confined in its precise

R. Perin (✉)
CERN (European Organization for Nuclear Research), Meyrin, Genève, Switzerland
e-mail: romeo.perin@cern.ch

D. Schoerling, A. V. Zlobin (eds.), *Nb_3Sn Accelerator Magnets*, Particle Acceleration and Detection, https://doi.org/10.1007/978-3-030-16118-7_4

shape during the whole fabrication process, and extreme care in all handling operations after the reaction heat treatment to not damage the brittle superconductor.

In view of the considerable potential of Nb_3Sn, notably its high critical current density at the field levels required at 4.2 K, and the higher specific heat and temperature margins as compared to Nb-Ti at 1.9 K, CERN decided to build a 1 m long dipole model using this superconductor. This decision was taken to ensure that the best choice for the LHC would be adopted, although the construction of this Nb_3Sn dipole model came with numerous challenges, which had to be overcome. An additional argument to test Nb_3Sn technology was the potential need for high-strength quadrupoles in the interaction regions with large apertures and in the presence of increased heat and radiation load.

The strategy for the construction of the Nb_3Sn dipole magnet was like that adopted for a Nb-Ti model (Perin et al. 1989), i.e., to design and build the magnet in close collaboration with a qualified industrial partner (Perin 1990). The reason for this approach was twofold: (1) to involve industry in the technological development in order to prepare it for the future production of the LHC magnets; and (2) the clearly insufficient resources available at that time in the CERN laboratory for LHC magnet research and development (R&D), as CERN was heavily committed to the construction of the Large Electron–Positron Collider (LEP).

A CERN delegation paid visits to several European electromechanical companies to invite them to collaborate with CERN in the development of the LHC magnets. Thorough discussions took place with the firms that expressed an interest: ELIN-UNION at Weiz, Austria, was selected as the most suitable for building a short model magnet using Nb_3Sn superconductor.

4.2 The CERN–ELIN Collaboration Agreement

A Collaboration Agreement for the design, manufacture, and testing of a 50 mm aperture, 1 m long single bore dipole model magnet using Nb_3Sn conductor was signed in August 1986 with the Austrian company ELIN-UNION.

In summary, the agreement foresaw the following.

- The design of the magnet and all necessary tooling were to be done at CERN by a common CERN–ELIN team. For this purpose, ELIN, at its own cost, sent to CERN one graduate electromechanical engineer, one production engineer, and one designer, who were integrated into the CERN project team.
- The superconducting cable was specified, procured, and qualified by CERN and then supplied to ELIN.
- Manufacturing of tooling, coils, mechanical parts, test assemblies, final assembly of the magnet, and factory tests were ELIN's responsibility with the assistance of members of the CERN team in the most delicate phases.
- Final tests were made at the CERN cryogenic laboratory under CERN's responsibility with the assistance of members of the ELIN team.
- The dipole magnet remains the property of ELIN.

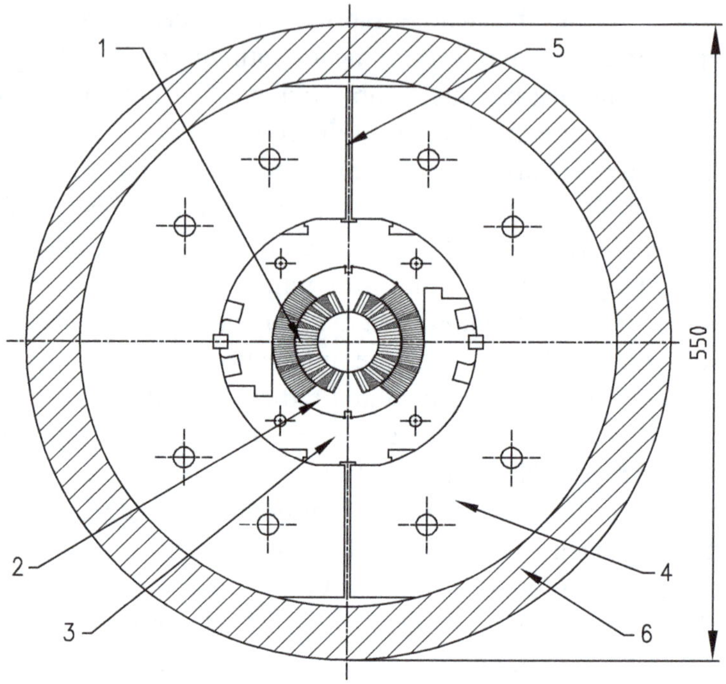

Fig. 4.1 Simplified cross-section of the magnet: *1* – coil; *2* – coil central post; *3* – collar; *4* – iron yoke; *5* – gap; *6* – shrinking cylinder

After the successful conclusion of this project and CERN's decision to adopt the Nb-Ti line for the LHC magnets, the collaboration of CERN with ELIN continued with the construction of short and long Nb-Ti models until the end of 1994.

4.3 Magnet Design Concept

The design, construction, and performance of the Nb_3Sn 1 m long model magnet are described in Wenger et al. (1989) and Asner et al. (1990). Transverse dimensions, coil cross-section, and the magnetic and mechanical structures were designed to be compatible with the 1985 preliminary design of the two-in-one dipole magnet for the LHC (Hagedorn et al. 1985). A simplified cross-section of the dipole cold mass is shown in Fig. 4.1. The main components are the two layer coils, a set of aluminum alloy (Al-Mg-4.5Mn) collars locked by stainless-steel keys, a laminated iron yoke vertically split in two halves, and an outer aluminum alloy (Al-Mg-4.5Mn) shrinking cylinder. The main parameters of the magnet are summarized in Table 4.1.

4.4 Superconducting Strands and Cables

Each coil consists of two layers composed of insulated Nb_3Sn Rutherford-type, keystoned cables having the same width but different thickness to grade the current density to peak field in each of the two layers. The superconducting cables were made from multifilamentary bronze route Nb_3Sn wires: their characteristics are

Table 4.1 Parameters of the magnet

Parameter	Value
Coil aperture diameter (mm)	50
Shrinking cylinder outer diameter (mm)	550
Magnetic length (m)	~1
Overall length (m)	~1.4
Overall mass (kg)	~2000
Nominal central field (T)	10
Peak field in inner layer cable (T)	~10.5
Nominal current I_{nom} (kA)	16.3
Stored energy at I_{nom} (kJ)	360
Lorentz forces F_x on first coil quadrant (MN/m)	2.3
Lorentz forces F_y on first coil quadrant (MN/m)	−1.2
Axial Lorentz force F_z per magnet end (MN)	0.32

Table 4.2 Characteristics of the superconducting cables

	Inner layer		Outer layer	
Parameter	Bronze	PIT	Bronze	PIT
Cable width (mm)	16.81	16.40	16.81	16.40
Cable thickness min/max (mm)	2.19/2.69	2.40/2.70	1.47/1.79	1.45/1.77
Strand diameter (mm)	1.38	1.37	0.92	0.90
Cu/non–Cu ratio	0.38 (0.22[a])	0.55	0.36	0.55
Strand number	24	24	36	36
Filament number	50,000	192	20,000	192
Nb_3Sn filament diameter (μm)	2.6	47	2.6	33
I_c(11 T, 4.2 K) (kA)	~18.1 (~19.6[a])	~23	~11.6	~16.5
I_c(12 T, 4.2 K) (kA)	~16.4 (~17[a])	~17.5	~10	~13.5

[a]These values pertain to the "mirror" coil inner layer. PIT powder-in-tube

reported in Table 4.2. These wires and cables (bronze route) were produced by Vacuumschmelze GmbH, Germany.

Cross-sections of inner and outer coil layer cables are shown in Figs. 4.2 and 4.3. A thin tantalum diffusion barrier (about 4% of the strand cross-section) separates the bronze/Nb/Nb_3Sn part from the outer stabilizing copper ring.

Samples, some of which were bent around a 5.5 mm radius before reaction, and a cable-to-cable splice about 125 mm long, were measured at the Fermi National Accelerator Laboratory (FNAL), which was at that time the only place having a suitable test station. No sample could be quenched up to 10.8 T and 19.7 kA at 4.2 K. The contact resistance of the splice measured at this field and current was about 1.25 nΩ.

An early coil (mirror coil), tested in the Mirror Coil Test Facility (see Sect. 4.7), had an inner shell made with a higher Nb_3Sn content cable (Cu/non–Cu ratio of 0.22 instead of 0.38).

Cable $I_c(B)$ lines and calculated coil load lines are shown in Fig. 4.4. The cable $I_c(B)$ lines were defined by adding the measured I_c of reacted round strands (not

Fig. 4.2 The inner layer cable's narrow edge showing strongly deformed strands. (Courtesy of Vacuumschmelze GmbH, Germany)

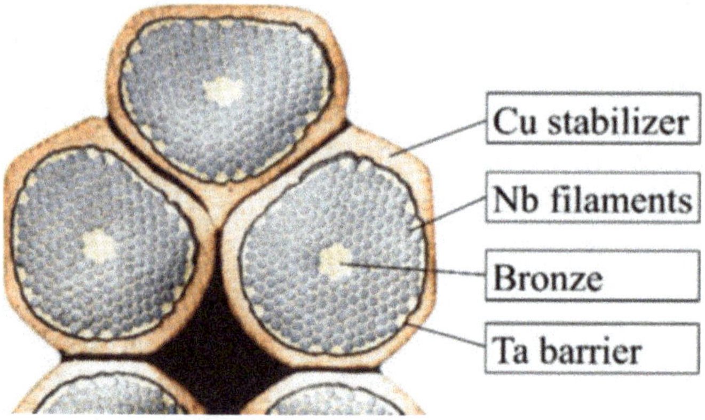

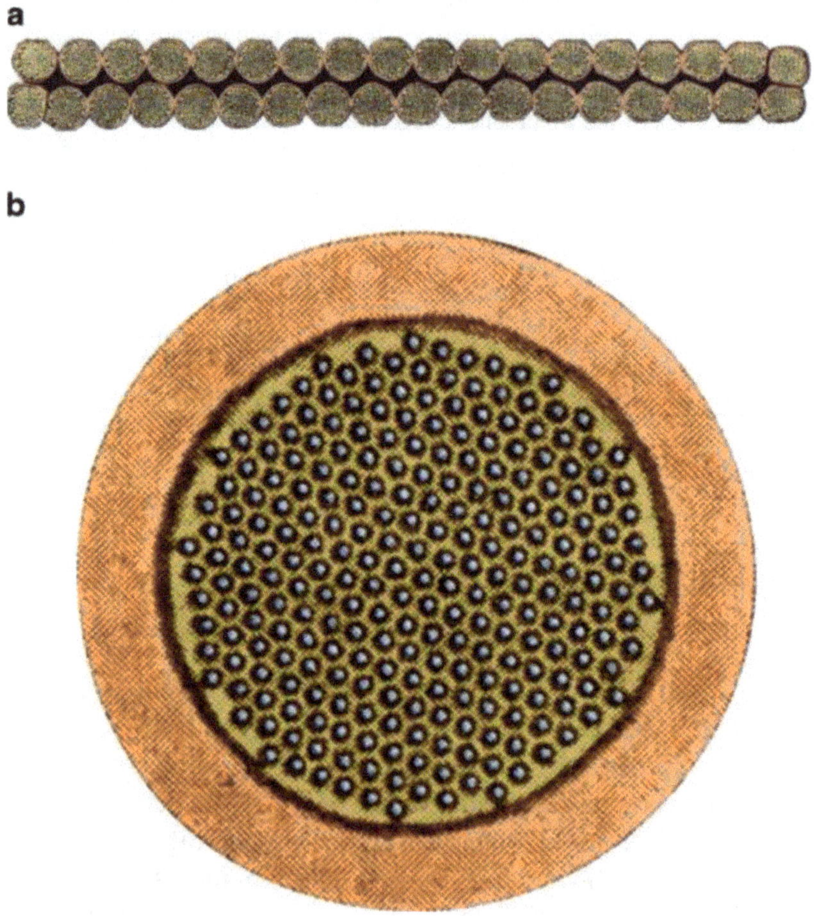

Fig. 4.3 Cross-section of: (**a**) outer layer cable; and (**b**) strand. (Courtesy of Vacuumschmelze GmbH, Germany)

extracted from the cables) and applying a reduction of 2% to take into account the expected degradation in cabling and a further 3% reduction due to degradation in bending around a 5 mm radius. The strand measurements were performed at the Vienna Technical University, Austria. These data showed that the magnet design parameters could be achieved with bronze route Nb_3Sn cables.

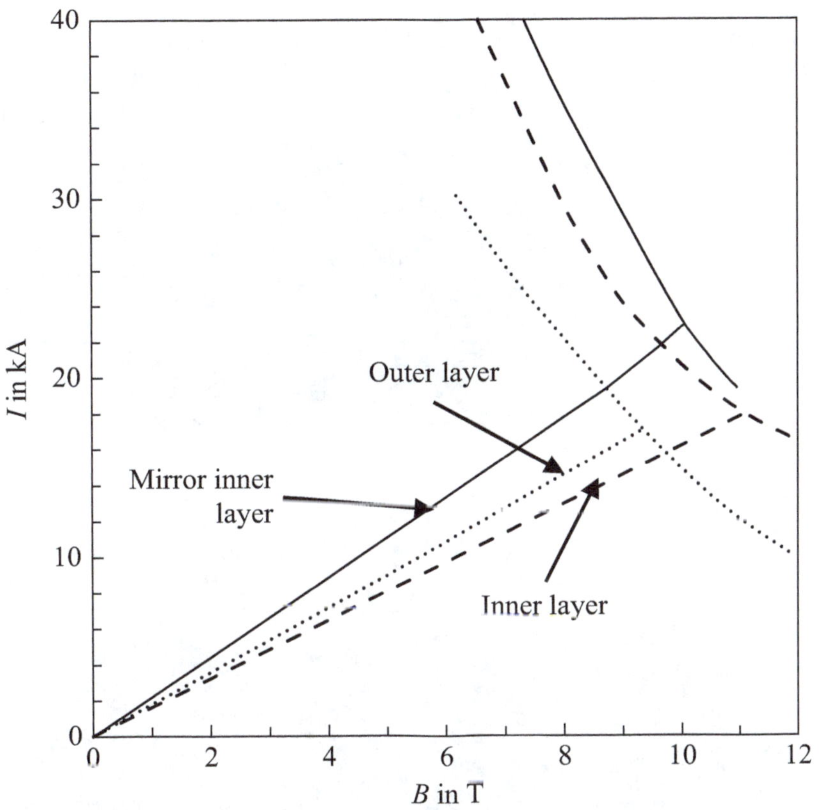

Fig. 4.4 Cable I_c vs. B dependence and computed coil load lines

Considering the limits in current density of the bronze route cables and the difficulty encountered in winding due to their stiffness, CERN with the University of Twente, Netherlands, had promoted the use of the innovative powder-in-tube (PIT) conductor under development by the Energy Research Centre of the Netherlands (ECN) (Hornsveld et al. 1988).

For this purpose, CERN ordered from ECN a small quantity of a pilot production of this conductor. A micrograph of a reacted filament is shown in Fig. 4.5, with voids seen in the core. The critical currents reported in Table 4.2 result from measurements performed at the University of Twente using a superconducting transformer (ten Kate et al. 1989). These high critical currents showed the potential of PIT conductors for the construction of high-field magnets. For particle accelerator magnets, however, considerable development was needed to reduce the filament diameter.

4.5 Magnetic and Mechanical Design

4.5.1 Magnetic Design

The conductors are arranged around the aperture in blocks separated by copper wedges to approximate the cos-theta current distribution. There are four blocks of

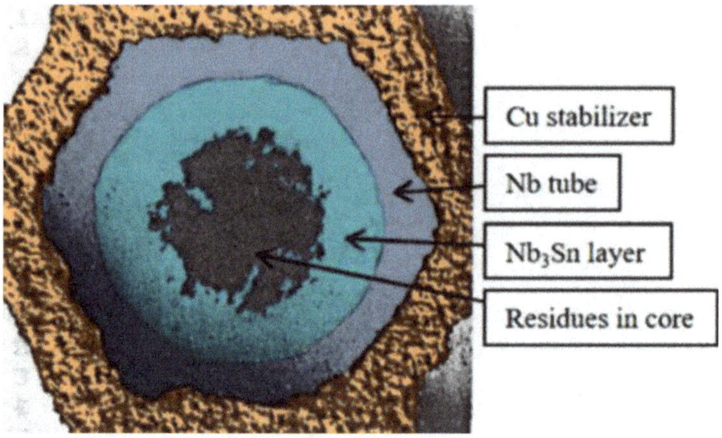

Fig. 4.5 Detail of a hexagon in the ECN conductor. (Modified from ten Kate et al. 1989)

Fig. 4.6 Coil cross-section model. (Courtesy of ELIN, Austria)

conductors in the inner layer and two in the outer. The electromagnetic design started with an analytical approach in two dimensions (Perin 1973, 1995), taking into account the presence of a circular iron screen considered as an equipotential surface ($\mu = \infty$). The coil was subdivided into a number of small circular sectors of uniform current density, and their contribution to the field added up by means of a simple computer program. In this case, the coil cross-section, shown in Fig. 4.6, was determined to eliminate the first higher order multipoles.

The real iron with its variable permeability was then taken into account, using the MARE (Perin and van der Meer 1967) and POISSON (Halbach and Holsinger 1972) programs, in order to optimize yoke thickness, yoke–coil distance and yoke shape. At intermediate fields of about 7 T the computed $\Delta B/B_0$ was within 1×10^{-4} at $R_{ref} = 10$ mm.

The coil end shapes follow the constant perimeter configuration designed by ELIN using a program based on the Biot–Savart law. The conductor blocks are separated by insulated bronze spacers. The ends of the coil, for a dummy coil made from Cu-Ni cable, are shown in Fig. 4.7. This coil was used to practice coil winding and to test the entire manufacturing process, without wasting superconductors. The Cu-Ni cables had mechanical properties similar to the bronze route cables.

Fig. 4.7 View of the ends of inner and outer coil layers after reaction (Cu-Ni dummy). (Courtesy of ELIN, Austria)

4.5.2 Mechanical Design

A hybrid structure was chosen for retaining the Lorentz forces and providing the necessary compressive pre-stress on the coils (Perin 1986). This was known as a hybrid structure, as it produces a moderate compression at room temperature, to avoid creep of the coil materials, and the higher pre-stress required only under cold conditions. In addition, it greatly reduces the deformation of the coil–collar assembly when the magnet is powered.

The yoke dimension is chosen such that a ~0.5 mm gap between the yoke halves is maintained at room temperature. This gap closes between 150 K and 100 K during cool-down depending on the precise dimensions of the collars and yoke. The outer cylinder is shrink-fitted. Figure 4.1 provides an overview of the parts of the structure.

The ANSYS software program (Swanson Analysis System, USA) was used for the structural analysis for which the Young's moduli E of the windings had been determined by measurements on reacted and impregnated cable stacks. Measurements up to 100 MPa pressure were performed in azimuthal and radial directions by pressing cable stacks in ad hoc made fixtures. Behavior close to linear was found above 40 MPa. The azimuthal E between 50 MPa and 80 MPa was about 19 GPa for the inner layer cable stacks and about 20 GPa for the outer layer. The radial E, almost equal for the inner and outer layer cable stacks, was about 24 GPa. The computed stresses at the pole and mid-plane of the coil at magnet assembly, cool-down and powering are reported in Fig. 4.8.

Results of measurements on Nb_3Sn strands and cables, available when the magnet was built, indicated a strong degradation of current density under transverse compression. For this reason, the pre-compression on the coil was limited to the minimum level to maintain contact at the interface with the central post in all operating conditions. Only 2 years later, new measurements made on assemblies

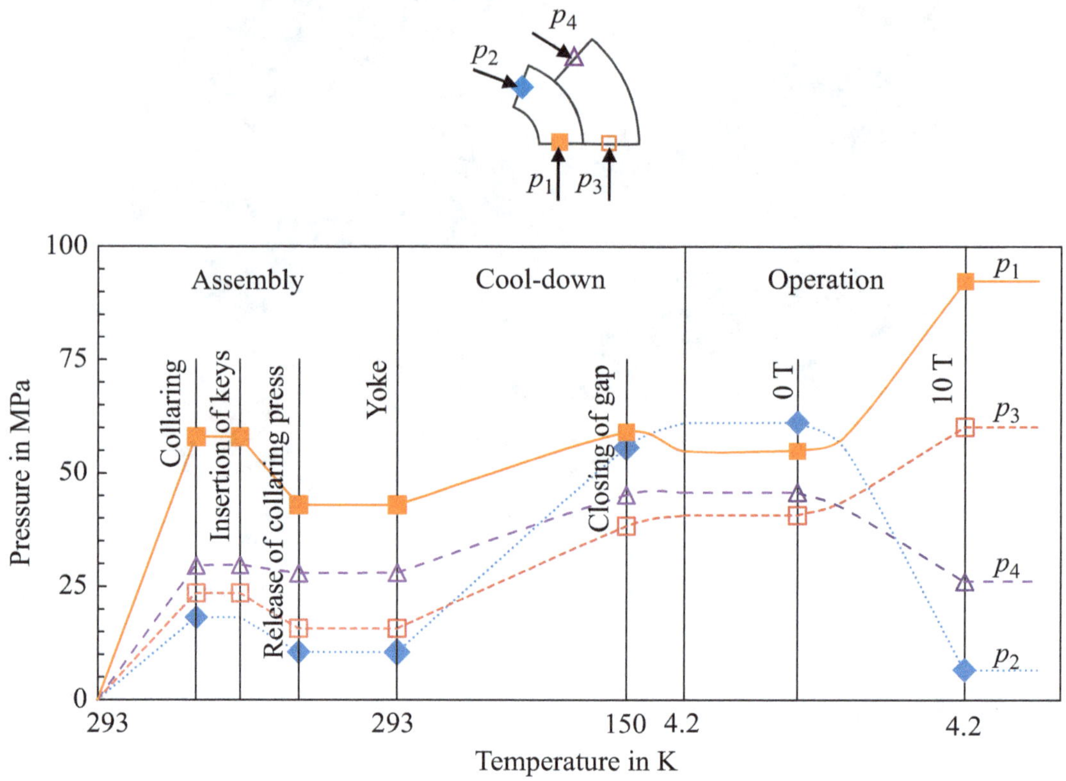

Fig. 4.8 Computed coil stresses at assembly, cool-down, and at 10 T bore field. (Courtesy of F. Zerobin)

of samples of the magnet cables reacted and epoxy-resin impregnated as in the magnet, showed much smaller reversible degradation (Bona et al. 1991): at $B = 10$ T less than 2% and about 4% under transverse compressions of 100 MPa and 150 MPa, respectively; at $B = 11.5$ T, 4% and 6% under the abovementioned compressions. This would have allowed a higher coil pre-compression of up to 150 MPa, which was the testing limit.

The axial Lorentz forces are taken at the coil ends by thick stainless-steel plates bolted to eight tie-rods passing through holes in the yoke.

4.6 Coil Manufacture and Magnet Assembly

The main components of a coil layer are:

- A central post, made of copper;
- Superconducting cables and their insulation;
- Longitudinal copper wedges;
- End spacers made of bronze.

A precise winding machine (Fig. 4.9) and sophisticated tooling were designed and built by ELIN, to shape the conductors and hold them in place during winding

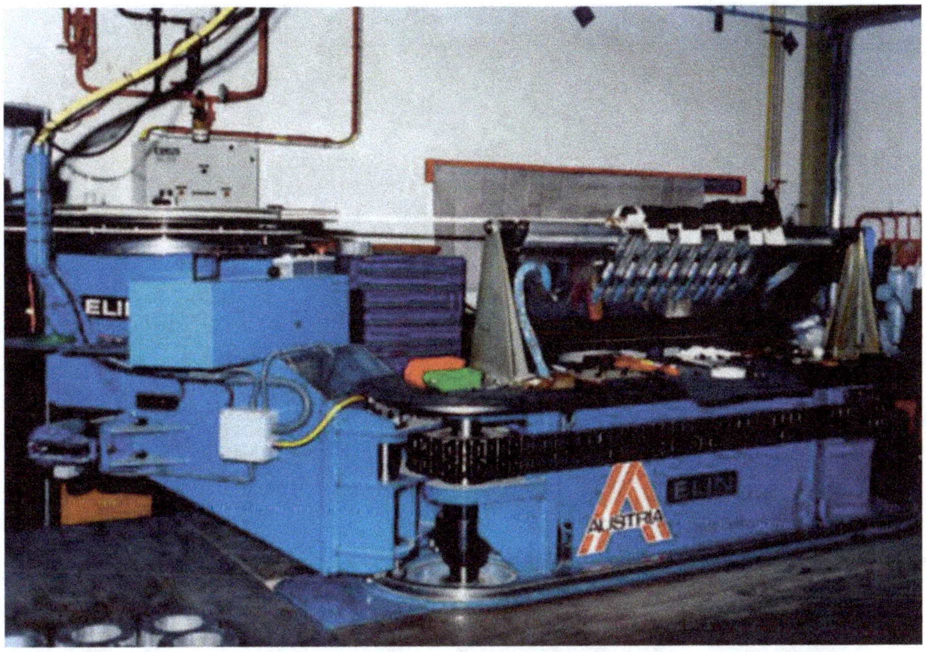

Fig. 4.9 Coil winding machine. (Courtesy of ELIN, Austria)

and the subsequent fabrication phases, since the bronze content made the conductors stiff and springy. This stiff and springy conductor required robust tools capable of applying large forces. No separate heat treatment mold was used, and therefore a part of the tooling used during winding had to be able to withstand the high temperature during reaction heat treatment for several hours. Stainless steel was the material generally used for various tooling.

The inter-turn insulation was made of half-overlapped 0.12 mm thick, 15 mm wide mica-glass tape (Cablosam® 366.21–10, Von Roll Isolawerke, Switzerland). This tape is composed of mica paper backed on one side by woven glass, and is entirely pre-impregnated with a flexible silicon elastomer. To ensure sufficient mechanical stability of the tiny mica flakes after firing of the organic elastomer, the tape was wound with the mica placed on the cable side, so that the external glass fiber (the mechanically stronger component) would maintain them in place and compress them against the cable. The same type of insulation tape was employed for the central post, the copper wedges, and the spacers at the coil ends.

The insulated wedges and end spacers were inserted during winding at the foreseen places. The end spacers were manufactured by casting and milling from bronze. The rough shape was obtained by casting. The final profile was made by inclining the milling cutter of a pre-determined angle and synchronically controlling the advance and rotation of the piece. Edges and corners were carefully rounded off to safeguard the insulation. An improvement of the insulation was tried by flame-spraying alumina, Al_2O_3, but mica-glass tape was absolutely needed.

The winding of a coil layer is shown in Fig. 4.10. The insulated cables were wound under a tension of 500 N for the inner layer and about 300 N for the outer layer. After winding, each single-layer coil with its clamping and confining fixtures

Fig. 4.10 Winding a coil layer. (Courtesy of ELIN, Austria)

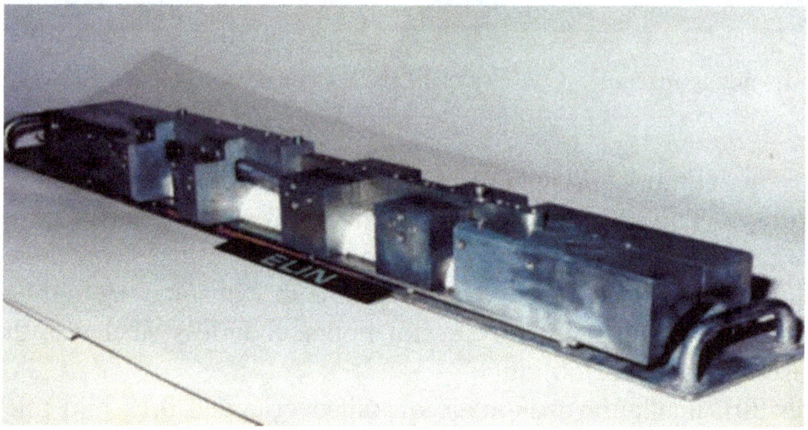

Fig. 4.11 A coil layer ready for reaction clamped by stainless-steel tooling blocks. (Courtesy of ELIN, Austria)

(Fig. 4.11) was placed in the reaction oven where the following operations were carried out: (1) evacuation at room temperature; (2) heating up to 600 °C in an argon atmosphere; (3) firing at 600 °C with dry air circulation for about half an hour in order to eliminate the organic component of the insulation; (4) evacuation; (5) heat treatment at 675 ° C under flowing argon for about 140 h; (6) slow cool-down to room temperature under an argon atmosphere. The end view of a reacted coil inner layer is shown in Fig. 4.12.

After reaction, an inter-layer insulation sheet (Nomex®, DuPont, USA) was placed around the inner layer, the two coil layers were assembled, and the internal Nb_3Sn-Nb_3Sn splice between the inner and outer layers was made. The making of the splice, placed in the straight part of the coil, can be described as follows: during winding, the first turn of the inner layer is bent (the jump) to reach the level of the outer layer where it continues straight for about 250 mm (about two transposition lengths). The corresponding part of the outer layer cable first turn, also about

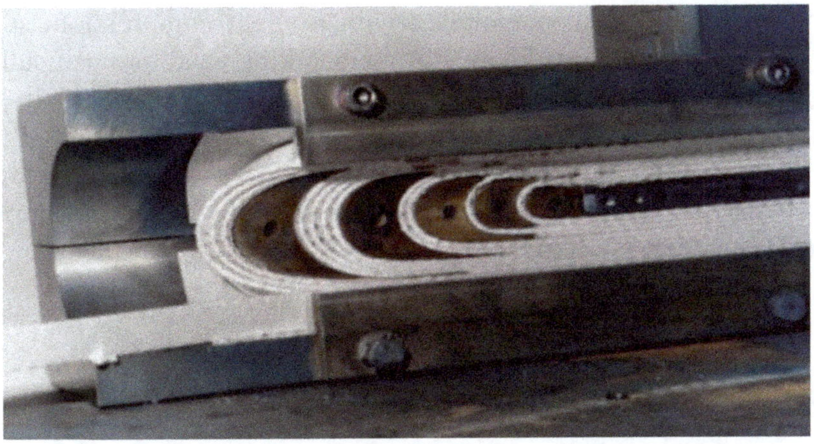

Fig. 4.12 End view of a reacted coil inner layer. (Courtesy of ELIN, Austria)

250 mm long, remains straight in its normal place. The soldering of the two corresponding parts is made by inserting between them a $Pb_{60}Sn_{40}$ foil, and then performing a pressing and heating operation. Two voltage taps are then soldered to the splice. For additional stability, the splice is then inserted and soldered in a U-shaped copper piece, which is insulated and lodged in a slot in the central post.

After placing the central post, the Nb_3Sn-Nb-Ti splices of the coil terminals were also made using $Pb_{60}Sn_{40}$ solder. The coil-to-ground insulation (Nomex®) was wrapped around the coil, which was then placed in the impregnation mold, whose closing plate was tightened by bolts with a force of 600 kN in order to apply a tangential compression of about 10 MPa to the windings.

After evacuation, impregnation with epoxy resin and curing were performed with a similar resin and procedure that had been used in the CERN ISR superconducting quadrupoles (Billan et al. 1976): Ciba-Geigy, Switzerland resin system composed of Araldite MY745 (100 pbw) epoxy, HY905 hardener (100 pbw), and DY072 (1 pbw) and DY073 (0.2 pbw) accelerators. The impregnation was performed for 8 h at 90 °C and a curing cycle for 24 h at 120 °C.

After impregnation, the coil was extracted from the mold, cleaned of resin and demolding agent residues, and replaced into the mold to measure its elastic modulus. Measurements were made in the central part of the coil over 100 mm, paying attention not to introduce shear. After some pressing cycles to verify the reproducibility of force vs. displacement behavior, the E modulus was measured at pressures of around 50 MPa: the result for the coil azimuthal E modulus was about 28 GPa. This value was in relatively good agreement with the E measured on cable stacks (see Sect. 4.5.2). This result was used to determine the thickness of the shims inserted between the coils in the magnet median plane to obtain the pre-stress required at collaring.

The two coils were then assembled, and the collars were placed around them. After some pressing cycles to settle the assembly, the collars were pressed to the prescribed force and the locking keys inserted.

The yoke halves were then mounted with a slight compression, sufficient to guarantee integrity of the assembly during handling, and provisionally kept in place with belts. The stiff end plates were fixed to the tie rods. Finally, the outer aluminum alloy cylinder was shrink-fitted by warming it up to 200–250 °C. An insulating plate to support the external connections was fixed to the magnet end plates. The two coils were then soldered together with two Nb-Ti cable terminals and the lead Nb-Ti cables were fixed to their support.

4.7 Mirror Coil Test Facility

A 1 m long, 5 cm diameter half-bore mirror dipole was built at ELIN to test single coils. Its cross-section is shown in Fig. 4.13. In the mirror dipole, one coil is replaced by an iron insert. In this set-up no collars were used: the necessary pre-stress on the coil is provided by the yoke halves surrounded by the aluminum shrinking cylinder.

4.8 Tests and Performance

The magnet prepared for installation in a Dewar is shown in Fig. 4.14. At various steps of magnet construction, voltage taps were placed: at the coil inter-layer splices, at the coil Nb_3Sn/Nb-Ti terminal splices, and at the magnet leads. All voltage taps

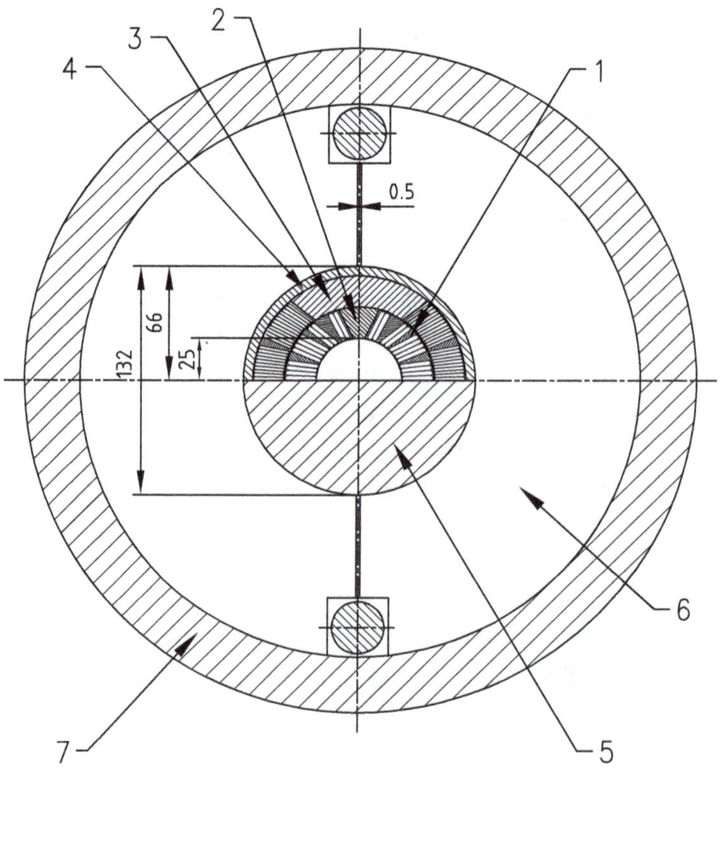

Fig. 4.13 Simplified cross-section of the Mirror Test Facility: *1* – coil; *2, 3* – central post parts; *4* – copper shield; *5* – iron mirror; *6* – iron yoke; *7* – shrinking cylinder

Fig. 4.14 The magnet ready to be inserted into a vertical cryostat for tests. (Courtesy of CERN)

were doubled for redundancy, so in total 12 voltage taps were installed. An elaborate protection system, with heaters for rapidly quenching the entire magnet, was not necessary for this short magnet and, therefore, not installed. Of course, for longer magnets such a system would be mandatory.

In this short model magnet, protection was ensured by two resistors inserted in the circuit. One resistor was placed across the magnet leads inside the cryostat, to provide protection in case one or both cryogenic current leads failed. The second resistor was placed outside the cryostat and was normally short-circuited by a circuit breaker. When a quench was detected (by comparing voltages across the two coils), the circuit breaker was opened, thus forcing the magnet current through the two resistors.

An early coil, mounted in the Mirror Coil Test Facility, had been tested in February 1989. It reached a central field of 10.2 T at 17,430 A, 4.3 K, after eight quenches (Fig. 4.15). The magnetic field was measured by means of a Hall plate magnetometer. After full thermal cycle, the same field was attained without retraining. When tested at 1.9 K the quench field was the same, an indication that the mechanical limit was reached. All quenches started in the inner layer of the coil.

The full magnet was tested in June 1989. It reached a central bore field of 9.5 T at a current of 15,600 A at 4.3 K, corresponding to about 10.05 T in a turn of the inner layer. Thermal cycling to room temperature did not affect the field level and negligible retraining occurred (Fig. 4.16). As in the mirror configuration, all

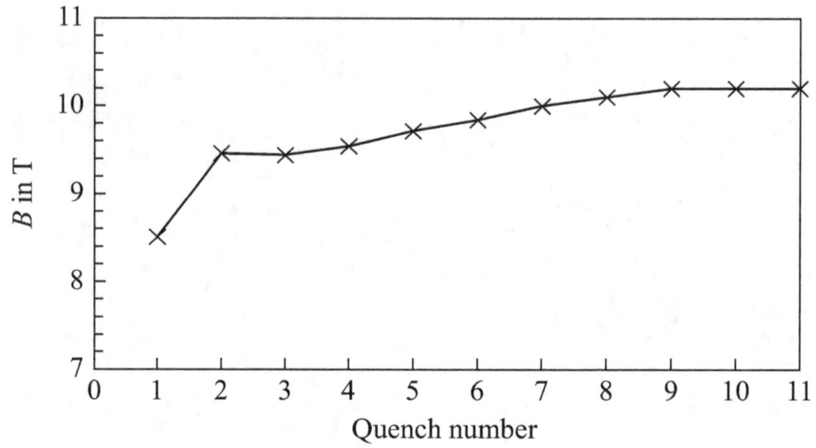

Fig. 4.15 Quench history of the mirror (half-dipole) at 4.3 K

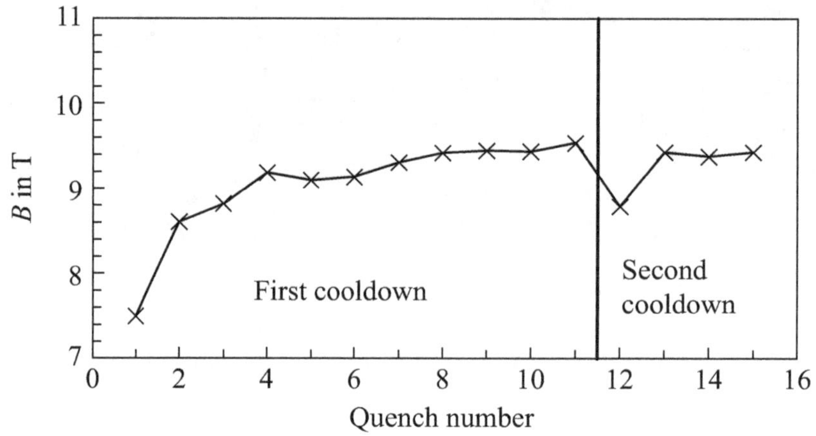

Fig. 4.16 Quench history of full dipole at 4.3 K

quenches started in the inner layer of the coils. The field attained in the dipole model was slightly lower than in the mirror configuration due to the lower Nb_3Sn content in the dipole inner-layer cable and the larger distance to the iron yoke.

4.9 Coil with Powder-in-Tube Cables

Upon delivery of the prototype cable lengths, a coil was manufactured by ELIN and mounted in the Mirror Coil Test Facility. Its construction followed the same procedure as the previous bronze route coils. It was tested in December 1991. The central field reached at the first quench was 7.43 T at a current of 11,270 A at 4.3 K, a remarkable result for a first experimental production of a novel conductor. More than 20 excitations were made. At small ramp rates up to 100 A/s all quenches happened at the same field level, about 7.4 T. A possible explanation could be that they were caused by a localized defect in a conductor. Higher ramp rates affected the quenching field: 7.1 T at 200 A/s, 6 T at 250 A/s, and 5 T at 300 A/s. This could be due to the large Nb_3Sn filament size.

This was the last construction activity at that time at CERN on the Nb_3Sn option, as all CERN resources had to be concentrated on the Nb-Ti line. In Europe activity on Nb_3Sn magnets continued, with CERN's involvement, by a collaboration with the Netherlands Institute for Nuclear & High Energy Physics (NIKHEF), the Netherlands Foundation for Fundamental Research on Matter (FOM) and the University of Twente. This culminated in the successful construction of the Model Single of the University of Twente (MSUT) dipole model, which was tested at CERN in 1995, reaching 11.03 T at 4.4 K at its first quench.

4.10 Conclusion

In 1989 the CERN–ELIN model magnet made a real breakthrough in accelerator magnet technology. A moment of celebration of this success is shown in Fig. 4.17. It was the first successful dipole of particle accelerator type made with Nb_3Sn superconductor. It reached record field levels of 9.5 T in the full dipole and 10.2 T in the magnetic mirror configuration. Its design and fabrication were in an industrial style, showing that the Nb_3Sn option was a realistic one for attaining field levels not accessible to Nb-Ti. Considerable technological advancement was achieved, in particular concerning the electrical insulation of the cables, the coil winding, and the reaction process. The manufacturing of the coil was, however, complicated and delicate, requiring maximum attention at all steps.

Fig. 4.17 Project leaders congratulate the team for the successful magnet test (17 June 1989)

It was recognized that important improvements were necessary to obtain better performing and less stiff conductors. The development of processes other than the bronze route such as jellyroll and PIT, still in their initial phases, had to be strongly supported because they promised higher current density and easier winding. One key challenge for PIT process development for use in particle accelerators was towards smaller filaments.

The cable insulation also needed great improvements in mechanical strength, flexibility, and resistance to high temperatures. At that time there was a consensus that insulation materials other than mica and glass should be investigated, for instance ceramics. These improvements on the basic materials, superconductors, and insulation, would lead to safer and more economical coil fabrication.

In 1991 CERN decided to adopt Nb-Ti for the LHC magnets, and consequently all CERN resources had to be concentrated on developing this technology. This decision was taken after thorough discussions at all levels. It was based on the following main arguments:

- The technology of Nb_3Sn superconductors was still in an experimental stage, while Nb-Ti was an industrial product manufactured and commercialized at market prices by several companies.
- Only one Nb_3Sn accelerator type short magnet and none of the length required for the LHC (10 m or more) had successfully been built.
- The cost of Nb_3Sn magnets was estimated to exceed by far the foreseen budget.
- Industrialization of superfluid cryogenics was deemed easier to achieve, less expensive, and less risky than that of Nb_3Sn magnets.

References

Asner A, Perin R, Wenger S et al (1990) First Nb_3Sn, 1 m long superconducting dipole model magnets for LHC break the 10 Tesla field threshold. In: Sekiguchi T, Shimamoto S (eds) Proceedings 11th international conference on magnet technology, Tsukuba, 28 Aug–1 Sep 1989, vol 1. Springer, Dordrecht, pp 36–41, CERN LHC Note 105, Aug 1989

Billan J, Perin R, Resegotti L et al (1976) Construction of a prototype superconducting quadrupole magnet for a high-luminosity insertion at the CERN intersecting storage rings. CERN Yellow Report 76–16, CERN, Geneva

Bona M, Jakob B, Pasztor G et al (1991) Reduced sensitivity of Nb_3Sn epoxy-impregnated cable to transverse stress. Cryogenics 31(5):390–391; CERN LHC Note 141, Apr 1991. https://doi.org/10.1016/0011-2275(91)90117-F

Haebel EU, Wittgenstein F (1971) Big European Bubble Chamber (BEBC) magnet progress report. In: Derrick M (ed) Proceedings international conference on bubble chamber technology, June 1970, vol 2. Argonne National Laboratory, Argonne, pp 1126–1149

Hagedorn D, Leroy D, Perin R (1985) Towards the development of high field superconducting magnets for a hadron collider in the LEP tunnel. In: Marinucci C, Weymuth P (eds) Proceedings of the 9th international conference on magnet technology, Swiss Institute for Nuclear Research (SIN), Villigen, pp 86–91, Zurich, 9–13 Sep 1985; CERN LHC Note 31, Aug 1985

Halbach K, Holsinger R (1972) Poisson user manual. Technical Report, Lawrence Berkeley Laboratory, Berkeley

Hornsveld EM, Elen JD, Van Beijnen CAM et al (1988) Development of ECN-type niobium-tin wire towards smaller filament size. Adv Cryo Eng 34:493–498

Morpurgo M (1970) The design of the superconducting magnet for the OMEGA project. Particle Accelerators 1:255–263. https://cds.cern.ch/record/350951/

Perin R (1973) Calculation of magnetic field in a cylindrical geometry produced by sector or layer windings. Internal note CERN ISR-MA/RP/cb, 15 Nov 1973. CERN, Geneva

Perin R (1986) Magnet research and development for the CERN Large Hadron Collider. In: Dahl PF (ed) Proceedings ICFA workshop on superconducting magnets and cryogenics, Brookhaven National Laboratory, Upton, NY, 12–16 May 1986, BNL 52006 pp 25–33; CERN LHC Note 41, CERN SPS/86–9 EMA

Perin R (1990) Superconducting magnets for the LHC: a report of CERN's collaboration with industry, vol 21. CERN, Europhysics News, Geneva, pp 90–92

Perin R (1995) Field, forces and mechanics of superconducting magnets, In: Proceedings CAS superconductivity in particle accelerators, 14–24 May 1995, Hamburg. CERN, Geneva pp 71–92 CERN Yellow Report 96–3

Perin R, Van Der Meer S (1967) The program MARE for the design of two-dimensional static magnetic fields. CERN Yellow Report 67–7

Perin R, Leroy D, Spigo G (1989) The first industry made model magnet for the CERN Large Hadron Collider. IEEE Trans Magn 25(2):1632–1635. https://doi.org/10.1109/20.92612. CERN LHC Note 83 Aug 1988

Ten Kate HHJ, Ten Haken B, Wessel S (1989) Critical current measurements of prototype cables for the CERN LHC up to 50 kA and between 7 and 13 Tesla using a superconducting transformer circuit. In: Sekiguchi T, Shimamoto S (eds) 11th international conference on magnet technology (MT-11), Tsukuba, 28 Aug–1 Sep 1989. Springer, Dordrecht, pp 60–65

Wenger S, Zerobin F, Asner A (1989) Towards a 1 m long high field Nb_3Sn dipole magnet of the ELIN-CERN collaboration for the LHC-project-development and technological aspects. IEEE Trans Magn 25(2):1636–1639. https://doi.org/10.1109/20.92613. CERN LHC Note 82 Aug 1988

Wilson RR (1978) The Tevatron. FERMILAB-TM-0763

Wolff S (1988) Superconducting HERA magnets. IEEE Trans Magn 24(2):719–722. https://doi.org/10.1109/20.11326

Chapter 5
The UT-CERN Cos-theta LHC-Type Nb_3Sn Dipole Magnet

Herman H. J. ten Kate, Andries den Ouden, and Daniel Schoerling

Abstract This chapter reports on the University of Twente (UT)–European Organization for Nuclear Research (CERN) cos-theta LHC-type Nb_3Sn dipole magnet program. In this program an experimental 1 m long two-layer cos-theta dipole magnet was developed using Rutherford cables made with powder-in-tube Nb_3Sn composite wires. The magnet reached a magnetic field of 11.3 T in a bore of 50 mm.

5.1 Introduction

To continue the successful development of Nb_3Sn dipoles started with the CERN–ELIN collaboration (see Chap. 4), the Applied Superconductivity Centre at the University of Twente, the Netherlands, initiated a collaboration with the National Institute for Subatomic Physics (Nikhef), the Energy Research Centre of the Netherlands (ECN), the European Organization for Nuclear Research (CERN), and the companies HOLEC and SMIT WIRE. The aim of the collaboration was to design and build an experimental 1 m long Large Hadron Collider (LHC)-type dipole magnet to convincingly break through the 10 T barrier experienced so far. In this program specific novel ideas and high current density powder-in-tube (PIT) Nb_3Sn composite wires were envisaged.

The main funding for this project was granted by the Netherlands Technology Foundation STW, with some contributions by companies, and was supported by

H. H. J. ten Kate (✉)
University of Twente, Enschede, Netherlands

CERN (European Organization for Nuclear Research), Meyrin, Genève, Switzerland
e-mail: Herman.TenKate@cern.ch

A. den Ouden
Radboud University, Nijmegen, Netherlands
e-mail: a.denouden@science.ru.nl

D. Schoerling
CERN (European Organization for Nuclear Research), Meyrin, Genève, Switzerland
e-mail: Daniel.Schoerling@cern.ch

D. Schoerling, A. V. Zlobin (eds.), *Nb_3Sn Accelerator Magnets*, Particle Acceleration and Detection, https://doi.org/10.1007/978-3-030-16118-7_5

CERN through the magnet test effort. The magnet parameters were chosen such to meet the LHC requirements at that time, i.e., a two-layer design with a minimum magnetic field of 10 T at 4.4 K within a free aperture of 50 mm (ten Kate et al. 1991).

Back then, the race between Nb-Ti dipole magnets operating at 1.9 K and Nb_3Sn magnets operating at 4.4 K was decided in favor of Nb-Ti technology. Nb_3Sn conductor technology was considered not sufficiently mature to be used reliably and efficiently for large-scale production. Moreover, the cost of Nb_3Sn conductor and magnet production in combination with the lack of experience with large-scale high current density Nb_3Sn superconductor and coil manufacturing in industry added to the decision to use Nb-Ti dipole magnet technology operating at 1.9 K for the LHC. Operation at 1.9 K was considered feasible following the progress demonstrated in cryogenic technology (Claudet and Aymar 1990).

In 1989, however, it was also clear that for a large number of Nb-Ti magnets connected in series it would hardly be possible to attain reliable 10 T operation, and indeed the operating magnetic field of the LHC was later reduced to 8.33 T. This limitation retained interest in further research towards Nb_3Sn LHC-type dipole magnets.

In the Nb_3Sn dipole magnet landscape of 1987, the program known as the Model Single of the University of Twente (MSUT), showed very few magnets with a real bore, and all performed well below 10 T. The latest was the CERN–ELIN magnet, which achieved only 9.7 T due to the limited current density in the bronze route Nb_3Sn wires used. In order to break through the 10 T barrier, a new project was launched with the ambition:

1. To exploit the potential of the Powder-in-Tube (PIT) Nb_3Sn conductor by targeting a magnetic field well above 10 T at 4.4 K (the LHC target in 1988);
2. To meet the LHC dipole parameters;
3. To perform research on PIT wires and cables for increasing the wire engineering critical current density $J_c(B, T)$, reduction of filament size, thereby reducing magnetization effects, and the degradation of I_c as a function of strain; and
4. To develop a dedicated design and technology to realize a 1 m long model single-aperture dipole magnet following a few new ideas to overcome problems with traditional designs.

As the requirements for Nb_3Sn conductor are fundamentally different from those for Nb-Ti conductor, the design has been re-worked from scratch and dedicated solutions addressing the specific challenges of using Nb_3Sn conductor and the wind-and-react (W&R) route were developed. In addition, based on a review of the manufacturing and performance problems of earlier Nb_3Sn dipole magnets, in particular concerning degradation and training, it has been decided to try out new solutions for achieving a better performance. Major investigations in the frame of this program included the development of the following:

1. A two-layer graded coil using 22 and 17 mm wide Rutherford cables with 1.26 and 0.98 mm strand diameters for the inner and outer layers, respectively. A two-layer design was chosen for two reasons: firstly to limit the coil winding cost,

and secondly to demonstrate the beneficial effect of stiff cables on the coil's training behavior.

2. New cable insulation based on a folded glass-mica tape wrapped with glass-fiber ribbon. The idea of using a dielectric mica film was motivated by reducing the risk of electrical shorts in the coil windings, in particular in full-size long magnets, and by reducing the risk of micro-cracking in windings.
3. A continuous support of the conductors at the transition from the straight part into the coil ends using strip-like end spacers brazed to the longitudinal wedges. Usually in magnets, spacers are not connected to the wedges, and a discontinuity or locally softer cable support is present, which may cause training quenches in magnets. Indeed, in this type of dipole magnet many training quenches have their origin in this area. Therefore it was considered necessary to avoid such a discontinuity.
4. A natural bending of the conductor over the coil heads to achieve a minimum stress layout even before cable heat treatment. The disadvantage of this method is that the cross-section of the coil head looks irregular since the cables are not forced to follow the shape of the predetermined end spacers.
5. A coil heat treatment without any constraint in the end spacer area. During heat treatment and roughly 3% Nb_3Sn volume expansion, the conductor can freely move and settle.
6. After the heat treatment, all local voids in the coil heads were securely filled with glass fibers. This filling was considered important to avoid any substantial voids filled with pure resin that may cause micro-cracks and, thus, training quenches.
7. A layer-to-layer joint using a superconducting shunt. In this way a layer jump can be avoided leading to a more uniform mechanical winding pack. The superconducting shunt was inserted and soldered after coil heat treatment but before resin vacuum impregnation.
8. Vacuum impregnation of the coil windings with Ciba-Geigy, Switzerland resin system composed of Araldite MY740 epoxy, HY906 hardener, and DY062 accelerator.
9. Closed, shrink-fit aluminum-alloy ring-collars for achieving optimal coil support without overstressing the Nb_3Sn. This type of collar and material were chosen to eliminate the risk of conductor degradation when using split collars and keys, and avoiding the risk of using a collaring press.

The PIT Nb_3Sn wire was developed and manufactured by ECN while cabling was performed at the Lawrence Berkeley National Laboratory (LBNL) in Berkeley, USA. After wire production, ECN closed its Nb_3Sn wire facilities. All wire-making equipment and knowledge were transferred, first to the University of Twente for a few years, and later in 1991 to the university spin-off company Shape Metal Innovation (SMI) in Enschede, The Netherlands. SMI further developed PIT conductors for many years while supported by Twente University, and delivered PIT Nb_3Sn wires for use in high-field magnets, in particular in various dipole magnets at the Fermi National Accelerator Laboratory and CERN. Finally, in 2006 the PIT technology was acquired from SMI by Bruker-EAS.

The 1 m, single-aperture dipole model magnet was manufactured in a five-year project from 1990 to 1995 at the University of Twente and tested first in 1995 and again in 1997 at CERN. The details of magnet design, technology development, system manufacturing, and magnet performance are discussed below.

5.2 Magnet Design

5.2.1 Electromagnetic Design

The initial twin-aperture electromagnetic design was optimized according to the following criteria (ter Avest et al. 1991):

- To minimize the amount of conductor, a shell-type (also known as cos-theta) design was chosen with graded current density, implying a higher current density in the outer layer than in the inner layer by reducing the cable cross-section.
- To ease operation, the same operating current ($\leq$20 kA) and similar load-line margins were foreseen.
- In view of the large scale of production, maintaining simplicity (considered to be important), costs, and labor, the number of layers was kept to two.
- To reach a field quality on the level of 10^{-4}, the design and fabrication tolerances needed to be small.
- To produce a stable cable, the number of strands per cable and the diameter of the strands needed to remain within a practical range.

These constraints are partly contradictory, and for practical reasons the following criteria were used: a cable thickness < 3 mm, a maximum angle of the layers $\leq$70°, and $\leq$ 36 strands per cable.

For a field of 10 T the overall current density was assumed to be 453 A/mm^2 and the non-Cu current density 1373 A/mm^2. These numbers include 13% overall void fraction in the cable, 16% area for the insulation thickness, and 10% operational margin on the load line for both layers.

The first round of optimization focused on a 2D optimization of a single-aperture magnet, assuming an iron yoke with a round inner diameter of 200 mm, infinite outer diameter and permeability, and the same cable properties for all layers. Figure 5.1 shows the simplified model used for the first optimization.

The field distribution in such a configuration can be calculated with analytical equations (Perin 1973; ter Avest et al. 1991). After setting the magnetic field to 11.5 T and the current density in the inner layer, the geometry was optimized by eliminating the two multipoles b_3 and b_5. This optimization was fast and produced two classes of solutions: one with maximum angles of about 70°/45° (inner/outer layer) and one with maximum angles of about 50°/80°. The last design option, however, requires more superconductor, and a difficult mechanical support structure, and was, therefore, abandoned.

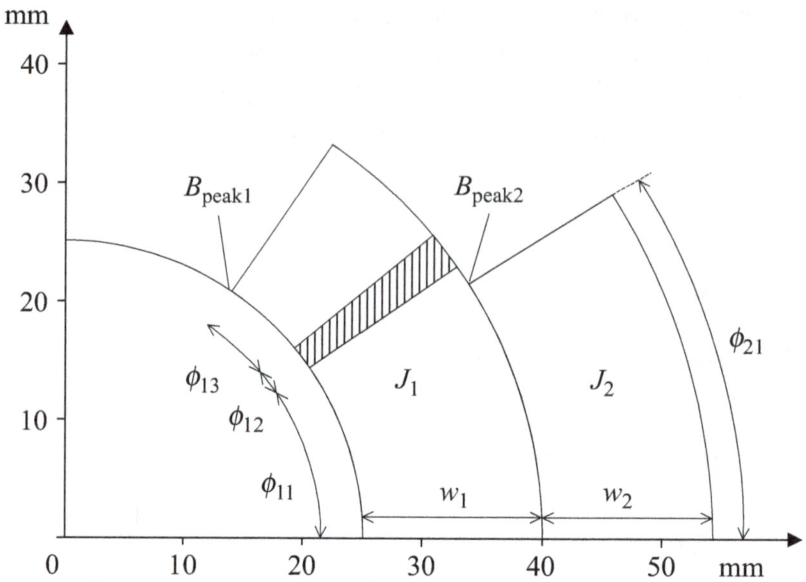

Fig. 5.1 The two-layer coil cross-section: B_{peak} – peak field in a layer, w – width of a layer, J – overall current density, and ϕ angle of a circular segment. The unit on both axes is mm. (Modified from ter Avest et al. 1991)

As the next step, the grading ratio between the inner and outer layers was optimized under the assumption that the transport current and the number of strands in the outer layer are fixed. It can be shown that, for an increasing ratio of w_2/w_1, the maximum angle of the coil increases. The total amount of superconductor is almost invariant to a w_2/w_1 variation for $w_2/w_1 \geq 0.8$. Therefore, a grading ratio of $w_2/w_1 = 0.8$ ($J_2/J_1 = 1.6$) was selected in order to minimize the top angle of the inner layer. Then, the higher multipoles b_7 and b_9 were minimized. Since more degrees of freedom were required, two additional wedges were added.

In the final step, the round shell segments were converted into keystoned cable segments and were optimized around the previously determined point. The geometric design multipoles (expressed relative to B_1) are $(b_3, b_5, b_7, b_9) = (-0.06, -0.5, -0.6, 0.3)\cdot 10^{-4}$. With a maximum angle $\phi_1 = 70.5°$ and $\phi_2 = 58.4°$, the required operating current is 17.7 kA. The cable parameters are presented in Table 5.1. The coil cross-section is shown in Fig. 5.2. A detailed treatment of the electromagnetic design optimization is given in ter Avest et al. (1991).

5.2.2 *Conductor Choice and Parameters*

To limit the coil width by two typical maximum widths of the Rutherford cable and enabling a twin-aperture design with an inter-beam spacing of 194 mm, the coil width was limited to around 35 mm. This limitation imposed for the reference design field level of 10 T a critical current density J_c of about 1580 A/mm^2 at 10 T in the non-copper part of the conductor, assuming a Cu/non–Cu ratio of 1.

Table 5.1 Nb_3Sn Rutherford cables and strand parameters for the inner and outer layers of the coils in twin-aperture configuration for the 10 and 11.5 T versions

	10 T design		11.5 T design	
Parameter	Inner layer	Outer layer	Inner layer	Outer layer
Thin edge (mm)	2.19	1.47	1.98	1.54
Thick edge (mm)	2.69	1.79	2.47	1.93
Bare cable width (mm)	16.8	16.8	21.7	17.4
Strand diameter (mm)	1.35	0.98	1.26	1.00
Cu fraction (%)	50	50	56	56
Filament diameter (μm)	–	–	42	32
Number of filaments	–	–	192	192
Filament twist pitch (mm)	–	–	30	30
Number of strands	24	36	33	33
RRR value	≥100	≥100	≥100	≥100
Cable pitch (mm)	≥120	≥120	150	150
Cable unit length (m)	4 × 25	4 × 40	4 × 25	4 × 40
Insulation thickness (mm)	0.14	0.14	0.14	0.14
Nominal current (kA)	16.0		17.7	
B_{peak} at I_{nom} (T)	10.2	8.7	11.9	9.6

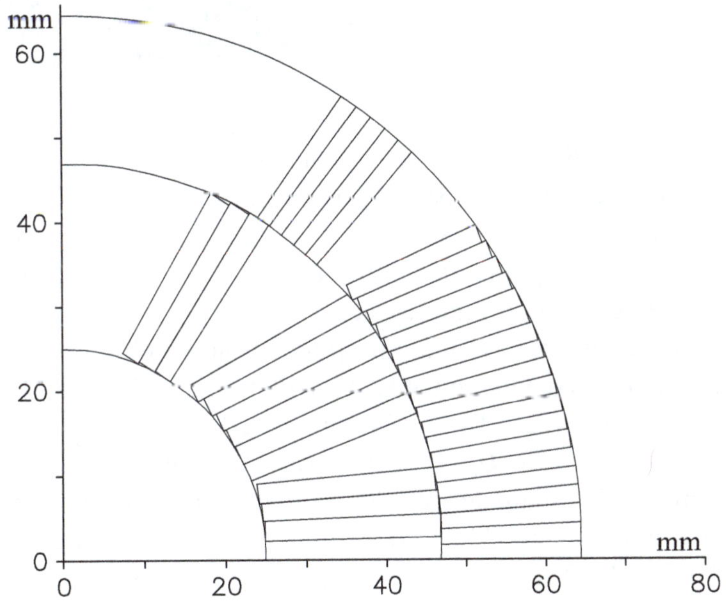

Fig. 5.2 Optimized cross-section of the 11.5 T design coil. The unit on both axes is mm. (Modified from ter Avest et al. 1991)

This high current density had to be achieved with small filaments in order to minimize magnetization effects, to improve field quality, and to minimize heat loss. First estimates called for an effective filament diameter below 10 μm. The combination of high critical current density and sufficiently small filaments was not available at the time. In the MSUT project, the focus was laid, therefore, on the conductor with the best performance available: the ECN PIT multi-filamentary wire. The non-copper critical current density J_c measured on short samples for these wires

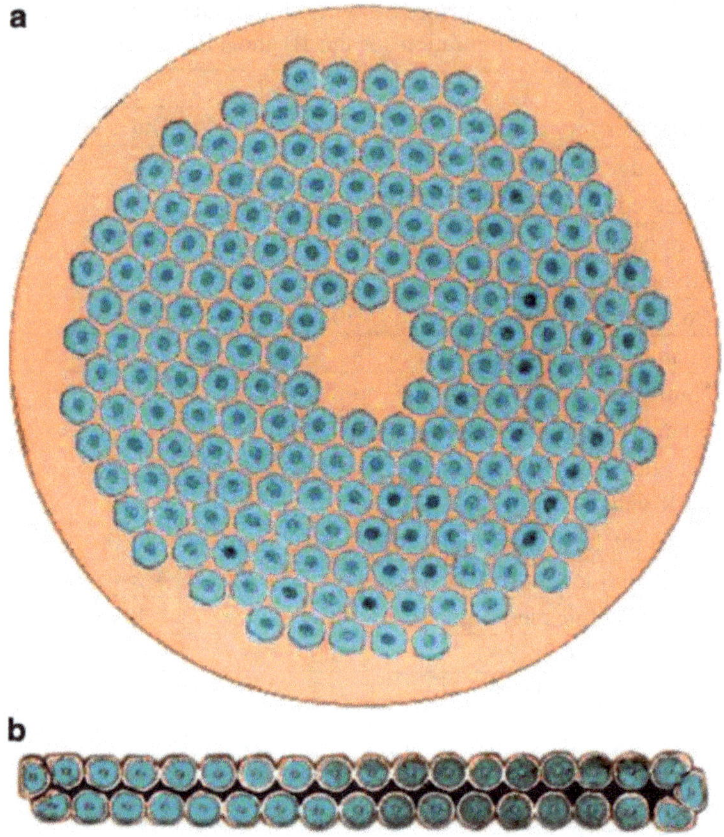

Fig. 5.3 (**a**) Cross-section of a 192-filament PIT Nb_3Sn wire with a pure copper fraction of 55%; and (**b**) a Rutherford cable made from this wire featuring 36 strands of 0.90 mm diameter, cable size = 1.45/1.77 × 16.70 mm. (Modified from ten Kate et al. 1991)

was about 2000 A/mm^2 at 10 T and 4.2 K, about a factor of 2.5 higher than in bronze conductors: at the expense, however, of a larger filament size. The program subsequently focused on decreasing the filament diameter, maintaining the same critical current density, increasing the unit lengths, and producing Rutherford cables with limited critical current degradation.

In an effort to reduce the filament size below $D_{\text{eff}} \leq 20$ μm a double-stacking method was developed at ECN. Billets with 18, 19, and 37 spokes were stacked into billets with 18 × 18 = 324 ($D_{\text{eff}} = 29$ μm), 36 × 19 = 684 ($D_{\text{eff}} = 21$ μm), and 36 × 37 = 1332 ($D_{\text{eff}} = 15$ μm) sub-elements. After assembly, they were drawn into wires with a diameter of 0.85 mm, and short sample measurements were performed (Hornsveld et al. 1988), confirming critical current densities of about 1900 A/mm^2 at 10 T and 4.2 K. The process, however, appeared rather costly, took a long time, and it was impossible to reliably draw sufficiently long lengths without breakage. Therefore, the single-stack wire design with 192 filaments was taken as a baseline and was used to produce about 75 km of wire, weighing about 500 kg, making up sufficient unit lengths. Tests of keystoned cables (Fig. 5.3) were performed at ECN and later at LBNL. Critical current measurements revealed little degradation, rendering a critical current of 30 kA at 8.7 T.

Based on this success, the magnet design was reiterated and the target magnetic field was increased from 10 T to 11.5 T for a twin-aperture design and 11.0 T for a single-aperture design, using a slightly larger cable. The cable parameters for the two

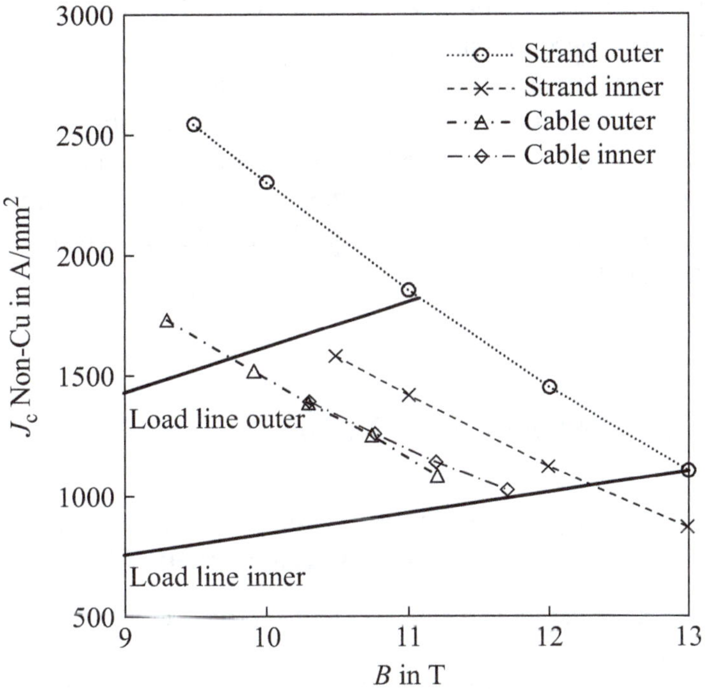

Fig. 5.4 Measured non-Cu J_c vs. magnetic field before (strand) and after cabling (cable) for inner and outer layer cables, averaged over at least five samples per case. The load lines for the inner and outer layers are also shown. (Modified from den Ouden et al. 1994)

design proposals for a bore field of 10 T (abandoned) and 11.5 T (pursued) are summarized in Table 5.1. At that time the cables for the 11.5 T design were the largest Rutherford cables ever made for use in an accelerator dipole magnet.

After the study of various designs, it was decided to construct a single-aperture dipole configuration. Due to a smaller yoke and the absence of magnetic field enhancement by a second aperture, the field factor of the model magnet was strongly affected. Where the twin-aperture system could generate 11.5 T at 17.7 kA, the single-aperture magnet would reach only 11.0 T at about 18.7 kA. This still impressive target field, however, appeared well reachable with the conductor properties. Figure 5.4 shows the results of critical current measurements on wire and cable samples together with the load lines for inner and outer conductor layers. In both cases, an I_c criterion of 5 μV/m was used. The strands produced by ECN for the final coils were their last production before their facilities were closed. This impeded stable production control and rejection of less-well performing batches if required. The data presented in Fig. 5.4 therefore represent an average of different batches with an unusual large scatter in critical current values of typically 10%.

Despite the abovementioned scatter between different batches of a single type of strand, the degradation due to the cabling for the inner and outer cables was about 15% and 30%, respectively. The degradation in the inner layer strand was less severe. Due to a non-optimized strand layout, however, the initial critical current density on virgin strands was already quite low.

Initially, the load-line margin was designed around 10% in both the inner and outer layers. As a result of the large I_c degradation the maximum short sample bore field was reduced to around 11.4 T, with a peak magnetic field of 11.8 T and 9.5 T in the inner and outer layers, respectively. The load-line margin in the inner and outer layer is therefore similar.

Time constraints in the project did not allow further optimizations of strand layout and cable production, but were considered imperative for a larger scale project. Similar PIT strand-based Rutherford cables made at a later stage after the MSUT, with a more adapted cable design (optimized keystone angle and width/thickness relation) and improved control of cabling settings, did not show such severe degradation.

5.2.3 *Mechanical Design*

To properly constrain the Lorentz force, a suitable twin-aperture structure was developed with the following design criteria (den Ouden et al. 1992):

- Displacements of the conductor blocks shall be less than 20 μm to minimize field distortions;
- Contact between the yoke halves shall be maintained during excitation;
- The azimuthal compressive stress in the coils shall be preserved after excitation;
- The initial 0.2 mm thickness of the conductor insulation is reduced before the coil heat treatment to 0.14 mm by pressing the wound coil to its final dimensions;
- The equivalent stress shall never exceed the yield stress in any of the used materials;
- The maximum equivalent stress in the coils shall be kept at warm and cold below 150 MPa.

A twin-aperture magnet was analyzed using ANSYS finite element software (ANSYS Inc., Canonsburg, PA). As a starting point, the collar system developed at that time for the LHC dipoles was used. This mechanical system appeared to be insufficient for a 11.5 T dipole magnet. At the location where the two parts of the collar meet, a large stress (Fig. 5.5a) leads to plastic flow of the locking rods and to a loss of pre-stress in the coils. Moreover, the coils are incompletely enclosed by the

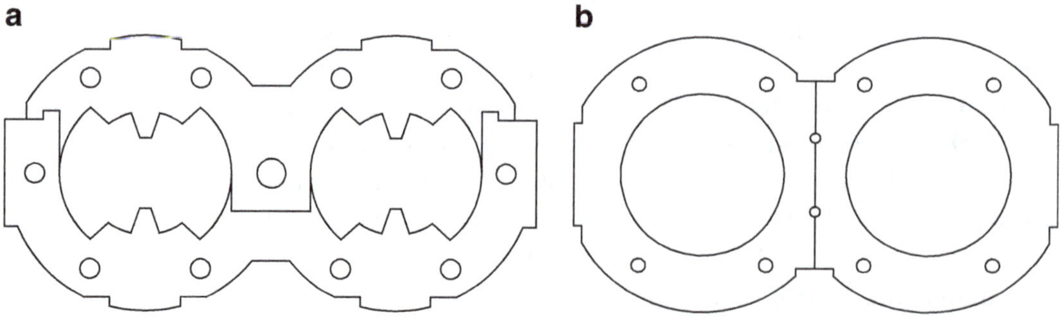

Fig. 5.5 Collar systems for LHC dipole magnets: (**a**) split collar according to the CERN-LHC reference design; (**b**) Al-alloy rings. (Modified from den Ouden et al. 1992)

collars in the same region, thus reducing the support area between the coils and collars. To overcome these limitations a ring-shaped shrink-fitted aluminum collar with an increased outer diameter of 110 mm was proposed (den Ouden et al. 1992) (Fig. 5.5b). For the shrink-fit collars Al-alloy 5083 plates with a thickness of 3 mm were chosen for reasons of workability, cryogenic stability, and yield strength.

To achieve pre-stress in the coils at room temperature, stacks of collar plates are heated up to 225 °C and shrink-fitted onto the polyimide ground-insulated coils. Tight tolerances are required for succeeding with the assembly and for achieving a homogenous pre-stress longitudinally along the magnet. The radial space for assembly is limited to about 0.2 mm. In order to better cope with the tight overall tolerances, a cylindrical outer shape of the coil was preferred.

The coils are ring-collared individually and then aligned with central rods to prepare for the yoke assembly (Fig. 5.5b). The yoke plate material is the standard CERN type with a plate thickness of 5 mm. The yoke assembly follows a standard LHC procedure where the two halves are firmly clamped together with an open gap and locked by welding a clamping piece to the yoke halves. During cool-down the gap between the two halves closes and pre-stress in the vertical plane builds up.

A mechanical model of the magnet is shown in Fig. 5.6. A distinct difference from the LHC type of collared coil is the incorporation of separated stainless-steel 316 L pole pieces, which are allowed to slide along the coil pole face. It is important to note that these pole pieces cover both the inner and the outer layers arranged in the

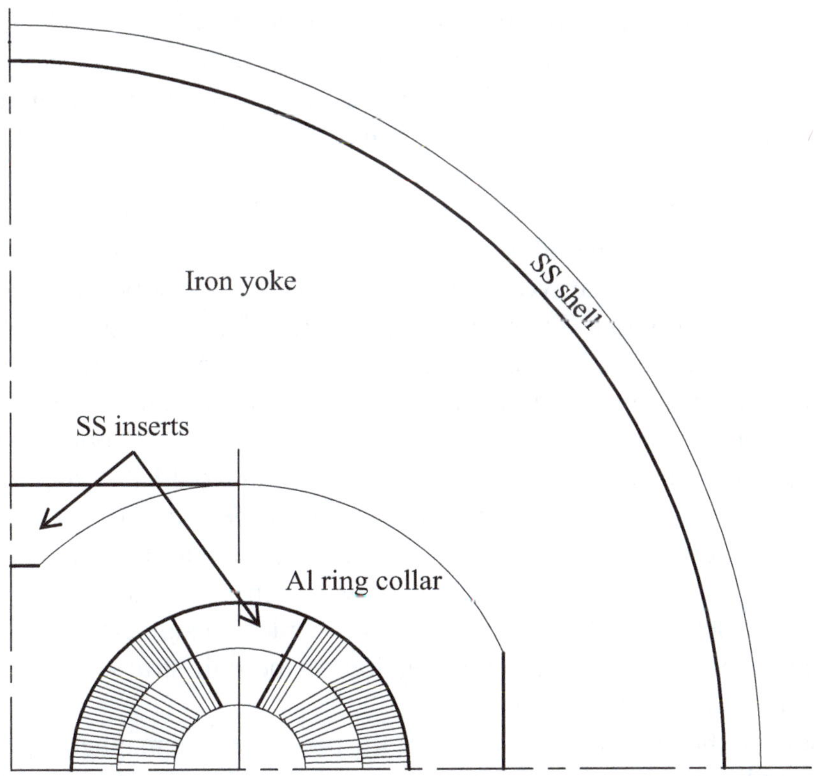

Fig. 5.6 Structural model of the twin-aperture configuration. Bold lines represent the interference layers where friction is included. (Modified from den Ouden et al. 1992)

Table 5.2 Peak equivalent stress in coil and support structures (MPa) for a friction coefficient of $\mu = 0$ (data from den Ouden et al. 1992)

	300 K (I)	300 K (II)	300 K (III)	4.2 K at 0 T	4.2 K at 11.5 T
Coil	63	61	108	116	138
Collar	103	92	124	161	223
Yoke	–	20	192	121	180
Shell	–	–	130	217	219

I = collaring, II = clamping, III = shell shrink-fitting

same plane. The contact between the Al-alloy collars and iron yoke has been optimized such that only the vertical outer planes and the horizontal planes below the triangular yoke insert pieces are in contact. By this approach, the horizontal stiffness of the yoke and shrinking cylinder is transferred efficiently to the Al-alloy collars. Additional pre-compression of the collared coil is realized by shrink-fitting a heated 13.5 mm thick stainless-steel 304 L cylinder around the yoke assembly.

As the sliding planes are metal–metal planes and the uncertainty in the friction coefficient for the metal pairs considered is high, a sensitivity analysis for friction coefficients $\mu = 0$, 0.3, 0.4, and 0.6 was performed. The equivalent stress in the coil and support structure with a friction of $\mu = 0$ is presented in Table 5.2.

The results show negative effects on the structural dynamics for realistic friction coefficients μ between 0.3 and 0.6. The yoke halves lose contact during excitation, and the inner layer loses contact with the pole insert. The shear stress between the layers increases to unacceptable values (larger than 35 MPa). A reduction of the friction coefficient using Teflon or phosphor-bronze sheets covered with molybdenum-sulfide (MoS_2) powder was therefore considered to impose small friction coefficients even under cryogenic conditions in the order of $\mu = 0.1$ (Tobler 1979; Lizon 1990).

The highest von Mises stress in the coils occur in the mid-plane at the innermost radius. On average, the E modulus for the winding pack was assumed to be in the order of 20 GPa, a value derived from measurements performed on impregnated stacks of PIT-Nb_3Sn-based ECN-SULTAN cable. These measurements showed E moduli in the range 16–21 GPa. A sensitivity analysis revealed that for E moduli below 15 GPa the pre-stress buildup would become insufficient. Moreover, plastic deformation of the copper matrix, which has a yield strength of about 35 MPa after the heat treatment, can be expected. Plastic deformation would also yield to a further and possibly unacceptable reduction of pre-stress in the winding pack. Therefore, work-hardening of the impregnated coils, by pressing them to their final dimensions, before assembly of the Al-alloy ring-collars was seen as a mandatory step during assembly. More details on the twin-aperture mechanical design are presented in den Ouden et al. (1992). The overall cross-section of the coil structure in the straight section is shown in Fig. 5.7.

The results of the mechanical analysis of the twin-aperture magnet appeared to be fairly well applicable to the single-aperture configuration (Fig. 5.8). For the single aperture, closure of the gap between the yoke halves appeared not to be a strict requirement anymore, which allowed for a simplified yoke configuration. A

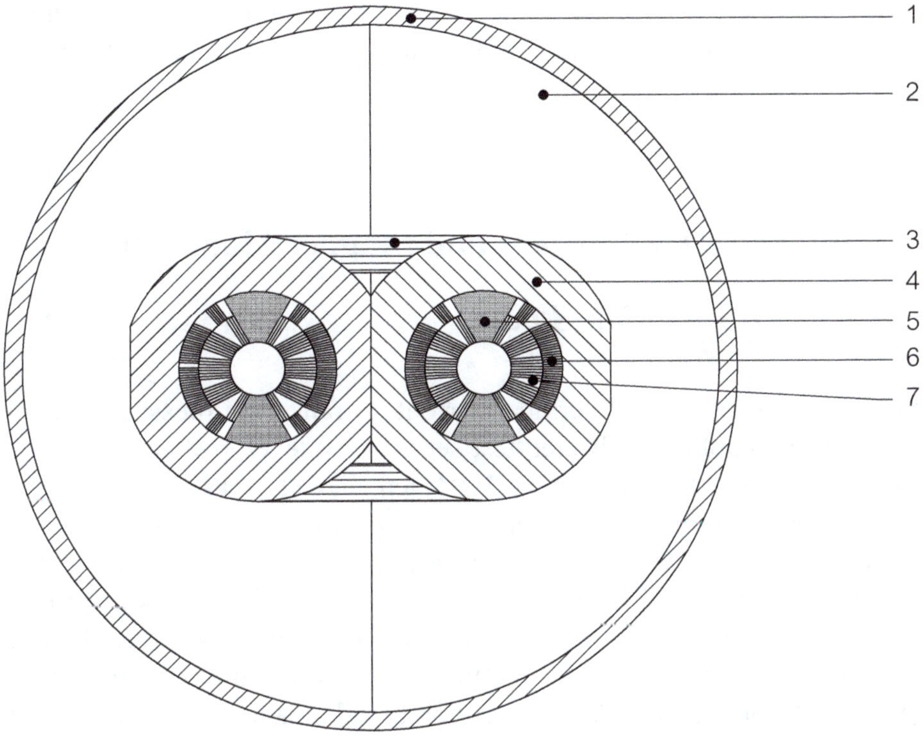

Fig. 5.7 Proposed structural design of the twin-aperture version of the 11.5 T Nb_3Sn LHC dipole magnet (Modified from den Ouden et al. 1992): *1* – outer cylinder; *2* – yoke; *3* – yoke insert; *4* – Al-alloy ring-collars; *5* – pole insert; *6* – conductor block; *7* – copper wedge

Fig. 5.8 Cross-section of the single-aperture magnet showing the coil support system (Modified from den Ouden et al. 1992): *1* – outer cylinder; *2* – yoke; *3* – Al-alloy ring-collars; *4* – pole insert; *5* – windings; *6* – copper wedge

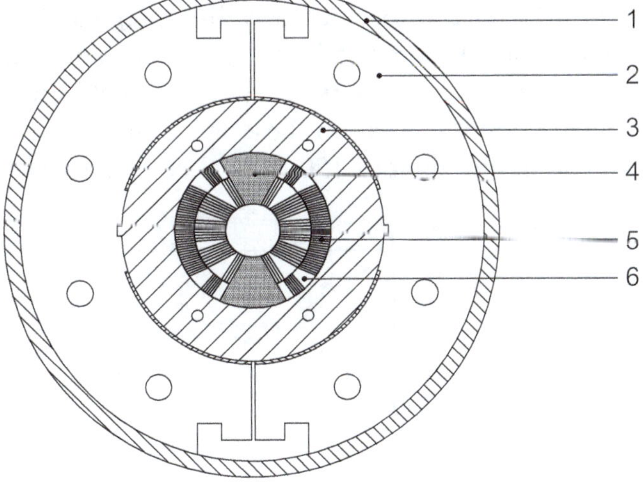

U-shaped aluminum clamp was engineered to hold together the two yoke halves during assembly but it has no structural purpose during cool-down and magnet powering. To prevent excessive stress in the collars due to bending of the yoke halves, the yoke–collar interference was also modified into a circular surface instead of a flat surface. This interference surface is explicitly defined by an arc-shaped zone at the inner radius of the yoke plates around the mid-plane. To simplify assembly and axial pre-stress control, it was decided to not connect the outer stainless-steel cylinder to the end plates. Instead, eight 20 mm thick tie rods (Fig. 5.8) were used to take up the axial loads and at the same time allowed for a controlled axial preload during assembly.

5.3 Technology Development and Magnet Manufacturing

5.3.1 *Cable Degradation Studies*

To measure the response of the critical current to transverse pressure applied to an Nb_3Sn Rutherford cable, a dedicated test set-up was developed and installed at the University of Twente (Boschman et al. 1991; ten Kate et al. 1992). In this set-up the wide side of the cable is exposed over a length of 40 mm to transverse mechanical pressures of up to 250 MPa. The critical current at 4.2 K is measured in a background magnetic field of up to 11 T oriented perpendicularly to the wide side of the cable. Currents up to 50 kA are supplied by a superconducting transformer. Impregnated samples of various versions of Nb_3Sn Rutherford cables employing different types of Nb_3Sn strands (bronze, modified jellyroll, internal tin and PIT) were prepared, and their critical current was measured as a function of transverse stress. Some samples of each kind of strand degraded severely and permanently under stress, others showed acceptable behavior.

Figure 5.9 shows the critical current as a function of transverse pressure up to 200 MPa and a field of 11 T for a cable made from ECN PIT strand. The decrease in I_c of 10% ± 2% for such a well-impregnated and moderately compacted Nb_3Sn cable is nearly completely reversible, meaning that during re-testing of the cable without applied transverse pressure the cable reached more than 98% of its initial critical current. Based on these and similar results on many other Nb_3Sn cables a general design limit of 150 MPa at cold and warm was formulated, at which only a

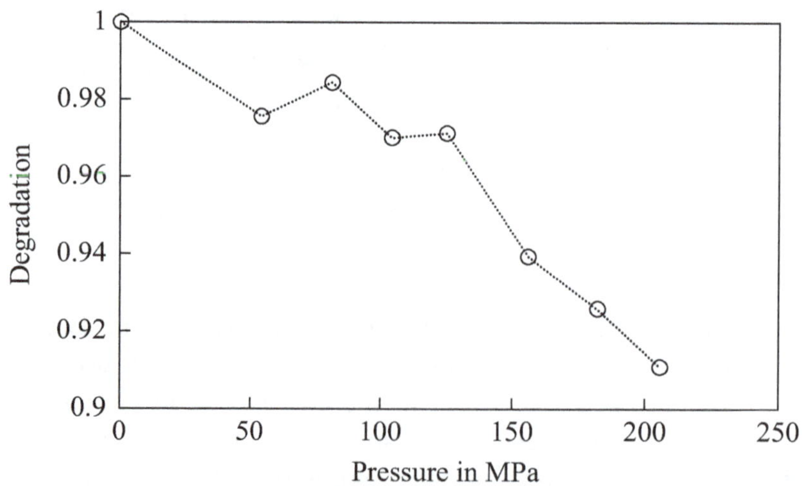

Fig. 5.9 Reduction of critical current density at 11 T vs. applied transverse pressure (voltage criteria 5 μV/m). (Modified from ten Kate et al. 1992)

reversible decrease in I_c of less than 10% at 4.4 K is to be expected. It is emphasized, however, that perfect resin impregnation on a strand level of the Rutherford cables in coil windings is essential to guarantee a more-or-less hydrostatic support of the Nb_3Sn layers in the strands of the cables to the imposed stress.

5.3.2 Mechanical Model

To verify the ANSYS simulation with respect to pre-stress build-up during the collaring process, a 100 mm long mechanical mock-up was built. Because at the time no representative coil blocks were available, the winding layers were replaced by custom-made G10 glass-fiber reinforced cylindrical pieces with a Young's modulus of 25 GPa. The pole inserts were equipped with bridge-type stress gauges. Low friction phosphor-bronze sheets covered with MoS_2 powder were inserted between the pole inserts and the G10 poles. A copper-beryllium tape of 0.4 mm thickness was wound around the coil halves, and phosphor-bronze sheets covered with MoS_2 powder were locked between coil and collars. The aluminum alloy ring-shaped collars were stacked, heated to about 200 °C, and shrink-fitted around the mock-up assembly. The mechanical model confirmed that the required pre-stress could be achieved and that the procedure is suited for at least a 1 m long coil.

5.3.3 Cable Insulation

The first trial cable insulation system that could withstand the reaction heat treatment at 675 °C consisted of a direct wrapping of a glass-mica ribbon, applied with 50% overlap. Due to insufficient penetration of epoxy resin through and between the overlapping tape areas, this insulation system resulted in incomplete resin penetration into the voids between the strands, a very low shear strength, and very low thermal conduction between adjacent conductors (den Ouden et al. 1991).

To improve the insulation system, a single sheet of the same glass-mica tape was folded parallel to the cable covering about 70% of its circumference. An S2-glass ribbon was then wrapped manually without overlap around the cable-mica assembly, as shown in Fig. 5.10. The wrapped glass ribbon holds the glass-mica tape in position, improves epoxy penetration into the cable and winding pack, enhances the shear strength between adjacent conductors, and therefore increases the effective thermal conductivity of the insulation layer at 4.4 K by at least a factor of 2. Before applying the Rutherford cable, the S2-glass tape was heat-treated in air at 300 °C to minimize carbon residuals in the coils during the high-temperature reaction heat treatment.

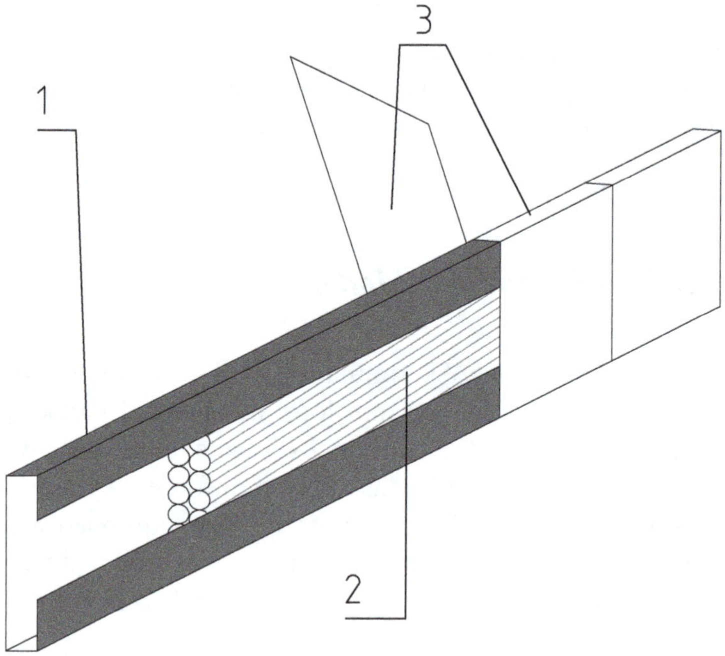

Fig. 5.10 Cable insulation (Modified from den Ouden et al. 1991): *1* – folded glass-mica tape; *2* – Rutherford cable; *3* – S2-glass tape

5.3.4 *Coil Wedges and End Spacer Design*

To meet the required tolerances, the copper wedges between the winding blocks in coil straight sections have been cut by electrical discharge machine into 140 mm long pieces and afterwards sandblasted to enhance epoxy resin adhesion.

Each conductor block is supported in the coil head by its own end spacer. No further subdivision was foreseen; the spacing between conductors was chosen based on winding tests. The innermost end spacer was made by placing a stainless-steel filler on the mandrel, leaving some space for the block just wound. A 1 mm thick stainless-steel strip was then brazed to a copper wedge and bent a few mm away from the filler into its natural shape. This shape appeared to be a proper inner support for the next turns. The space between the strip and the last turn for the second spacer was filled using a mixture of castable alumina and sand that can easily be removed after the heat treatment. After heat treatment and before resin impregnation, all open spaces were filled tightly with fiberglass. These end spacers were easy to make. They are flexible and form a continuous support path from the straight section into the coil ends, which favors cable mechanical support and prevents damage to the cable insulation and, last but not least, may prevent coil quenching originating from this area.

5.3.5 Coil Winding

Prior to the winding of the Nb_3Sn coils a complete two-layer dummy Nb-Ti pole was wound, heat-treated, stacked, and impregnated. To prevent collapsing of the cables, the winding tension had to be kept within 150–200 N. During coil winding each wound layer was firmly clamped tangentially against the winding poles along the straight parts, in particular close to the ends.

To reduce the effective insulation thickness from 0.20 mm to 0.14 mm the wound coils were inserted in their heat treatment mold and pressed to a precise, pre-defined size with a pressure of about 50 MPa. The coil ends were intentionally left unsupported during all further process steps. Figure 5.11 shows the end region of an outer layer coil after winding (Fig. 5.11a) and after reaction heat treatment (Fig. 5.11b). Figure 5.12 shows a cross-section of the end region of a completely processed Nb-Ti practice coil. The Rutherford cable in the ends shows a remarkable deformation. After some modifications of the end spacer shape this deformation has been suppressed nearly completely during winding of the final Nb_3Sn coils.

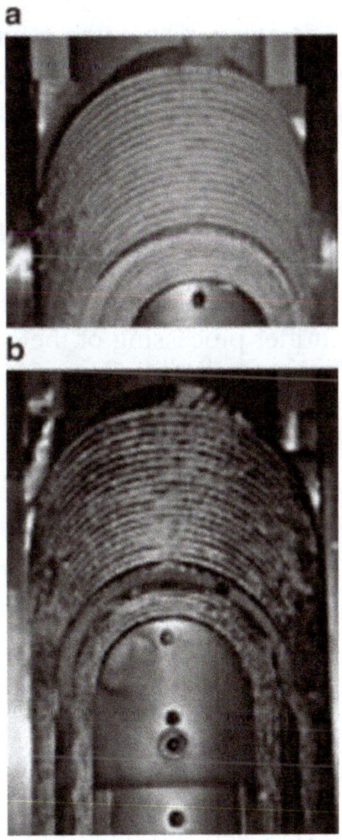

Fig. 5.11 Coil fabrication: (**a**) coil end after winding; (**b**) coil end after reaction heat treatment

Fig. 5.12 Impregnated Nb-Ti practice coil: (**a**) top view of the coil end; and (**b**) lead end longitudinal cross-section

5.3.6 Coil Reaction Heat Treatment

The reaction heat treatment of each single layer placed in its compressed mold took place under vacuum at 675 °C for 14 h. Cable ends were not sealed. During this process, most of the binder material from the glass-mica tape evaporated at around 300 °C and was deposited upon cold surfaces. The remainder did not cause problems with respect to the electrical insulation or binding of epoxy resin. Due to compressing the layer tangentially without axial support, gaps as large as 1 mm between the end spacers and the first conductor of each winding block appeared. As mentioned, these gaps were manually tightly filled with glass-fiber cloth before further processing of the coil layer.

5.3.7 Electrical Connection Between the Coils

To allow for a connection plane between the two coil layers, the outer layer coil was started at the same angle as the pole plane of the inner layer. After separate heat treatment of the coil layers, they were stacked, and a pre-manufactured connection piece consisting of a copper plate was wrapped with Nb_3Sn wires and reacted independently such that it could connect both coil leads. Figure 5.13 shows a cross-section of the splicing area of the dummy Nb-Ti coil. The length of the splice was 450 mm (three twist pitch lengths). The connection piece and coil leads in both layers were thoroughly cleaned with alcohol, brushed with Scotch-Brite® pads (3M, Maplewood, MN) and soldered with Ag-Sn solder. To prevent contamination of adjacent windings temporary glass-mica sheets were inserted between the pole and other turns. The resulting resistance at 20 kA ranged from 0.3 nΩ at 0 T to 1.5 nΩ at 10 T.

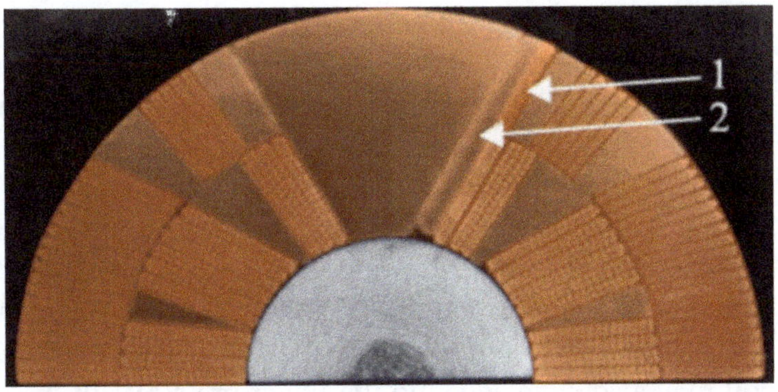

Fig. 5.13 Dummy coil cross-section: *1* – additional outer layer pole turn (length 450 mm) to match the azimuthal position of the inner layer pole turn; *2* – joint piece jumper, a Cu plate wrapped with heat-treated Nb_3Sn wires

The generated heat of about 0.5 W at nominal operating current and magnetic field was conducted away to the helium bath by an extension of the copper plate, which was installed after impregnation inside the aperture of the final Nb_3Sn coils. This design was acceptable for this test magnet, but would need reconsideration for an accelerator magnet.

5.3.8 *Impregnation*

Both layers were equipped with G10 end pieces (coil saddles), and temporary Teflon-coated pole inserts. The coil assembly was then placed into an impregnation mold and covered by using the respective parts of the heat treatment mold. Steel foils with a thickness of 50 μm, covered with a non-adhesive layer, were used to ensure release between coils and mold parts.

The resin was Ciba-Geigy MY740 epoxy mixed with HY906 hardener and DY062 accelerator. This resin system was chosen for its demonstrated performance in superconducting magnet systems and for its acceptable pot life of 6 h at 55 °C. The training performance of the resin system was not very well investigated. After vacuum resin impregnation, the final stainless-steel pole inserts were mounted. Two MoS_2 powder-coated phosphor-bronze sliding foils were inserted between the pole inserts and the coils' pole faces to ensure a low friction coefficient at all stages.

5.3.9 *Coil Assembly and Collaring*

After installation of the instrumentation (see below), the two half-coils were stacked and wrapped with four layers of 0.075 mm thick polyimide foil to ensure sufficient ground insulation. Finally, two 0.20 mm thick, coil-long MoS_2 powder-coated phosphor-bronze sheets were subsequently tightly wrapped around the coils and locked to each other at the mating faces of the second layer by a soft soldered connection.

The precisely aligned and vertically stacked aluminum alloy ring-collar laminations were heated up to 200 °C, which created a 0.20 mm radial gap for the coils, which were quickly inserted vertically into this stack within 10 s. At room temperature, the Al-alloy collars shrunk by 0.08 mm into the coils, providing initial coil pre-stress. After insertion, the heated collars could not be cooled fast enough and heated up the coils to about 90 °C, a temperature at which the solidified epoxy becomes soft and probably yielded. After cool-down, the stress transducers indicated a pre-stress, which amounted to about 30 MPa, 50% of the target value of 60 MPa. Apart from the softened epoxy, the lower-than-expected Young's modulus of the coil may also have added to the lower pre-stress in the coil pack.

5.3.10 Collared Coil Yoking and Skinning

The mounting procedure for the yoke halves and clamping parts was straightforward. Starting in the straight section a few centimeters from the beginning of the coil ends, slightly radially oversized stainless-steel yoke pieces were applied. In the straight section iron blocks were applied. This approach effectively lowered the maximum magnetic field in the coil ends. The magnetic field maximum in both layers is located in the straight section. The yoke was also enclosed by two MoS_2 covered sliding phosphor-bronze foils.

The heated but undersized stainless-steel outer cylinder around the yoke-collared coil assembly should result in a 40 MPa increase of the coil's tangential pre-stress. Despite this increase in pressure, the shrinking process appeared straightforward and fairly easy to control. After axial insertion of eight 20 mm diameter tie rods through the holes in the yoke, 50 mm thick end plates were mounted. The rods were tensioned to supply about 10 MPa axial pre-compression through the end plates to the coil layers alone. For ease of assembly, these end plates were deliberately not welded to the outer cylinder. Figure 5.14 shows a cross-section of the assembled coil and support structure.

5.3.11 Instrumentation and Quench Heaters

Before stacking the two two-layer poles together, each inner layer block was equipped at the inner side with miniature germanium resistors (diameter 1.6 mm, length 5 mm), which were used as thermometers. The sensitivity is of a few kΩ/K with a resistance of typically 10 kΩ at 4.2 K and a relatively low magnetoresistance of less than 3% at 8 T (Zarubin et al. 1990). The thermometers were inserted in small copper tubes that were soft soldered onto a single strand of a particular coil turn. The tubes and about 10 mm of the sensor wires were thermally insulated by 1 mm thick polyimide domes that were glued onto the tubes and sealed by epoxy resin.

Fig. 5.14 Cross-section of the assembled coil and support structure

Strain gauges (100 Ω) were mounted on similarly mounted copper tubes at the inner layer, which served as a spot heater for the study of normal zone propagation.

Furthermore, many voltage taps were soldered to individual coil turns by locally removing the insulation after resin impregnation.

For measurement of the tangential pre-stress at the poles, bridge-type strain gauges were integrated in the pole inserts.

At the bore side of the inner layer, four meander-shaped, polyimide insulated quench stainless-steel heaters were glued to the coil, covering all windings. These heaters, however, appeared to be ineffective for quench protection. No inter-layer or outer layer quench heaters were foreseen.

5.4 Test Results

The magnet was tested in 1995 for 3 weeks at the model magnet test facility at CERN at a temperature of 4.4 K. After a rather long thermal cycle, the magnet was then retested in 1997. Since the quench heaters appeared ineffective for protection, at least 50% of the stored energy was extracted during quenches. A detailed summary of the test results is presented in den Ouden et al. (1997a, b).

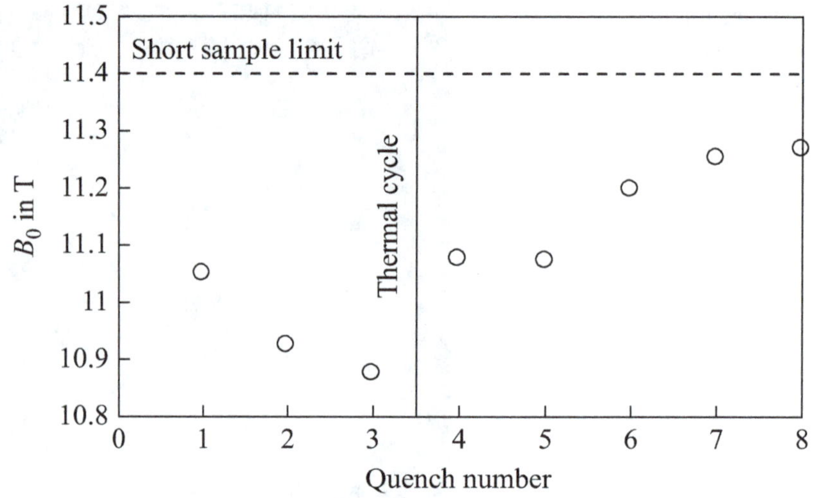

Fig. 5.15 Magnet training history. (Modified from den Ouden et al. 1997b)

5.4.1 *Magnet Training*

After first cool-down, the magnet was ramped without a quench to 15 kA and, after a flat top of a few minutes, the ramp was continued with a ramp rate of 2 A/s until quench. The first high-field quench occurred at a current of 18.7 kA, corresponding to a record central dipole field of 11.03 T.

A second run immediately after the first quench resulted in a quench current of 18.5 kA (10.92 T), about 1.5% lower than the first quench. After a 3 week measurement program, a third (and for this run, final) high current ramp resulted in a quench at 18.4 kA (10.86 T). A thermal cycle was then performed, and re-testing 2 years later resulted in a first quench at 18.8 kA at a bore field of 11.07 T. The magnet trained in the four subsequent quenches to a maximum current of 19.1 kA and a field of 11.27 T. Note that in these tests the ramp rate was slightly increased to 5 A/s. Figure 5.15 shows the magnet training history.

For quench localization, a static pick-up coil set was used (Bottura 1995). All quenches of the first test occurred in the outer layer of the same pole in the splice region, where the field current margin is the lowest. At many ramps during the entire test campaign, large voltage spikes were observed at low currents, indicating large-scale flux jumps, although without causing the magnet to quench.

5.4.2 *Magnetic Measurements*

The MSUT transfer function (TF) vs. the magnet current is shown in Fig. 5.16. Due to the relatively small iron yoke outer diameter, its saturation starts at low currents, and at 12 kA the TF reduction reaches ~13%. At currents above 8 kA the curve flattens, indicating that the yoke is practically fully saturated.

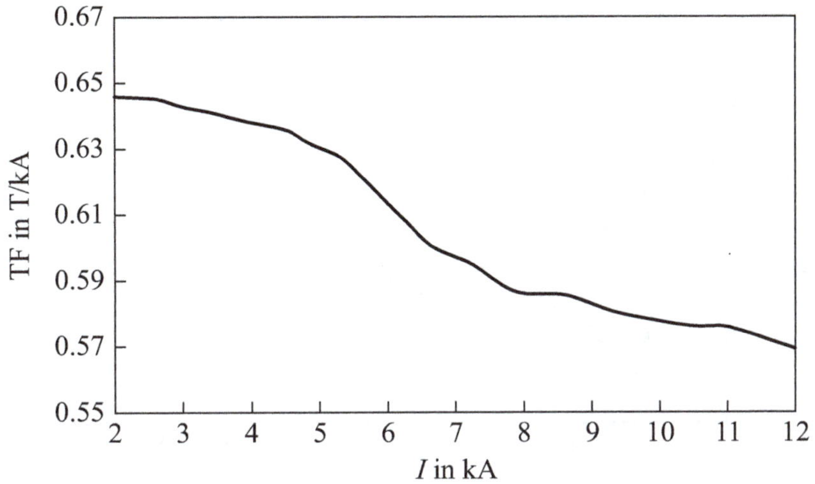

Fig. 5.16 Transfer function vs. magnet current measured at a ramp rate of 10 A/s in the straight part of the magnet. (Modified from Bottura 1995)

Table 5.3 Normalized multipoles $b_n = B_n/B_1$ and $a_n = A_n/B_1$ at constant magnetic field, measured in the straight part (10^{-4}) at a reference radius of 10 mm

B_0, (T)	b_2	a_2	b_3	a_3	b_4	a_4	b_5	a_5
0.4	−8.4	−21.0	55.8	−4.2	0.7	−3.3	−2.1	−1.6
9.0	−3.3	0.9	−4.4	−1.0	−0.3	−0.6	−0.3	−0.3

In view of possible use in the LHC, the field quality of this magnet was measured at a low field of 0.4 T, which is slightly below the LHC injection field of 0.54 T, at a high field of 9 T, and during magnetic field ramping. At low magnetic fields the most significant field errors originate from persistent filament magnetization currents, whereas at high fields they are affected by iron saturation. During ramping of the magnetic field, inter-strand coupling currents (ISCCs) and boundary induced coupling currents (BICCs) cause additional field errors. Note that the two Rutherford cables used in MSUT did not have a stainless-steel core, which already at that time was considered necessary to limit magnetization effects and magnet ramp losses.

The results of static field measurements are summarized in Table 5.3. The large normal b_2 and skew a_2 quadrupole components may be caused by a combination of small misalignments of the coils and asymmetry in the permanent vertical gap between the yoke halves. The positive sign for b_3 at 0.4 T indicates that at this field the coil re-magnetization process is not complete (see Fig. 5.17).

Figure 5.17 presents the normal sextupole b_3 vs. the magnet bore field measured with a current ramp rate of 10 A/s (corresponding to ~6 mT/s) in the magnet straight part and a reset current of around 250 A. The persistent current effect at low fields below ~4 T is the main contributor to b_3 due to the 40 μm effective filament diameter and the high J_c in superconducting filaments at low magnetic field. The b_3 reaches its minimum of −15 units at a bore field of ~1.25 T. Note that these values are affected by the reset current. The asymmetry of the two branches at high fields is due to iron saturation. The iron saturation effect in b_3, however, is much smaller than in the TF.

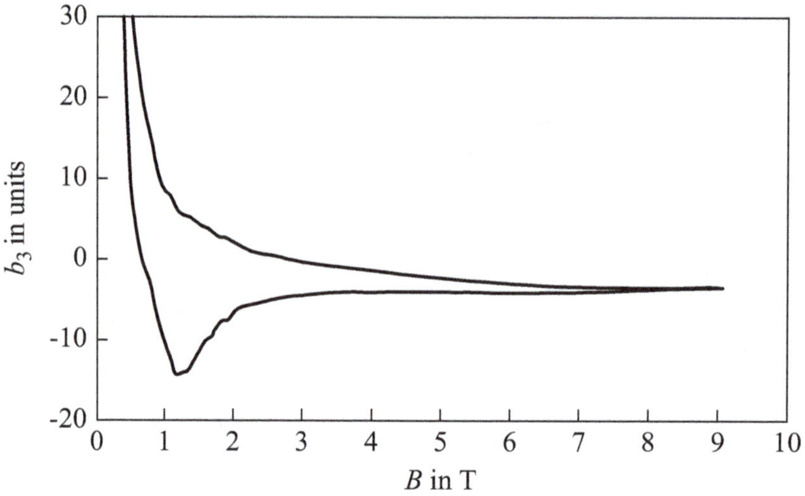

Fig. 5.17 The sextupole component b_3 over the magnetic field measured with a ramp rate of 10 A/s in the straight part of the magnet at $R_{ref} = 10$ mm. (Modified from Bottura 1995)

5.4.3 Losses During Ramping

The integrated hysteresis and coupling losses during magnetic field cycles were measured electrically (Siemko et al. 1995). Figure 5.18 shows the integrated total hysteresis (extrapolated for a theoretical ramp rate of 0 A/s) and coupling losses during field cycles between 1.9 and 3.8 T for both poles. From these measurements, an average value for the resistance between crossing strands $R_c = 1.2$ μΩ and a time constant of $\tau = 11$ s were calculated. During the heat treatment at 675 °C the uncoated Cu surfaces of the strands are probably sintered, which results in this very low R_c. The systematic difference between the different coils points to a small difference in the average R_c (R_{c1}, R_{c2}).

5.4.4 Ramp Rate Sensitivity

Figure 5.19 shows the quench currents at different ramp rates for linear ramps from 0 A until quench. Two different types of coil quenching were identified based on the analysis of the voltage tap measurements. At moderate ramp rates ($dI/dt \leq 75$ A/s), the normal and boundary-induced coupling currents are responsible for the relatively fast decrease of the quench current with increasing ramp rate (quench type 1). AC loss measurements show a very low average inter-strand resistance of 1.2 μΩ. Together with the large cable width, the long cable twist pitch, and the number of strands, this results in relatively high coupling currents. At fast ramp rates ($dI/dt \geq 75$ A/s), large voltage differences between the two poles were observed just before the quench occurred (60–120 mV, duration 10–15 ms). Analysis of the quench location showed that such quenches (labeled quench type 2), exclusively

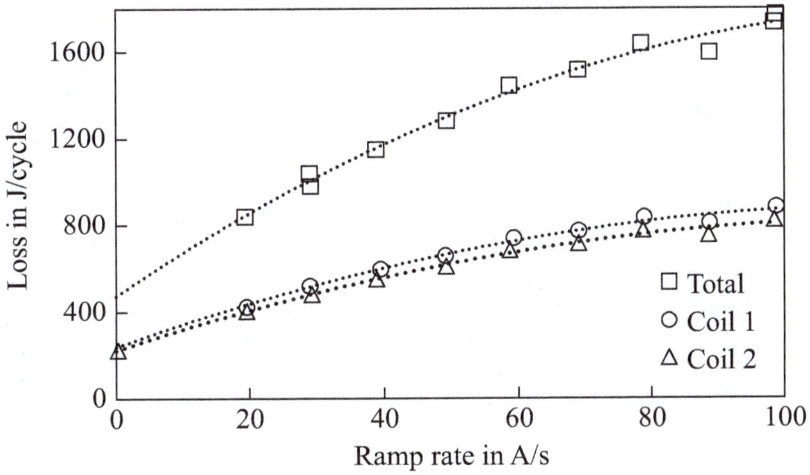

Fig. 5.18 Measured integrated hysteresis and coupling losses as a function of current ramp rate during a 1.9–3.8–1.9 T cycle for both poles separately and the total loss

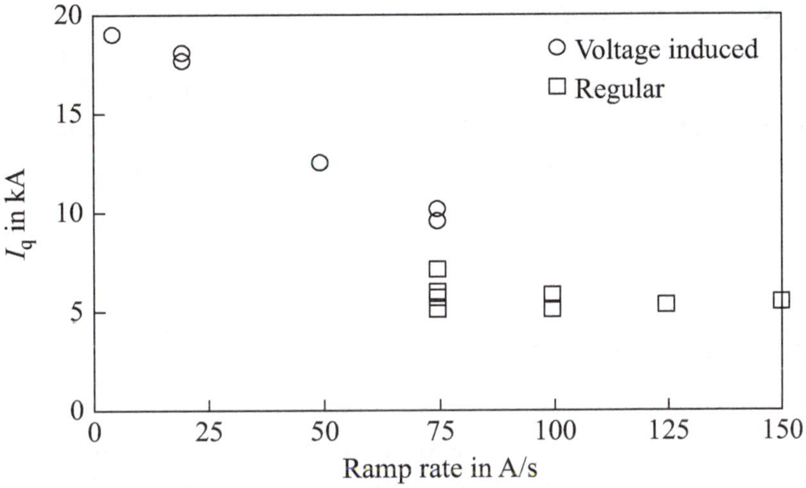

Fig. 5.19 Quench current as a function of ramp rate (circles: regular type 1; squares: type 2). Each ramp started from 0 T. (Modified from den Ouden et al. 1997b)

originated in or near the splice region of one particular pole. For ramp rates at $dI/dt \approx$ 75 A/s both types of quenches occur.

5.4.5 *Temperature Development*

During all magnetic field sweeps, the temperature development in the inner layer conductor blocks was measured with the temperature sensors. Figure 5.20 shows the temperature in the mid-plane block during ramps from 0 to 5 T with ramp rates of 10 to 60 A/s. Due to the high magnetization loss a large temperature rise at low fields

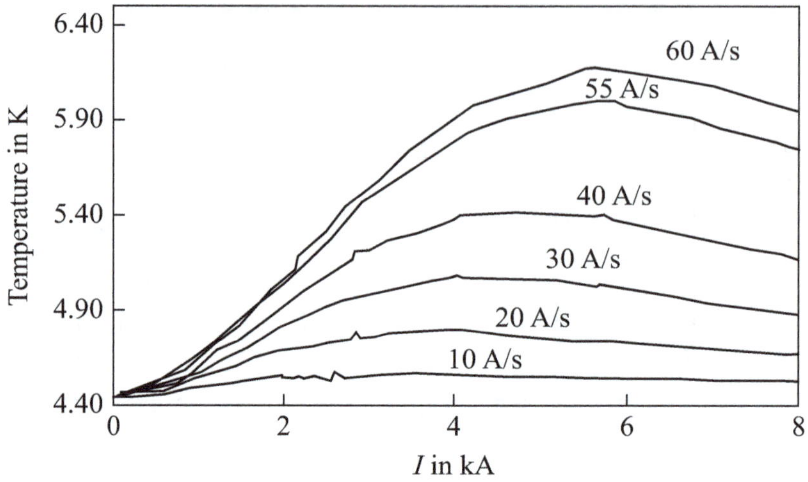

Fig. 5.20 Measured temperature development in the mid-plane block during a field sweep from 0–5 T at different ramp rates

occurs. At the end of the ramp, the magnetization contribution is negligible and only the coupling loss determines the local temperature. Given the measured value for R_c, the calculated temperature increase and the location of the maximum agree very well with the measurements. These unique in situ experiments show the reliability of the thermometer layout and are a valuable tool in the study of the thermal behavior and stability at cable level.

5.4.6 Quench Propagation Velocity

Strain gauges installed on a single strand of the inner layer pole turn were used as a spot heater. Quenches induced in this location spread under quasi-adiabatic condition and cause a quench of the entire pole turn cable within a few milliseconds. The quench propagation in the inner layer was recorded with both the voltage taps and the thermometers. Results were found to be consistent. The difference in the measured voltage between the poles was used to estimate the turn-to-turn propagation, which was cross-checked with measurements from the voltage turn signals. Figure 5.21 shows the turn-to-turn propagation times (Fig. 5.21a) and the normal zone propagation velocity (Fig. 5.21b), obtained from the series of experiments described above.

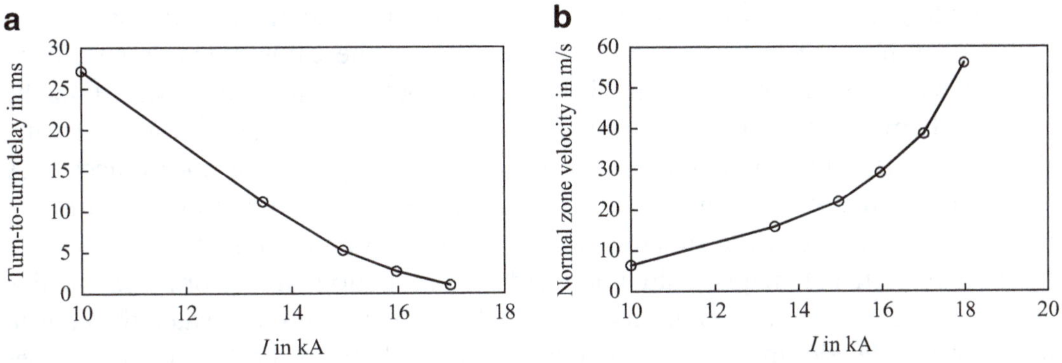

Fig. 5.21 (**a**) Turn-to-turn propagation times for the pole blocks; and (**b**) normal zone propagation velocity. (Modified from den Ouden et al. 1997b)

5.5 Conclusion

In the framework of the LHC research and development program, twin- and single-aperture dipole magnets were designed based on PIT-Nb_3Sn high current density conductors. In 1995 the dipole magnet was successfully tested, reaching a record bore magnetic field of 11.03 T at 4.4 K. This result was achieved on first cool-down and first quench of the magnet, which is, even today, a remarkable result given the usual long training curves experienced in other similar Nb_3Sn dipole magnets. After only seven training quenches the magnet reached a bore field of 11.27 T, around 99% on the load line of the cable taking into account the cable degradation during cabling.

The project addressed successfully the development of high current density and high-performing PIT-type Nb_3Sn composite wires, their application in wide cables, and glass-mica cable insulation. The program also brought forward the first systematic electromagnetic optimization of the cross-section of dipole magnets; the application of mechanical structure based on a shrink-fit Al-alloy round collars; coil end spacer designs without discontinuities between the coil ends and the straight section; the introduction of sliding planes based on phosphor-bronze sheets covered with MoS_2 powder; and an innovative layer-to-layer joint using a superconducting shunt. Finally, the uttermost relevance of being able to fill gaps in the winding pack after the reaction heat treatment with glass-fiber fillers to prevent crack formation and propagation cannot be underestimated, as shown by the very good training behavior of this magnet.

The manufacturing and testing experience also revealed a number of challenges, which confirmed that Nb_3Sn accelerator magnet technology remained tedious, costly, and not yet ready for industrialization at that time, as required for series construction of LHC dipoles. The main points, which were seen as being critical towards better performance, were the significant and hard-to-predict conductor performance degradation during cabling as well as the magnetization and coupling current effects. For accelerator operation, the magnet had too large magnetization and coupling currents with a negative impact on both static and dynamic field quality. The use of a resistive stainless-steel core in the cable helps drastically to reduce the magnetization effects and is nowadays routinely applied. At that time, the filament diameter was seen as a major issue, and development towards smaller filaments while maintaining a high critical current density was considered necessary to reduce drastically the magnetization effects. Dedicated correction schemes are now studied to mitigate the superconductor magnetization effects.

The stress in the magnet did not yield any reduction of performance, which confirmed the mechanical reliability of the PIT conductor and motivated the continuation of the development of this conductor technology in the following decades (den Ouden et al. 2001; Boutboul et al. 2009).

Finally, it is important to realize that one of the primary goals of the MSUT magnet was to achieve 11 T field level. Bearing this in mind, the UT-CERN MSUT program was a great success. It is noted that most of the new ideas (such as wide and stiff high-current cables, continuous end spacers, layer joint jumper, shrink-fit collars etc.) were never all reproduced together in later magnets. It is therefore very hard to correlate without speculation the remarkable MSUT test results—achieving 11 T design field at first cool-down without any training, not yet shown again in later magnets—to one or more of the newly applied current technical solutions.

References

Boschman H, Verweij AP, Wessel S et al (1991) The effect of transverse loads up to 300 MPa on the critical currents of Nb_3Sn cables. IEEE Trans Magn 27(2):1831–1834. https://doi.org/10.1109/20.133551

Bottura L (1995) Results of magnetic measurements on the MSUT (Nb_3Sn) model dipole. Technical Note, AT-MA Note 95–104, July, CERN, Geneva

Boutboul T, Oberli L, den Ouden A et al (2009) Heat treatment optimization studies on PIT Nb_3Sn strand for the NED project. IEEE Trans Appl Supercond 19(3):2564–2567. https://doi.org/10.1109/tasc.2009.2019017

Claudet G, Aymar R (1990) Tore Supra and He II cooling of large high field magnets. In: Fast RW (ed) Advances in cryogenic engineering, vol 35. Springer, Boston, pp 55–67

Den Ouden A, ten Kate HHJ, Wessel S et al (1991) Thermal conduction in fully impregnated Nb_3Sn windings for LHC type of dipoles. Adv Cryo Eng 38B:635–642. https://research.utwente.nl/en/publications/thermal-conduction-in-fully-impregnated-nb3sn-windings-for-lhc-ty

Den Ouden A, ten Kate HHJ, ter Avest D et al (1992) Analysis of the mechanical behaviour of an 11.5 T Nb_3Sn LHC dipole magnet according to the ring collar concept. IEEE Trans Magn 28 (1):331–334. https://doi.org/10.1109/20.119878

Den Ouden A, Wessel S, Krooshoop E et al (1994) An experimental 11.5 T Nb_3Sn LHC type of dipole magnet. IEEE Trans Magn 30(4):2320–2323. https://doi.org/10.1109/20.305740

Den Ouden A, Wessel S, Krooshoop E et al (1997a) Application of Nb_3Sn superconductors in high-field accelerator magnets. IEEE Trans Appl Supercond 7(2):733–738. https://doi.org/10.1109/77.614608

Den Ouden A, ten Kate HHJ, Siemko A et al (1997b) Quench characteristics of the 11 T Nb_3Sn model dipole magnet MSUT. In: Liangzhen L, Guoliao S, Luguang Y (eds) Proceedings of fifteenth international conference on magnet technology (MT-15), 20–24 Oct 1997 Beijing, pp 339–342

Den Ouden A, Wessel WAJ, Kirby GA et al (2001) Progress in the development of an 88-mm bore 10 T Nb_3Sn dipole magnet. IEEE Trans Appl Supercond 11(1):2268–2271. https://doi.org/10.1109/77.920312

Hornsveld EM, Elen JD, van Beijnen CAM et al (1988) Development of ECN-type niobium-tin wire towards smaller filament size. Adv Cryo Eng 34:493–498

Lizon JL (1990) Comparison between various lubricants at cryogenic temperatures in a vacuum. Adv Cryo Eng (Materials) 36B:1209–1215

Perin R (1973) Calculation of magnetic field in a cylindrical geometry produced by sector or layer windings. Internal note CERN: ISR-MA/RP/cb, CERN, Geneva

Siemko A, Billan J, Gerin G et al (1995) Quench location in the superconducting model magnets of the LHC by means of pick-up coils. IEEE Trans Appl Supercond 5(2):1028–1031. https://doi.org/10.1109/77.402726

Ten Kate HHJ, den Ouden A, ter Avest D et al (1991) Development of an experimental 10 T Nb_3Sn dipole magnet for the CERN LHC. IEEE Trans Magn 27(2):1996–1999. https://doi.org/10.1109/20.133597

Ten Kate HHJ, Weijers H, Wessel S et al (1992) The reduction of the critical current in Nb_3Sn cables under transverse forces. IEEE Trans Magn 28(1):715–718. https://doi.org/10.1109/20.119979

Ter Avest D, ten Kate HHJ, van de Klundert LJM et al (1991) Optimizing the conductor dimensions for a 10–13 T superconducting dipole magnet. IEEE Trans Appl Supercond 27(2):2000–2003. https://doi.org/10.1109/20.133598

Tobler RL (1979) A review of antifriction materials and design for cryogenic environments. Adv Cryo Eng (Materials) 26:66–77. https://doi.org/10.1007/978-1-4613-9859-2_5

Zarubin LI, Nemish IY, Szmyrka-Grzebyk A et al (1990) Germanium resistance thermometers with low magnetoresistance. Cryogenics 30(6):533–537. https://doi.org/10.1016/0011-2275(90)90055-h

Chapter 6
LBNL Cos-theta Nb_3Sn Dipole Magnet D20

Shlomo Caspi

Abstract In 1996 the LBNL D20 Nb_3Sn dipole magnet reached 12.8 T at 4.4 K and 13.5 T at 1.8 K, and caught the accelerator magnet community by surprise. Not only did it achieve a record field for an accelerator dipole magnet but it also demonstrated a breakthrough in Nb_3Sn conductor technology. This chapter summarizes the technical aspects of D20 development and test, addresses lessons learned, and includes suggestions relevant to developing high-field magnets for future accelerators.

6.1 Introduction

In the 1980s and early 1990s the US Department of Energy (DOE) considered developing magnets using Nb_3Sn conductor at Lawrence Berkeley National Laboratory (LBNL) (Taylor et al. 1983, 1985). Nb_3Sn, a brittle, strain-sensitive superconductor, was the only superconductor that at that time had practical current density for fields up to 16 T. It was also well understood that the next high-energy hadron collider would likely require higher operating fields well beyond the present 10 T capabilities of Nb-Ti. Prior improvements in Nb_3Sn magnets had been few and progress slow, but recent results at that time had been encouraging.

D20 was a magnet set to demonstrate the technology needed for future high energy colliders by building and testing a 1 m long accelerator-quality 14 T dipole. At that time, the risk of failure using brittle Nb_3Sn conductor was not totally unfounded and very real—the untested technology, untested structure, and uncommon conductor were just part of the uncharted technology so that, by the time issues were resolved, 6 years had passed. Compared with how long it took to design and build Nb-Ti magnets, such a long-extended period was beyond what was commonly expected. Also compounding the technical difficulties were differences of opinion that weighed heavily on decisions. In the end, the two coils, each a double-layer Nb_3Sn coil, went beyond the 10 T "wall," and today such technology has matured to a point where Nb_3Sn magnets are included in the Large Hadron Collider (LHC)

S. Caspi (✉)
LBNL (Lawrence Berkeley National Laboratory), Berkeley, CA, USA
e-mail: s_caspi@lbl.gov

D. Schoerling, A. V. Zlobin (eds.), *Nb_3Sn Accelerator Magnets*, Particle Acceleration and Detection, https://doi.org/10.1007/978-3-030-16118-7_6

Hi-Luminosity upgrade. In hindsight, 20 years after D20, some valuable lessons from that time are still relevant.

6.2 Magnet Design

6.2.1 Design Approach

The design approach towards D20 (Dell'Orco et al. 1993a) was "think-outside-the-box," a concept that followed a similar approach previously tried with a unique Nb-Ti magnet (Dell'Orco et al. 1993b). D19, a 50 mm aperture Nb-Ti dipole magnet, was built and tested as a candidate for the US Superconducting Super Collider (SSC) program. With a record field of 7.6 T at 4.4 K and 10 T at 1.9 K, that magnet specifically addressed issues of high fields and Lorentz forces. By separating the traditional concept of using self-supporting collars that combine assembly and pre-stress, D19 used thin non-supporting collars for assembly alone, leaving the functionality of pre-stress to be applied during the final structural assembly. Thin elliptical collars (with a width of 3 mm on the mid-plane) were used for coil assembly and initial alignment. Pre-stress was split between a "rings and collets" loading system and a cool-down shrinkage of aluminum shell over iron. The close proximity of the iron to the coils contributed to the dipole field, and saturation harmonics were handled by the elliptically shaped collars. Aluminum bars between the yokes controlled the final assembly.

The successful test of D19, performing without training at 4.4 K, became a model for the D20. The D20 design (Dell'Orco et al. 1995; Scanlan et al. 1995) had a considerable number of additional steps with a complexity that required an integrated design approach. A multi-phased heat treatment, thermal expansion of materials, protection heaters, epoxy impregnation, assembly, and stress pre-loading were mostly still to be understood. This chapter describes the D20 design as it was done in the early 1990s: a time when many presently available tools and programs, especially 3D, were not available or had just started to emerge. Therefore, certain design aspects, routinely used today, were missing and were not implemented in D20.

6.2.2 Magnetic Design Optimization

D20 was designed to extend accelerator magnet technology to high fields by using Nb_3Sn cables with sufficient current density to deliver fields over 13 T. The graded design used two double layers with different wire and cable sizes, placing the larger strand size in the inner layer. The magnetic design underwent three successive iterations:

1. Infinite permeability, sector coils, no wedges;

2. Infinite permeability, stacked turns, and wedges;
3. Elliptical inner yoke with finite permeability, stacked turns, and wedges (the elliptical yoke was later replaced by a circular one).

In the first design the coils were modeled as circular sectors, no wedges, and the number of strands, diameter, and the Cu/non-Cu ratio optimized for both layers. This step led to a design of two double layers, eliminating the use of single layers with wide cables and a lead emerging out of the pole. The design with the field computed analytically was an attempt to maximize the central dipole field and reduce the sextupole and decapole components of the field. The initial graded two double-layer coils assumed a keystone cable, and used the same current margin and copper current density. The resulting magnet configuration had a short sample field of 14.3 T with a current of 6280 A/turn.

In the second design an infinite permeable boundary was placed 9 mm away from the coils, leaving room for a circular collar. Design variables were the number of block and turns keeping them close to a radial position in order to reduce wedges with sharp tips. The results yielded 13.15 T and 5 kA/turn; a lower field and current, respectively.

In the third design, permeable iron was used to size and shape the collar. The challenge was to reduce changes in the sextupole harmonic from low to high field. Up to 13 T that change was reduced to 1.8 units (Fig. 6.1). The thin collars placed the iron structure closer to the coils, resulting in a high transfer function of 2.22 T/kA that was only 14% lower than that with unsaturated iron.

6.2.3 *Conductor Development*

The four-layer D20 cosine-theta magnet (two double layers) was wound with two types of Rutherford cable over a 50 mm bore. The inner double layer (layers 1 and 2)

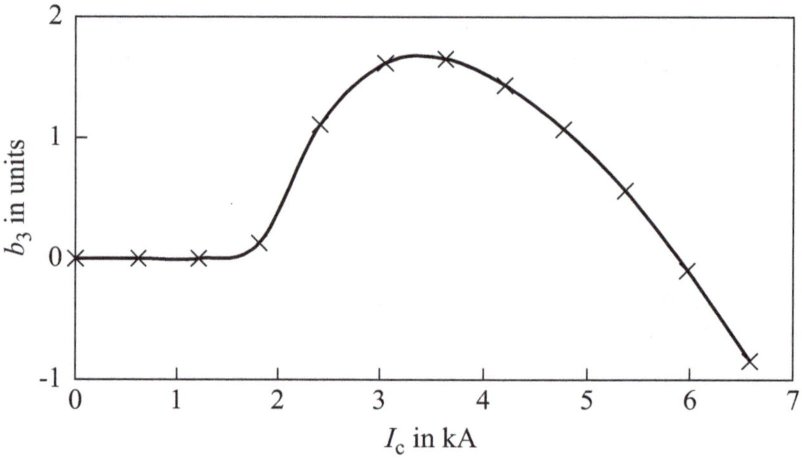

Fig. 6.1 Sextupole variation due to iron saturation

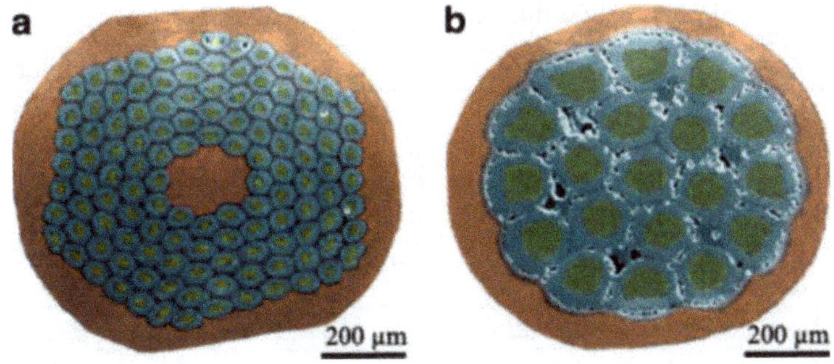

Fig. 6.2 Cross-sections of: (**a**) reacted TWCA-MJR; and (**b**) IGC-IT wires. (Dietderich and Godeke 2008)

used a thicker 37-strand cable with a 0.75 mm diameter strand, while the graded outer layer design used a thinner 47-strand cable with 0.48 mm diameter strand.

Two sources of Nb_3Sn wire, both based on the internal tin (IT) process, were used. The construction details of these conductors were different, although the critical currents were similar. Teledyne Wah Chang Albany (TWCA) provided a modified jellyroll (MJR) Nb_3Sn composite wire with a high volume fraction of tin, niobium, and Nb diffusion barrier in order to produce a high critical current density J_c. Intermagnetics General Corporation (IGC) provided a composite wire with a low copper to non-copper ratio, a moderately high volume fraction of niobium and tin, and a tantalum diffusion barrier. Two different wires, one for the inner layer cable and one for the outer layer cable, were ordered from each supplier. Both cables initially had a keystone cross-section with keystone angles of 1.11° and 0.87° for the inner and outer double layers, respectively. The cable cross-section was later changed to rectangular. Cross-sections of reacted MJR-TWCA and IT-IGC strands in soldered cables are shown in Fig. 6.2.

To optimize the cables' parameters, strands were extracted from the cables and their critical current was measured. In addition, the critical currents of these cables were measured as a function of applied stress, indicating a large degradation in critical current due to the cabling operation and a large dependence of critical current on the applied transverse stress. Subsequently, cables were made without a keystone and repeated tests showed little cabling degradation and improved stress dependence (Dietderich and Godeke 2008). As a result, the magnet cross-section was redesigned using rectangular cables.

Micrographs of strands removed from keystone cables showed the cause of the J_c degradation to be severe deformation of the filament bundles on the cables' narrow edge. Micrographs of strands removed from the rectangular cables showed no sheared bundles of filaments. Strands of the TWCA inner layer material still, however, showed some breaks in the niobium diffusion barriers at the edge of the rectangular cable. Since the inner strands of the IGC material did not show any breaks in the diffusion barriers, this material was chosen for the inner coils.

The geometry and final cables used in this magnet are listed in Table 6.1. The cables were insulated using a 0.12 mm thick S2-glass sleeve, which is discussed below.

Table 6.1 D20 cable final parameters

	Inner coil[a]	Outer coil
Layer	1 and 2	3 and 4
Conductor type	IGC/TWCA	TWCA
Strand diameter (mm)	0.75	0.48
Cu/non-Cu ratio	0.43	1.06
Number of strands	37	47
Insulated cable width (mm)	14.66	11.88
Insulated cable thickness (mm)	1.58	1.12
Keystone angle	0	0
Twist pitch length (mm)	93.5	81.28
Residual resistivity ratio (RRR) (upper/lower coil)	201/47.5	44.9/51.6

[a]One of the two inner coils was made with TWCA wire

6.2.4 Mechanical Design and Analysis

The mechanical structure of D20 (see Fig. 6.3) was similar to that of D19. The use of elliptical collars was later abandoned as it was realized that they were unnecessary with four-layer-thick coils. The initial design considered the use of an external structural shell. This was later revised and the shell replaced by 18 layers of a rectangular stainless-steel wire wound over the yoke with a tension of 500 N. The high tension wire winding forced the yoke gap to close during cool-down. A coil pre-stress of 110 MPa on the inner layer and 90 MPa on the outer layer was needed to prevent coil separation at 13 T.

The vertically split iron yoke had a measured tapered gap that varied from 0.56 mm inside to 0.76 mm at the yoke outer radius of 381 mm. An aluminum bar, 280 mm long with a 0.25 mm clearance, was placed between the yokes to control the initial assembly and cool-down. This clearance allowed initial compression to take place during assembly and the yoke gap to close tightly during cool-down. During cool-down, the coils lost on average 11 MPa of pre-stress. When the magnet was energized to 13 T, the Lorentz force partially unloads the yoke gap without completely opening it. This approach of keeping the yoke gap closed at all times made the structure very stiff. Winding a rectangular wire around the yoke until the coil pre-stress was achieved had a real advantage compared with the alternative use of a 25 mm thick welded shell. The wire-wound method was easier and more controllable during that phase of research and development.

The mechanical analysis assumed that all materials are isotropic, linearly elastic, the coils have no hysteresis, the coils and the copper wedges are bonded, there is no friction, and the plane-stress analysis is valid. The initial coil Young's modulus was 17 GPa; it was, however, later revised according to measurements (Chow and Millos 1999) (Table 6.2). The calculated average stress distributions in the D20 design with an outer welded shell structure and an outer winding structure are shown in Tables 6.3 and 6.4. Three load cases have been examined: (a) full assembly at

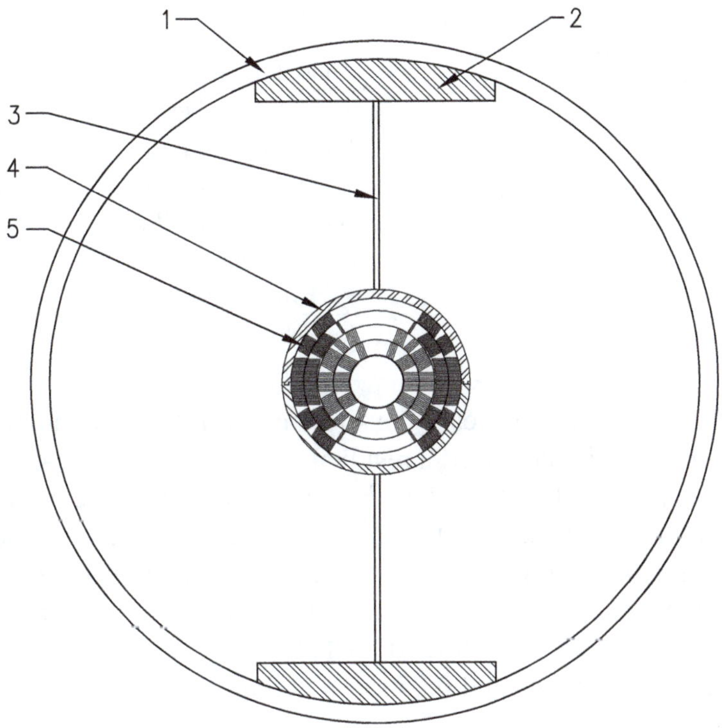

Fig. 6.3 Final cross-section: *1* – stainless-steel wire wrap; *2* – aluminum block; *3* – split yoke gap; *4* – circular spacer; *5* – coil

Table 6.2 Mechanical properties of D20 coils (Chow and Millos 1999)

	E modulus at 300 K measured/calculated (GPa)	*E* modulus at 77 K measured (GPa)	Contraction at 293–77 K (%)
Inner coil	38/40	42	0.29
Outer coil	33/37	36	0.26

Table 6.3 Average stress distribution in D20 design with outer welded shell structure only

	300 K	4.3 K	13 T
Mid-plane inner coil (MPa)	107	93	100
Mid-plane outer coil (MPa)	90	81	122
Pole inner coil (MPa)	109	96	25
Pole outer coil (MPa)	90	81	22
Force half gap (N/mm)	0	2826	1240
Force Al bar (N/mm)	516	0	0
Stress shell (MPa)	280	405	406

room temperature; (b) magnet cool-down to 4.2 K; and (c) a final load to a field of 13 T.

The room-temperature coil pre-stress did not change much during the cool-down. With Lorentz forces and a bore field of 13 T, the mid-plane stress on the inner coils increased by 12 MPa and on the outer coils by 56 MPa (Table 6.4). At the poles,

Table 6.4 Stress distribution in D20 design with outer winding structure

	300 K	4.3 K	13 T
1 and 2 mid-plane (MPa)	79	65	77
3 and 4 mid-plane (MPa)	85	75	131
1 and 2 pole (MPa)	83	76	18
3 and 4 pole (MPa)	86	75	15
Outer wire wrap (MPa)	240	360	360
F_x (N/mm/quad.)	–	–	5622
F_y (N/mm/quad.)	–	–	−1984
F_z (kN/end)	–	–	820

Lorentz forces decrease the stress on both layers by approximately 60 MPa, leaving an approximate 18 MPa residual compression on the coils. With no pole separation, the possibility of motion that may cause training was not expected. The loaded aluminum spacer between the yokes shrank considerably during cool-down, unloading completely and closing the yoke gap. When the field was raised to 13 T the yoke at the gap became partially unloaded but it did not open, thus keeping the coil displacements low. At 13 T, the radial displacement of the collar was 0.063 mm at the mid-plane and 0.058 mm at the pole.

6.2.5 D20 Final Design

The final magnet design was based on four layers of graded cos-theta coils, rectangular cable, a 50 mm aperture, a thin coil–yoke radial spacer, and a vertically split iron yoke supported by stainless-steel wire wrap. The final magnet cross-section is shown in Fig. 6.3 and the coil cross-section using rectangular cables is shown in Fig. 6.4. The coil parameters are given in Table 6.5.

Calculated magnet design parameters are summarized in Table 6.6. The strand short sample current density measurements went through several tests and revisions, and remained uncertain even as the magnet test commenced. At best the D20 design suggested it could reach 14 T at 4.5 K and 7 kA. While “doom and gloom” predictions persisted, they did not materialize and D20 succeeded beyond expectations.

6.3 Magnet Fabrication

Each double-layer coil was wound with an internal ramp in order to avoid an inter-layer splice in the high-field region (Fig. 6.5). The cable was insulated using a 0.12 mm thick S2-glass sleeve. The coils were fabricated using the wind-and-react (W&R) method following the final choice of heat treatment at 210 °C for 100 h, 340 °C for 48 h and 650 °C for 180 h, followed by vacuum impregnation with epoxy.

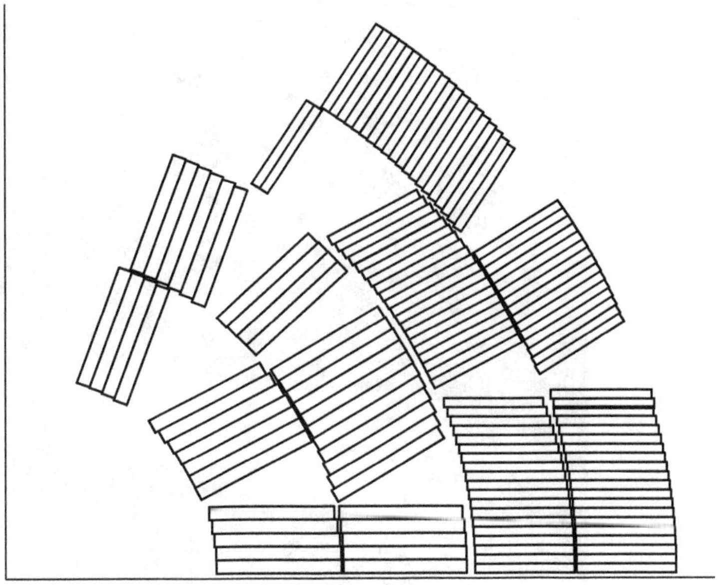

Fig. 6.4 D20 final coil cross-section using rectangular cables

Table 6.5 D20 coil final parameters

	Inner coil		Outer coil	
Layer	1	2	3	4
Number of turns	16	26	41	55
Number of wedges/quadrant	2	3	2	2
Inner radius (mm)	24.75	40.15	55.82	67.95
Outer radius (mm)	39.41	54.81	67.73	79.86
Coil pole corner (°)	70.56°	67.65°	57.89°	56.42°
Coil pole tilt (°)	21.16°	21.16°	35.03°	35.03°

Table 6.6 Magnet design parameters

Parameter	Value
Magnet aperture (mm)	50
Yoke inner/outer diameter (mm)	178/762
Coil outer diameter (mm)	160
End shoe-to-shoe length (mm)	1250
Magnet end physical length (mm)	246
Magnet end magnetic length (mm)	150
Transfer function (T/kA)	2.22
Current at 13 T (kA)	6.4
Stored energy at 13 T (MJ/m)	1.1
Magnet inductance (mH)	45.6

6.3.1 Coil Components

The requirements for the coil component parts were metallic non-magnetic, *E*-modulus, and a thermal expansion coefficient matching that of the windings, and capable of surviving the reaction temperature. An aluminum-bronze alloy was

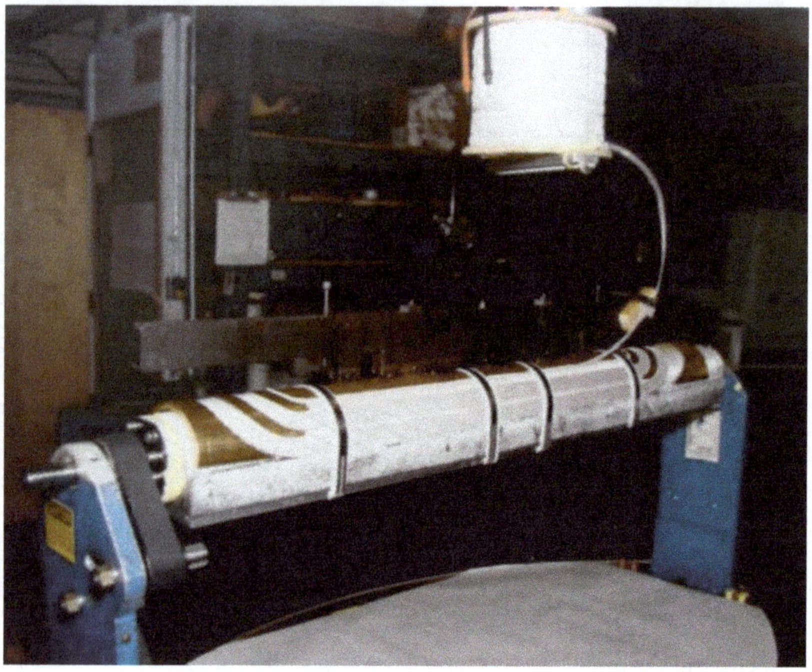

Fig. 6.5 Split spool windings across the internal layer jump, ready to wind layer 2

chosen to meet these requirements, and several different fabrication approaches for these pieces were evaluated. Wedges were made by conventional machining, as were the pole pieces. End spacers could not, however, be machined using conventional methods due to their complex shapes.

The design of the end spacers (Fig. 6.6) was based on surface shapes according to the rulings obtained by the BEND software program (Cook 1990, 1991; Brandt et al. 1991; Caspi 1993a, 1993b; Caspi et al. 1995). Surfaces were constructed from straight lines formed by geodesic curves, which ensured minimum strain energy in the cables wound around them (Fig. 6.7). Early attempts to manufacture the spacers required the use of numerically controlled five-axis machines and data information in the form of rulings. At that time the technology was limited, and spacers were cut from machined wax cylinders and used, in a lost wax process, for casting from aluminum bronze. The casting operation proved to be very unsatisfactory, yielding parts with unacceptable tolerances and voids. A lot of handwork was required to improve the spacers' quality before they could be used in the coils. Figure 6.8 shows a selection of islands and end spacers for inner layers 1 and 2. Figures 6.9 and 6.10 show the nested turns and spacers before and after reaction. An axial cut through the inner double layer ends (Fig. 6.11) shows how turns nest as they turn over the pole. Turns near the pole clearly nest much better compared with those near the mid-plane by the shoe. Despite their distortion and de-cabling, no quenches initiated at these locations.

The technique was later improved considerably when a fully automated electrical discharge machine (EDM) became available. Shape information in the form of rulings was sent electronically to a computer-controlled EDM, which was used to

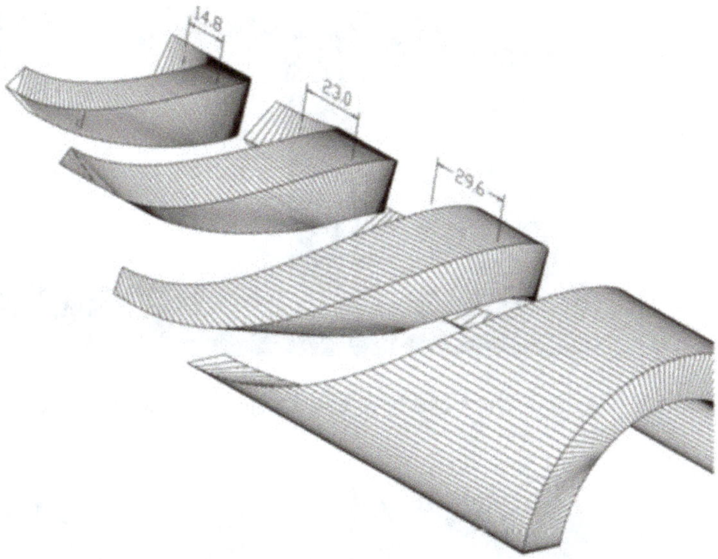

Fig. 6.6 CAD view of layer 2 end spacers generated by the program BEND

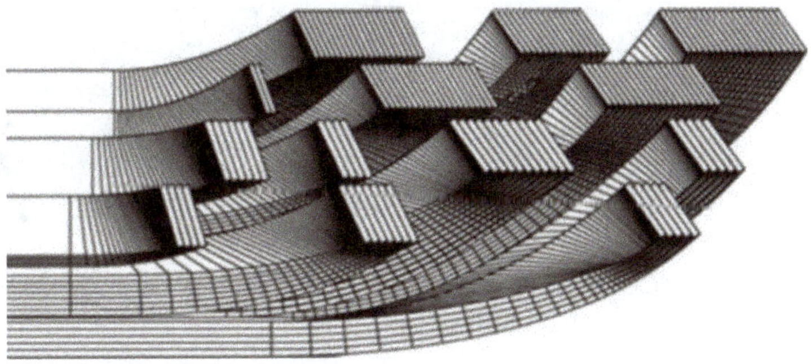

Fig. 6.7 View of the D20 end-turn design

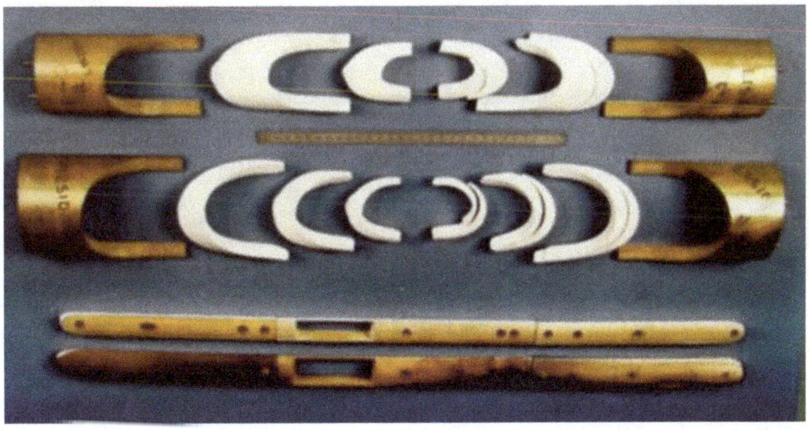

Fig. 6.8 Layers 1 and 2 pole blocks (islands), end spacers, and saddles (shoes). White alumina coating was added to the end spacers for extra insulation

cut all the parts of a given layer from a single prefabricated bronze cylinder. This process required little human intervention, was cost-effective, and yielded high-quality parts.

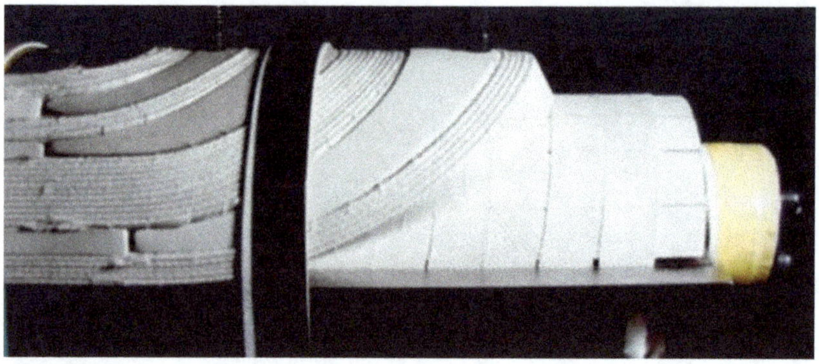

Fig. 6.9 Layer 2 return end before reaction (end spacers coated)

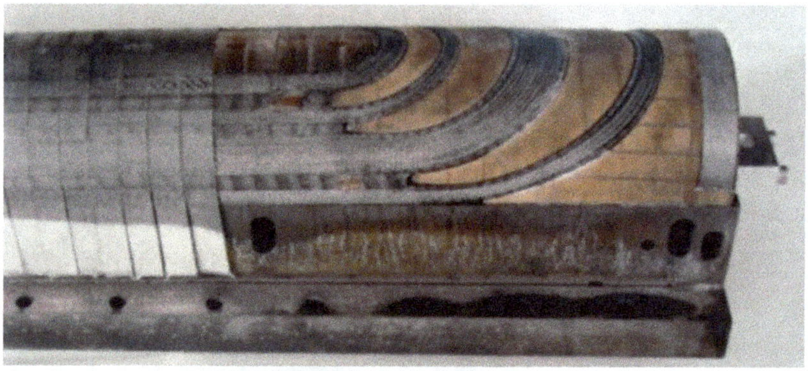

Fig. 6.10 Prototype coil with tape insulation added (note the carbon residue)

a

b

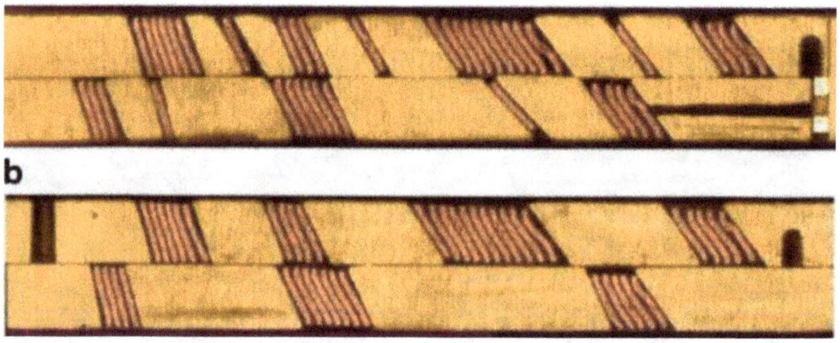

Fig. 6.11 A longitudinal cut through (**a**) the inner coil lead-end; and (**b**) the return-end (compare with Fig. 6.7)

6.3.2 Coil Insulation and Reaction

Glass-fiber insulation was readily available. It was inexpensive and could withstand a high-temperature reaction. Initial reaction tests by the conductor manufacturers showed a somewhat surprising result. At 680 °C, the insulation remained intact and could be handled to some extent. Above 740 °C, however, the heat treatment recommended by IGC, the insulation became extremely fragile and crumbled under the slightest handling or abrasion. Alternatives such as similar ceramic or quartz fibers were also considered, but both suffered from limited availability and relatively high cost. The solution adopted was to use a less than optimum heat treatment for the IGC conductor and keep the original glass-fiber insulation. Two types of glass-fiber insulation were evaluated: a braided tape, which was formed around the cable, and a sleeve that was made and then slipped onto the cable. The sleeve was adapted following the development of a process to show that it could be applied to a cable length of over 500 m.

The sizing used by the manufacturer left a heavy residue of carbon after reaction in an argon atmosphere. Removal of the sizing with a partial atmosphere of oxygen has been previously used in W&R Nb_3Sn magnets. This approach was, however, abandoned due to problems with oxidation of the copper matrix and incomplete carbon removal in less accessible areas of the coils. Instead, the sizing was removed by pre-heating, and replaced with palmitic acid, which volatizes in argon below 400 °C and did not leave a heavy carbon residue. A high gas flow through the coils was, however, needed to flush it from the coils. The palmitic acid was applied by dip-coating the sleeve through a solution of 1 part palmitic acid to 20 parts ethanol by weight. After dip-coating, the sleeve was passed through a drying tower to evaporate the ethanol.

Initial winding tests with the S2-glass insulation showed that occasional shorts could still be developed due to abrasion by end spacers and wedges. This problem was solved by applying a coat of plasma-sprayed aluminum oxide ceramic insulation 0.1 mm thick (Fig. 6.9). The process took some time to develop, but the ceramic coating had good adherence to the metal pieces, good bonding to epoxy, and good insulating properties.

6.3.3 Splice Joints

Each of the eight Nb_3Sn leads was ramped slightly away from the mid-plane surfaces, and axially extended underneath the coil end into a low field region ($B < 2$ T). It was then cleaned, fluxed, and sandwiched between two solder foils: Nb-Ti cables within a rectangular high purity copper box. The box was pressed towards the wide face of the cable and resistively uniformly heated until flux and excess solder flowed out both ends. The temperature was monitored to maintain reproducible splices. A typical joint took less than 2 min to fuse after initiating heating. The finished joint was insulated, mechanically supported, and vacuum-potted within the bronze end-shoe during impregnation. Typical Nb-Ti/Nb_3Sn joints were measured between 0.5 and 1 nΩ at 6 T.

6.3.4 Coil Epoxy Impregnation

The wind-and-react (W&R) approach made it necessary for the coils to be reinforced by a combination of glass fiber and epoxy. The strength of the reinforcing matrix, in combination with the reacted Nb_3Sn cable, required a series of tests to measure its mechanical properties. For this application, the epoxy had to have a low viscosity, long pot life, and be compatible with glass-fiber reinforcement. In addition, it was required to withstand cryogenic conditions. After preliminary screening, the epoxy chosen for more extensive testing was CTD-101 (Composite Technology Development, Lafayette, CO).

Both inner and outer layer cables were tested as well as samples of pure epoxy and epoxy plus glass fiber. Compression tests were performed at ambient temperature and at 77 K. Test data provided by the epoxy manufacturer were used for properties at 4.2 K. The values obtained for *E*-modulii and thermal expansion for both inner and outer cable samples are shown in Table 6.2. The values agreed well with values calculated using a rule of mixtures; the measured coefficient of expansion, however, did not. Additionally, tests were performed to find the maximum compressive load the composite stacks could withstand before the epoxy began to show cracking; loads of at least 100 MPa could be applied before the epoxy cracked. The cross-section autopsy of the four-layer impregnated coil is shown in Fig. 6.12.

6.3.5 Quench Protection

With a relatively high magnet inductance of 45 mH, quench protection issues had to be addressed. Protection heaters were designed to dissipate the energy over a large volume to reduce the coil's "hot-spot" temperature. These heaters were fabricated from a polyimide/stainless-steel composite sheet, and the traces were formed by a photoresist/etching process. After etching, a top sheet of polyimide was bonded to the package to produce an insulated unit 0.12–0.15 mm thick that included 0.025 mm thick stainless-steel heaters. These heater traces became bonded to the

Fig. 6.12 D20 coil cross-section

Fig. 6.13 End view of a double-layer impregnated coil showing the end-pole, end spacer, end-shoe, and inner-layer quench protection heater

coils during the epoxy impregnation, covering over 80% of the coil turns (Fig. 6.13). With a quench detection level of 0.25 V, the temperature of the coil hot-spot was limited to less than 300 K.

6.3.6 Magnet Assembly

After potting, each double layer was placed back into the potting fixture, replacing one or more insulation layers on the appropriate loading surfaces with a pressure-sensitive film. The coils were then loaded to a few MPa to determine the surface profiles as well as the coil size. This sizing information along with design numbers and component measurements were used to determine the assembly details of the entire magnet. During an intermediate final assembly test, pressure-sensitive films once again replaced insulation layers on mating surfaces to check for the appropriate fit and the maximum load applied.

Initially during the first sub-assembly test the coils were placed within two aluminum-bronze half-shells and wrapped with 304 stainless-steel wires to about 50% of the calculated target tension. Pole strain gauges were used to monitor the entire operation. During the final assembly test, two yokes were placed over the bronze spacers and held together over the gap by aluminum blocks that formed a continuous circular feature.

Eighteen layers of 304 stainless steel 3 mm × 1 mm round edge wire, 14 km long were wrapped around the entire yoke assembly (Fig. 6.14). With a tension of 850 N, the stress applied to the coils was about 100 MPa and 65 MPa for the inner and outer layers, respectively. These results then led to the loading numbers of the final assembly within the yokes. End bolts held by end plates were attached to the magnet

Fig. 6.14 Close-up of the wire wrap

Fig. 6.15 D20 non-lead end before installing the end plate

ends providing axial compression up to 30% of the total Lorentz force (820 kN). Bolts with pre-calibrated strain gauges were used to monitor the end force throughout the assembly and testing. Figure 6.15 shows D20 during the final assembly placed on a delivery cart.

Part theory and part reality, it turned out that during the applied pre-stress by the outer wire wrap the gap between the yokes did not remain parallel, and closed, first near the outer diameter (OD) (near the wire wraps). That required a revision of the yoke interfaces to be tapered, with a larger gap near the OD, as mentioned above. Controlling the way the iron acted during assembly and cool-down remained difficult and resulted in only partial closing of the yokes. The impact therefore was uncertainty of the coils' pre-stress.

6.4 Test Results and Discussion

The D20 magnet was tested at the LBNL test facility in a horizontal cryostat with an ambient pressure of liquid helium at 4.4 and 1.8 K (Fig. 6.16). The facility used an external dump resistor to reduce the stored energy absorbed by the magnet coils during a quench.

6.4.1 Training and Operation

The magnet was heavily instrumented with voltage taps and strain gauges. The first quench originated in the layer 1 pole at a bore field of 10.2 T. It was then followed by 13 quenches before reaching 11.34 T (Fig. 6.17, cy1). The rate of training improved when the magnet was cooled to 1.8 K, immediately raising the quench field by 9% to 12.3 T (cy1). Prior to the D20 test such jumps were seen in Nb-Ti magnets (Gilbert et al. 1989). Upon a warm-up cycle (cy2) the magnet retained a quench level of 11.9 T, and a second thermal cycle (cy3) resulted in the highest field achieved.

Fig. 6.16 LBNL personnel during the first cycle of D20 testing. The sign reads “D20 Magnet a World Record Dipole Field of 13.3 T Reached at LBNL on March 13 1997.” The highest field record was later revised to 13.5 T as the test continued

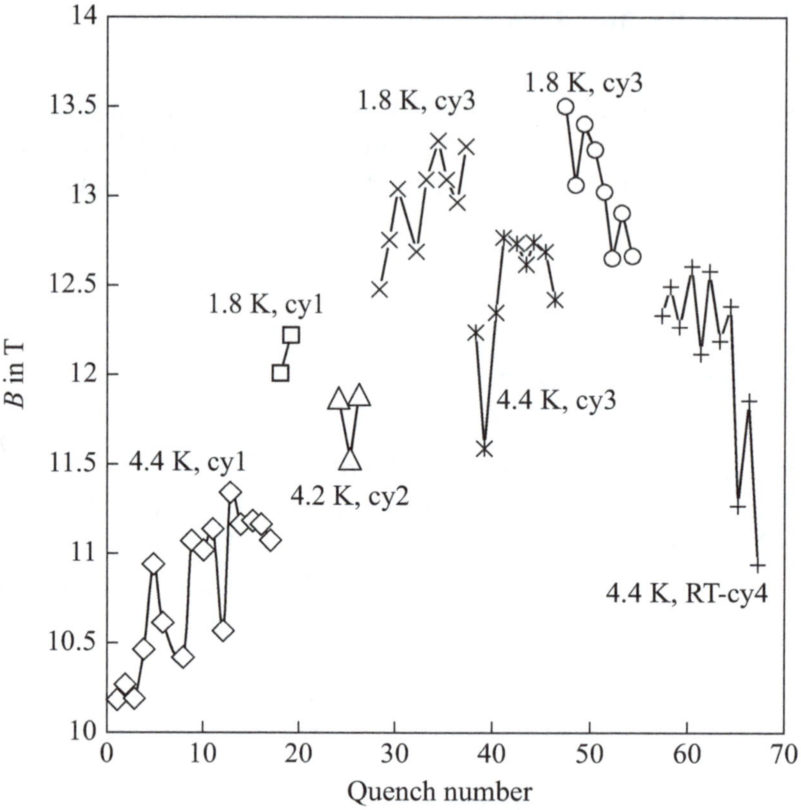

Fig. 6.17 D20 training history

Out of the first 65 training quenches, 17 originated in layer 1, 2 in layer 2, 6 in layer 3 and 15 in layer 4 (at the low field ramp). Most quenches started at the pole turn, with later inner-layer quenches originating in the ramp region between the layer 1 and 2 poles. Six quenches originated somewhere between the pole and mid-plane. Only one quench originated in the end region. The mid-plane quenches that were anticipated from conductor degradation under the accumulated Lorentz load (130 MPa expected) never materialized. A full thermal cycle to 300 K and cool-down was also completed. Although there is considerable doubt that there was enough pre-stress when the magnet was cold, the basic mechanical structure was apparently adequate for 13.5 T. A more detailed description of test results can be found in Lietzke et al. (1997), McInturff et al. (1997), Scanlan et al. (1997) and Benjegerdes et al. (1999).

From the training curve (Fig. 6.17), D20 reached a 12.8 T limit at 4.4 K. Data suggested that the onset of this quench was thermally triggered (no spike as the quench started) indicating a possible higher limit, perhaps as high as 13.1 T.

At 1.8 K, D20 reached 13.5 T followed by a continuous slow and unstable quench performance suggesting a potential problem in the low field splice ramp of layer 4 (Fig. 6.17, cy4). This ramp had been modified before reaction in order to adjust for some earlier conductor damage. The modification is believed to have compromised the cable mechanical support, causing the quench to start with a major voltage jump

Table 6.7 D20 highest values reached during testing

	IGC/TWCA (inner)		TWCA (outer)	
	4.35 K	1.8 K	4.35 K	1.8 K
I_{max} (A)	6300	6712	6300	6712
B_{bore} (T)	12.8	13.5	12.8	13.5
B_{coil} (T)	13.1	13.8	10.3	10.8
$J_{non\text{-}Cu}$ (A/mm^2)	550	587	1528	1527
J_{Cu} (A/mm^2)	1283	1367	1441	1535

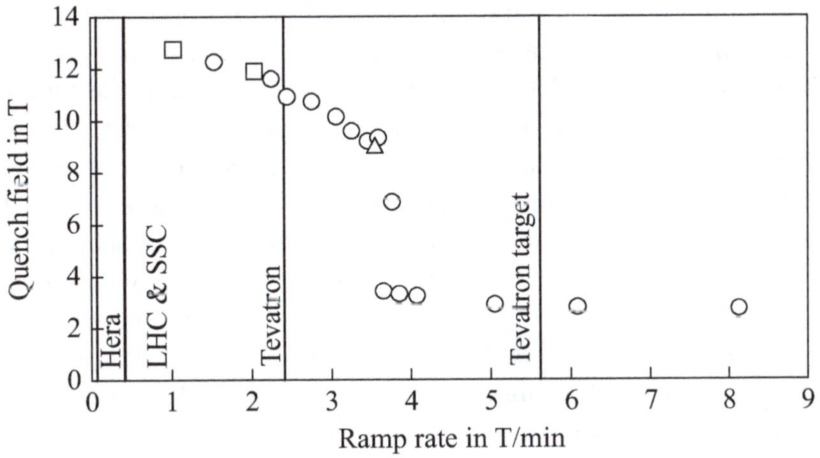

Fig. 6.18 Ramp rate dependence at 4.4 K. Circles mark ramps starting at zero current; squares and triangles mark those started at a higher current

and then propagating very slowly before entering the high-field region. Due to the limitation in the layer 4 field ramp, training was terminated. An autopsy following disassembly showed that part of the mid-plane structure at that location did indeed collapse. It was not clear whether the conductor near the joint has been damaged irreversibly.

The highest values reached during D20 testing at 4.35 and 1.8 K are summarized in Table 6.7.

Protection heaters were generally fired with the smallest possible delay after a quench was detected, using a nominal power of 25–30 W/cm^2. The measured average temperature rise was either 120 K (inner coil quench), or 185 K (outer coil quench). Typically, quench integrals of 4.3 million A^2 s (MIITS) were absorbed following a 13 T quench.

The ramp-rate sensitivity of D20 was measured at 4.4 K and the results compared with the required ramp-rates limits of Nb-Ti magnets of past projects (Hadron-Elektron-Ring-Anlage, LHC, Tevatron) (Fig. 6.18). Below 1 T/min the magnet was insensitive to the ramp rate, but at 3.5 T/min the field dropped abruptly from a maximum of 9 T to 3 T. We speculate that overcoming an inter-strand voltage threshold caused the sharp decline in quench current as the coil cooling transitioned from nucleate to film boiling.

6.4.2 Magnetic Measurements

The transfer function (Fig. 6.19) was measured with both a Hall probe and a rotating coil. Hall probe measurements were made at 8 A/s going up and 16 A/s going down. The spread in the current ramp-up vs. ramp-down measurements due to conductor magnetization is significantly reduced above 7 T. At low field of 1 T the up vs. down ramp reaches 10% with an average of 2.33 T/kA. We have checked and verified that the source of undulation in the transfer function measurements arises from the characteristics of the Hall probe and may be due to a non-linear response.

The D20 harmonics were measured with two rotating-coil systems. A 6 Hz rotating "Morgan coil" (R = 10 mm, L = 460 mm), placed at the radial and axial center of the magnet, measured the direct field harmonics b_n and a_n (for n = 1, 2, 3) during a stepped power cycle up to 2 T. A wait period of several minutes was imposed after the field was changed to a ramp rate of 1 T/min to allow the harmonics to settle to steady values before measurements were recorded. A second system used a tangential and dipole-bucking coil set (R = 11.2 mm, L = 424 mm, f = 1 Hz), for time-dependent measurements that were integrated from the magnet center through its lead end. Both the tangential and bucked signals were integrated and sampled.

The field in the magnet straight section is represented in terms of harmonic coefficients defined by the expansion

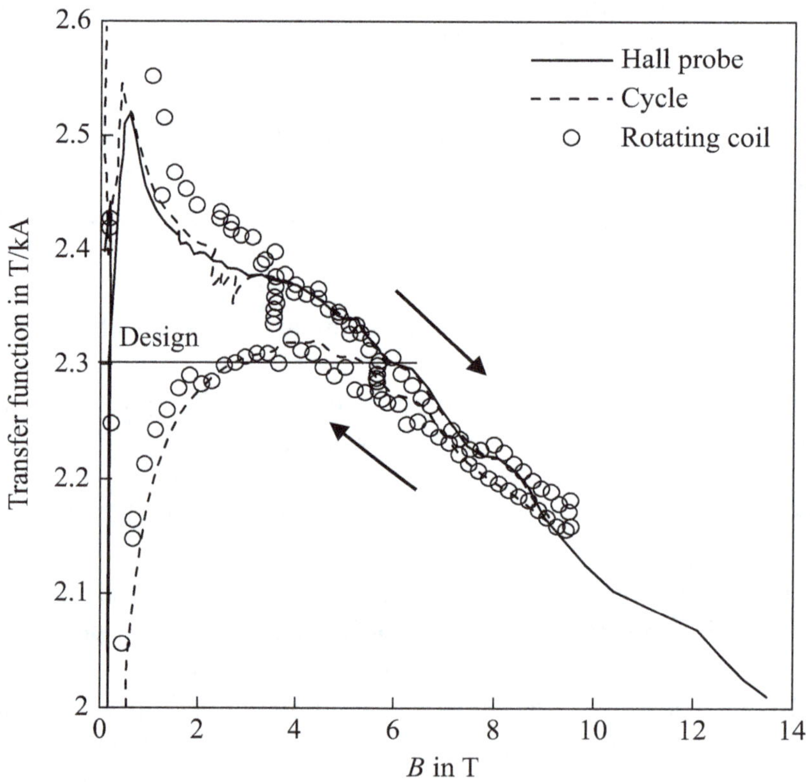

Fig. 6.19 D20 measured transfer function vs. the bore field. The horizontal line is the computed bore field value without the coil magnetization effect

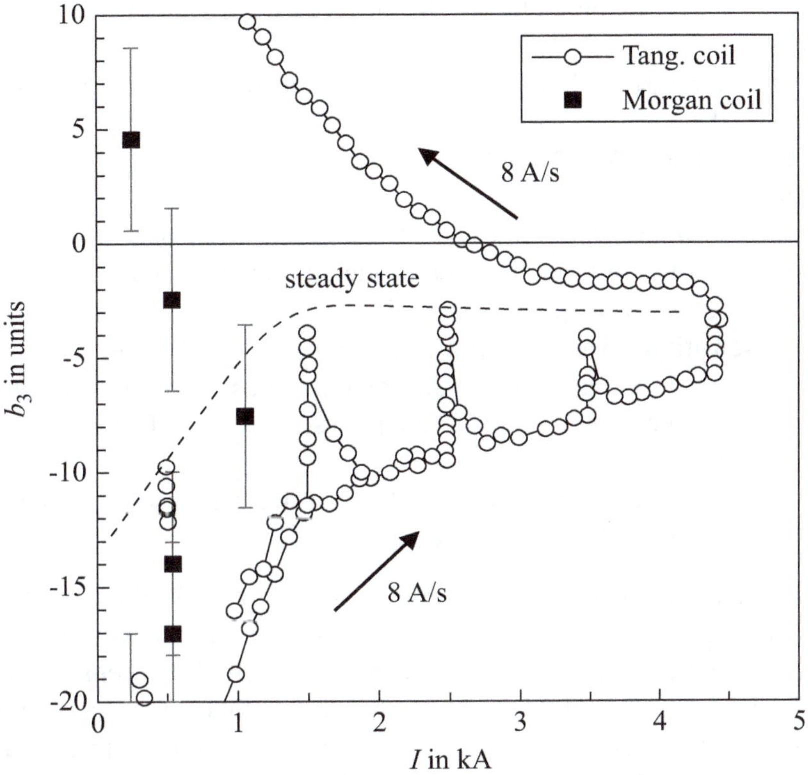

Fig. 6.20 Measured sextupole vs. the current at low field

$$B_y + iB_x = B_1 \sum_{n=1}^{\infty} (b_n + ia_n) \left(\frac{x + iy}{R_{\text{ref}}} \right)^{n-1}$$

where B_x and B_y are the horizontal and vertical transverse field components, B_1 is the dipole field component, and b_n and a_n are the $2n$-pole coefficients at a reference radius $R_{\text{ref}} = 10$ mm.

The normal sextupole b_3 expressed as "units" (10^{-4} of the main field component) is plotted in Fig. 6.20. Up to 10 T the sextupole b_3 is quite small, approaching an asymptotic value of 4 units. All higher order harmonics were measured to well below 0.5 units except for the non-allowed skew quadrupole a_2, which was measured at 2.5 units.

6.4.3 Discussion

The successful test of D20 and the record field it achieved revealed several important points that suggested a potential use and improvements of Nb_3Sn technology in future high-field accelerator magnets.

- The iron yoke closed prematurely during cool-down, thereby yielding a lower cold pre-stress;
- Initial training was slow but was accelerated substantially when the magnet was cooled to 1.8 K;
- Neither the IGC nor TWCA wires seemed as stress-sensitive as short sample tests had predicted;
- Cooling was sufficient to operate at 12.5 T and 2 K when a steady-state heat input of 24 W was applied by the distributed quench heater on the inner edge of layer 1 or the outer edge of layer 2 of one coil;
- Ramp-rate sensitivity was exceptionally low for a W&R impregnated magnet;
- Except for one low field ramp, the magnet was very robust, withstanding 60 quenches above 10 T, with 37 quenches above 12 T, including a second thermal cycle;
- The magnet could be protected despite its relatively high quenching current density in the copper, $J_{Cu} = 1535$ A/mm^2.

The favorable performance of D20 did not, however, erase the previous concerns about Nb_3Sn magnets. The magnet design and construction was long and time-consuming due to the brittle nature of Nb_3Sn, and resulted in a sense of a need for a future change by trying something other than a cosine-theta cross-section. Following the D20, two other types of magnet cross-section were built and tested at LBNL (Chiesa et al. 2003): the common coil and the window frame block. Both were explored to determine if a simpler geometry and construction technique and strain management might lead to reduced training and offset the conductor cost. As of the time of writing (2018), the gain in field has been less than 2 T and new attempts are being made to reconsider a cos-theta design like D20.

Following the thermal cycle D20 was used to provide a background field to test an insert Nb_3Al cos-theta dipole constructed by The High Energy Accelerator Research Organization and Japanese industry (Wake et al. 1997). The insert coil was placed snugly within the D20 bore. During the test D20 could be held at 12.5 T, 4.4 K for an extended period (hours) without quenching. After warm-up the Nb_3Al insert could not be removed without taking the D20 magnet apart. The insulation between the insert and D20 was left with deep imprint marks, suggesting the insert acted as a ridged bore supporting D20 as it tried to go elliptical.

Following the D20 test, recommendations and suggestions made by members of LBNL design team remain part of the lessons learned (not implemented):

- "...Proposals considered using D20 as a background field for testing small sample coils. A test facility capable of measuring short-sample at the 13 T level would have been, at that time, complementary to CERN's 10 T FRESCA-I test facility."
- "...Repair or replace the outer coil, re-assemble with a larger iron gap (to reduce cool-down pre-stress loss), and install it in a permanent Dewar."
- "...The magnet reached a 4.4 K limit below the expected wire short-sample value (1500 A/mm^2 at 12 T and 4.4 K). For a magnet like D20 to reach 16 T the current density would have to be doubled (3000 A/mm^2 at 12 T). With today's conductor,

this may not be as difficult as it appears. Recent short-sample tests on newly available Nb_3Sn wire have delivered such non-copper current densities."

6.5 Conclusion

It has been close to 25 years since the D20 program. My personal perspective following that period is that one magnet test does not tell all of the story. We should also equally recognize successes and failures and learn from both before a design changes course. The present current density of Nb_3Sn composite wires is now available beyond 3000 A/mm^2 at 12 T. As already suggested at that time, a magnet like D20 made with today's conductor could reach a central field above 16 T. Figure 6.21 extends D20 load lines of the central dipole field (circle) and the maximum field at the conductor (square) to a short sample of today's conductor (J_{sc} = 3000 A/mm^2 at 12 T and 4.2 K).

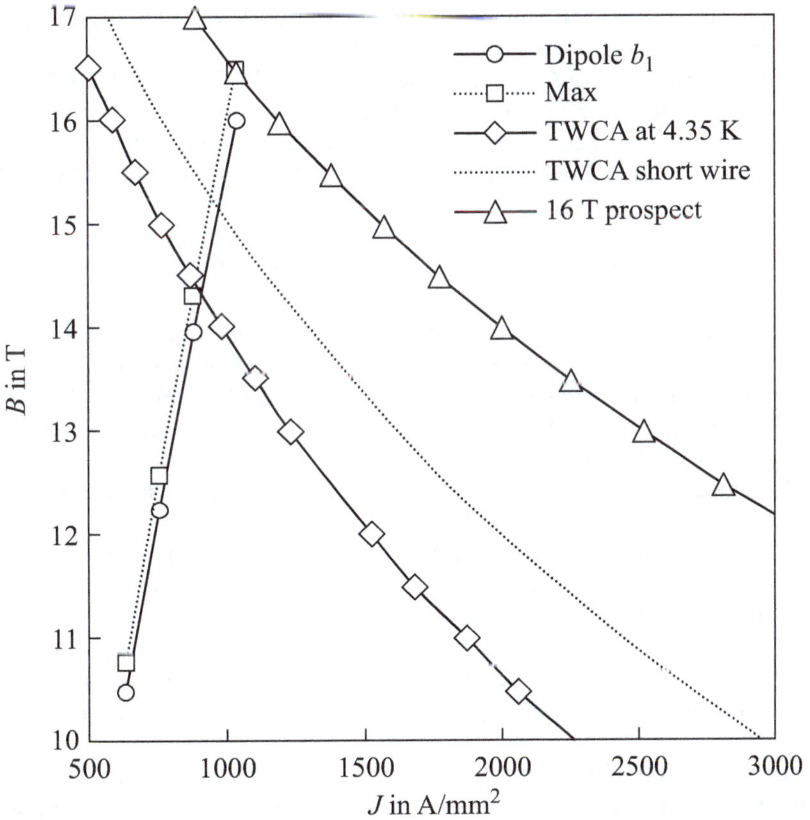

Fig. 6.21 By extending D20 load lines and using today's high J_c Nb_3Sn wires D20 could reach a central field over 16 T

References

Benjegerdes R, Bish P, Caspi S et al (1999) Operational characteristics, parameters, and history of a (13T) Nb_3Sn dipole. In: Luccio A, MacKay W (eds) Proceedings of the 1999 particle accelerator conference (Cat. No.99CH36366). New York, 27 Mar–2 Apr 1999, vol 5. IEEE, Piscataway, pp 3233–3235

Brandt JS, Bartlett NW, Bossert RC et al (1991) Coil end design for the SSC collider dipole magnet. In: Conference record of the 1991 IEEE particle accelerator conference, vol 4. San Francisco, 1991, pp 2182–2184

Caspi S (1993a) Conceptual end design of dipole D20. LBNL SC-MAG-424. Lawrence Berkeley National Laboratory, Berkeley

Caspi S (1993b) End design of dipole D20—version d. LBNL, SC-MAG-430. Lawrence Berkeley National Laboratory, Berkeley

Caspi S, Dell'Orco D, Ghiorso WB et al (1995) Design and fabrication of end spacers for a 13 T Nb_3Sn dipole magnet. IEEE Trans Appl Supercond 5(2):1004–1007. https://doi.org/10.1109/77.402720

Chiesa L, Caspi S, Coccoli M et al (2003) Performance comparison of Nb_3Sn magnets at LBNL magnet. IEEE Trans Appl Supercond 13(2):1254–1257. https://doi.org/10.1109/tasc.2003.812651

Chow KP, Millos GA (1999) Measurements of modulus of elasticity and thermal contraction of epoxy impregnated niobium-tin and niobium-titanium composites. IEEE Trans Appl Supercond 9(2):213–215. https://doi.org/10.1109/77.783274

Cook JM (1990) An application of differential geometry to SSC coil end design. SSC Laboratory SSCL-N-720, Fermilab TM-1663 internal report

Cook JM (1991) Strain energy minimization in SSC magnet winding. IEEE Trans Magn 27 (2):1976–1980. https://doi.org/10.1109/20.133592

Dell'Orco D, Scanlan RM, Taylor CE (1993a) Design of the Nb_3Sn dipole D20. IEEE Trans Appl Supercond 3(1):82–86. https://doi.org/10.1109/77.233677

Dell'Orco D, Caspi S, O'Neill J et al (1993b) A 50 mm bore superconducting dipole with a unique iron yoke structure. IEEE Trans Appl Supercond 3(1):637–641. https://doi.org/10.1109/77.233783

Dell'Orco D, Scanlan RM, Taylor CE et al (1995) Fabrication and component testing results for a Nb_3Sn dipole magnet. IEEE Trans Appl Supercond 5(2):1000–1003. https://doi.org/10.1109/77.402719

Dietderich DR, Godeke A (2008) Nb_3Sn research and development in the USA—wires and cables. Cryogenics 48(7–8):331–340. https://doi.org/10.1016/j.cryogenics.2008.05.004

Gilbert WS, Althaus R, Benjegerdes R et al (1989) Training of LBL-SSC model dipole magnets at 1.8 K. In: Proceedings of the 1989 IEEE particle accelerator conference on accelerator science and technology, vol 3. Chicago, pp 1780–1782

Lietzke AF, Benjegerdes R, Caspi S et al (1997) Test results for a Nb_3Sn dipole magnet. IEEE Trans Appl Supercond 7(2):739–742. https://doi.org/10.1109/77.614609

McInturff AD, Bish P, Benjegerdes R et al (1997) Test results for a high field (13 T) Nb_3Sn dipole. In: Comyn M, Craddock MK, Reiser M et al (eds) Proceedings of the 1997 particle accelerator conference (Cat. No.97CH36167), Vancouver, 12–16 May 1997, vol 3. IEEE, Piscataway, pp 3212–3214

Scanlan RM, Dell'Orco D, Taylor CE et al (1995) Fabrication and preliminary test results for a Nb_3Sn dipole magnet. IEEE Trans Appl Supercon 5(2):1000–1003. https://doi.org/10.1109/77.402719

Scanlan RM, Benjegerdes RJ, Bish PA et al (1997) Preliminary test results of a 13 Tesla niobium tin dipole. In: Rogalla H, Blank DHS (eds) 3rd European conference on applied supercon. Netherlands, EUCAS-97, 30 Jun–3 Jul 1997, Konigshof, The Netherlands. Institute of Physics Publishing, Bristol, Philadelphia, 158(2):1503–1506

Taylor C, Meuser R, Caspi S et al (1983) Design of a 10-T superconducting dipole magnet using niobium-tin conductor. IEEE Trans Magn 19(3):1398–1400. https://doi.org/10.1109/tmag.1983.1062261

Taylor C, Scanlan R, Peters C et al (1985) A Nb_3Sn dipole magnet reacted after winding. IEEE Trans Magn 21(2):967–970. https://doi.org/10.1109/tmag.1985.1063680

Wake M, Jaffery TS, Shintomi T et al (1997) Insertion Nb_3Sn coils for the magnet development. In: Liangzhen L, Guoliao S, Luguang Y (eds) Proceedings of the 15th international conference on magnet technology, Beijing, 20–24 Oct 1997. Science Press, Beijing, pp 103–106

Chapter 7
Cos-theta Nb_3Sn Dipole for a Very Large Hadron Collider

Alexander V. Zlobin

Abstract A series of 1 m long Nb_3Sn dipole models with a magnetic field of 10–11 T in a 43.5 mm bore was developed at FNAL as part of a research and development effort for a Very Large Hadron Collider. This chapter describes the magnet design in single-aperture and twin-aperture configurations, details of the magnet short model fabrication, and summarizes the results of magnetic, mechanical, and quench protection analyses, and model test results.

7.1 Introduction

In 1998 the Fermi National Accelerator Laboratory (FNAL, also known as Fermilab), in collaboration with the Lawrence Berkeley National Laboratory (LBNL) and the High Energy Accelerator Research Organization (KEK), Japan, started a new high-field accelerator magnet research and development (R&D) program with the goal of developing a cost-effective and robust magnet design and technology for a post-LHC Very Large Hadron Collider (VLHC). VLHC studies showed that a nominal operating field between 10 and 12 T is optimal to provide adequate radiation beam damping without significant complication of the machine's cryogenic and vacuum systems (Fermilab 2001). The VLHC target operating fields excluded using the traditional Nb-Ti magnet technology due to the low upper critical magnetic field for this superconductor. Alternative superconductors for high-field accelerator magnets are the A15 materials, primarily Nb_3Sn. Ternary Nb_3Sn composite wires with an upper critical magnetic field of ~24 T at 4.2 K and a critical temperature of ~18 K were already being produced by industry on a fairly large scale.

Although the work on Nb_3Sn accelerator magnets started in the 1960s, just a few years after the discovery of this material, it was only in the 1990s that some Nb_3Sn short dipole magnets exceeded the 10 T magnetic field threshold (see Chap. 3). All of those magnets were shell-type (also known as cos-theta) dipoles with a bore diameter

A. V. Zlobin (✉)
Fermi National Accelerator Laboratory (FNAL), Batavia, IL, USA
e-mail: zlobin@fnal.gov

D. Schoerling, A. V. Zlobin (eds.), *Nb_3Sn Accelerator Magnets*, Particle Acceleration and Detection, https://doi.org/10.1007/978-3-030-16118-7_7

of 50 mm. Two 1 m long dipole models, Model Single of the University of Twente (MSUT), the Netherlands and D20 by LBNL, reached 11.4 T and 12.5 T, respectively, at 4.5 K (den Ouden et al. 1997; McInturff et al. 1997) (see also Chaps. 5 and 6). None of those magnets, however, were of accelerator quality, focusing mostly on reaching high fields.

Progress in raising the critical current density of commercial Nb_3Sn composite wires in the late 1990s made it possible to design cost-effective Nb_3Sn accelerator magnets with a nominal field of 10–12 T. Extensive studies of shell-type and block-type dipole designs with small apertures, various current block arrangements and cable parameters, etc., were carried out at FNAL with the goal of finding optimal magnet parameters and robust, cost-effective designs and technologies for high-field dipoles suitable for a VLHC (Ambrosio et al. 2000a; Sabbi et al. 2000). This chapter summarizes the magnetic and mechanical design studies, describes basic magnet design and fabrication technology, as well as specific features and parameters of the cos-theta Nb_3Sn dipole developed at FNAL for the VLHC in the framework of the High-Field Magnet (HFM) R&D program. Tests results for the first series of Nb_3Sn accelerator magnets are presented and discussed.

7.2 Design Studies

The work started with conceptual design studies of cos-theta dipoles for VLHC. These studies were first performed to select optimal magnet aperture, nominal operating field and operating margins, conductor and iron yoke parameters as well as to estimate field quality, Lorentz forces, stored energy, and magnet inductance (Ambrosio et al. 2000a). These studies used the parameters of superconducting composite wires developed and produced at the time by Intermagnetics General Corporation based on the internal tin (IT) process. The wire critical current density at 12 T and 4.2 K was 1.9 kA/mm^2, the Cu/non-Cu ratio was 0.85, and the copper matrix residual resistivity ratio (RRR) was 100. Two types of Rutherford cable were used in the analysis. Cable 1 consisted of 28 strands, 1 mm in diameter, had 14.23 mm width, 1.8 mm mid-thickness, and 1° keystone angle. Cable 2 consisted of 38 strands, 0.808 mm in diameter, had 15.4 mm width, 1.46 mm mid-thickness, and 0.5° keystone angle. The thickness of the cable insulation was 0.125 mm, which corresponded to the commercially available S2-glass tape and to the ceramic insulation developed at that time at Composite Technology Development, Inc. (CTD).

Several opposing requirements were considered to select the magnet aperture size. A large magnet aperture helps to achieve good field quality, and simplify the design of the beam screen and the coil ends. A small magnet aperture reduces the magnet stored energy, the inductance, and the mechanical stresses. It also decreases the coil mass size and, thus, the magnet's cost.

To choose the preliminary coil cross-sections, the following constraints were imposed: (a) a range of coil aperture of 30–50 mm; (b) a target design field of 12 T; (c) no coil grading; and (d) low-order geometrical harmonics below 1 unit

Table 7.1 Magnet design comparison

	Cable 1			Cable 2
Design option	I	II	III	IV
Bore diameter (mm)	50	45	40	40
Turns per dipole	64	60	52	64
Max. bore field B_{ss} (T)	12.4	12.4	12.5	12.5
Max. magnet current I_{ss} (kA)	16.8	16.8	18.5	15.4
Stored energy at 11 T (kJ/m)	289	256	221	230
Inductance (mH/m)	2.75	2.32	1.67	2.53
Coil area (cm^2)	32.8	30.7	26.6	28.8
Min. pole width (mm)	17.5	16.2	15.0	14.6

(10^{-4} parts of the main field). A collarless structure with a 9 mm gap between coil and yoke was chosen, and no iron saturation effect was anticipated. To simplify coil fabrication, coil turns were positioned radially, and the minimal width of the coil inner-layer pole was 13 mm. The latter requirement was based on previous coil winding experience and on special cable winding tests.

Magnet cross-sections were analyzed using the ROXIE code (Russenschuck 1995). Pre-selected designs were compared based on transfer function (TF), stored energy, inductance, coil mechanical stress, and some other parameters. Designs I–III used Cable 1 and had a coil aperture of 50, 45, and 40 mm respectively. Design IV used Cable 2 and had a 40 mm aperture. For each case it was possible to achieve field quality to the level of the field quality requirements for the Superconducting Super Collider (SSC) dipoles (Jackson 1986). Coil cross-sections with 30 mm and 35 mm aperture and rather good field quality were also studied, but they were rejected due to potential coil winding problems.

Table 7.1 summarizes the main parameters of the pre-selected dipole designs. For all four magnets the maximum bore field exceeds 12 T. While the number of turns decreases by 20% as the bore diameter reduces from 50 to 40 mm, the magnet current increases only by 9%, the stored energy reduces by 25%, and the inductance decreases by 35%. A significant saving of superconductor was achieved in magnets with the smallest coil aperture. The minimal pole width and the cable block position for each preselected design ensured easy cable windability.

The maximum stress in each of the four coils due to Lorentz forces was less than 100 MPa at 11 T. It is much less than the threshold value of 150 MPa, at which the critical current degradation of Nb_3Sn strands becomes significant and irreversible. Since Lorentz forces and stresses become smaller when the coil diameter decreases, a lower coil azimuthal pre-stress at room temperature is required.

The analysis showed that there were many good reasons to reduce the magnet aperture diameter. On the other hand, the analysis of the persistent current effect revealed some benefits of having larger apertures. The sensitivity of field harmonics to block displacements also decreases as the aperture increases. Iron saturation effects did not add any strong restrictions to the cross-section choice, although a smaller iron yoke diameter would have helped reducing the magnet's weight and cost.

In addition to the pros and cons above, the final choice for the magnet cross-section was made by also taking into account the availability of the magnet fabrication infrastructure and tooling at FNAL, and specifically the availability of equipment from the high gradient quadrupole program for the LHC interaction regions. To make robust coils and reduce the risk of turn-to-turn shorts, the cable insulation thickness was increased to 0.25 mm. With these additional considerations, a coil cross-section with 43.5 mm bore diameter and 400 mm outer diameter yoke was eventually selected.

7.3 Magnet Design and Parameters

The geometrical parameters of the Rutherford cable used to optimize the coil cross-section were similar to Cable 1, described in the previous section. The cable consists of 28 1 mm strands, has a width of 14.24 mm, a thin edge of 1.687 mm, and a thick edge of 1.913 mm. The cable packing factor is 0.884. The cable insulation is 0.25 mm thick.

The optimized coil cross-section of the cos-theta dipole is shown in Fig. 7.1. Each coil consists of 24 turns: 11 in the inner layer and 13 in the outer layer. The thickness of the inter-layer insulation is 0.28 mm, and the thickness of the mid-plane insulation spacers is 0.125 mm per quadrant for both layers. Pole blocks are integrated into the coil inner and outer layers. The optimized width of the inner-layer pole is 15.09 mm, which is convenient for coil winding. Each coil has four wedges per side, two for each layer, which are used to minimize the low-order geometrical harmonics and ensure the radial turn position. To decrease the maximum field and improve the field quality in the magnet ends, the coil end design also has a block-wise arrangement with the same number of blocks and turns per block as in the magnet body.

Fig. 7.1 Coil cross-section

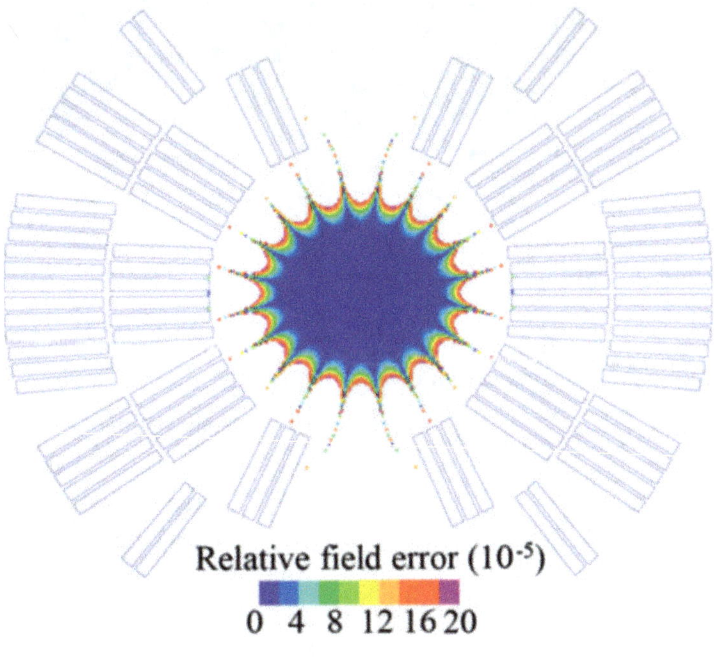

7.3.1 Single-Aperture Design and Parameters

The field distribution diagram and the cross-section of the single-aperture dipole magnet are shown in Fig. 7.2. To reduce magnet cost, the traditional collars are replaced with 8 mm thick coil–yoke spacers. The iron yoke has an inner diameter of 120 mm and an outer diameter of 400 mm. The yoke length of 600 mm is shorter than the coil length, which allows reducing the coil end fields and maximizing the length of uniform field in the magnet aperture. The coils are supported by a vertically split iron yoke, locked with two aluminum clamps, and by the stainless-steel skin.

The coil–yoke spacers protect the coil during magnet assembly. The coils are aligned relative to the spacers via two vertical extensions in the outer-layer pole posts, whereas the spacers are centered inside the yoke via two mid-plane keys. The interference of spacer and pole extension guarantees their contact during magnet assembly and operation.

Calculated magnet design parameters are shown in Table 7.2. A noticeable reduction of coil area was achieved in this design compared with earlier Nb_3Sn magnets with similar design fields. The coil cross-section area is smaller than that in the MSUT (design field of 11.4 T) (den Ouden et al. 1997) by a factor of ~2 and than that in D20 (design field of 13.4 T) (McInturff et al. 1997) by a factor of 3. Notice that all those magnets have a slightly larger bore of 50 mm.

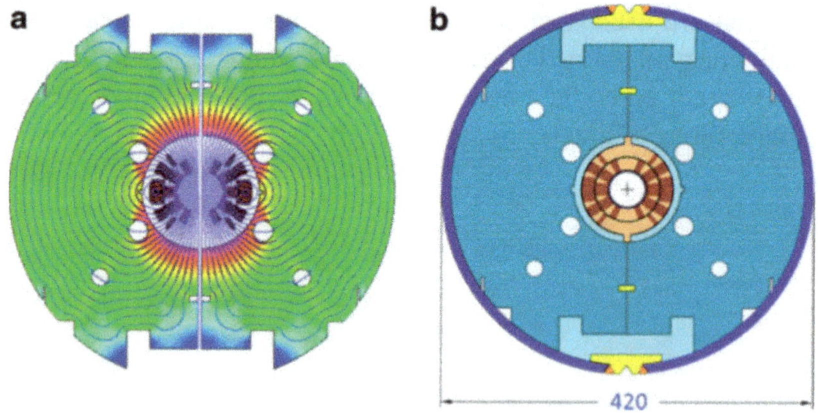

Fig. 7.2 Single-aperture dipole cross-section: (**a**) magnetic flux distribution and (**b**) mechanical structure

Table 7.2 Magnet design parameters

Parameter	Value
Transfer function (T/kA)	0.555
Current at 11 T in the bore (kA)	19.82
Stored energy at 11 T (kJ/m)	241
Magnet inductance (mH/m)	1.50
Total cable area (mm^2)	2461

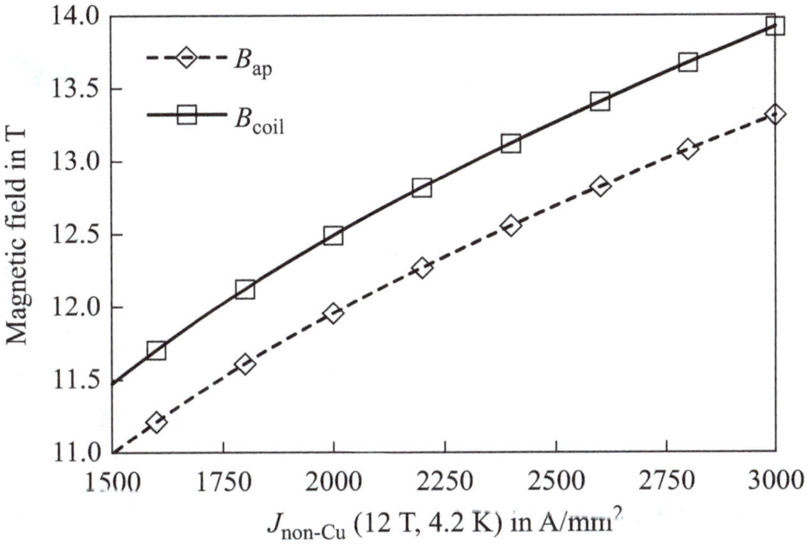

Fig. 7.3 Peak coil and bore fields vs. wire J_c(12 T, 4.2 K)

The calculated values of the coil peak field and the magnet bore field at quench vs. the critical current density J_c(12 T, 4.2 K) of Nb_3Sn cable are shown in Fig. 7.3. To reach a bore field of 11 T the critical current density of round wires at 12 T and 4.2 K has to be 1650 A/mm^2, assuming 10% of critical current degradation during cabling.

7.3.1.1 Magnetic Analysis

The field in the magnet aperture is represented in terms of harmonic coefficients defined by

$$B_y + iB_x = B_1 \sum_{n=1}^{\infty} (b_n + ia_n) \left(\frac{x + iy}{R_{ref}} \right)^{n-1},$$

where B_x and B_y are the horizontal and vertical transverse field components, B_1 is the dipole field component, and b_n and a_n are the $2n$-pole coefficients at a reference radius $R_{ref} = 10$ mm.

Due to the coil cross-section optimization the low-order geometrical harmonics are small, less than 10^{-5} of the main dipole field (0.1 unit). The calculated effect of iron saturation on field quality is shown in Fig. 7.4. For the yoke without special correction holes the b_3 variations are within 7 units, reaching a maximum at ~8 T. It was found that adding one or two correction holes decreases this effect below 2 units for bore fields up to 11 T (Ambrosio et al. 2000b). The iron saturation effect on higher order harmonics is small.

Analysis shows that the coil magnetization effect at low fields in Nb_3Sn magnets is large. For the Nb_3Sn wires available at the time, with an effective filament

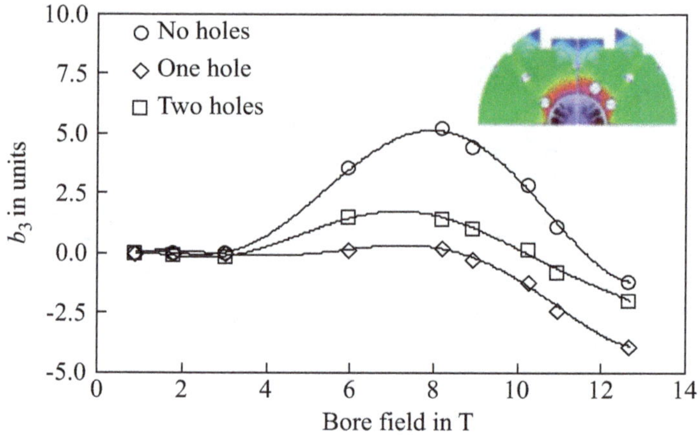

Fig. 7.4 Effect of iron saturation on the sextupole field component b_3 with and without correction holes. (Ambrosio et al. 2000d)

diameter D_{eff} of ~0.10 mm, b_3 in the magnet aperture at 1 T reaches −25 units. Some decrease of this effect was expected by reducing the D_{eff} in Nb_3Sn wires. The eddy current effects were expected to be suppressed by using Nb_3Sn wires with a small twist pitch and cables with a resistive inter-layer core.

7.3.1.2 Mechanical Analysis

A finite element analysis (FEA) using ANSYS software (ANSYS Inc., Canonsburg, PA) was performed to optimize the coil stress during magnet assembly and operation, and to evaluate the maximum stresses in key elements of the magnet support structure (Ambrosio et al. 2000c). The results are summarized in Table 7.3.

After clamping and skin welding at room temperature, the area with a highest stress of 80 MPa is at the inner radius of the coil inner layer near the pole wedge. The clamps provide ~40% of the total coil pre-stress during magnet assembly. The coil bore is almost round with a difference of less than 0.004 mm along the horizontal and vertical radii.

At liquid helium temperatures, the stress distribution on the coil inner surface becomes non-uniform due to the coil horizontal deformation. The difference between the aperture horizontal and vertical radii increases to approximately 0.10 mm. The coil stress ranges from 121 MPa in the inner-layer pole turn to 11 MPa in the inner-layer mid-plane turn.

At the design field of 11 T the coil is still under compression and the maximum stress of 100 MPa is at the mid-plane of inner and outer layers. At 12 T the stress distribution is the same, although some small fractions of the coil–pole interfaces are under a small tension. The coil bore shape returns back to an approximate circle, with a difference along the horizontal and vertical radii of ~0.01 mm.

The data show that the magnet coil is under compression under all conditions, and that the maximum stress in the coil is always less than 125 MPa. The maximum stress in key elements of the magnet support structure is less than the material's yield stress. The shear stress on the major interfaces does not exceed 30 MPa. All these values are acceptable.

Table 7.3 Peak equivalent stress in coil and support structure

Stage	Peak equivalent stress (MPa)				
	Coil	Spacer	Yoke	Clamp	Skin
Assembly (300 K)	80	166	110	135	200
Cool-down (4.2 K)	121	125	110	124	330
Nominal field (11 T)	100	97	133	128	350

7.3.1.3 Quench Protection

Coil protection in case of a quench in accelerator magnets is typically provided by internal quench heaters (Ambrosio et al. 2000d). In a two-layer coil, quench heaters can be placed inside the thin inter-layer spacer, to quench practically all of the turns in both top and bottom coils, or on the coil outer layer. Analysis shows that in both cases quench heaters provide reliable magnet protection, ensuring a low coil temperature and low voltages.

7.3.2 Twin-Aperture Designs and Parameters

The Nb_3Sn coil described above was used to design twin-aperture dipoles for VLHC (Kashikhin and Zlobin 2001). The dipole cross-sections with cold and warm iron, and with horizontal and vertical arrangements of apertures are shown in Fig. 7.5.

In the design shown in Fig. 7.5a, the two coils are placed side by side inside a cold iron with an aperture separation of 180 mm. This distance can be varied within the 160–200 mm range for the same iron size without noticeable impact on the field quality. The iron is split into three pieces. The coil pre-stress and support is provided by clamped iron blocks and an external shell (also known as skin). Since the horizontal components of the Lorentz force between coils are compensated inside the iron, a 10 mm thick stainless-steel skin is adequate for coil support. An open vertical gap between the iron pieces minimizes the decrease of coil pre-stress after cool-down. To minimize the impact of gap size variations on field quality, the iron split is partially parallel to the magnetic flux lines. The iron saturation effect is suppressed by special holes and by optimizing the iron size.

In the design shown in Figs. 7.5c, d, the warm iron is sufficiently distanced from the coils to provide room for the cold mass support system, the thermal shield, and the vacuum vessel. The two coils are placed inside the iron with an aperture separation of 180 mm as in the design in Figs. 7.5a, b. The coil pre-stress and mechanical support is provided by thick aluminum rings, stainless-steel inserts and, if necessary, the cold mass skin. The iron inner radius and thickness were optimized considering field quality, fringe fields, iron cross-section, and the cryostat design requirements. This design allows for a significant reduction of magnet cross-section with respect to the cold iron design. The iron saturation effect in the warm iron design is small. Magnetic coupling between the two apertures, however, produces a

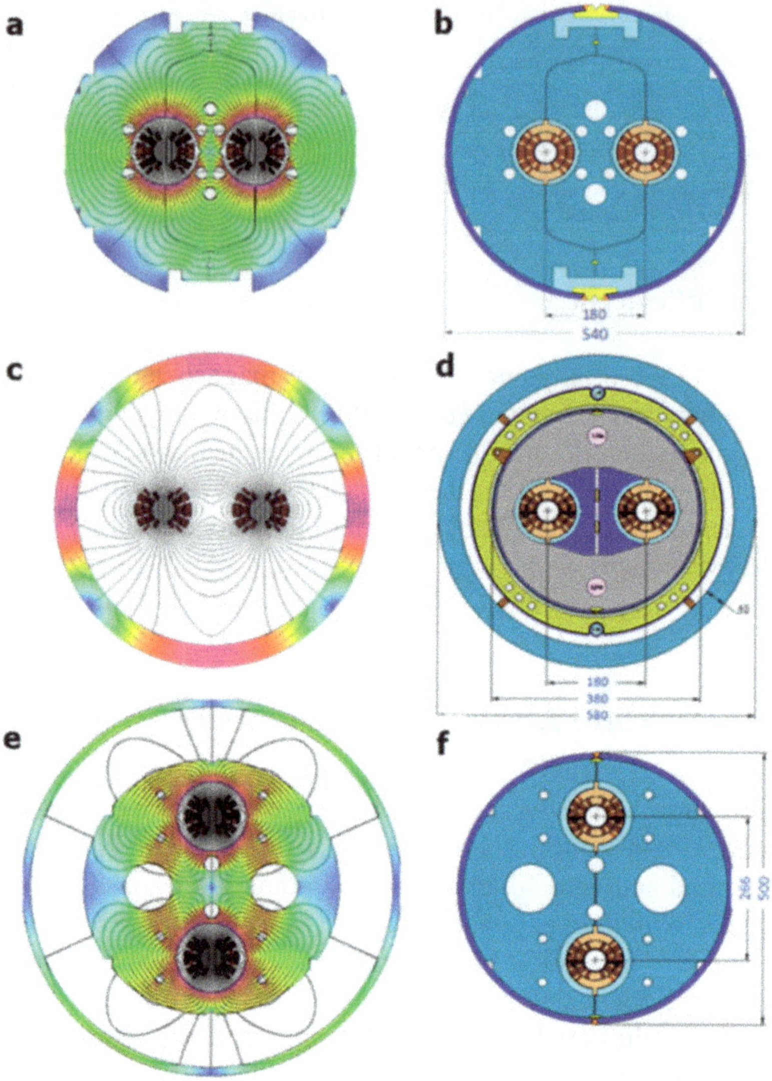

Fig. 7.5 Twin-aperture dipole magnets based on the cos-theta Nb_3Sn coil. (**a**, **b**, **e**, **f**) Cold; and (**c**, **d**) warm iron. (**a**, **b**, **c**, **d**) Horizontal; and (**e**, **f**) vertical bore arrangement. (Kashikhin and Zlobin 2001)

large quadrupole field component. It is cancelled in each aperture by the geometrical quadrupole component produced by the small left–right asymmetry in the coil block position. To reduce the force imbalance and the effect on the field quality, the warm iron design requires a proper alignment of the cold mass inside the iron. Analysis shows that these alignment requirements can be easily met.

In the design shown in Figs. 7.5e, f, the two coils are placed vertically inside the cold iron. To reduce the negative magnetic coupling, which reduces the dipole field in each aperture, the aperture separation should be at least 266 mm. To reduce the size of the cold mass, the iron is divided into two parts—cold and warm. The iron cold part is vertically split into two pieces for coil assembly and preload. To provide an adequate coil pre-stress and support in this design, a 20 mm thick stainless-steel skin is needed. Since the horizontal components of the Lorentz force in each coil are added, the skin in this design is a factor of two thicker than in the design in Figs. 7.5a, b. To minimize the pre-stress decrease after cool-down, the cold iron gap remains open. The warm part of the iron is properly distanced to accommodate the cryostat elements as in the design in Figs. 7.5c, d. The iron saturation effect in this design is corrected by using holes in the cold iron, and optimizing the inner and outer radii of the cold and warm iron parts.

The mechanical analysis shows that in all the above designs the coil is under compression and the stress is always less than 150 MPa, which is acceptable for brittle Nb_3Sn cable. The coil bore deformations are less than 0.10 mm, and all structural elements work in elastic mode.

The main parameters of the dipole magnets described above are summarized in Table 7.4. The calculated geometrical harmonics are small; they are shown in Table 7.5.

The maximum bore field and quench current were calculated for Nb_3Sn wires with a Cu/non-Cu ratio of 0.85 and a critical current density J_c(12 T, 4.2 K) of 2 kA/mm^2 assuming a 10% critical current degradation due to wire plastic deformation during cabling. These designs could provide a nominal field of

Table 7.4 Calculated magnet parameters

	Bore arrangement		
	Horizontal		Vertical
Yoke design	Cold	Warm	Cold/warm
Magnet aperture (mm)	43.5	43.5	43.5
Aperture separation (mm)	180	180	266
Iron yoke diameter (mm)	520	580	564/710
Iron yoke area (cm^2)	1722	679	1378/327
Maximum bore field (T)	12.05	11.34	11.59
Maximum quench current (kA)	21.9	24.0	22.3
Stored energy at 11 T (kJ/m)	520	588	554
Inductance at 11 T (mH/m)	2.68	2.16	2.46

Table 7.5 Geometrical harmonics b_n at R_{ref} = 10 mm (10^{-4})

	Horizontal		Vertical
Harmonic number	Cold yoke	Warm yoke	Cold yoke
2	–	0.000	–
3	0.000	0.000	0.000
4	–	0.000	–
5	0.000	0.001	−0.000
6	–	−0.012	–
7	0.000	−0.011	−0.006
8	–	0.031	–
9	−0.091	−0.130	−0.067

10–11 T with a margin within 10–15% using Nb_3Sn wires with J_c(12 T, 4.2 K) of 3 kA/mm^2 and a Cu/non-Cu ratio of 1.2 to ensure quench protection. Nb_3Sn composite wires with such a high J_c became available around 2005 from Oxford Superconductor Technologies (OST) (Parrell et al. 2005). Notice that the magnet design with a warm yoke allows for a substantial reduction in size without noticeable decrease in performance.

7.4 Fabrication Technology

Due to the small bending radii defined by the size of the inner-layer pole block, the coil manufacturing technology requires using the wind-and-react (W&R) method. In this approach the superconducting Nb_3Sn phase is formed after coil winding during its high temperature heat treatment.

The W&R technique imposes demanding requirements upon the coil components, which must survive a long heat treatment at a high temperature of ~700 °C, in the presence of thermo-mechanical stresses. Despite the relatively high cost, ceramic insulation was selected for the dipole models of this series. This insulation, developed by CTD in the late 1990s, did not use organic ingredients and showed excellent mechanical and electrical properties before and after the high-temperature heat treatment (Rice et al. 1999; Chichili et al. 2000). Alternative, less expensive S2-glass tape, traditionally used in Nb_3Sn magnets, was initially declined, primarily due to the presence of organic sizing, which involved additional processing. Later S2-glass tapes without organic components were successfully used in short and long coils of this series. To reinforce insulation after reaction, Nb_3Sn coils are impregnated with epoxy.

Cos-theta coils use complex 3D end parts. In the case of the W&R approach they also must withstand the coil heat treatment without noticeable deformations, which could cause shorts to the coils. An optimization method to design 3D metallic end parts was developed and successfully tested at FNAL (Yadav et al. 2001). Newly developed rapid prototyping techniques were used for the first time to reduce the time and cost of end part design. Emerging technologies, such as water-jet cutting, which reduced the end part fabrication cost by a factor of 2 and the processing time by a factor of 10, were also successfully used (Zlobin et al. 2005).

New features were introduced in the magnet fabrication process to simplify coil manufacturing and handling, and to reduce magnet production cost.

- To increase the cable mechanical stability and reduce the risk of turn-to-turn shorts during winding, the ceramic tape was impregnated with a liquid ceramic binder and pre-cured at 120 °C.
- To obtain a solid coil structure and the desired coil size, each coil layer was impregnated again after winding with a ceramic binder and cured at 120 °C in a precise mold.

- To reduce coil fabrication time and allow easy coil handling and warm field quality control before magnet assembly, two coils were assembled, reacted, and epoxy impregnated to form a thick, round, solid pipe.
- To reduce assembly time and costs, expensive collars and delicate and time-consuming collaring procedures were eliminated in this design. Later, safe collaring procedures for brittle Nb_3Sn coils were also developed and successfully demonstrated at FNAL (Bossert et al. 2010), and used in the 11 T dipole for the LHC upgrades (see Chap. 8).

The main steps of the developed coil fabrication technology, magnet assembly, and coil preload were verified using mechanical and technological models (Andreev et al. 2000; Chichili et al. 2001). This approach proved to be very useful in complicated accelerator magnet R&D programs.

7.4.1 Mechanical Model

A 200 mm long mechanical model was assembled and tested to verify the results of the magnet mechanical analysis and to choose the most appropriate coil shim plan. For this purpose a special coil was wound, cured, reacted, and epoxy impregnated. The Nb_3Sn cable for this coil was made from IT composite wire developed for the International Thermonuclear Experimental Reactor and insulated with S-2 glass. The coil straight section was cut into two halves, covered with spacers, and assembled inside the iron blocks. The assembly was clamped with aluminum clamps in a press and preloaded by a welded skin (Fig. 7.6). The mechanical model was heavily instrumented with strain gauges to monitor stress evolution during coil clamping, skin welding, and model cooling-down with liquid nitrogen to a temperature of 77 K.

Table 7.6 presents the strain gauge data at various stages of the model assembly. The experimental data are in good agreement with the FEA predictions shown in parentheses. The outer diameter of the epoxy impregnated coil "pipe" was 0.25 mm

Fig. 7.6 Instrumented short mechanical model. (Andreev et al. 2000)

Table 7.6 Data from the mechanical model and FEA

	Coil	Spacer	
	Pole (MPa)	Mid-plane (MPa)	Pole (MPa)
Yoking in press	154 (145)	88 (122)	152 (156)
After spring-back	32 (40)	40 (43)	51 (50)
After skin welding	66 (72)	68 (65)	84 (81)
At 77 K	61 (73)	46 (45)	68 (56)

FEA predictions are shown in parentheses

larger than the nominal coil size, leading to high stresses in the coil during yoking in press. After spring-back, the stress in the coil decreased to ~32 MPa. The final coil pre-stress was provided by the skin through weld shrinkage, achieving the required pre-stress within 10%.

7.4.2 *Technological Model (HFDA01)*

The main goal of the 1 m long technological model was to check the tooling, optimize the coil fabrication and magnet assembly processes, and develop the instrumentation and quality control procedures. In the case of successful assembly, it was expected to proceed with cryogenic testing of this model.

7.4.2.1 Strand and Cable

The Rutherford cable used in this model had 28 strands, each of 1.0 mm diameter, a 0.025 mm thick stainless-steel core, a width of 14.23 mm, a mid-thickness of 1.82 mm, and a keystone angle of 0.927°. The cable was made at LBNL. The Nb_3Sn composite wire was produced by OST using the modified jelly roll (MJR) process. The round wire had a filament diameter of ~0.115 mm and a Cu/non-Cu ratio of 0.92. The measured critical current density at 12 T and 4.2 K for the round wire was ~2000 A/mm^2 and the RRR was ~30. The I_c degradation due to cabling, measured using round wires and extracted strands, was less than 10% for this cable, which had a quite large packing factor of 89%. The quality of the cable was not completely satisfactory. It was found that at several locations the core stuck between the strands, although it did not protrude out of the cable surface, except for one location.

The cable was cleaned with ABZOL VG solvent and wrapped with CTD-CF100 ceramic tape with a 50% overlap. The nominal thickness of the cable insulation was 0.25 mm. Inorganic CTD-1002x ceramic binder was applied to the insulated cable using rollers. The entire spool of wet insulated cable was cured at 80 °C for about 20 min to achieve a strong cable insulation system.

7.4.2.2 Coil Winding and Curing

Both coil layers were wound using a single cable piece. Coil end parts, wedges, and pole blocks were made of aluminum-silicon bronze C642. The end parts were fabricated at LBNL using five-axis computerized numerical control machining and coated with a 0.125 mm thick ceramic layer.

Small axial gaps were introduced between the pole blocks and between the wedges and end parts to account for differential thermal expansion during reaction in the axial direction. Voltage taps made of thin stainless-steel strips were inserted between the cable and the insulation during winding around the end parts. After the winding of the inner coil, the ceramic binder was applied and the wet coil was prepared for curing. After preliminary cycling at low pressures, a final azimuthal pressure of 45 MPa and a radial pressure of 10 MPa were applied to achieve the target coil geometry. The curing temperature of 150 °C was held for 30 min to provide good turn bonding.

The inter-layer insulation was made of three layers of 0.125 mm thick ceramic cloth. The middle layer contained two quench protection heaters made of 0.025 mm thick stainless-steel strips (Fig. 7.7). The insulation assembly was impregnated with ceramic binder and cured in a special fixture. The coil outer layer was wound on top of the cured inner layer, which was covered by the inter-layer insulation. The outer layer was then filled with ceramic binder and cured along with the previously cured inner layer. A cured coil for the technological model is shown in Fig. 7.8.

The azimuthal size of the cured coils was measured at four locations along the coil straight section at an azimuthal pressure of 2 MPa. The average azimuthal size for the first coil was 0.2 mm over its nominal size. To correct the coil size of the second coil, the wedge insulation was changed from 50% overlap to butt lap. The average azimuthal size of the second coil at 2 MPa was only 0.01 mm over the nominal coil size. The average measured azimuthal modulus of elasticity for the cured coils was about 20 GPa. Electrical measurements, taken on the coils to check for turn-to-turn shorts, did not detect any problems at that stage.

Fig. 7.7 The middle layer of the inter-layer insulation with strip-heaters. (Chichili et al. 2001)

Fig. 7.8 Dipole coil with coated parts after impregnation with ceramic binder and curing in a precise mold. (Chichili et al. 2001)

7.4.2.3 Coil Reaction

Two coils with ground insulation were mounted around a stainless-steel mandrel and placed in the reaction fixture. The ground insulation consisted of three layers of 0.125 mm thick ceramic cloth preformed into shape using the ceramic binder. The cable ends were welded to prevent tin leakage during reaction. The retort was pumped for several hours and then purged with argon with a flow rate of ~1 cm^3/s. The reaction cycle had two steps: 575 °C for 200 h followed by 700 °C for 40 h with a ramp rate of 25 °C/h. The coil temperature was monitored at several locations throughout the reaction. The difference between the coil and the furnace temperature above 500 °C was less than 2 °C.

Both mechanical and electrical measurements were also taken on the coils after reaction. It was found that after reaction the coils expanded axially by 9 mm (!) whereas the length of an annealed MJR cable in free state expanded by only 1 mm/m. This unexpectedly large increase of the coil length was related to the restricted radial and free axial coil expansion in the reaction fixture.

The azimuthal coil size was measured using a reacted practice coil. It was found that it increased after reaction by about 0.5 mm at an applied pressure of ~15 MPa. Therefore, the size of the next coils during curing was reduced by 0.5 mm to allow the coils to grow azimuthally to the nominal size during reaction, thereby avoiding high pressure in the mold and coil axial extrusion.

Tin leaks were found in both coils after reaction. This defect was attributed to the lack of a low temperature step at 200–210 °C, when tin diffuses as a solid phase, as well as to the high coil compression during reaction. Even though the coils were clearly damaged, magnet fabrication was continued to study all aspects of the assembly procedure.

7.4.2.4 Splice Joints

Brittle Nb_3Sn coil leads were spliced to flexible Nb-Ti cables using special tooling and Pb-30Sn solder with Kester 44 Rosin Flux. Splicing was performed while coils were still inside the reaction fixture. Each Nb_3Sn lead was placed between two Nb-Ti cables and encased in a 0.55 mm thick, 125 mm long copper cage with a copper top plate. The Nb_3Sn leads and the Nb-Ti cables were always supported to protect the brittle Nb_3Sn cable from any possible damage. Both splice joints were heated to ~230 °C by heaters placed in tooling slots. The bolts on the tooling were tightened with the increase in temperature until the right splice geometry was obtained.

7.4.2.5 Epoxy Impregnation

Two coils were impregnated together with CTD-101K epoxy at a temperature of 60 °C. The stainless-steel mandrel used during reaction was replaced with a Teflon mandrel. The mandrel diameter was chosen such that at the curing temperature it expanded to fill the coil aperture. Thanks to the large thermal expansion of Teflon, after impregnation it was easily removed from the coil aperture. A 0.125 mm thick mold-released polyimide film was placed between and around the coils in the impregnation fixture. In principle, this polyimide layer allowed splitting the coils after impregnation. Pictures of impregnated coils are shown in Fig. 7.9.

7.4.2.6 Model Assembly and Instrumentation

The impregnated coil assembly and various components of the magnet support structure were instrumented with gauges to measure stress values during magnet assembly and operation. To measure the azimuthal stress, four capacitance gauges were installed between the outer pole blocks and the coil. Capacitance gauges were

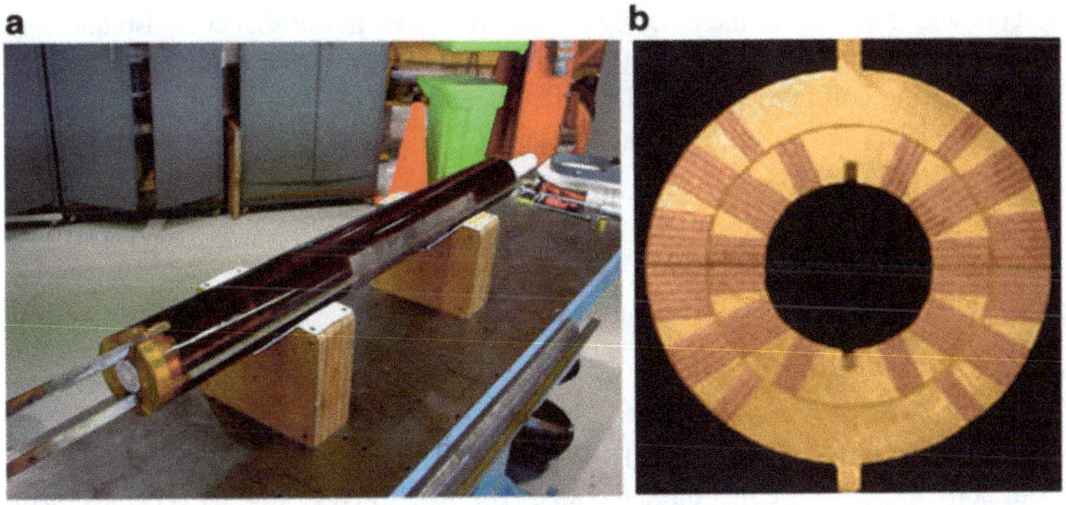

Fig. 7.9 (**a**) Epoxy impregnated coil assembly with Teflon mandrel in the aperture; and (**b**) cross-section of the impregnated coil. (Chichili et al. 2001)

Fig. 7.10 Coil-yoke assembly inside the yoking and clamping tooling ready for installation in the vertical press. (Chichili et al. 2001)

fabricated in-house and calibrated at room and helium temperatures. Resistive strain gauges were installed on spacers, clamps, and the skin.

The coils and yoke were assembled in a dedicated tooling and then compressed in a vertical press (Fig. 7.10). The press load was increased incrementally until the gap between the yoke halves, which was controlled by dial indicators, reached the nominal value. At this pressure the aluminum clamps were easily inserted using a separate set of pusher blocks. Then, the press load was released and the assembly sprang back to an equilibrium position. The measured maximum stress in the pole region of the coil outer layer was about 100 MPa, while the stress in the coil–yoke spacers was 200 MPa in the pole region and 100 MPa in the spacer mid-plane.

At this stage the assembly of the technological model was complete. Lessons learned during coil fabrication and magnet assembly were implemented in the next dipole models.

7.5 Short Models

7.5.1 Short Dipole Models

Six 1 m long dipole models of the HFDA series were fabricated and tested at FNAL from 2001 to 2005. The first three models, HFDA02 to HFDA04, used cables made from MJR Nb_3Sn composite wire produced by OST. The next three models,

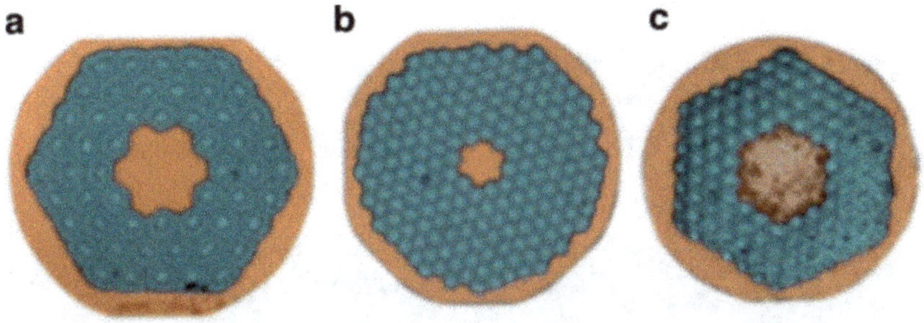

Fig. 7.11 Cross-sections of Nb_3Sn round wires: (**a**) MJR54/61; (**b**) PIT192; and (**c**) RRP108/127. (Zlobin 2011)

HFDA05 to HFDA07, used cables made from powder-in-tube (PIT) Nb_3Sn wire produced by Shape Metal Innovation. The MJR round wire had a higher critical current density J_c(12 T, 4.2 K) of 2.0–2.2 kA/mm^2, but a larger sub-element diameter D_{eff} of ~0.100 mm, whereas the PIT round wire had a lower J_c of ~1.6–1.8 kA/mm^2 and a smaller D_{eff} of ~0.050 mm.

In the early 2000s, a new improved R&D composite wire based on the restack rod process (RRP) was developed by OST (Parrell et al. 2005). This wire had a high J_c(12 T, 4.2 K) up to 3 kA/mm^2, and a larger number of smaller sub-elements, which significantly improved the stability of this wire with respect to "flux jumps." The cross-sections of the round Nb_3Sn wires used in the HFDA dipole models are shown in Fig. 7.11.

The cable in HFDA01 to HFDA03 had a 0.025 mm thick stainless-steel core to control the inter-strand resistance, while the cable in HFDA04 to HFDA07 had no core. Cables used in HFDA01–05 had 28 strands; the number of strand was then reduced to 27 to optimize the cable packing factor. Cross-sections of the 28-strand and 27-strand Rutherford cables without core made from PIT composite wire are shown in Fig. 7.12.

The cable for the first eleven 1 m long coils was produced at LBNL. The cable for 1 m long coils 12–19, for the 2 m long coil 20, and for the 4 m long coil 21 was made in-house using FNAL cabling machine. The cable for coils 1–11 was cleaned with ABZOL VG to remove any oil residue left from the cabling process. Later, it was found that cable cleaning does not affect coil properties, thus cable cleaning for the subsequent coils was not used.

The main features of the 1 m long dipole models HFDA02–07 are summarized in Table 7.7. The magnet design and fabrication procedures were similar to those used in the technological model HFDA01. Some design and technological changes were introduced throughout the short model R&D to address the problems found during the technological model fabrication, and to incorporate the feedback from dipole model fabrication and tests. The most important changes are described below.

Fig. 7.12 Cross-sections of (a) 28-strand and (b) 27-strand cables

Table 7.7 HFDA model features

Magnet HFDA	Coil number	Strand type	Cable core	Coil ends	Coil impregnation	Skin design
02	3, 4	MJR	Yes	v. 1	Glued	Welded
03	5, 6	MJR	Yes	v. 1	Glued	Welded
04	7, 8	MJR	No	v. 2	Glued	Welded
05	12, 13	PIT	No	v. 2	Separate	Bolted
06	14, 15	PIT	No	v. 2	Separate	Bolted
07	12, 14	PIT	No	v. 2	Separate	Bolted

7.5.1.1 HFDA02

The cable insulation had 30% overlap in coil 3 and 40% in coil 4. The coil azimuthal size during curing was reduced by 0.5 mm with respect to the nominal coil size, such that after reaction the coils were at the nominal size without excessive pressure on the conductor during reaction. Coil 3 was about 0.2 mm larger than coil 4 due to the difference in mid-thickness of the used bare cable.

The coil end parts (v. 2) were re-optimized to better match the cable positions in the end areas (Fig. 7.13). The end parts were produced using water-jet machining, which is more cost-effective than five-axis CNC machining. No end-part coating was used in this and the following magnets.

The coil reaction cycle was revised to avoid the tin leaks observed in the technological model coils. The new heat treatment cycle included three steps: 210 °C for 100 h, 340 °C for 48 h, and 650 °C for 180 h. This heat treatment schedule was used for all the short models made of the MJR cables.

Ground insulation was made of two layers of 0.25 mm thick ceramic cloth. Since it was found that the heater strips between the two coil layers in the technological model had been jammed during coil processing, which increased the risk of heater-to-coil shorts, quench protection heaters in this and the next models of this series were installed between the two layers of ground insulation. Voltage taps were installed only on the coil leads.

Fig. 7.13 Longitudinal sections of the coil lead end with (**a**) first; and (**b**) second generation end parts

7.5.1.2 HFDA03

The HFDA03 fabrication process was similar to that of HFDA02. The two coils in this magnet had virtually the same azimuthal size. To provide more flexibility in adjusting the tooling and to improve splice quality, a new splicing fixture was made to splice each Nb_3Sn lead individually. The splicing procedure was adjusted to eliminate possible strand damage, which was believed to be one of the main reasons for the poor quench performance of HFDA02. To provide visual control of the splice quality, copper boxes were not used.

The ground insulation consisted of three layers of 0.125 mm thick ceramic cloth. Strip heaters were weaved (see Fig. 7.7) into the insulation middle layer. Additional voltage taps were placed on the outer layer of each coil, adjacent to the layer jump. Special holes were made in the iron yoke to correct the iron saturation effect in the normal sextupole b_3. This iron yoke was used for all following dipole models.

7.5.1.3 HFDA04

Based on the HFDA03 test results, the coil lead end was redesigned to keep the coil lead cables in the mid-plane (Fig. 7.14). The entire splice joint was placed in the lead end saddle to ensure reliable splice support during splicing and at all stages of magnet assembly and operation. The length of the coil straight section in this and the next short models was reduced by ~200 mm to house the lead splices within the coil end saddle and still use the same tooling. To allow better control of the coil mid-plane position and provide the possibility of splicing each coil lead individually, the two coils were separated in the reaction fixture by a stainless-steel plate.

a

b

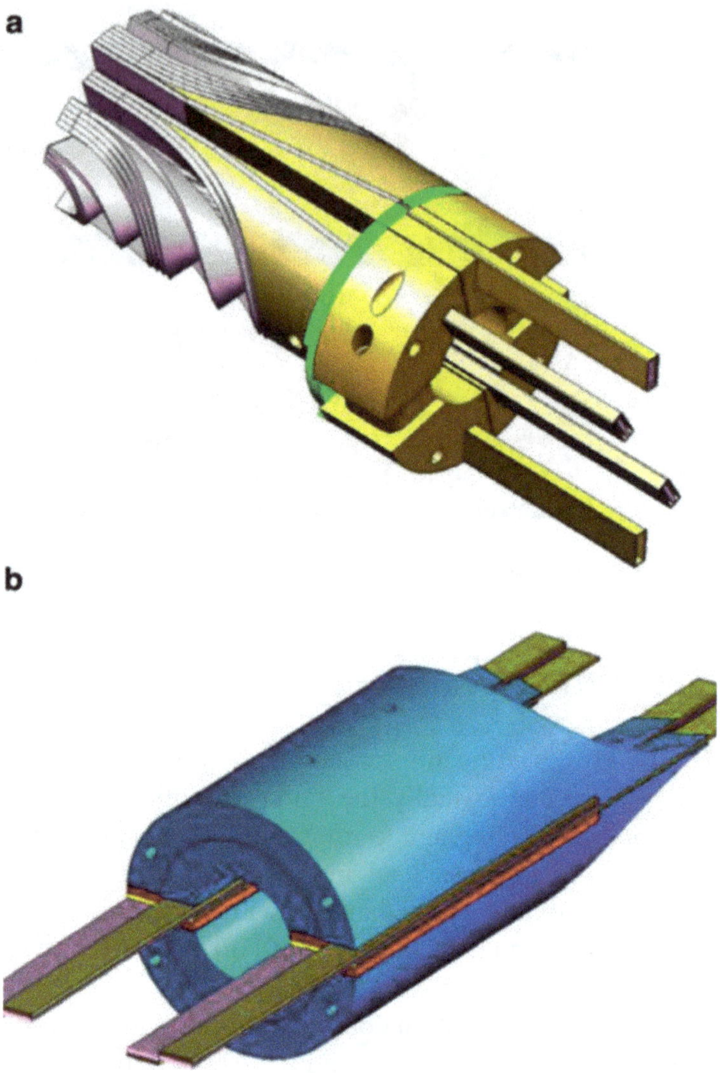

Fig. 7.14 Coil end design: (**a**) version 1; and (**b**) version 2 with straight coil leads

Coil 7 was impregnated first alone, and then coil 8 was impregnated together with coil 7. The first step ensured an accurate mid-plane position and the second step provided proper matching of the two coils in the mid-plane areas.

Coil alignment inside the iron yoke was done using only one outer pole block with an extension. Special spacers were used to prevent large variations of the azimuthal stress in the left and right sides of the coil during yoking. The stainless-steel end blocks extended to cover part of the splice block, thus moving the stress discontinuity away from the Nb_3Sn lead.

HFDA04 was the last model that used the cable with MJR strands. By that time, it had been recognized and experimentally confirmed that strong flux jump instabilities in these strands with a large Nb_3Sn sub-element size were the main cause of the poor quench performance of HFDA02–04 (Zlobin et al. 2005).

7.5.1.4 HFDA05

The HFDA05 dipole was similar to HFDA04, except for the cable, which was made of PIT Nb_3Sn composite wires. Before insulation, the cable was annealed at 200 °C for 30 min to reduce residual stresses accumulated during cable manufacturing. The cable was insulated using a new insulation type. It was wrapped with two layers of 0.125 mm thick ceramic tape. The first layer was dry ceramic tape wrapped with a 0.75 mm gap. The second layer consisted of the same ceramic tape impregnated by the manufacturer with CTD-1008 binder. It was also wrapped with 0.75 mm gaps overlapping the gaps from the first layer.

The reaction cycle was adapted for the PIT wire and had only one step with a temperature ramp rate of 25 °C/h to a reaction temperature of 655 °C for 170 h. Both coils were reacted and impregnated separately to allow testing of the first PIT coil in a dipole mirror structure (see Sect. 7.5.1.7).

None of the outer pole blocks had alignment extensions. The coil alignment inside the yoke was done by using "scale" measurements. To eliminate large differences in azimuthal stress between the left and right sides of the coil assembly, special spacers were used, as in HFDA04. Unlike previous HFDA models, the HFDA05 yoke gap was, by design, closed after cool-down.

7.5.1.5 HFDA06

The cable type and preparation procedures were the same as in HFDA05. The cable was wrapped with 0.75 mm gaps with two layers of 0.125 mm thick and 12 mm wide ceramic tape pre-impregnated with binder. The outer layer was wrapped to overlap the gaps in the inner layer.

Each coil was impregnated with binder and cured in a closed cavity mold at 150 °C for 0.5 h with a 0.125 mm azimuthal polyimide shim in the mid-plane.

Coils 14 and 15 were reacted individually in a single-coil reaction fixture. The intended reaction cycle for both coils had three steps: 210 °C for 100 h, 331 °C for 48 h, and 675 °C for 64 h (coil 14) or 100 h (coil 15). The reaction of coil 14 was interrupted during ramping at 331 °C and then restarted several times due to malfunctioning of the oven. Nonetheless, coil 14 was the best performing PIT coil. Coils 14 and 15 were impregnated with epoxy separately after splicing with Nb-Ti leads in the same fixture that was used for reaction.

7.5.1.6 HFDA07

This magnet was assembled using the best coils, 12 and 14, previously tested in HFDA05 and HFDA06.

7.5.1.7 Dipole Mirror Magnets

Single dipole coils were tested using a special coil test structure (CTS), also known as dipole mirror, under operating conditions similar to those of real magnets (Chichili et al. 2004). CTS noticeably reduced the turnaround time of coil fabrication and evaluation, as well as material and labor costs. The dipole mirror used the same mechanical structures and assembly procedures as the complete dipole magnets, and allowed advanced instrumentation to be used.

The HFDM dipole mirror is shown in Fig. 7.15. The mirror's mechanical structure is similar to the HFDA dipole structure, except for the iron yoke, which is split horizontally, and one of the two coils is substituted with half-cylinder iron blocks. The transverse coil pre-stress and support are provided in the same way as in the dipoles by a combination of aluminum clamps and a bolted stainless-steel skin.

The main parameters of the coils tested in the dipole mirror structure are summarized in Table 7.8. The most important details of the mirror magnets are described below.

7.5.1.8 HFDM01

The first mirror magnet HFDM01A (also called HFDA03A) had two main goals—to test a mirror magnet structure with bolted skin, and to study the effect of lead splices on the magnet quench performance. The magnet used coil 5, which was tested previously in HFDA03. The goal of the HFDM01B test (also called HFDA03B) was to assess the splice joints in a configuration similar to a real magnet, because the

Fig. 7.15 Return end of short dipole mirror structure. The coil is paired with an iron semi-cylinder inside the iron yoke and the bolted skin half-shell. (Zlobin 2011)

Table 7.8 Dipole mirror design features

Mirror magnet	Coil number	Coil length (m)	Wire type	Wire diameter (mm)	Skin type
HFDM01	5	1	MJR54/61	1	Bolted
HFDM02	10	1	MJR54/61	1	Bolted
HFDM03	12	1	PIT192	1	Bolted
HFDM04	16	1	RRP54/61	0.7	Bolted
HFDM05	17	1	RRP54/61	0.7	Bolted
HFDM06	19	1	RRP108/127	1	Bolted
LM01	20	2	PIT192	1	Welded
LM02	21	4	RRP108/127	1	Welded

splices were initially suspected as the main cause of poor quench performance of the first HFDA dipole models. Additionally, HFDM01B allowed checking whether the conductor could carry currents above 20 kA under conditions similar to those found in a magnet. For this purpose, the inner- and outer-layer mid-plane turns on each side of coil 5 were cut from the rest of the coil, spliced to flexible Nb-Ti cables, and connected in series.

7.5.1.9 HFDM02

HFDM02 was fabricated and tested with a new coil 10, made from 28-strand MJR cable. The cable was insulated for the first time with pre-impregnated ceramic tape, which would later be used in HFDA05–07. Each end saddle had holes underneath the lead splice joints to improve splice cooling. The coil was reacted without the inner mandrel to allow for cable expansion inside the bore. This approach was chosen to reduce possible cable stress/strain degradation during reaction and improve gas removal from the coil before and during reaction. This trial coil demonstrated poor quench performance similarly to previously tested MJR coils. Therefore, it was not used in the next HFDA magnets.

7.5.1.10 HFDM03

Coil 12, the first coil made of PIT cable, was tested first in a dipole mirror configuration to verify the effect of cable stability on the magnet performance. For the coil design and fabrication features, see HFDA05 described above.

7.5.1.11 HFDM04, HFDM05

These two mirror magnets were fabricated and tested to increase the dipole field to 11–12 T by using the newly developed high-J_c RRP strands. The cable had 39 strands, each of 0.7 mm diameter, a width of 14.34 mm, a mid-thickness of

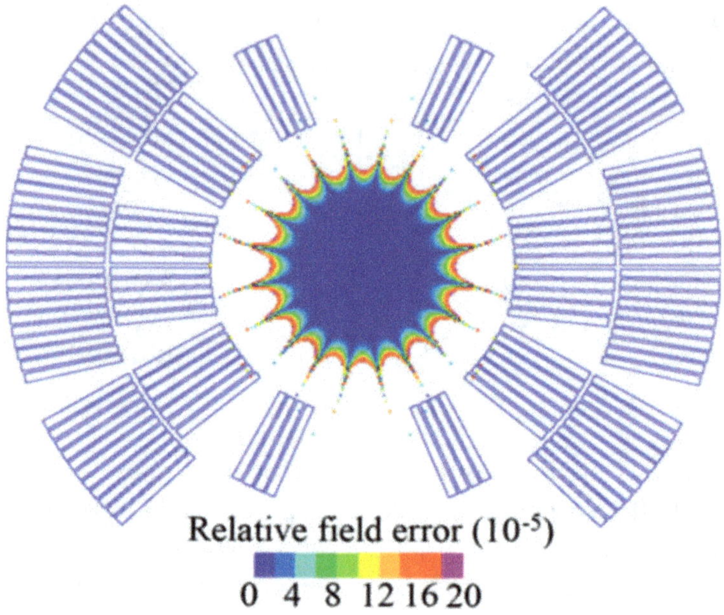

Fig. 7.16 Coil cross-section based on 39-strand cable

1.258 mm, and a keystone angle of 0.972°. The cable for HFDM04 was produced at LBNL. The cable for HFM05 was made in two steps. First, the rectangular cable was made at LBNL. Then it was annealed and re-rolled to its final keystone geometry at FNAL. A high-J_c 0.7 mm diameter RRP54/61 strand with a sub-element size of ~0.085 mm was used to mitigate the instability problems and thereby increase the achievable field. The coil cross-section was modified for this thinner 39-strand cable without changing the coil layer width and outer coil radius (Fig. 7.16). Three coils of this design were fabricated and two were tested in a mirror configuration.

7.5.1.12 HFDM06

This mirror magnet used the coil made of RRP108/127 Nb_3Sn strand 1 mm in diameter with a larger number of sub-elements and increased sub-element spacing. To reduce the strand deformation on the cable edges, the number of strands in the cable was reduced from 28 to 27. The Rutherford cable was made at FNAL in two steps. First a low-compaction rectangular cable was produced, and it was then re-rolled to its final keystoned cross-section, after a short intermediate annealing of the rectangular cable at 190 °C in air.

7.5.1.13 LM1 (HFDM07)

This mirror magnet was used to test the first 2 m long coil 12, which was made of 27-strand PIT cable. The coil fabrication procedure was similar to coils 14 and 15. The magnet assembly was similar to HFDM03.

7.5.1.14 LM2 (HFDM08)

This mirror magnet was used to test the first 4 m long Nb_3Sn dipole coil as part of the Nb_3Sn coil technology scale-up program. The coil was made of 27-strand Rutherford cable with 1 mm Nb_3Sn RRP108/127 strands. The total cable length in the coil was about 166 m. The coil design and the magnet fabrication procedure were similar to HFDM06.

7.6 Dipole Model Tests

Six short dipoles of the HFDA series were built and tested at FNAL from 2002 to 2006. It was the first in the world series of nearly identical Nb_3Sn accelerator magnets, which provided the first data on magnet quench performance, field quality, and especially on the reproducibility of magnet technology and performance.

7.6.1 Quench Performance

The HFDA dipole models were tested in liquid helium at 4.5 K and at lower temperatures. Quench performance of HFDA02 to HFDA07 is shown in Fig. 7.17. The first three models HFDA02 to HFDA04, made of the MJR wire, were limited by

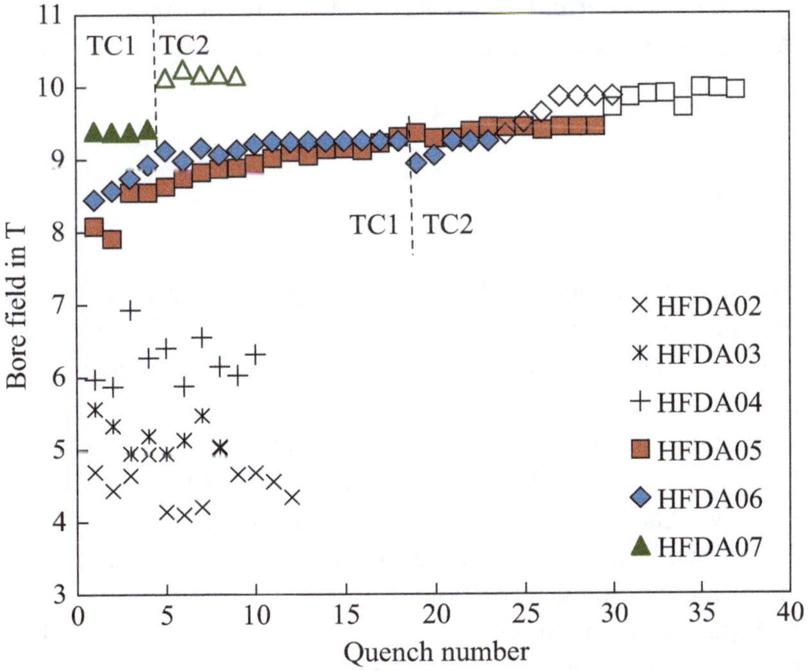

Fig. 7.17 HFDA02–07 training at 4.5 K (solid markers) and 2.2 K (open markers). TC1/TC2 represents thermal cycles

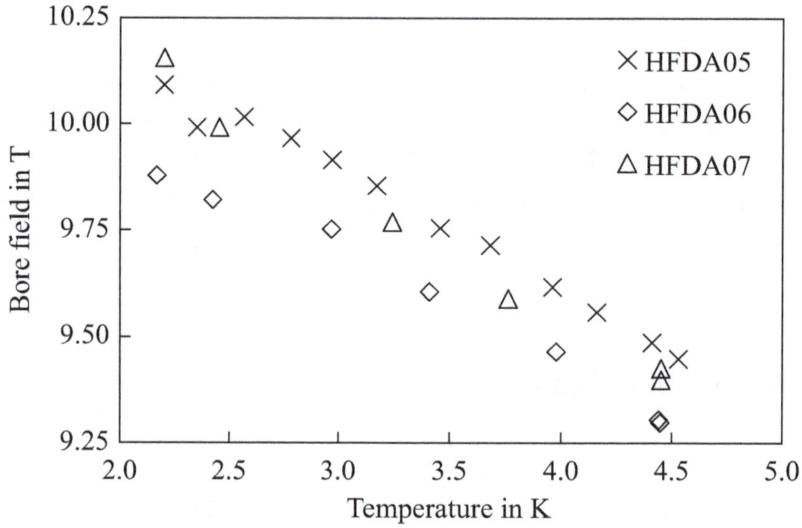

Fig. 7.18 Maximum bore field vs. coil temperature

flux jumps in the superconductor and only reached 4–7 T (Zlobin et al. 2005). The three last magnets were made of the more stable PIT wire, and at 4.5 K reached their short sample field of 9.3–9.4 T. The field level in these models was limited by the relatively low J_c of the PIT wire. At 2.2 K, the maximum field in PIT models increased to ~10 T thanks to the increase of the superconductor J_c at lower temperatures (Fig. 7.18).

Ramp rate dependences of the HFDA05–07 quench field normalized to the B_{max} value measured at 4.5 K and $dI/dt = 20$ A/s are shown in Fig. 7.19. The shape of the ramp rate dependences at current ramp rates above 125–150 A/s suggests that they are dominated by the large eddy current losses in the cable without a stainless-steel core.

7.6.2 *Field Quality*

The field harmonics were measured at 4.5 K using a 250 mm long (43 mm long in HFDA05) 25 mm diameter probe. The probe had a tangential coil to measure high-order harmonics, as well as dedicated dipole and quadrupole coils to measure low order harmonics.

The transfer function TF, normal sextupole b_3 and decapole b_5 harmonics measured in HFDA07 vs. the bore field are shown in Fig. 7.20.

Analysis shows that the iron saturation effect in the TF is in a good agreement with the calculations. In b_3 the iron saturation effect was minimized at high fields by using special correction holes in the yoke.

Figure 7.21 shows the measured ramp rate sensitivity of the normal sextupole b_3 in HFDA02–07. In this figure the width Δb_3 of the sextupole loop at $B = 2$ T is plotted vs. the current ramp rate (Zlobin et al. 2007). HFDA02 to HFDA04 models

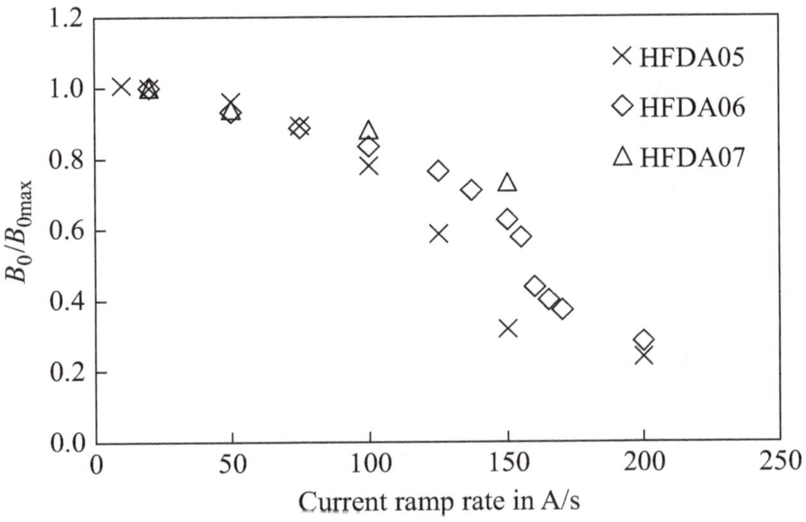

Fig. 7.19 Ramp rate sensitivity of magnet quench field

demonstrated a very small and reproducible eddy current effect, due to large crossover resistances in the cable with stainless-steel core and due to the high resistance of the strand matrix (RRR ~10). The Δb_3 in HFDA05 to HFDA07 rapidly changed with the ramp rate. The negative slope of the Δb_3 ramp rate dependences indicates that the effect was due to the eddy currents in the cable rather than in the strands in spite of the lower copper matrix resistivity in the PIT wires (RRR ~50). These results prove that the eddy current magnetization effect could be suppressed using cored cables and strands with a small twist pitch.

The width of sextupole loops Δb_3 extrapolated to $dI/dt = 0$ corresponds to the persistent current component of a coil magnetization, which is proportional to $J_c \cdot D_{eff}$. The persistent current effect is reproducible in HFDA models made of the same wire type. The b_3 loops in HFDA02 to HFDA04 were larger than in HFDA05 to HFDA07 due to the higher J_c and larger D_{eff} in the MJR wires. These larger values also caused noticeable b_3 fluctuations at low fields in these models, associated with flux jumps in the superconductor (Zlobin et al. 2006). It was realized that the large persistent current effect in Nb_3Sn accelerator magnets cannot be reduced to an acceptable level by reducing the sub-element size because of technological limitations for high-J_c Nb_3Sn wires based on the IT or PIT processes. It was shown, however, that the main component can be compensated using a simple passive correction based on thin iron strips developed and tested at FNAL (Kashikhin et al. 2003).

Measurements of b_3 decay and "snap-back" effects, important for accelerator magnets, were done at a 3 kA plateau for 30 min. It was found that the b_3 decay was very small relative to that seen in Nb-Ti accelerator magnets.

Geometrical field harmonics in HFDA02 to HFDA07 are shown in Table 7.9. They were determined as average values between current up and down ramps at 3 kA. The average values of the low-order geometrical harmonics for HFDA series

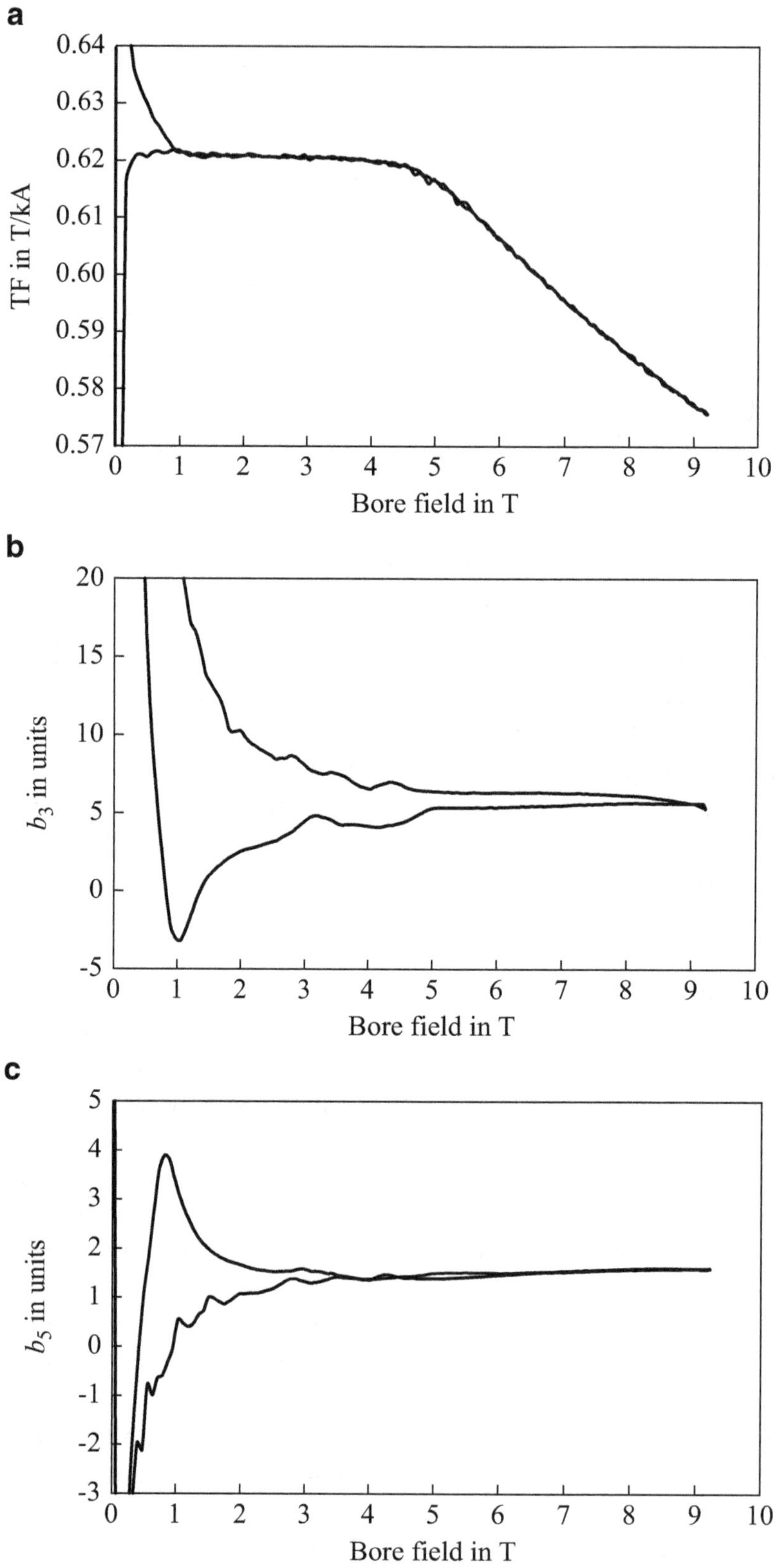

Fig. 7.20 Measured: (**a**) transfer function; (**b**) b_3; and (**c**) b_5 in HFDA07 vs. bore field

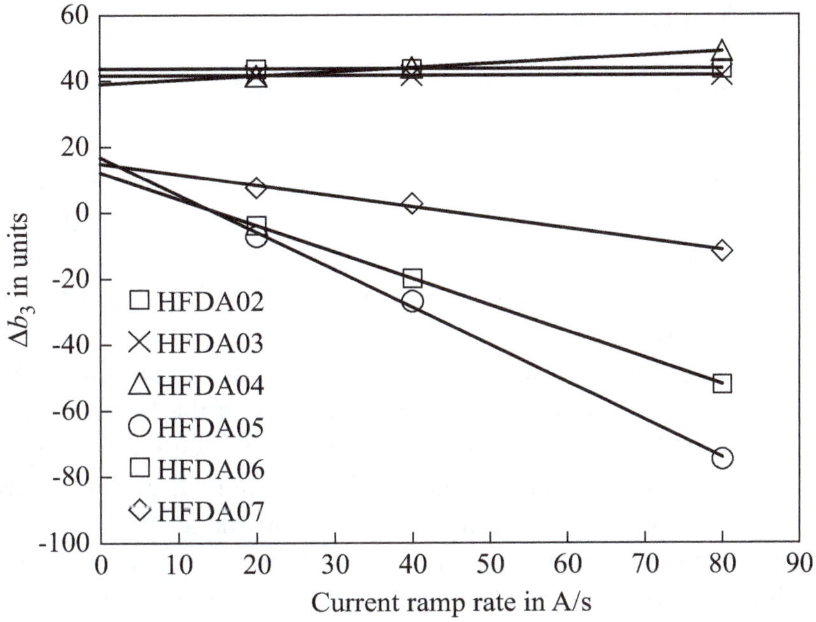

Fig. 7.21 Ramp rate sensitivity of the normal sextupole b_3. (Zlobin et al. 2007)

Table 7.9 HFDA02–07 geometrical field harmonics, 10^{-4}

n	HFDA				SSC-40 mm	
	Average		SD		SD	
	a_n	b_n	a_n	b_n	a_n	b_n
2	−0.37	−0.15	7.3	4.1	2.77	0.79
3	0.55	2.06	0.6	4.3	0.22	1.24
4	−0.73	−0.06	0.9	0.7	0.29	0.15
5	0.17	0.60	0.1	0.9	0.12	0.30
6	−0.04	0.00	0.2	0.2	0.11	0.03
7	0.01	0.20	0.1	0.2	0.03	0.06
9	−0.02	−0.05	0.1	0.1	0.08	0.05

are rather small, below one unit, except for the normal sextupole b_3. Standard deviations (SD) of the measured normal b_2 and skew a_2 gradients, and the normal b_3 sextupole, are relatively large with respect to the other harmonics.

The standard deviations of normal and skew harmonics measured in six HFDA models are compared in Table 7.9 with the results of the first six 40 mm aperture SSC dipole models made with Nb-Ti superconductor (Jackson 1986). The variation of skew harmonics in the Nb_3Sn and Nb-Ti models is rather close. The variation of normal harmonics is larger in the former, since it includes both fabrication errors of coil components and size variations of the shims used to adjust the coil pre-stress.

7.7 Dipole Mirror Tests

7.7.1 *Conductor and Coil Technology Study*

Six short dipole mirror magnets of the HFDM series were built and tested at FNAL between 2002 and 2006. The first tests were performed to validate the mirror structure and to develop and demonstrate the coil technology (HFDM01–02). Then the focus moved towards understanding and improving conductor and magnet flux jump stability (HFDM03–06).

Training data of dipole coils tested using the dipole mirror structure are plotted in Fig. 7.22. The coils made of 1 mm MJR54/61 wire with the largest D_{eff}, and those made with the first high-J_c 0.7 mm RRP54/61 wire with relatively low RRR demonstrated erratic quench performance and large quench current degradation at 4.5 K, as the corresponding dipole models. The coil made of 1 mm PIT192 wire showed stable training performance and reached its short sample limit (SSL) at 4.5 K. A similar performance was later confirmed by the PIT dipole models (Fig. 7.17). The coil with the high-J_c 1 mm RRP108/127 wire reached at 4.5 K the highest quench current, which corresponds to ~97% of its SSL limit. Noticeable variations of quench current at the current plateau, however, pointed to magnetic or, perhaps, mechanical instabilities in the coil.

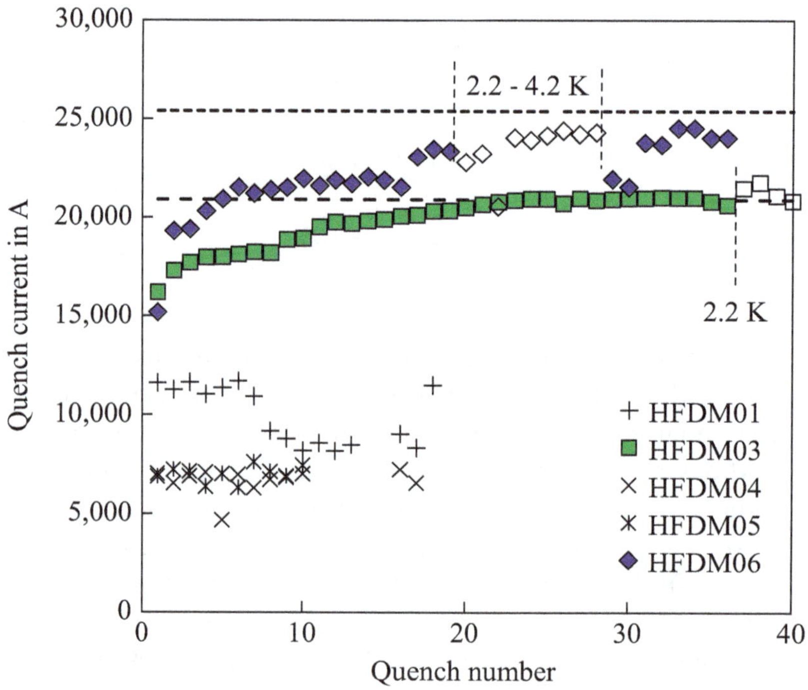

Fig. 7.22 Dipole mirror training at 4.5 K (solid markers) and lower temperatures (open markers). (Zlobin 2011)

7.7.2 *Technology Scale-Up*

Nb_3Sn technology scale-up is a key phase for magnet implementation in accelerators. It addresses problems related to winding, curing, reaction, impregnation, and handling of long Nb_3Sn coils, as well as long magnet assembly and performance. The scale-up was done in two steps, starting in 2006–2007 with fabricating and testing of a 2 m long Nb_3Sn dipole coil made of PIT wire (Zlobin et al. 2007). Then, a year later, the first 4 m long dipole coil made of RRP108/127 Nb_3Sn wire was fabricated and tested (Chlachidze et al. 2009). Pictures of the 4 m long Nb_3Sn coil and dipole mirror are shown in Fig. 7.23.

Training quenches of the 2 m long PIT coil (LM01) and the 4 m long RRP coil (LM02) at 4.5 K are shown in Fig. 7.24, where they are compared with the corresponding 1 m long coils that were also tested in dipole mirror magnets.

The 2 m PIT coil reached its SSL and a field level of 10 T at 4.5 after a short training, similar to the corresponding 1 m long PIT coil tested in dipole mirror HFDM03. The 4 m long RRP coil, unlike its short version, was limited at 4.5 K by strong flux jump instabilities in the coil outer layer (perhaps caused by conductor damage during coil fabrication or magnet assembly). After suppressing these instabilities by heating the coil outer layer using quench heaters, however, it reached ~90% of its SSL at 4.5 K. The maximum quench current was limited by quenches in the coil inner-layer mid-plane turns, which were impacted by the heat flux from the heaters.

7.8 Conclusion

Nb_3Sn accelerator dipole magnets based on shell-type coils and the W&R method were developed at FNAL. The twin-aperture magnet designs with horizontal and vertical aperture arrangements and cold and warm iron yokes met the VLHC technical requirements and offered substantial cost reductions for the collider magnet system.

The R&D program experimentally demonstrated the main magnet parameters (maximum field, quench performance, field quality) and their reproducibility using a series of 1 m long single-aperture models, as well as demonstrated the technology scale-up using longer coils. As part of the technology development, nineteen 1 m long two-layer dipole coils were fabricated and tested in six dipole and six dipole mirror models. The last three dipoles and two mirrors reached their design fields of 10–11 T. All six short dipole models showed good, well-understood, and

Fig. 7.23 (**a**) First in the world 4 m long Nb_3Sn dipole coil. (**b**) Dipole mirror LM02 with 4 m coil prepared for transportation to the FNAL magnet test facility. (Zlobin 2011)

reproducible field quality. The quench performance and field quality data confirmed the good reproducibility and robustness of the magnet design and technology. It was the first time that Nb_3Sn technology scale-up was performed by building and testing

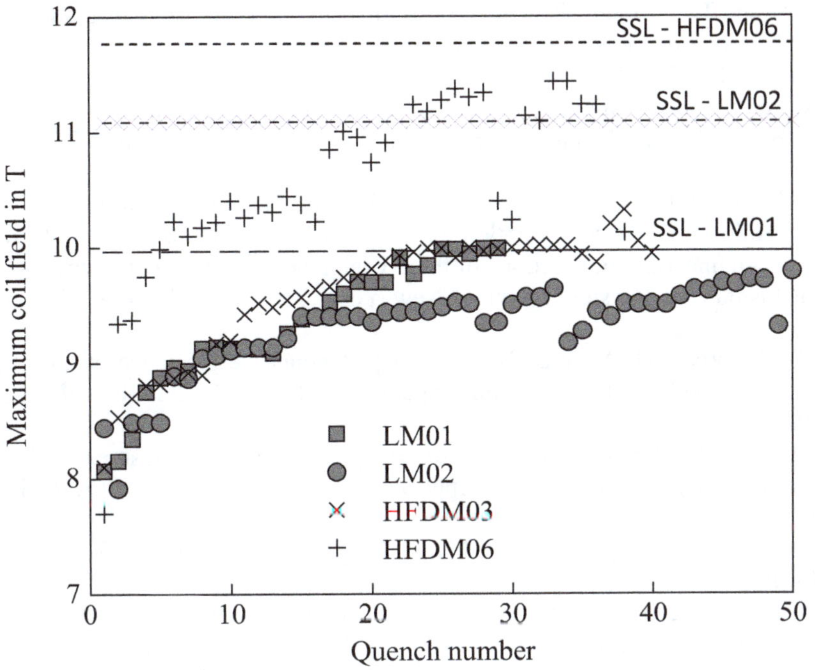

Fig. 7.24 Mirror models training summary at 4.5 K. (Chichili et al. 2004)

2 m and 4 m long dipole coils. The first positive results of the Nb_3Sn technology scale-up phase reinforced high expectations for the practical use of this technology in particle accelerators.

References

Ambrosio G, Kashikhin VV, Limon PJ et al (2000a) Conceptual design study of high field magnets for very large hadron collider. IEEE Trans Appl Supercond 10(1):310–313. https://doi.org/10.1109/77.828236

Ambrosio G, Andreev N, Caspi S et al (2000b) Magnetic design of the Fermilab 11 T Nb_3Sn short dipole model. IEEE Trans Appl Supercond 10(1):322–325. https://doi.org/10.1109/77.828239

Ambrosio G, Andreev N, Chichili DR et al (2000c) Mechanical design and analysis of the Fermilab 11 T Nb_3Sn dipole model. IEEE Trans Appl Supercond 10(1):306–309. https://doi.org/10.1109/77.828235

Ambrosio G, Andreev N, Arkan TT et al (2000d) Development of the 11 T Nb_3Sn dipole model at Fermilab. IEEE Trans Appl Supercond 10(1):298–301. https://doi.org/10.1109/77.828233

Andreev N, Arkan TT, Chichili DR et al (2000) Fabrication and testing of a high field dipole mechanical model. IEEE Trans Appl Supercond 10(1):314–317. https://doi.org/10.1109/77.828237

Bossert R, Ambrosio G, Andreev N et al (2010) Development and test of collaring methods for Nb_3Sn quadrupole magnets. In: Weisend JG II, Barclay J, Breon S et al (eds) AIP conference proceedings, 28 June–2 July 2018, Tuscon, vol 1218. American Institute of Physics, Melville, New York, pp 507–514

Chichili D, Arkan TT, Ozelis JP et al (2000) Investigation of cable insulation and thermo-mechanical properties of Nb_3Sn composite. IEEE Trans Appl Supercond 10(1):1317–1320. https://doi.org/10.1109/77.828478

Chichili DR, Ambrosio G, Andreev N et al (2001) Fabrication of the shell-type Nb_3Sn dipole magnet at Fermilab. IEEE Trans Appl Supercond 11(1):2160–2163. https://doi.org/10.1109/77.920285

Chichili DR, Ambrosio G, Andreev N et al (2004) Design, fabrication and testing of Nb_3Sn shell type coils in mirror magnet configuration. In: Waynert J, Barclay J, Breon S et al (eds) Advances in cryogenic engineering: transactions of the cryogenic engineering conference—CEC, vol 49. American Institute of Physics, Melville, New York, 710:775–782. https://doi.org/10.1063/1.1774754

Chlachidze G, Ambrosio G, Andreev N et al (2009) Quench performance of a 4-m long Nb_3Sn shell-type dipole coil. IEEE Trans Appl Supercond 19(3):1217–1220. https://doi.org/10.1109/tasc.2009.2018276

Den Ouden A, Wessel S, Krooshoop E et al (1997) Application of Nb_3Sn superconductors in high-field accelerator magnets. IEEE Trans Appl Supercond 7(2):733–738. https://doi.org/10.1109/77.614608

Fermilab (2001) Design study for a staged Very Large Hadron Collider. Fermilab, Batavia, IL, TM-2149, June 4

Jackson JD (1986) Conceptual design of the superconducting super collider. SSC-SR-2020, SSC Central Design Group, Berkeley

Kashikhin VV, Zlobin AV (2001) Magnetic designs of 2-in-1 Nb_3Sn dipole magnets for VLHC. IEEE Trans Appl Supercond 11(1):2176–2179. https://doi.org/10.1109/77.920289

Kashikhin VV, Barzi E, Chichili D et al (2003) Passive correction of the persistent current effect in Nb_3Sn accelerator magnets. IEEE Trans Appl Supercond 13(2):1270–1273. https://doi.org/10.1109/tasc.2003.812642

McInturff AD, Benjegerdes R, Bish P et al (1997) Test results for a high field (13 T) Nb_3Sn dipole. In: Comyn M, Craddock MK, Reiser M et al (eds) Proceedings of the 1997 particle accelerator conference (PAC1997), 12–16 May 1997, Vancouver. IEEE, Piscataway, pp 3212–3214

Parrell JA, Field MB, Zhang Y et al (2005) Advances in Nb_3Sn strand for fusion and particle accelerator applications. IEEE Trans Appl Supercond 15(2):1200–1204. https://doi.org/10.1109/tasc.2005.849531

Rice JA, Fabian PE, Hazelton CS et al (1999) Mechanical and electrical properties of wrappable ceramic insulation. IEEE Trans Appl Supercond 9(2):220–223. https://doi.org/10.1109/77.783276

Russenschuck S (1995) A computer program for the design of superconducting accelerator magnets. CERN, Sep 1995. CERN, Geneva

Sabbi G, Ambrosio G, Andreev N et al (2000) Conceptual design of a common coil dipole for VLHC. IEEE Trans Appl Supercond 10(1):330–333. https://doi.org/10.1109/77.828241

Yadav S, Chichili DR, Terechkine I (2001) Coil end parts design and fabrication issues for the high field dipole at Fermilab. IEEE Trans Appl Supercond 11(1):2284–2287. https://doi.org/10.1109/77.920316

Zlobin AV (2011) Status of Nb_3Sn accelerator magnet R&D at Fermilab. CERN Yellow Report CERN-2011-003. CERN, Geneva, pp 50–58 [arXiv:1108.1869]

Zlobin AV, Ambrosio G, Andreev N et al (2005) R&D of Nb_3Sn accelerator magnets at Fermilab. IEEE Trans Appl Supercond 15(2):1113–1118. https://doi.org/10.1109/tasc.2005.849507

Zlobin AV, Kashikhin VV, Barzi E (2006) Effect of flux jumps in superconductor on Nb_3Sn accelerator magnet performance. IEEE Trans Appl Supercond 16(2):1308–1311. https://doi.org/10.1109/tasc.2006.870557

Zlobin AV, Ambrosio G, Andreev N et al (2007) Nb_3Sn accelerator magnet technology R&D at Fermilab. In: Petit-Jean-Genaz C (ed) Proceedings of 2007 IEEE particle accelerator conference (PAC), 25–29 June 2007. Albuquerque, pp 482–484

Chapter 8
Nb_3Sn 11 T Dipole for the High Luminosity LHC (FNAL)

Alexander V. Zlobin

Abstract This chapter describes the design and parameters of the 11 T dipole developed at the Fermi National Accelerator Laboratory (FNAL) in collaboration with the European Organization for Nuclear Research (CERN) for the High Luminosity LHC project, and presents details of the single-aperture and twin-aperture dipole models that were constructed and tested. Magnet test results including magnet quench performance, magnetic measurements, and quench protection studies performed using the dipole mirror and single-aperture and twin-aperture dipole models are summarized and discussed.

8.1 Introduction

The operation of the Large Hadron Collider (LHC) at higher luminosities requires the installation of additional collimators in the dispersion suppression (DS) regions (Bottura et al. 2012; de Rijk et al. 2010) (see also Chap. 9). The free warm longitudinal space of ~3.5 m that is required for additional collimators can be provided by substituting some regular 14.3 m long 8.33 T LHC main dipoles (MB) with a pair of 5.5 m long 11 T dipoles (Bottura et al. 2012). These twin-aperture dipoles will operate at 1.9 K in series with the main dipoles, and deliver the same integrated strength of 119 T m at a nominal operating current of 11.85 kA. The operating field level of ~11 T calls for magnets based on Nb_3Sn superconductor.

To demonstrate feasibility of such magnets and study their performance parameters, in 2011 the European Organization for Nuclear Research (CERN) and the Fermi National Accelerator Laboratory (FNAL) started a research and development (R&D) program with the goal of developing and testing a 5.5 m long twin-aperture

A. V. Zlobin (✉)
Fermi National Accelerator Laboratory (FNAL), Batavia, IL, USA
e-mail: zlobin@fnal.gov

D. Schoerling, A. V. Zlobin (eds.), *Nb_3Sn Accelerator Magnets*, Particle Acceleration and Detection, https://doi.org/10.1007/978-3-030-16118-7_8

Nb_3Sn dipole prototype. The original FNAL–CERN R&D plan comprised three phases:

- Phase 1 (2011–2012): development and testing of a single-aperture 2 m long dipole demonstrator, first at FNAL and then, after technology transfer, at CERN;
- Phase 2 (2013–2014): development and testing of two 2 m long twin-aperture dipole models at each laboratory to study the magnets' performance parameters and their reproducibility, and to select the magnet's final design;
- Phase 3 (2014–2015): development and testing of a 5.5 m long twin-aperture dipole prototype to demonstrate the technology scale-up and production readiness. It was assumed that one 5.5 m long collared coil would be produced by FNAL and the other one by CERN.

In 2013 the FNAL plan was modified due to the reduction of the FNAL program budget as well as CERN's priorities and the schedule for US contributions to the LHC Luminosity Upgrade (HL-LHC) project. In the new plan the third phase of the FNAL plan was cancelled, and the scope of the second phase was modified. The model length was reduced from 2 m to 1 m, to minimize the magnet cost.

The design and technology of the 11 T dipole was influenced by the design of the LHC Nb-Ti main dipole and by the results of the Nb_3Sn magnet R&D program at FNAL (Zlobin 2010) (see also Chap. 7). To meet the tight project schedule within the available budget, the magnet was designed to make maximum use of the existing tooling, infrastructure, and magnet components at both laboratories.

8.2 Magnet Design Concept

8.2.1 Design Considerations

The main design goals included achieving a dipole field above 11 T at a current of 11.85 kA with 20% margin on the load line at an operating temperature of 1.9 K, and providing geometrical field harmonics below the 10^{-4} level at the reference radius of 17 mm (Karppinen et al. 2012; Zlobin et al. 2011).

The design concept for the 11 T dipole features a two-layer shell-type coil, stainless-steel collars, and a vertically split iron yoke, surrounded by a stainless-steel outer shell. The 60 mm coil aperture, slightly larger than the 56 mm aperture of the LHC main dipole, was selected to accommodate the beam sagitta in the 11 T dipoles and to avoid the manufacture of curved Nb_3Sn coils. The aperture separation is 194 mm, as in the LHC main dipole. The size and location of the heat-exchanger and the slots for the bus-bars inside the iron yoke of the 11 T dipole are identical to those of the MB yoke.

The parameters of the Nb_3Sn Rutherford cable were selected using the following considerations. The maximum number of strands in the cable was limited to 40 by

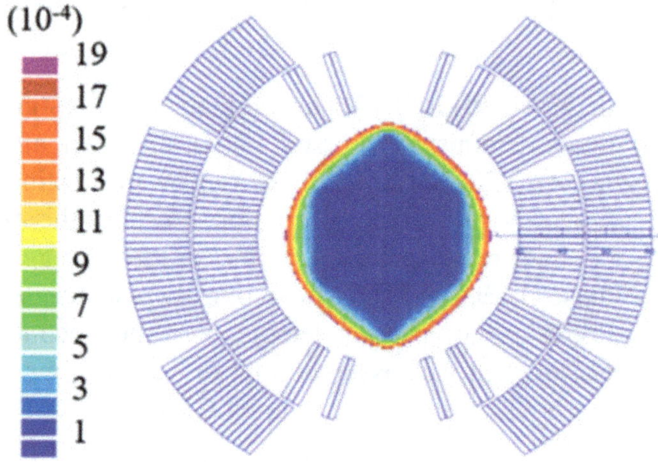

Fig. 8.1 Coil cross-section with geometrical field errors in units

the capability of CERN's cabling machine (FNAL cabling machine allows for 42 strands). The maximum strand diameter was restricted by 0.7 mm to limit the cable thickness and, thus, to achieve the required field of 11 T or more at the nominal current of 11.85 kA. The acceptable critical current degradation due to cabling was limited to 10%. Magnet design studies started with the following cable geometrical parameters: a width of 14.85 mm, a thin edge of 1.20 mm, a thick edge of 1.41 mm, and a keystone angle of 0.81°. The cable insulation thickness was 0.1 mm.

Several two-layer coil cross-sections with five, six, and seven blocks per quadrant and 60 mm aperture were analyzed using the ROXIE program (Russenschuck 1995). A six-block design with 56 turns, which satisfied the above criteria, was selected for further optimization. A cross-section of this coil, optimized for the twin-aperture layout with a round iron yoke separated from the coil by 30 mm, is shown in Fig. 8.1.

To simplify the magnet assembly and reduce the risk of coil damage during collaring, it was decided to use separate collars for each aperture. An additional advantage of this approach is the possibility of testing collared coils in both single-aperture and twin-aperture configurations. The collar width in this magnet is limited to 30 mm by the available space between the two apertures. This space is not enough for using a freestanding collar design. With support from the yoke and skin, however, the collar width can be less than 30 mm. The minimum collar width that satisfies the stress limits in the collar and key materials is about 20 mm.

Analysis of the stress distribution in the coil showed that during collaring the stress in the coil pole regions is significantly smaller than the stress in the mid-plane regions due to the relatively large width and high rigidity of the coil. The low pre-stress in the pole region, limited by the maximum allowed stress in the coil mid-plane during collaring, can be increased to the required level by coil bending using the mid-plane collar–yoke shim or by using removable poles with shims. Both methods were adopted for the mechanical structure of the 11 T dipole. CERN focused on the removable pole design and wide round collar, whereas FNAL pursued the integrated pole design and a narrower elliptical collar to provide coil bending. The results of magnetic and mechanical analyses for the FNAL 11 T dipole models are presented in Auchmann et al. (2012), Karppinen et al. (2012), and Zlobin et al. (2012a).

8.2.2 Mechanical Designs and Analysis

The integrated pole design approach stands upon previous FNAL experience with Nb_3Sn dipole and quadrupole coils. Cross-sections of the collared coil and the 11 T single-aperture and twin-aperture dipoles that were developed at FNAL are shown in Figs. 8.2 and 8.3.

The collar has a slightly elliptical shape, with a minimum width in the mid-plane of 25 mm. In the single-aperture configuration, the collared coil is placed inside the 400 mm diameter iron yoke used previously in FNAL HFDA series dipole models (see Chap. 7). The inner contour of the yoke was adapted to the collared coil design. The two yoke halves are connected by Al clamps and a 12 mm thick stainless-steel skin. In the twin-aperture configuration, two collared coils, separated by an iron

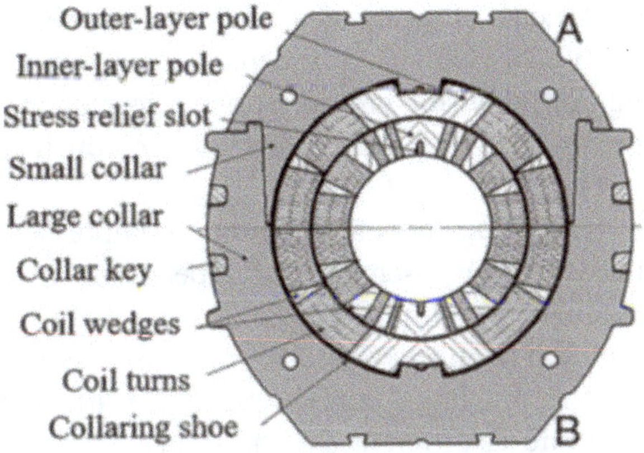

Fig. 8.2 Collared coil cross-section

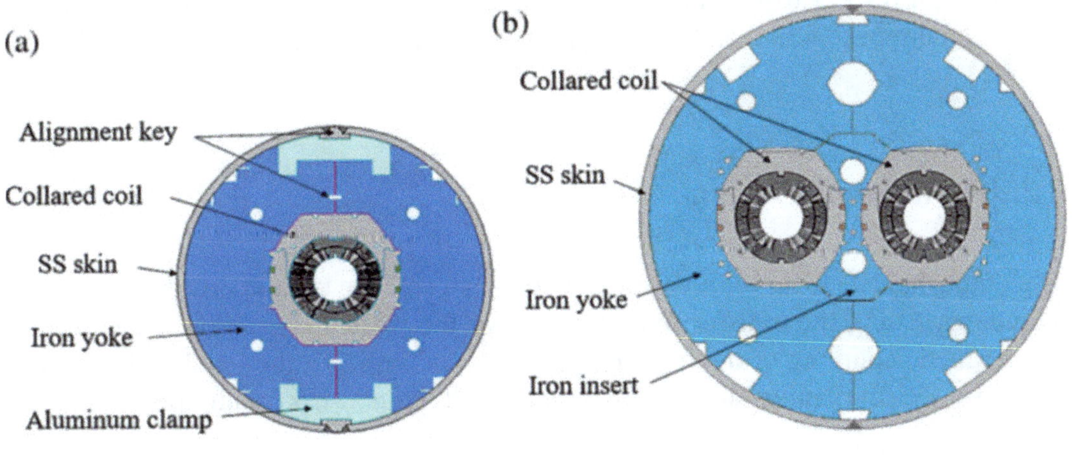

Fig. 8.3 (**a**) Single-aperture and (**b**) twin-aperture dipole cross-sections

insert, are placed inside the 550 mm diameter iron yoke between the two yoke pieces surrounded by a 12 mm thick stainless-steel skin.

The required coil pre-stress is applied in several steps. The initial coil preloading is provided by adding 0.1 mm mid-plane (or by using coils with a 0.1 mm larger azimuthal size) and 0.025 mm radial shims during the collaring of the coils. The maximum achievable pre-stress in the pole region is limited at this stage by the maximum allowed stress near the coil mid-plane. During cold mass assembly the coil pole pre-stress is increased to the nominal level by the horizontal deformation of the collared coil, using 0.125 mm horizontal collar–yoke shims. These shims also provide the collar–yoke contact in the mid-plane area after cool-down.

The vertical gap between the two yoke halves is open at room temperature. It is controlled by the precise collar dimensions near the top and bottom horizontal surfaces of the collared coil (areas A and B in Fig. 8.2). Due to the larger thermal contraction of the outer shell, the clamps, and the collared coil relative to the iron yoke, the gap is closed during the cool-down to 1.9 K, and compressed by the shell and clamps to stay closed up to the maximum design field of 12 T. The maximum tensile stress in the 12 mm thick stainless-steel shell is approximately 250 MPa. This large shell stress increases the compression of the iron halves with very small impact on the coil preload.

A detailed mechanical analysis was performed to optimize the stress in the coil and major elements of the magnet support structure, and to minimize the conductor motion and coil cross-section deformation at room and operating temperatures. Stress distribution diagrams for the coil in twin-aperture configuration are shown in Fig. 8.4.

The values for the maximum compressive coil stress in the pole and mid-plane turns during assembly and operation for the twin-aperture and single-aperture magnets are summarized in Table 8.1. In both cases the poles remain under compression at all steps, and the maximum coil stress remains below 165 MPa.

The maximum stress values for the major elements of the magnet support structure at different assembly and operating stages are below the yield stress of the structural materials chosen. The maximum radial deflection of the coil cross-section with respect to the magnet's nominal design decreases from 0.135 mm at injection current to less than 0.04 mm at the nominal current, which is acceptable.

8.2.3 *Magnetic Design and Analysis*

The main challenges for the electromagnetic design of the twin-aperture 11 T dipole include matching the transfer function (TF) of the main LHC dipoles, controlling the magnetic coupling of the two apertures, and minimizing the level and variation of unwanted multipoles in the operating field range. Iron saturation and coil

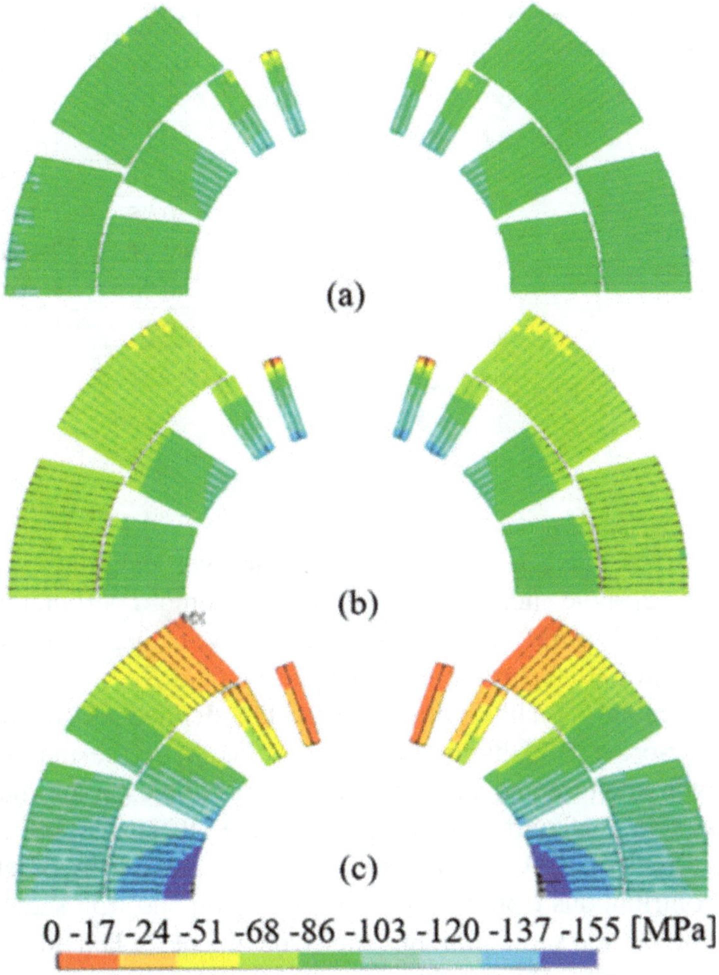

Fig. 8.4 Distributions of the azimuthal stress in the coil for twin-aperture configuration: (**a**) after assembly; (**b**) after cool-down; and (**c**) at 12 T

Table 8.1 Maximum compressive azimuthal stress in coil pole and mid-plane turns (MPa). The data in parentheses corresponds to the single-aperture design

Position in coil	Collared coil	Cold mass	Cool-down	Design field 12 T
Inner-layer pole	47 (44)	103 (120)	117 (136)	5 (2)
Outer-layer pole	81 (64)	85 (87)	59 (110)	25 (17)
Inner-layer mid-plane	117 (97)	77 (79)	80 (97)	165 (141)
Outer-layer mid-plane	59 (51)	85 (108)	66 (124)	116 (153)

magnetization are the two most important effects that contribute to the TF and low-order field harmonics. Both effects are non-linear, which makes their compensation rather challenging.

Field harmonic coefficients in a magnet aperture are defined as

Table 8.2 Transfer function and field harmonics (in units) at $R_{ref} = 17$ mm

Parameter	Twin-aperture		Single-aperture	
	0.757 kA	11.85 kA	0.757 kA	11.85 kA
B_1/I (T/kA)	1.01	0.95	1.01	0.92
b_2	0.13	−7.14	–	–
b_3	44.51	6.04	38.4	−0.8
b_4	0.05	−0.44	–	–
b_5	0.01	−0.02	5.3	0.1
b_7	0.04	0.02	0.0	0.0
b_9	0.19	0.96	1.0	1.0

$$B_y + iB_x = B_1 \sum_{n=1}^{\infty} (b_n + ia_n) \left(\frac{x + iy}{R_{\text{ref}}} \right)^{n-1},$$

where B_x and B_y are the horizontal and vertical field components, and b_n and a_n are the $2n$-pole normal and skew harmonic coefficients at the reference radius $R_{\text{ref}} = 17$ mm. The normal b_n and skew a_n harmonic coefficients are expressed in units of 10^{-4} parts of the main dipole field B_1.

Table 8.2 shows the magnet TF, defined as the ratio B_1/I, and the allowed low-order field harmonics b_n calculated for the magnet's straight section of the twin-aperture and single-aperture 11 T dipoles at an injection current of 757 A and at a nominal current of 11.85 kA. The LHC current pre-cycle with a minimal (reset) current of 350 A was used. The data include geometrical components and the contributions from the coil magnetization and iron saturation effects.

The iron saturation influences the magnet's TF and the low-order field harmonics b_2, b_3, and b_5. Due to this effect, the TF of the twin-aperture model is reduced by 6%. Two large holes in the yoke insert (see Fig. 8.3) were added to limit the b_2 variation by 11 units. The cut-outs above and below the collared coils, the size and position of the small holes around the collared coils, and the location of the holes for tie-rods in Fig. 8.3 were used to keep the b_3 and b_5 variations from injection to a nominal current below 0.1 unit. The iron cross-section of the single-aperture models was not optimized to suppress the iron saturation effect. Thus, the TF reduction in the single-aperture design reaches 9%, and the b_5 variation increases to 1.3 units.

The coil magnetization effect in the 11 T Nb_3Sn dipoles is much larger than in the Nb-Ti LHC MB dipoles due to the higher critical current density and the larger size of the superconducting filaments in the state-of-the-art Nb_3Sn composite wires. Beam dynamics studies have shown that b_3 errors of up to ±20 units at injection can be tolerated without compromising the LHC's dynamic aperture. The calculated b_3 due to the persistent currents in the 11 T dipole is ~44 units at the LHC injection current. Analysis has shown that the b_3 variations in the operating field range could be limited to ±20 units by reducing the LHC reset current to 100 A. Possible passive correction schemes were also studied to further reduce b_3 at low fields (Auchmann et al. 2012).

Table 8.3 Dipole design parameters at $I_{nom} = 11.85$ kA

Parameter	Twin-aperture	Single-aperture
Yoke outer diameter (mm)	550	400
Nominal bore field at I_{nom} (T)	11.23	10.88 (11.07)[a]
Short sample field B_{SSL} at T_{op} (T)	13.9	13.4 (14.1)[a]
Margin B_{nom}/B_{SSL} at T_{op} (%)	83	81 (79)[a]
Stored energy at I_{nom} (kJ/m)	969	424 (445)[a]
F_x/quadrant at I_{nom} (MN/m)	3.16	2.89
F_y/quadrant at I_{nom} (MN/m)	−1.59	−1.58

[a]1 m long model

8.2.4 Magnet Design Parameters

The 2D calculated design parameters for the twin-aperture and single-aperture dipole magnets at a nominal operating current of 11.85 kA and a temperature of 1.9 K are presented in Table 8.3. The calculation was performed for a wire J_c(12 T, 4.2 K) of 2750 A/mm^2, a Cu fraction of 0.53, and a cable I_c degradation of 10%. In the single-aperture configuration the calculated nominal central field is 10.88 T, whereas in the twin-aperture magnet it increases to 11.23 T due to field enhancement in the twin-aperture configuration. For a 1 m long model the nominal and short sample bore fields are also slightly larger due to contributions to the central field from the coil ends.

8.2.5 Quench Protection

The 11 T Nb_3Sn dipoles have a larger stored energy and inductance per unit length than the main LHC dipoles and, thus, require special attention to their protection during a quench. The preliminary quench analysis suggested that the quench protection scheme with efficient outer-layer (OL) heaters could provide adequate protection for the 11 T Nb_3Sn dipoles. The quench protection heaters are made of stainless-steel strips and placed on the coils' outer surface. The heater strips on one side of each coil are connected in series with the strips on the same side of the other coil, forming two symmetric heater circuits (see Fig. 8.5). The two circuits are connected in parallel for redundancy.

In the case of a magnet quench at the maximum operating current of 11.85 kA, the expected average temperature of the coil OL under the heaters is less than 140 K when both heater circuits operate, and less than 200 K in the case of one heater circuit failure. The maximum hot-spot temperature calculated for a 50 ms protection system delay does not exceed 240 K and 340 K, respectively.

Reliable quench protection for the 11 T dipoles with OL heaters requires high heater efficiency. Experimental studies and optimization of the protection heaters were a key part of the 11 T dipole R&D program at FNAL.

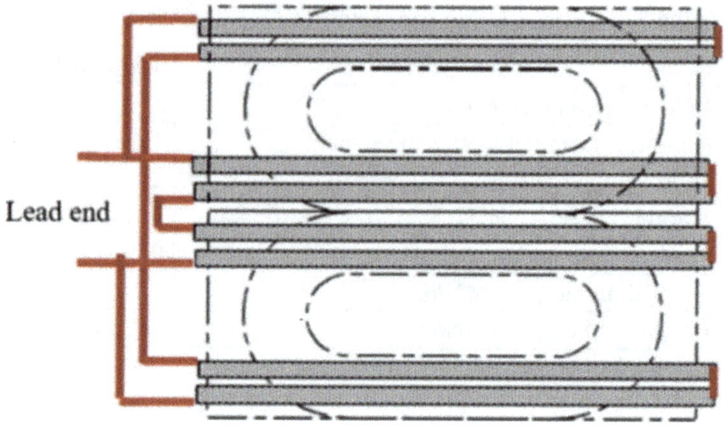

Fig. 8.5 Electrical connection of protection heaters

8.3 11 T Dipole R&D

8.3.1 Model Design and Fabrication

The 11 T dipole R&D at FNAL started with the development of the baseline technology and tooling, and the optimization of strand and cable parameters. The strand and cable parameters are the same for both the FNAL and CERN dipole designs. The design of the FNAL dipole including coil, collar, and iron yoke, as well as magnet assembly processes and preload procedures, were developed in parallel. Experimental studies of magnet quench performance, protection, field quality, and performance reproducibility were then performed.

8.3.1.1 Nb_3Sn Wire and Cable

The 11 T dipole uses Rutherford cable with 40 Nb_3Sn strands, 0.7 mm in diameter. The optimization of cable parameters included the selection of the cable cross-section geometry and compaction to achieve a good mechanical stability of the cable and acceptable critical current degradation, incorporating a stainless-steel core and preserving a high residual resistivity ratio (RRR) of the copper matrix. Two strand designs (baseline and R&D) with a different sub-element number, size, and distribution in the cross-section were used. The cross-sections of the round wires and the cored cable are shown in Fig. 8.6.

The main parameters and properties of the Nb_3Sn wires and Rutherford cables are discussed in Chap. 2 of this book. Results of the wire and cable study and optimization for the 11 T program at FNAL are reported by Barzi et al. (2012). The Nb_3Sn wires were produced using the Restack Rod Process® (RRP) by Oxford Superconducting Technologies (OST). The main parameters of RRP108/127 (baseline) and RRP150/169 (R&D) round wires are summarized in Table 8.4.

During the reaction, Nb_3Sn strands expand due to the phase transformation. The 11 T dipole cable measured in free conditions showed an average width expansion of

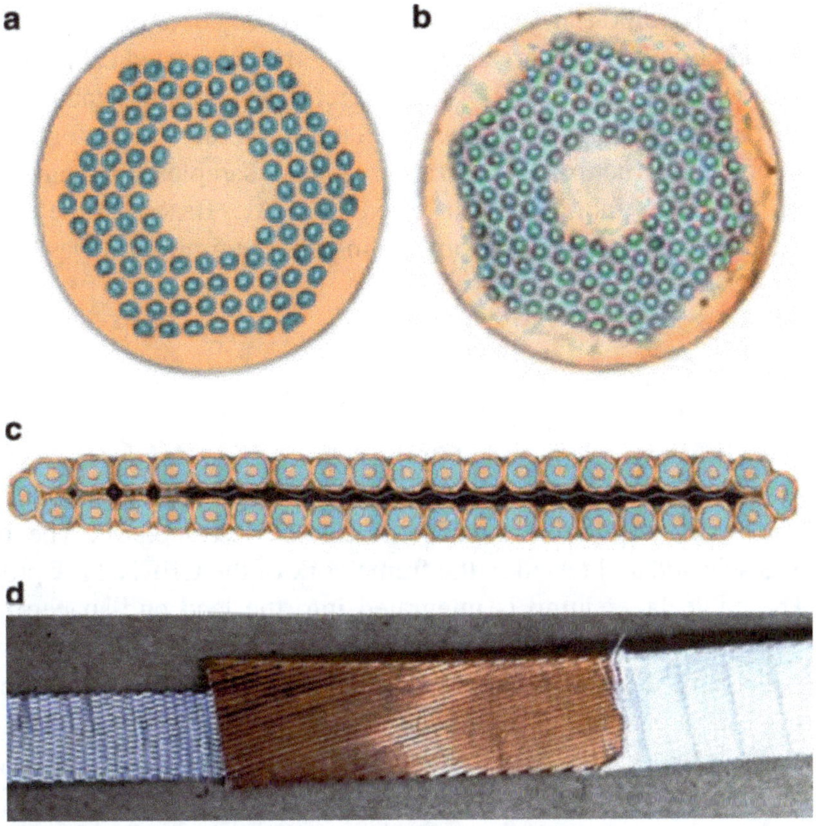

Fig. 8.6 (**a**) RRP108/127 and (**b**) RRP150/169 composite wires; (**c**) the 40-strand Rutherford cable with stainless-steel core; and (**d**) the cored cable insulated with E-glass tape

Table 8.4 Nb_3Sn round wire parameters

Parameter	RRP108/127	RRP150/169
Strand diameters (mm)	0.70	0.70
Average J_c(12 T, 4.2 K) (kA/mm^2)	2.68	2.65
Effective filament size D_{eff} (μm)	41	36
Twist pitch (mm)	14	13
Cu fraction (%)	55.5	51.8

Table 8.5 Cable geometrical parameters

Parameter	Un-reacted	Reacted
Mid-thickness (mm)	1.25	1.30
Thin edge (mm)	1.15	1.19
Thick edge (mm)	1.35	1.40
Width (mm)	14.70	15.08
Keystone angle (°)	0.79	0.81

2.6%, an average mid-thickness expansion of 3.9%, and an average length decrease of 0.3%. An explanation of the anisotropic cable expansion is presented by Andreev et al. (2002). The geometrical parameters of the un-reacted and reacted cable optimized for the 11 T dipole are listed in Table 8.5. The geometrical parameters of the coil winding and curing tooling are determined by the un-reacted cable cross-section, whereas the geometrical parameters of the coil reaction and impregnation tooling are based on the reacted cable dimensions. The latter were also used for the magnet's electromagnetic and structural optimization. Cable samples with and without a stainless-steel core were fabricated and tested at FNAL. Based on the experimental data, the critical current degradation from cabling after optimization was less than 4%.

8.3.1.2 Coil

Each coil consists of two layers and 56 turns. The coil is wound from a single piece of cable insulated with two layers of E-glass tape 0.075 mm thick and 12.7 mm wide (see Fig. 8.6d). This insulation is adequate for the R&D phase. The final cable insulation was selected and tested in the framework of the CERN 11 T program (see Chap. 9). The cable layer jump is integrated into the lead end-spacers. Both coil poles were made of Ti-6Al-4 V alloy, whereas the wedges, end-spacers, and saddles were made of stainless steel. The end-spacers were fabricated using the selective laser sintering (SLS) process and provided by CERN for all FNAL coils.

Coils are fabricated using the wind-and-react method, i.e., the superconducting Nb_3Sn phase is formed during the coil high-temperature heat treatment. The details of the coil fabrication process are reported by Zlobin et al. (2012a, 2013). After winding, each coil layer was impregnated with CTD-1202x liquid ceramic binder and cured under a small pressure at 150 °C for 0.5 h. During curing the coil inner and outer layers were shimmed in the mid-plane to a size of 1.0 and 1.5 mm, respectively, smaller than the layer nominal sizes, to provide room for the Nb_3Sn cable volume's expansion after reaction. Each coil was reacted separately in an argon atmosphere using a three-step cycle. The maximum temperature for coils with RRP108/127 wires was 640 °C for 48 h (HT1); for the last two coils (11 and 12) it was increased to 645 °C for 50 h (HT2). The coils with RRP150/169 wires were reacted at 665 °C for 50 h (HT3). Before impregnation, the brittle Nb_3Sn coil leads were spliced to flexible Nb-Ti cables, and the coils were wrapped with a 0.125 mm thick layer of E-glass or S2-glass cloth. The coils were impregnated with CTD-101K epoxy and cured at 125 °C for 21 h. Pictures of a coil after curing, reaction, and impregnation, and a coil cross-section are shown in Fig. 8.7.

The radial and azimuthal sizes of each coil were measured in free condition in several cross-sections using a 3D Cordax machine. An example of the coil size measurements is shown in Fig. 8.8. One can see that the coil outer radius is smaller than nominal by ~0.05 mm. This difference was compensated for during magnet assembly by adding an appropriate layer of Kapton film. An oversizing of the coil mid-plane by ~0.1 mm was introduced to achieve the target coil pre-stress without using special mid-plane shims. It was found that the radial and azimuthal sizes of impregnated coils are reproducible from coil to coil. Thus, they can be adjusted using appropriate radial and azimuthal shims in the impregnation mold.

Fig. 8.7 (**a**) Coil after winding and curing (lead end); (**b**) reacted coil with v.3 end-spacers (G10 plugs are inserted in special holes in end spacers before impregnation); (**c**) epoxy impregnated coil and splice block; (**d**) coil cross-section illustrating the accuracy of the turn and spacer positions

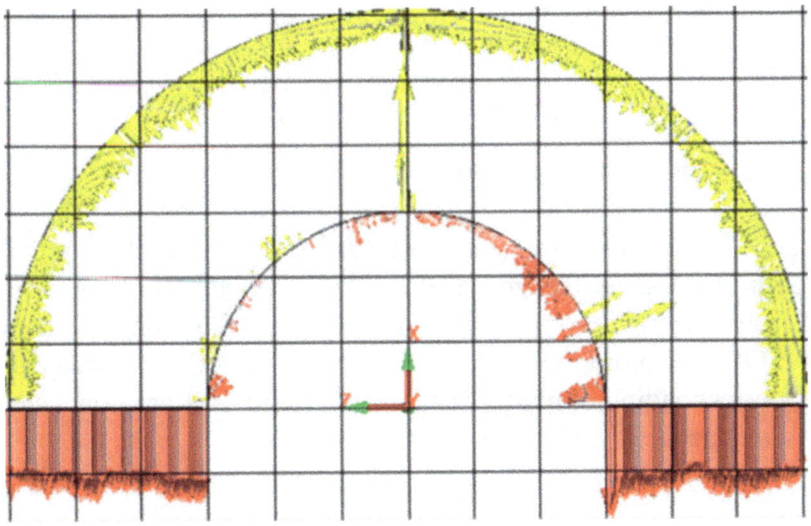

Fig. 8.8 Example of coil size measurements (coil 10). The side of each square corresponds to 0.1 mm

Four 2 m long and eight 1 m long coils were fabricated at FNAL during 2012–2014 using various wires, cables, coil end parts, and reaction cycles. Three end-spacer designs were used: the original design (v.1), the design with shortened legs (v.2), and the design with flexible legs and round holes (v.3) filled with

Table 8.6 Coil design features

Coil number	Length (m)	Strand design	Cable core	End-spacers	Reaction cycle
1[a]	2	108/127	No	v.1	HT1
2, 3, 4[b]	2	108/127	No	v.1	HT1
5, 6[b], 7	1	150/169	Yes	v.1	HT3
8	1	108/127	Yes	v.1	HT1
9, 10	1	108/127	Yes	v.2	HT1
11, 12	1	108/127	Yes	v.3	HT2

[a]Practice coil
[b]Coil damaged during fabrication

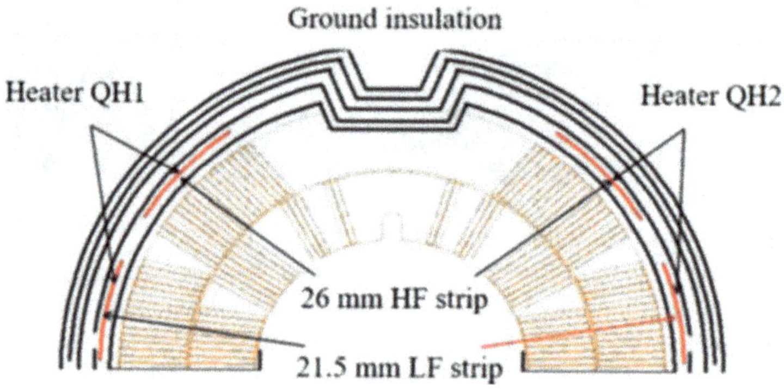

Fig. 8.9 Position of high-field (HF) and low-field (LF) protection heaters

G10 rods after reaction (see Fig. 8.7b). Coil design features are summarized in Table 8.6.

8.3.1.3 Ground Insulation and Quench Protection Heaters

The coil ground insulation consists of five layers of 0.125 mm thick polyimide film. Two quench protection heaters made of 0.025 mm thick stainless-steel strips are placed on each side of the coil between the first and second insulation layers, covering the OL coil blocks (Fig. 8.9). The resistance of each heater at 300 K is 5.9 Ω. The corresponding strips on each side of each coil are connected in series forming two independent heaters, as shown in Fig. 8.5.

8.3.1.4 Collared Coil

The collared coil assembly consists of two coils, a multilayer polyimide ground insulation, 316 L stainless-steel protection shells (collaring shoe), and collar laminations made of Nirosta high-Mn stainless steel. Collar blocks are locked on each side by two bronze keys. The models were assembled first with laser-cut collars (v.1) and later with stamped collars (v.2). The stamped collars had a slightly larger inner radius to accommodate a thicker protective shell.

Assembly and preload procedures for brittle Nb_3Sn coils with a collar structure were developed and successfully demonstrated at FNAL using quadrupole coils (Bossert et al. 2011). Since the collaring of Nb_3Sn coils always requires great care and process control, a 0.6 m long mechanical model with instrumented Nb_3Sn coils was assembled and used to optimize these procedures, as well as the coil final preload.

8.3.1.5 Short Dipole Models

Single-Aperture Models

In the single-aperture configuration, a collared coil is installed inside a vertically split yoke of 400 mm outer diameter made of SAE 1045 iron and fixed with Al clamps similar to the HFDA series dipoles (see Chap. 7). The collared coil inside the iron yoke is shown in Fig. 8.10. To assure uniform mechanical support of the collared coil, the yoke covers the entire coil length, including the Nb_3Sn/Nb-Ti lead splices. In this case, the maximum field in the coil ends is ~2% higher than the maximum field in the coil straight section.

The 12.7 mm thick 304 L stainless-steel skin is pre-tensioned during welding (welded skin) or by strong bolts (bolted skin) to provide the coil's final pre-compression. The required minimal stress in the skin at room temperature was

Fig. 8.10 Collared coil with collar-yoke shims inside iron yoke (lead end)

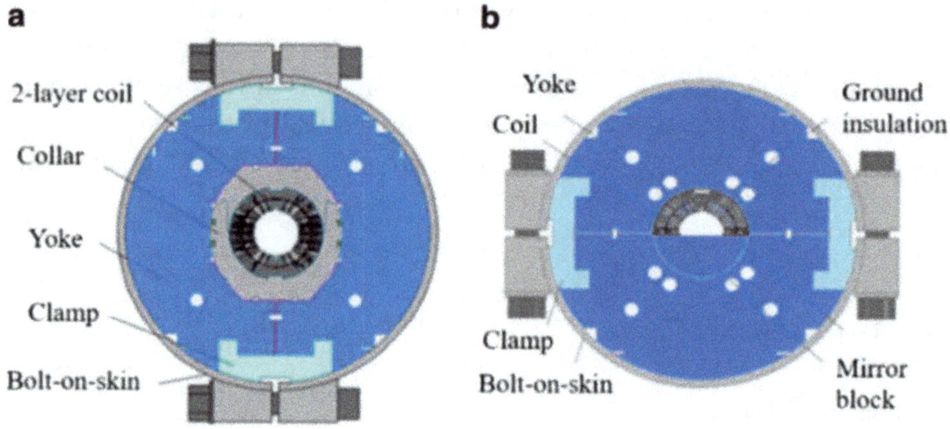

Fig. 8.11 Cross-sections of (**a**) 11 T dipole and (**b**) dipole mirror with a bolted skin

controlled using the data from strain gauges installed on the skin. Two 50 mm thick 304 L stainless-steel end-plates are attached to the shell to restrict the axial motion of the coil ends.

Some single 11 T dipole coils were tested using the dipole mirror structure and assembly procedure developed for the HFDM series (see Chap. 7). Due to the larger radial size of the 11 T coils, the 8 mm thick radial bronze spacer in HFDM structure was replaced by a 2 mm thick stainless-steel shell. Cross-sections of the single-aperture 11 T dipole and dipole mirror models with bolted skin are shown in Fig. 8.11.

The 11 T dipole demonstrator MBHSP01 used 2 m long coils 2 and 3. Single-aperture dipole models MBHSP02 and MBHSP03 used 1 m long coils 5, 7, 9, and 10. Coil 8 was heavily instrumented and was first tested in the dipole mirror model MBHSM01 to study the effect of coil pre-stress and to measure quench protection parameters (Zlobin et al. 2014). It was then assembled with coil 11 and tested in a single-aperture dipole MBHSP04 without collars, using the dipole mirror structure. Coil 11 was later tested again in the dipole mirror MBHSM02.

Twin-Aperture Model

In a twin-aperture configuration, two collared coils are placed inside a vertically split 550 mm outer diameter (OD) iron yoke with an iron spacer between them. The yoke is surrounded by a 12.7 mm thick welded stainless-steel skin. Two 50 mm thick stainless-steel end-plates, welded to the skin, restrict the axial motion of both collared coils. No axial pre-stress was applied to the coil ends.

The twin-aperture dipole model MBHDP01 was assembled using collared coils that were previously tested in MBHSP02 and MBHSP03. Based on the test results in a single-aperture configuration, the MBHSP03 collared coil was re-collared with a slightly larger radial coil–collar shim to increase the coil pre-stress. Both collared coils were installed inside the MBHDP01 iron yoke with the same collar–yoke mid-plane shims as in MBHSP03. These shims provided a collared coil bending

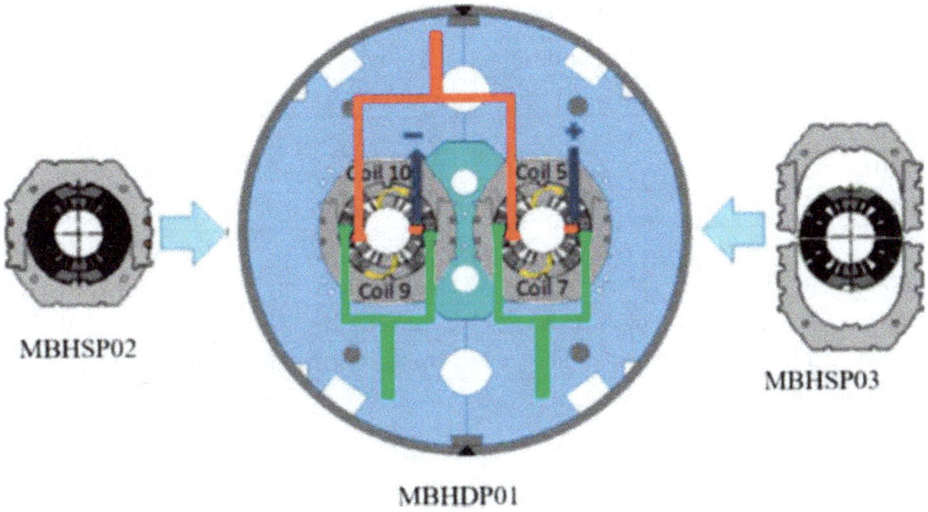

Fig. 8.12 MBHDP01 assembly and coil electrical connection schemes

of ~0.1 mm to ensure collar–yoke contact after cooling down. Figure 8.12 shows the twin-aperture dipole MBHDP01 assembly and the coil electrical connection scheme.

8.3.2 *Magnet Test*

The design features and test dates of the single-aperture dipole and dipole mirror models, and twin-aperture dipole model are summarized in Table 8.7.

All the models were tested at the FNAL Vertical Magnet Test Facility (VMTF) during 2012–2017. The magnets were instrumented with voltage taps and a quench antenna; strain gauges on the coils, shell, and end bullets; and temperature sensors to monitor the magnet parameters during assembly and test.

The typical test plan included magnet training, measurements of the ramp rate and temperature dependencies of the magnet quench current, magnetic measurements, and protection heater studies.

8.3.2.1 Quench Performance

Single-Aperture Dipoles and Mirror Models

The training quenches of single-aperture dipole models MBHSP01–04 and dipole mirror models MBHSM01–02 are summarized in Fig. 8.13. Typically, each magnet was trained first at 4.5 K with a current ramp rate of 20 A/s. When the training was slowing down or a plateau was reached, the magnet training was continued at 1.9 K.

Table 8.7 Design features of short models

Model	Coil number	Collar design	Skin type	Test date
MBHSP01	2, 3	v.1	Welded	June–July 2012
MBHSP02	5, 7	v.1	Bolted horizontal	March 2013
MBHSM01	8	No collar	Bolted horizontal	Dec 2013–Jan 2014
MBHSP03	9, 10	v.2	Bolted vertical	April–May 2014
MBHDP01	5, 7 & 9, 10	v.1. & v.2	Welded	Feb–March 2015 (TC1)[a]
MBHSP04	8, 11	No collar	Bolted horizontal	June–July 2015
MBHDP01	5, 7 & 9, 10	v.1. & v.2	Welded	June–July 2016 (TC2)[b]
MBHSM02	11	No collar	Bolted horizontal	March–April 2017

[a]Thermal cycle 1
[b]Thermal cycle 2

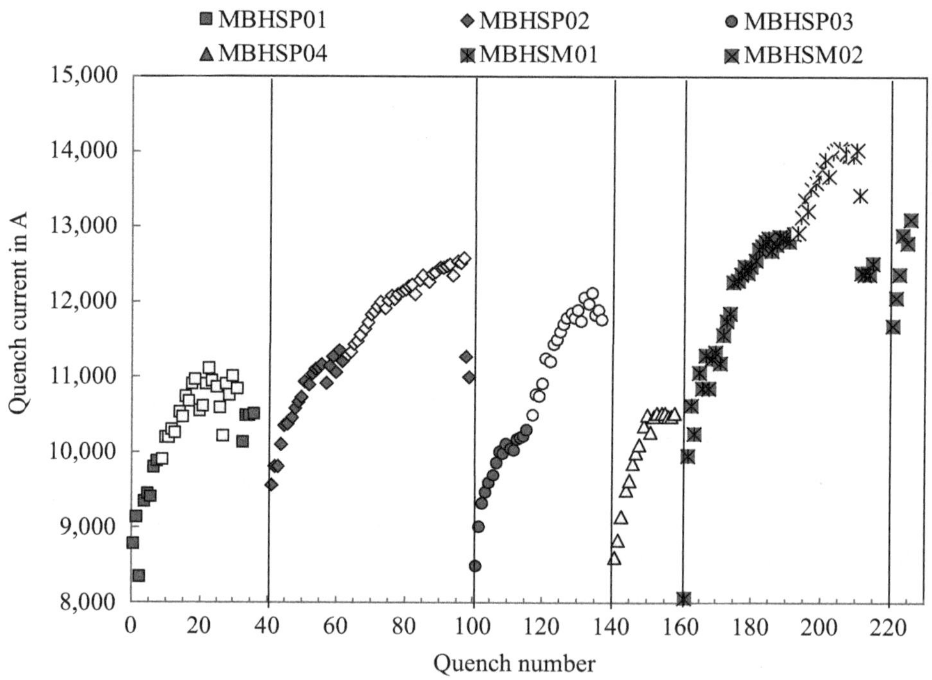

Fig. 8.13 Magnet training. The data for 4.5 K and 1.9 K are represented with filled and non-filled markers, respectively

The 2 m long, single-aperture demonstrator dipole, called MBHSP01, was tested in June–July 2012, only 18 months (!) after the start of the program. The goal of this test was to achieve the design field of 12 T and to check the magnet's design, technology, and performance. After a first quench at ~8.2 T and 4.5 K, the magnet only reached 10.4 T, or 87% of the magnet design field of 12 T at 1.9 K. The quench performance was erratic. The magnet also spontaneously quenched after 1–6 min at constant currents above 7.5 kA (7.0 T) at 4.5 K and above 9 kA (8.4 T) at 1.9 K.

Practically all of the low ramp-rate quenches and all the quenches at constant currents at 4.5 K, 1.9 K, and intermediate temperatures started in the mid-plane block of the coil outer layer. Only a few of the first training quenches occurred in the high-field region at the very beginning of the test. The analysis of the quench location data, ramp rate, and temperature dependencies of the magnet quench current and magnet operation at a constant current pointed to conductor damage in the outer coil mid-plane area. Later, the magnet autopsy confirmed this conclusion, and indicated that this damage occurred during coil reaction and collaring. The coil end parts, collars, and coil–collar shells were modified, and the coil fabrication and magnet assembly processes of the following 1 m long models were corrected based on the lessons learned from MBHSP01.

The first 1 m long dipole model, called MBHSP02, was tested almost a year later, in March 2013. The first quench current was 9.57 kA, which corresponds to ~9 T in the aperture. After 17 quenches in the inner-layer (IL) end-blocks of both coils, three consequent quenches were detected in the outer mid-plane blocks of coil 7. After 23 training quenches and ramp-rate dependence studies, magnet training was continued at 1.9 K. A maximum bore field of 11.7 T (97.5% of the magnet design field but only 83% of the magnet short sample limit (SSL) bore field) was reached after 57 training quenches. At this point the magnet training at 1.9 K was discontinued. All quenches at 1.9 K occurred in the IL end-blocks of both coils. After quench studies at 1.9 K, MBHSP02 was quenched several times again at 4.6 K. These quenches at ~11.4 kA (10.7 T in the aperture) showed that the magnet reached the conductor limit at 4.6 K, which is only 84.3% of the magnet SSL based on witness sample data. Spontaneous quenches at constant currents were also observed in this magnet, although they occurred at higher currents, above 9 kA (8.7 T) and 11 kA (10.4 T) at 4.5 K and 1.9 K respectively, than in MBHSP01.

The unexpectedly large level of quench current degradation, as well as spontaneous quenches at a current plateau below the maximum quench current in MBHSP02, were associated with the excessive coil pre-stress and the large radial deformation of coil mid-plane areas that was used to pre-stress the coil pole turns. These issues were studied by testing a single coil in a dipole mirror structure called MBHSM01.

The dipole mirror model MBHSM01, which was assembled with smaller mid-plane shims to reduce coil pre-stress, was tested in December 2013–January 2014. The magnet was trained to 80% of its SSL after only four quenches and to almost 100% of the SSL at 4.5 and 1.9 K after 25 and 15 quenches, respectively. The coil maximum field was 12.5 T at 1.9 K and 11.6 T at 4.5 K. All training quenches started in the high-field area of the coil inner layer, with only two quenches in the coil outer layer. The quenches after reaching a training plateau at both temperatures started in the blocks, next to the IL middle wedges. Unlike MBHSP01 and MBHSP02, dipole mirror MBHSM01 demonstrated stable performance during a 25 min long current plateau (no so-called ‘holding quenches’) at 13 kA (90% of SSL) at 1.9 K and 12 kA (92% of SSL) at 4.5 K. Since the design and fabrication processes for coil 8 in MBHSM01 were the same as for coils 5 and 7 in MBHSP02, the improved quench performance of coil 8 in the dipole mirror structure suggests that the large mid-plane collar-yoke shim was likely a major cause of the conductor degradation in the dipole model MBHSP02. This shim size in the next dipole model MBHSP03 was reduced to the level necessary to just compensate for the difference in collar and yoke thermal contraction.

The second 1 m long dipole model, called MBHSP03, with reduced coil preload, was tested in April–May 2014. Magnet training started at 4.5 K with a first quench at 8.49 kA, which corresponds to ~8.4 T in the aperture. After 16 quenches at 4.5 K, magnet training was continued in superfluid helium at 1.9 K. A maximum bore field of 11.6 T, which is 96.7% of the magnet design field, was reached after 35 training quenches. All of the training quenches at 1.9 K occurred in the coil IL high-field blocks. No holding quenches were detected in MBHSP03 over ~30 min at various steady currents up to the nominal LHC operating current of 11.85 kA. The observed variations of quench currents in MBHSP03 were likely to be due to epoxy cracking between the pole blocks and coil turns caused by the inadequate coil pre-stress. Therefore, to avoid possible conductor degradation, magnet training was discontinued.

It is interesting to note that, despite the different strand design and critical current density, and the coil pre-stress, the first 18 quenches at 4.5 K that were normalized to the corresponding magnet SSLs for both dipole models are very close. The training rates of the magnets at 1.9 K are, however, quite different. MBHSP03 (low pre-stress) was trained to 85% of its SSL bore field after 35 quenches, whereas MBHSP02 (high pre-stress) needed 65 quenches to reach 83% of its SSL bore field. Since MBHSP03 training at 1.9 K was not completed and the magnet was not quenched again at 4.5 K, it is unknown if the conductor degradation was reduced as well.

Another attempt to better understand the effect of the coil preload in the dipole structure on magnet training and degradation was made using coil 8, previously tested in the mirror structure MBHSM01, and new coil 11, which was built using modified end-spacers with flexible legs and reduced azimuthal and axial rigidity thanks to small radial holes. These two coils were assembled without collars in a dipole configuration using the dipole mirror structure with a thin stainless-steel shell between the coils and the iron yoke. The collarless dipole model MBHSP04 was tested in June–July 2015. Magnet training was only performed at 1.9 K. After 10 quenches the magnet quench current reached a stable plateau at ~10.5 kA (see Fig. 8.13), which corresponds to a bore field of 10.7 T. The magnet SSL at 1.9 K based on coil 8 witness sample data is 12.7 T. Thus, this magnet reached ~84% of its conductor limit, as did MBHSP02 and MBHSP03 with collared coils. All of the quenches were detected in the previously tested coil 8. This suggests that coil 8 could be degraded during the disassembly of MBHSM01 or during the assembly of MBHSP04.

To check if coil 11 was also degraded during MBHSP04 assembly, it was tested in the dipole mirror structure MBHSM02 in March–April 2017. The first six training quenches of this magnet at 4.5 K are shown in Fig. 8.13. One can see that coil 11 demonstrated even better performance than the virgin coil 8, confirming that it was not damaged during MBHSP04 assembly and disassembly, or during the assembly of MBHSM02.

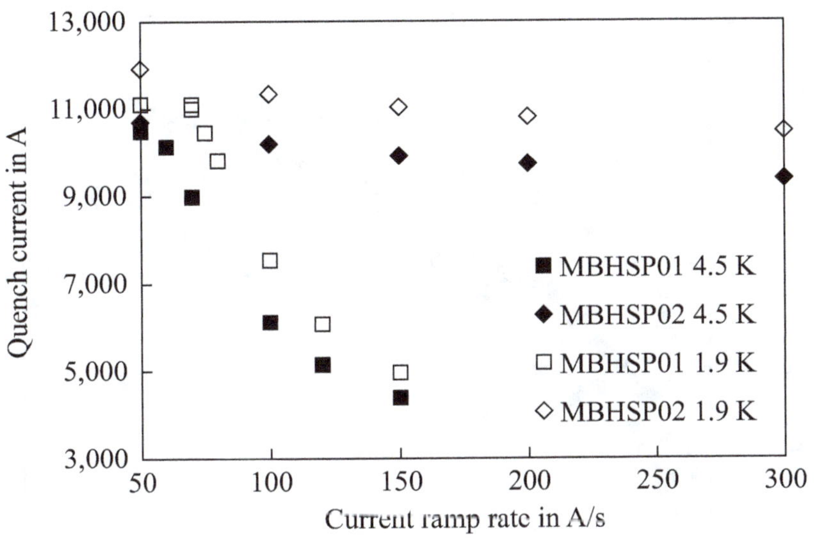

Fig. 8.14 Ramp rate dependence of magnet quench current. The data at 4.5 K and 1.9 K are represented with filled and non-filled markers, respectively

A good performance of coil 11 could be a result of its design improvements, of the new end-spacers in particular. The evaluation of the new design of end-spacers was planned by testing coils 11 and 12 in single-aperture dipole MBHSP05 (with or without collars). Due to a change of program priorities, however, this magnet was not assembled and tested.

All of the short models, except for MBHSP01, used the cable with a stainless-steel core. The positive effect of the core on the magnet ramp-rate sensitivity is shown in Fig. 8.14, where MBHSP02 ramp rate sensitivity is compared with MBHSP01, which was made from a cable without a core.

Twin-Aperture Model

The twin-aperture dipole model MBHDP01 was tested for the first time in February–March 2015 (thermal cycle TC1). A year later, in June–July 2016, it was re-tested (thermal cycle TC2) with a new instrumentation header that permitted the installation of an anti-cryostat for magnetic measurements in one of the two apertures. The anti-cryostat with magnetic measurement probes was placed in the aperture with coils 9 and 10 used in MBHSP03.

The main goals of the MBHDP01 test were: (a) comparison of collared coil performance in single-aperture and twin-aperture configurations; (b) observation of the effect of coils 9 and 10 disassembly and re-collaring with higher pre-stress, and the effect of the smaller bending of coils 5 and 7 on magnet training and conductor degradation.

The MBHDP01 training was performed in superfluid helium at 1.9 K. Quenches occurred in all four coils, which is not surprising since the mechanical stress was changed in both collared coils. The values of the bore field in MBHDP01 and, for comparison, in MBHSP02 and MBHSP03 during magnet training are plotted in

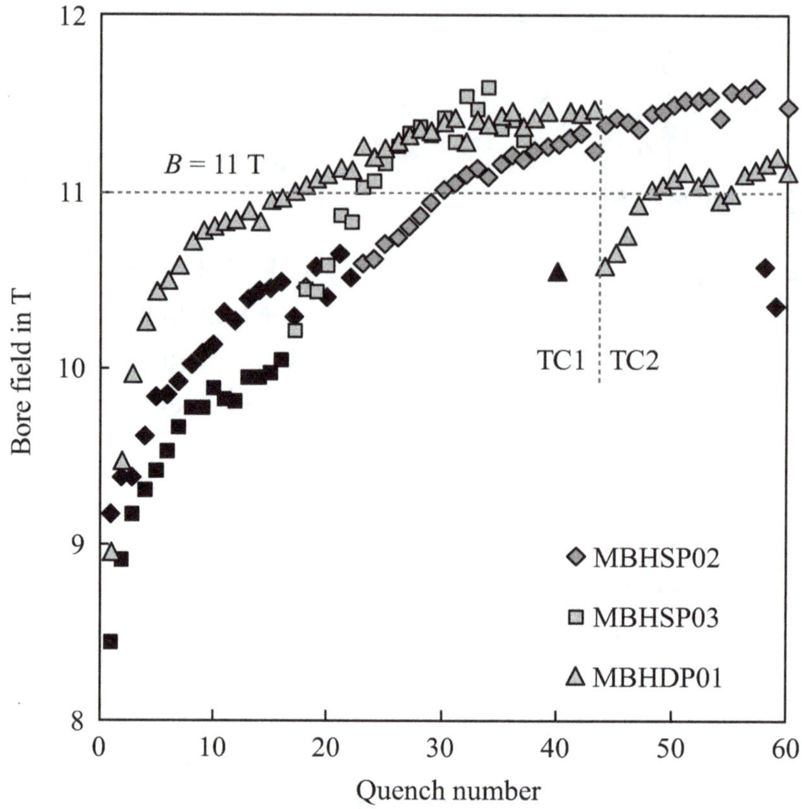

Fig. 8.15 Quench bore field during magnet training. The data at 4.5 K and 1.9 K are represented with filled and non-filled markers, respectively

Fig. 8.15. The bore field was calculated using the measured quench currents and the magnet TFs. The first low-current quenches and the quenches at the highest currents occurred in the IL blocks of coil 7 and coil 10 (the same location and same coils as in the corresponding single-aperture models). The training curve exhibits two regions: the first region with faster training rates, and the second with slower training rates.

The first quench in twin-aperture MBHDP01 occurred at a bore field of ~9 T, as in the single-aperture dipole models. A maximum bore field of 11.5 T was reached at a current of 12.1 kA, which is only 0.1 T lower than for the single-aperture models. It can be seen that MBHDP01 training slowed down but still continued.

MBHDP01 training in TC2 started ~9% lower than the maximum bore field achieved in TC1. Magnet re-training was also rather long; after 17 quenches the magnet still quenched below the bore field level reached in TC1. Most of the training quenches in TC2 occurred in coil 10, although each of the four coils quenched at least once, typically at the same locations as in TC1, around the IL middle wedges. The data from the voltage taps and the quench antenna also indicate that the training quenches started near the body-end transition regions.

The ramp-rate dependencies of the MBHDP01 and MBHSP02 bore fields at 1.9 and 4.5 K are plotted in Fig. 8.16. There is a very good correlation of the data at ramp

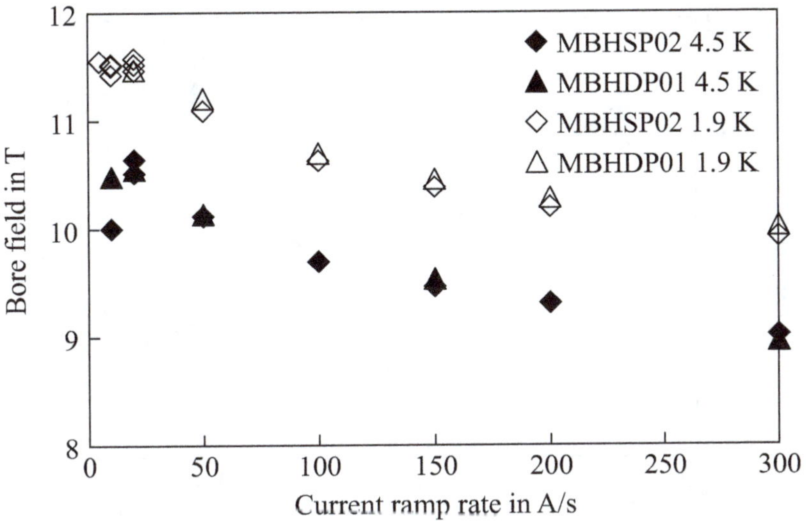

Fig. 8.16 Ramp rate dependence of quench bore field. The data at 4.5 K and 1.9 K are represented with filled and non-filled markers respectively

rates above 20 A/s at both temperatures. All ramp-rate quenches in MBHSP02 and MBHDP01 started in coil 7 at the same location as the training quenches. A relatively low ramp-rate sensitivity of the magnet quench current is due to the use of cables with a stainless-steel core, which suppresses the inter-layer eddy currents in the cable.

Quenches at ramp rates below 20 A/s indicate that the magnet training was not completed at 1.9 K or that the magnet performance was limited by some other effects, for example the current redistribution. The shape of the ramp-rate dependencies at high current ramp rates points to a non-uniform current distribution in the cable, which is also consistent with the axial harmonics variations observed in MBHSP03. Extrapolation of the ramp-rate curves at high ramp rates to zero gives a B_{max} ~11 T at 4.5 K and 12 T at 1.9 K.

Temperature dependencies of the quench bore field in twin-aperture MBHDP01 and single-aperture MBHSP02, measured in the temperature range 1.9–4.6 K, are shown in Fig. 8.17. There is a very good correlation between the data for both dipole models at all measured temperatures.

8.3.2.2 Magnetic Measurements

Magnetic measurements were performed in all MBH models using two 16-layer rotating coil probes manufactured with printed circuit board (PCB) technology (DiMarco et al. 2013). In MBHDP01 the coils were placed in the aperture with coils 9 and 10 used in MBHSP03. The measurement data were compared with magnetic measurements in the corresponding single-aperture model MBHSP03 and calculations (Strauss et al. 2016). Figure 8.18 shows the TF for both dipole

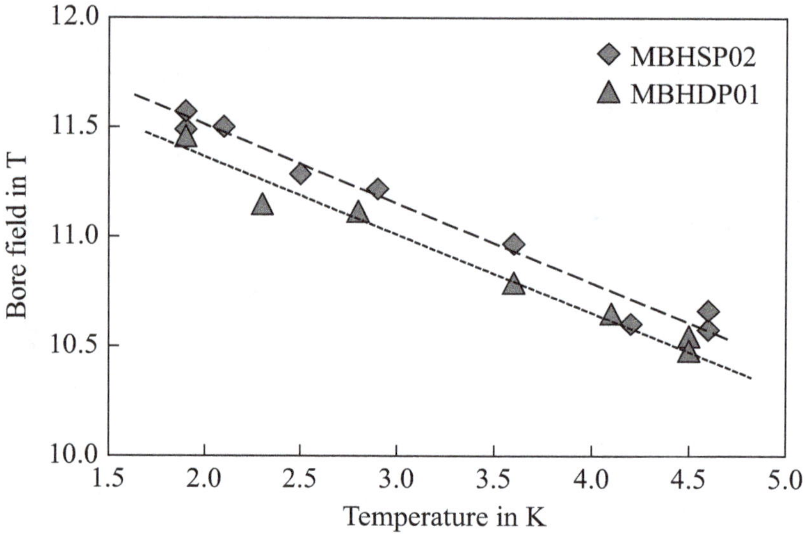

Fig. 8.17 Temperature dependence of the quench bore field in twin-aperture MBHDP01 and single-aperture MBHSP02

models. Figure 8.19 presents the evolution of b_2 and b_3 at $R_{ref} = 17$ mm vs. the magnet current at a current ramp rate of 20 A/s with a reset current of 100 A.

The iron saturation effect in the TF and b_3 at currents above 4 kA is, in general, consistent with calculations based on iron's magnetic properties and geometry (see Table 8.2). At high currents the difference between calculated and measured TF is less than 1.5%, and the difference for b_3 is less than 6 units. As expected, unlike the single-aperture dipole, the twin-aperture dipole has slightly smaller effects from iron saturation in the TF and b_3, whereas b_2 is significantly affected due to the aperture cross-talk.

The persistent current effect in the TF and b_3 is substantial in both models at low currents due to the large superconducting filament size and the high critical current density of the Nb_3Sn RRP wires used in both models (see Table 8.4). The ramp-rate effect is small, as expected for a cable with a resistive core.

There is a quite good correlation of the measured and calculated data for the persistent current effect in the TF and b_3 at currents above 1.5 kA when the coil re-magnetization is practically complete, whereas at lower currents there are large discrepancies (Andreev et al. 2013). Therefore, the coil magnetization effect at low currents was studied experimentally at various reset currents in the pre-cycle. The results are shown in Fig. 8.20.

The studies have shown that, due to the relatively long re-magnetization of the Nb_3Sn coils, b_3 at the LHC injection field strongly depends on the reset current. For the Nb_3Sn wires used in the MBH models and a reset current below 100 A, however, this process is practically complete prior to reaching the LHC injection current. It significantly simplifies b_3 correction in the 11 T Nb_3Sn dipoles at injection and at the beginning of acceleration. With the present reset current in the LHC of 100 A, RRP108/127 wire can be used in production magnets for the LHC collimation system upgrade.

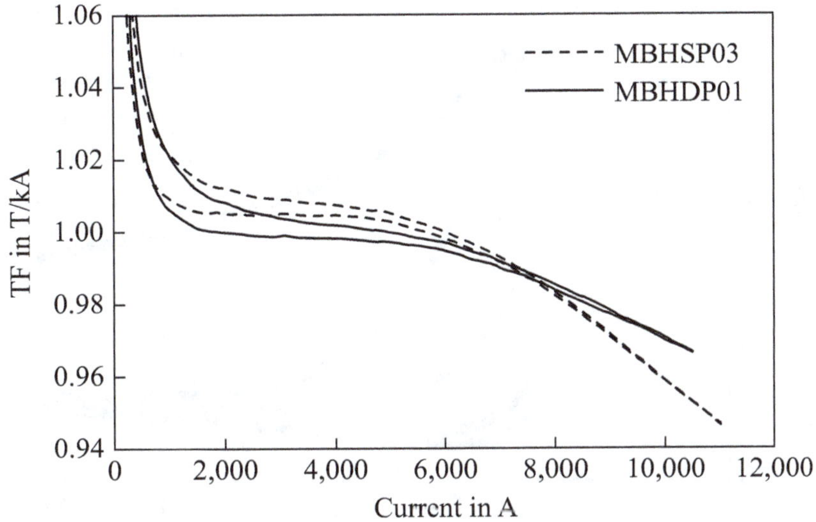

Fig. 8.18 Transfer function vs. current in single- and twin-aperture models

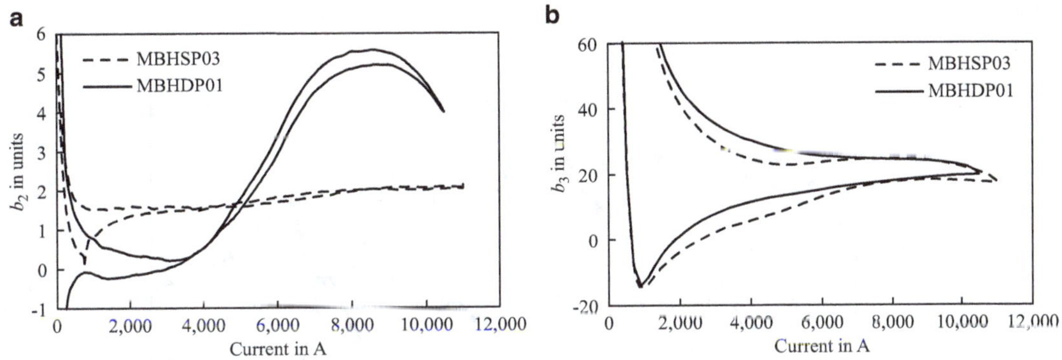

Fig. 8.19 (**a**) b_2 and (**b**) b_3 vs. current in single- and twin-aperture models

The b_3 decay was measured in all short models at an LHC injection current of 760 A. It is shown for various reset currents for MBHDP01 in Fig. 8.20. In all the 1 m long magnets, the b_3 decay is reproducible and quite large, within 4–7 units, unlike for the 2 m long MBHSP01 (Andreev et al. 2013) and previously tested Nb_3Sn dipoles (Barzi et al. 2002). The cause of the unexpectedly large b_3 decay in some MBH dipole models remains unknown. One possible explanation could be local core damage (e.g., in the coil ends where the cable experiences large and complex bending deformations), which leads to a local decrease of the inter-strand resistance in these areas.

Axial variations of the normal b_2 and skew a_2 quadrupole components were measured in MBHSP03 using a 26 mm long probe at a magnet current of 6 kA. The harmonic variations shown in Fig. 8.21 had a period comparable to the cable transposition pitch. This periodic variation may indicate a non-uniform current distribution in the cable cross-section, which could also be the cause of the large degradation of the magnet quench currents observed in the models discussed.

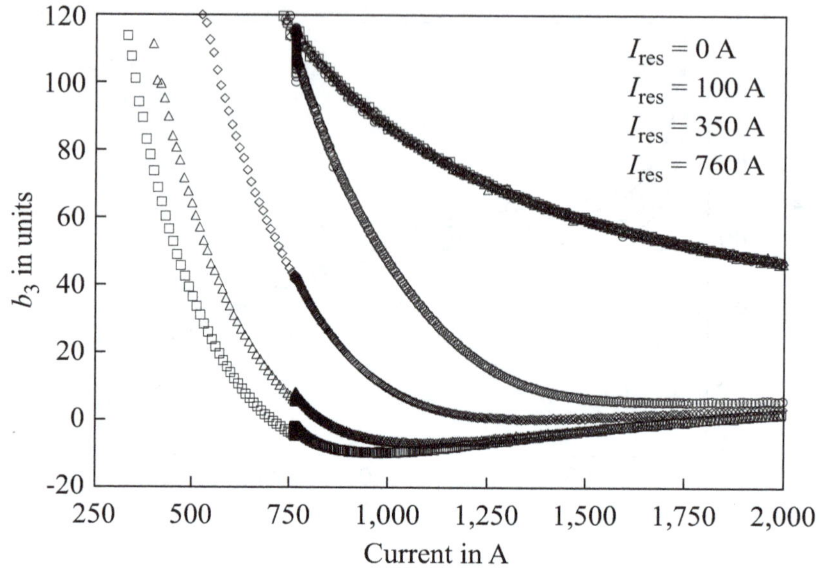

Fig. 8.20 Variations of b_3 at low currents at various reset currents I_{res} in the pre-cycle measured in MBHDP01

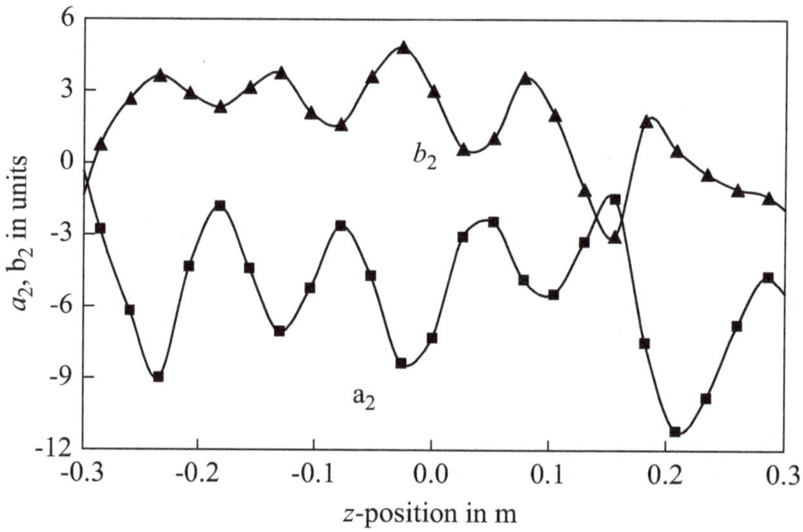

Fig. 8.21 Axial variations of b_2 and a_2 in MBHSP03

The geometrical harmonics at a magnet current of 3.5 kA for the single-aperture and twin-aperture models are summarized in Table 8.8. The resolution of the measurements is better than 0.5 units. The higher order harmonics ($n > 3$) in all of the models are small except for b_9, which is slightly larger than 1 unit in MBHSP03 and MBHDP01. On the other hand, shimming variations in the models to achieve the target pre-stress levels give rise to sizable differences in the lower-order harmonics.

Table 8.8 Field harmonics at $I = 3.5$ kA

n	MBHSP02		MBHSP03		MBHDP01	
	a_n	b_n	a_n	b_n	a_n	b_n
2	0.1	−4.9	−4.6	1.4	−3.5	0.6
3	−1.4	8.4	2.0	16.1	0.4	20.9
4	0.2	−0.2	−0.1	0.1	0.3	0.3
5	0.2	1.0	−0.1	0.8	−0.5	−0.2
6	0.0	−0.2	−0.3	−0.2	−0.1	0.4
7	−0.1	0.0	0.0	0.3	−0.5	−0.2
8	0.0	0.0	0.1	0.0	0.1	0.2
9	0.1	0.2	0.2	1.3	0.5	1.1

8.3.2.3 Quench Protection Studies

The quench protection problem of the 11 T dipoles was comprehensively studied at FNAL, including simulations and measurements of short dipole and dipole mirror models (Chlachidze et al. 2013a, b; Zlobin et al. 2012b, 2014). In dipole mirror MBHSM01 spot heaters made of stainless-steel strip were mounted on the IL and OL mid-plane turns of coil 8. Each spot heater was surrounded by two voltage taps. Two additional voltage taps, separated by 10 cm, were also installed next to the spot heater. Due to the damage to the IL spot heater wiring during magnet assembly, studies were only performed with the OL spot heater.

Quench Temperature Measurements

The coil maximum temperature after a quench is estimated based on the quench integral (*QI*) calculated over the current decay time using the adiabatic approach (see Chap. 1), and usually represented in MIITs (1 MIIT = 10^6 A^2s). Simulations of quench processes show that heat transfer from the quenched cable inside the magnet coil plays an important role. To study the effect of heat transfer from the cable, the cable temperature growth in the coil due to a quench was measured using quenches induced by the spot heater at fixed coil current. The temperature of the cable in coil was estimated using the known dependence of the copper matrix resistivity on the temperature and magnetic field.

The measured cable temperature vs. the time after quench at constant coil currents is plotted in Fig. 8.22. The dashed lines connect the temperature points corresponding to the same *QI* values. The temperature points on the vertical axis ($t = 0$) represent the adiabatic calculations for the corresponding bare cable. It can be seen that the cable temperature depends not only on the value of *QI*, but also on the time during which it is accumulated, which is consistent with efficient heat transfer from the quenched cable. Since the quench time of an accelerator magnet is usually longer than 0.2 s, traditional adiabatic calculations significantly overestimate the coil temperature after quench.

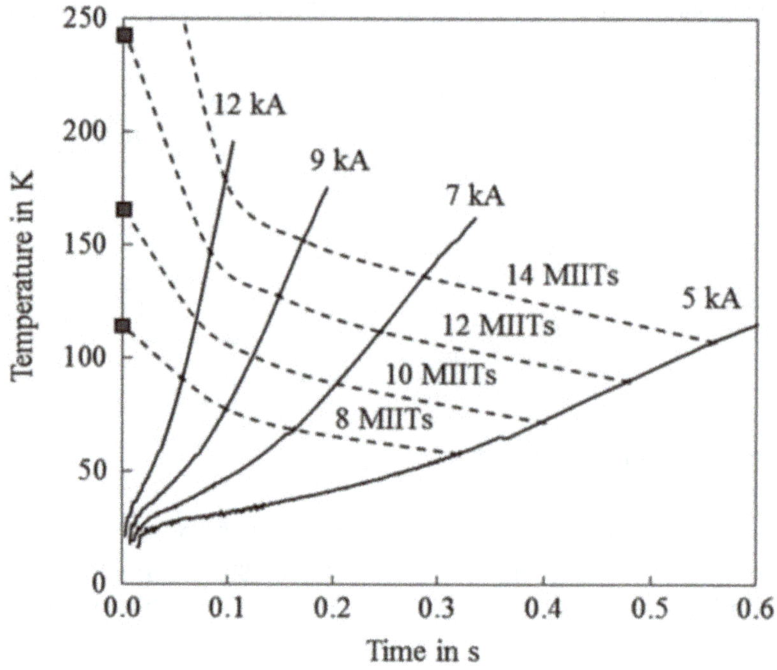

Fig. 8.22 Cable temperature after quench vs. time measured at various currents in coil (MBHSM01)

Longitudinal Quench Propagation Velocity

The quench propagation velocity along the cable in a coil is an important parameter for estimating the *QI* in the area of quench origin and for optimizing the protection heater design. The quench propagation velocity in the OL mid-plane turn was measured using the spot heater. The quench propagation velocity along the cable was determined using the slope of $V(t)$ dependence between the voltage taps next to the spot heater (method A), and the measured quench propagation time and the known distance between the two voltage taps (method B). The quench propagation velocity in the IL pole turn was estimated using the dV/dt slope during some training quenches. The measured data for the OL mid-plane turn and the IL pole turn are shown in Fig. 8.23. The experimental data in Fig. 8.23 are in agreement with calculations. These data provide an important input for the optimization of the quench detection voltage threshold and the signal discrimination time.

Radial Quench Propagation

Simulations and heater studies in 11 T dipole models revealed that a quench propagates quite rapidly in the radial direction from OL to IL coil blocks. It helps to distribute the magnet's stored energy over a larger coil volume and, thus, to reduce the coil maximum temperature. The quench delay time (the time between heater initiation and coil quench) was measured separately for the coil OL and IL blocks at 4.5 K and 1.9 K. The quench propagation time between the coil layers was estimated

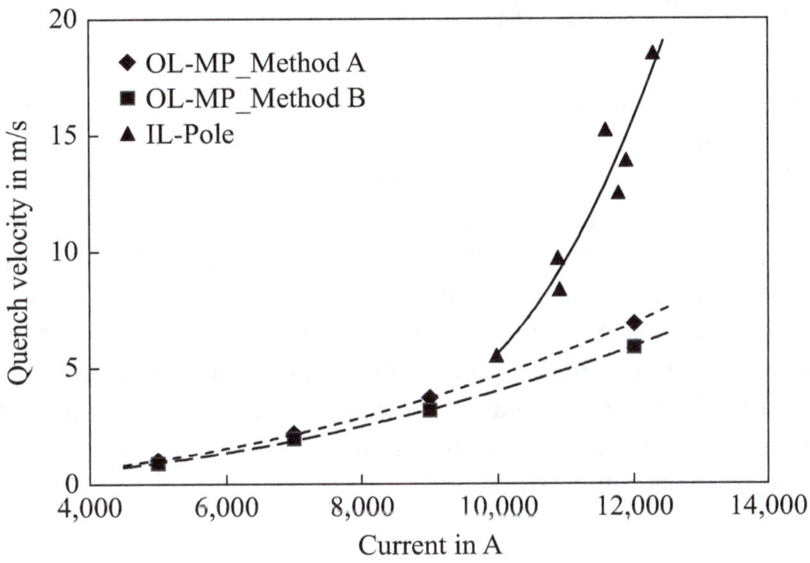

Fig. 8.23 Longitudinal quench velocity in coil (MBHSM01). MP, mid-plane; OL, outer layer; IL, inner layer

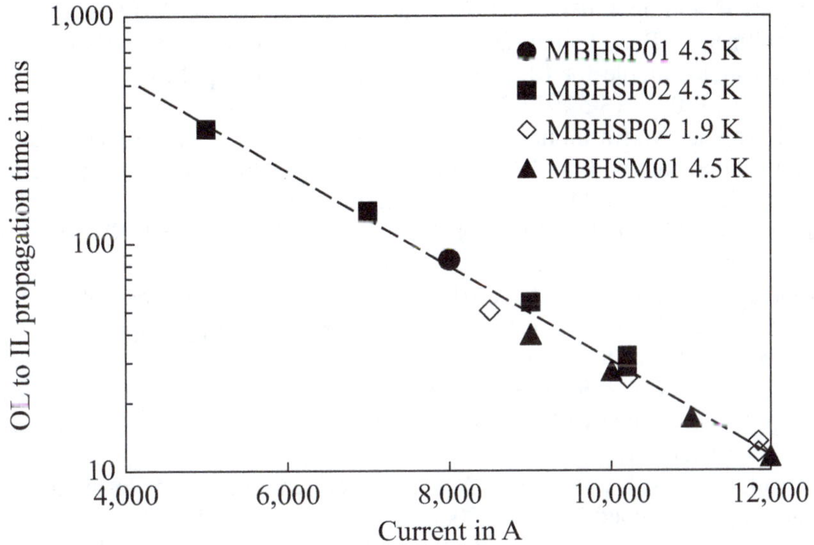

Fig. 8.24 Radial quench propagation time from outer layer to inner layer

as the time difference between the detection of a quench in the OL and IL of the coil. Figure 8.24 shows this time difference vs. the magnet current. Similar results for different coils in different magnets confirm the excellent reproducibility of this effect in Nb_3Sn coils. At currents close to the LHC nominal operating current the radial quench propagation time is smaller than 20 ms. It means that, after a heater-induced quench, the coil OL serves as an efficient quench heater for the IL coil.

8.4 FNAL 11 T Dipole R&D Summary

Dipole models with a field strength of 11 T for the LHC upgrade have been developed, fabricated, and tested at FNAL. Two single-aperture and twin-aperture models were trained to bore fields of 11.5–11.7 T, which are close to the magnet design field of 12 T. It demonstrated the viability of the magnet design and technology developed. At this point FNAL's part in the joint 11 T dipole R&D program for the LHC upgrade was completed. Some open questions, however, remain. The most important open questions are the observed long magnet training and noticeable re-training, large conductor degradation, and the non-uniform current distribution in the Rutherford cables in the magnet coils. The latter could be a cause of the spontaneous quenches at a current plateau observed in some models. These issues need to be further understood and addressed.

References

Andreev N, Barzi E, Chichili DR et al (2002) Volume expansion of strands and cables during heat treatment. Adv Cryo Eng 48(614):941–948. https://doi.org/10.1063/1.1472635

Andreev N, Apollinari G, Barzi E et al (2013) Field quality measurements in a single-aperture 11 T Nb_3Sn demonstrator dipole for LHC upgrades. IEEE Trans Appl Supercond 23(3):4001804. https://doi.org/10.1109/TASC.2013.2237819

Auchmann B, Karppinen M, Kashikhin VV et al (2012) Magnetic analysis of a single-aperture 11 T Nb_3Sn demonstrator dipole for LHC upgrades. In: Proceedings of 2012 international particle accelerator conference, New Orleans, May 2012, p 3596

Barzi E, Carcagno R, Chichili D et al (2002) Field quality of the Fermilab Nb_3Sn cos-theta dipole models. In: Proceedings of 2002 European particle accelerator conference, Paris, p 2403

Barzi E, Karppinen M, Lombardo V et al (2012) Development and fabrication of Nb_3Sn Rutherford cable for the 11 T DS dipole demonstration model. IEEE Trans Appl Supercond 22(3):6000805. https://doi.org/10.1109/TASC.2011.2180869

Bossert RC, Andreev N, Chlachidze G et al (2011) Fabrication and test of 90-mm Nb_3Sn model based on dipole type collar. IEEE Trans Appl Supercond 21(3):1777–1780. https://doi.org/10.1109/TASC.2010.2089418

Bottura L, de Rijk G, Rossi L et al (2012) Advanced accelerator magnets for upgrading the LHC. IEEE Trans Appl Supercond 22(3):4002008

Chlachidze G, Novitski I, Zlobin AV et al (2013a) Experimental results and analysis from 11 T Nb_3Sn DS dipole. FERMILAB-CONF-13-084-TD, and WAMSDO'2013, CERN-2013-006, CERN, Geneva

Chlachidze G, Andreev N, Apollinari G et al (2013b) Quench protection study of a single-aperture 11 T Nb_3Sn demonstrator dipole for LHC upgrades. IEEE Trans Appl Supercond 23 (3):4001205. https://dx.doi.org/10.1109/TASC.2013.2237871

de Rijk G, Milanese A, Todesco E (2010) 11 Tesla Nb_3Sn dipoles for phase II collimation in the Large Hadron Collider. sLHC Project Note 0019, CERN, Geneva

DiMarco J, Chlachidze G, Makulski A et al (2013) Application of PCB and FDM technologies to magnetic measurement probe system development. IEEE Trans Appl Supercond 23 (3):9000505. https://doi.org/10.1109/TASC.2012.2236596

Karppinen M, Andreev N, Apollinari G et al (2012) Design of 11 T twin-aperture Nb_3Sn dipole demonstrator magnet for LHC upgrades. IEEE Trans Appl Supercond 22(3):4901504, https://doi.org/10.1109/TASC.2011.2177625

Russenschuck S (1995) A computer program for the design of superconducting accelerator magnets. CERN AC/95–05 (MA), September 1995, CERN, Geneva

Strauss T, Apollinari G, Barzi E et al (2016) Field quality measurements in the FNAL twin-aperture 11 T dipole for LHC upgrades. In: Proceedings of 2016 North American particle accelerator conference, Chicago, 9–14 October 2016, p 158

Zlobin AV (2010) Status of Nb_3Sn accelerator magnet R&D at Fermilab. In: EuCARD – HE-LHC'10 AccNet mini-workshop on a High-Energy LHC, 14–16 October 2010, CERN Yellow Report CERN-2011-003, p 50 [arXiv:1108.1869], CERN, Geneva

Zlobin AV, Andreev N, Apollinari G et al (2011) Development of Nb_3Sn 11 T single aperture demonstrator dipole for LHC upgrades. In: Proceedings of 2011 particle accelerator conference, New York, March 2011, p 1460

Zlobin AV, Andreev N, Apollinari G et al (2012a) Design and fabrication of a single-aperture 11 T Nb_3Sn dipole model for LHC upgrades. IEEE Trans Appl Supercond 22(3):4001705. https://doi.org/10.1109/TASC.2011.2177619

Zlobin AV, Novitski I, Yamada R (2012b) Quench protection analysis of a single-aperture 11 T Nb_3Sn demonstrator dipole for LHC upgrades. In: Proceedings of 2012 International particle accelerator conference, New Orleans, May 2012, p 3599

Zlobin AV, Andreev N, Apollinari G et al (2013) Development and test of a single-aperture 11T Nb_3Sn demonstrator dipole for LHC upgrades. IEEE Trans Appl Supercond 23(3):4000904. https://doi.org/10.1109/TASC.2012.2236138

Zlobin AV, Chlachidze G, Nobrega F et al (2014) Quench protection studies of 11 T Nb_3Sn dipole coils. In: Proceedings of 2014 international particle accelerator conference, Dresden, 15–20 June 2014, p 2725

Chapter 9
Nb_3Sn 11 T Dipole for the High Luminosity LHC (CERN)

Bernardo Bordini, Luca Bottura, Arnaud Devred, Lucio Fiscarelli, Mikko Karppinen, Gijs de Rijk, Lucio Rossi, Frédéric Savary, and Gerard Willering

Abstract This chapter describes the design of, and parameters for, the 11 T dipole developed at the European Organization for Nuclear Research (CERN) for the High Luminosity Large Hadron Collider (HL-LHC) project. The results for testing the short models as well as the first 5 m long magnet are also presented. Finally, the production process for the six units, of which four are to be installed in the HL-LHC, is described, with a description of the process for planning the installation. In 2020, these dipoles should be the first Nb_3Sn magnets working in an accelerator.

9.1 Introduction

The Large Hadron Collider (LHC) will undergo a major upgrade to increase the luminosity of proton–proton collisions. The upgrade, called the High Luminosity LHC (HL-LHC or HiLumi LHC) (Apollinari et al. 2017; Brüning and Rossi 2015), calls for a more intense proton beam, with a circulating current of 1.1 A vs. the 0.56 A nominal current value in the LHC. The intensity of the ion beams (usually Pb ions) for ion–ion collisions will actually be increased by a factor of three: from 40×10^9 to 120×10^9 circulating particles. This intensity increase, both for protons and ions, will increase the diffractive losses at the primary collimators, located in LHC Point 7 (P7), which may drive the heat losses in the main dipoles in the dispersion suppressor (DS) region above the quench limit.

B. Bordini · L. Bottura · A. Devred · L. Fiscarelli · M. Karppinen · G. de Rijk · F. Savary · G. Willering
CERN (European Organization for Nuclear Research), Meyrin, Genève, Switzerland
e-mail: Bernardo.Bordini@cern.ch; Luca.Bottura@cern.ch; Arnaud.Devred@cern.ch; Lucio.Fiscarelli@cern.ch; Mikko.Karppinen@cern.ch; Gijs.Derijk@cern.ch; Frederic.Savary@cern.ch; Gerard.Willering@cern.ch

L. Rossi (✉)
CERN (European Organization for Nuclear Research), Meyrin, Genève, Switzerland

University of Milan, Physics Department, Milano, Italy
e-mail: Lucio.Rossi@cern.ch

D. Schoerling, A. V. Zlobin (eds.), *Nb_3Sn Accelerator Magnets*, Particle Acceleration and Detection, https://doi.org/10.1007/978-3-030-16118-7_9

To avoid limiting the machine due to this effect, various countermeasures have been studied, and the solution chosen was to intercept these diffractive losses via warm absorbers (also called *collimators*) placed in the cold region of the LHC. The most elegant and practical way to introduce a room-temperature zone in the DS, at a location corresponding to the middle of the second dipole of the DS cell, was to substitute an LHC dipole (8.33 T of central field and 14.3 m of magnetic length) with a magnet of 11 T with a length of approximately 11 m, yielding the same bending strength while saving about 3.5 m. The 11 T field in the magnet bore inevitably calls for Nb_3Sn technology. The space gained, thanks to the higher field and shorter length, is sufficient to insert a cold–warm–cold bypass on which to allocate all lines for cryogenic and electrical continuity of the circuits in series in the LHC arc, and the target collimator long dispersion suppressor (TCLD). For reasons of beam dynamics (reduction of the orbit excursion with respect to the ideal trajectory) and to reduce the technology risk associated with the innovative and relatively expensive Nb_3Sn superconductor, we decided to split the 11 T dipole into two magnets of 5.5 m length, with the bypass and collimator in the middle. A schematic layout of the assembly with its position in the LHC is shown in Fig. 9.1.

9.1.1 History of the 11 T Project

The possibility of using 11 T dipoles in the LHC was first envisaged at the LHC Performance Workshop in January 2010 (Carli 2010, 29). After a preliminary study (de Rijk et al. 2010), it was included in the HL-LHC project and in the main goals of the European Organization for Nuclear Research (CERN) High-Field Magnet (HFM) program (Bottura et al. 2012). The project was set up after the end of 2010 as a close collaboration between CERN and the Fermi National Accelerator Laboratory (FNAL). FNAL has carried out an extensive research and development (R&D) program on dipoles for reaching 10–11 T for the Very Large Hadron Collider (VLHC) (see also Chap. 7, Cos-theta Nb_3Sn Dipole), besides participating in the US LHC Accelerator Research Program (LARP) (Zlobin 2010). This was in a period when CERN was restarting Nb_3Sn R&D, with a main focus on the Inner Triplet program for the HL-LHC.

A parallel program was established, with two companion lines: one in FNAL and another in CERN, as described in detail in Chap. 8, The Nb_3Sn 11 T Dipole for the High Luminosity LHC (FNAL). The initial scope was to design and manufacture short models and one 5.5 m long prototype in each laboratory, allowing for small differences in design. In this program, intensive technology transfer from FNAL to CERN took place, especially in the domain of Nb_3Sn coil technology.

During its course the plan was modified, mainly because the funding profile at FNAL was reduced and shortened, so FNAL concentrated on the design, construction, and test of a few short models, initially with a length of 2 m, and then 1 m.

A detailed description of the design, and the similarities and differences between the FNAL and CERN designs, as well as the initial R&D, can be found in Chap. 8,

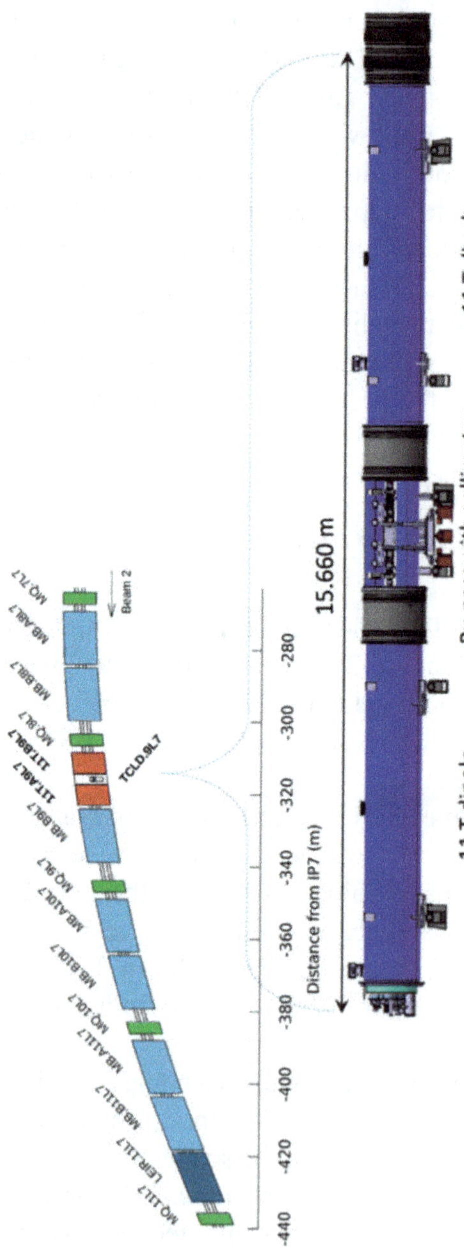

Fig. 9.1 Schematic of the 11 T cryo-assembly, with the bypass hosting the collimator at the center. In the dispersion suppressor (DS) cells (the array in the top where LHC dipoles are in light blue) one standard dipole is replaced by the 11 T cryo-assembly, indicated in red

The Nb_3Sn 11 T Dipole for the High Luminosity LHC (FNAL), and are not discussed here. In this chapter we concentrate on the specifics of the R&D at CERN, on the design's main features, as well as short model testing, both single and two-in-one apertures, and on the industrialization aspects related to long magnet production for the HL-LHC.

9.1.2 Present Scope of the 11 T Project

It is foreseen that most of the HL-LHC installation work, including the Nb_3Sn Inner Triplet quadrupoles, will take place during the period 2024–2026, known as Long Shutdown 3 (LS3). The LHC Injectors Upgrade (LIU) project, which will allow the production of more intense beams for the LHC (Damerau et al. 2014), will, however, already have been carried out during the period 2019–2020, known as Long Shutdown 2 (LS2). It is then foreseen that ion beams will be collided at their maximum intensity during LHC Run 3 (2021–2023), i.e., after LS2. In the same period, HL-LHC high-intensity proton beams will also be tested. For this reason, the diffractive losses in P7 described above could play a limiting role in LHC Run 3.

To avoid this limitation, it was decided in 2010 to prepare the 11 T dipole with its collimation system for installation during LS2, requiring that both 11 T magnet assemblies (one per side with respect to LHC P7) and their ancillary systems are ready by 2019. Originally, it was foreseen that further sets of 11 T dipoles would be installed in LS3: an additional set in both sides of the P7 DS and around the Point 2 (P2) interaction point (Alice detectors), for a total of 16 11 T dipoles, each 5.55 m in length. Because of a re-evaluation of heat deposition (in particular in P7), and because other solutions could be devised with modified beam optics, it was later decided that these additional 11 T dipoles would no longer be necessary. They have been removed from the HL-LHC baseline, which is hence presently limited to four 11 T dipoles, to be installed in two sets around Point 7 at LHC locations MB.AR7 and MB.AL7.

9.2 Magnet Design

9.2.1 Design Constraints

The main constraint is due to the fact that the 11 T dipole pair is a part of the LHC main circuit and, as such, is powered in series with the other 153 LHC two-in-one-aperture dipoles comprising one of the eight LHC main dipole circuits. Each 11 T dipole pair must have an integrated transfer function (ITF) as near as possible to that of the LHC main dipoles over the whole dynamic range, from the injection energy of 0.45 TeV (7.7 Tm of bending strength) to the collision energy of 7 TeV (119.2 Tm of bending strength). These values for the bending strength must be obtained at the nominal current of the main bending (MB) circuit, i.e., 760 A at the 0.45 TeV injection energy and 11.85 kA at the 7 TeV collision energy. A second condition was to respect the LHC's basic geometry: the distance between the center of the two apertures (194 mm at 1.9 K) and the position of the cryogenic and electrical lines passing through the yoke. Finally, the timescale did not allow for a long period of R&D. It was therefore decided to rely on a design concept as close as possible to that of the LHC MB dipoles, from which CERN has accumulated more than two decades of experience (Rossi 2004): the two-layer cos-theta coil layout, with the force supported by a classical collar structure, a vertically split iron yoke, and an external shrinking cylinder also serving as a helium vessel. Thanks to CERN's experience and to the FNAL program on the Nb_3Sn cos-theta layout with collars, the initial decision on the layout was taken.

The preliminary design study proved that all main design criteria, i.e., field strength with a reasonable ITF over all of the dynamic range, critical current margin, and field quality (both at injection and at high field), could be met with the available Nb_3Sn wire properties, despite the considerable difficulties given by the various constraints. Structural and protection issues were left to a more detailed, realistic design.

9.2.2 Basic Features

Different from the LHC MB, the 11 T dipoles have separate stainless-steel collars for each aperture, compared to common stainless-steel collars in the case of the LHC MBs. This design change was introduced to allow a better-controlled symmetric loading of the coils and make it possible to test the collared coils in a one-in-one configuration without the need for de-collaring prior to integration in the two-in-one cold mass. To make maximum use of the existing infrastructure and cold-mass assembly tooling, the outer contour of the cold mass was chosen to be identical to the LHC MBs. The location and the section of the slots' bus-bars was to be preserved, as well as the location of the heat exchanger in the iron yoke.

A nominal field of 11 T requires a magnet that is about 11 m long. To reduce the risks associated with the fabrication of brittle Nb_3Sn coils, the original 11 m long magnet was split into two units, each 5.5 m long and with straight coils, not bent as in the MBs. The sagitta of the beam trajectory in each 5.5 m long straight magnet is only around 2 mm compared to 9 mm in a standard MB. During the initial phase of the project, it was considered important to compensate for the effect of the sagitta on the free aperture by enlarging the coil aperture to 60 mm (this is 56 mm in the LHC dipoles). Later, it was decided to use existing spare beam screens from the LHC without an increased free aperture.

The field quality targets were similar to those of the LHC MB, and at the 10^{-4} level at the reference radius of 17 mm, and special attention was to be paid to the multipoles arising from persistent currents induced in the inherently larger filaments of the available Nb_3Sn strands and the higher yoke saturation (with a larger magnetic flux conveyed in the same yoke as the MB).

A further important advantage of this solution is the possibility of placing the collimator between the two 5.5 m 11 T dipoles, reducing the orbit excursion, as mentioned above.

To ensure reliable operation, the design goal was to provide an operational margin of 20% on the load line, as for other Nb_3Sn HL-LHC magnets.

The initial design of the 11 T dipole is described by Karppinen et al. (2012) and Zlobin et al. (2012).

9.2.3 Conductor Choice and Nominal Dimensions

The parameters of the strands and the Rutherford cable were selected based on the required number of ampere-turns to generate the requested ITF under the 20% operating margin, the available coil space, and the maximum number of strands possible in the cabling machine. For this last constraint, the most stringent limits were between CERN (40 strands), and FNAL (42 strands). The selected strand diameter was 0.7 mm, with an expected cable thickness in the range 1.2–1.3 mm (depending on the allowed compaction). The strand geometry and performance specification are summarized in Table 9.1.

The optimization of cable parameters was done jointly by FNAL and CERN (Zlobin et al. 2012), and included the selection of the cable cross-section geometry and compaction to achieve good mechanical stability of the cable and acceptable I_c degradation (less than 10%), incorporating a stainless-steel core (25 μm thickness), and preserving a high residual resistivity ratio (RRR) of the Cu matrix (RRR larger than 100 in extracted strands).

Various strand designs, with different technologies (Restack Rod Process® (RRP) and PIT), sub-element number, size, and distribution in the cross-section were considered for the optimization. Trial cabling runs at FNAL in the R&D phase were done using Nb_3Sn wires produced using the Restack Rod Process® by Oxford Superconducting Technologies (OST) (see Chap. 8, The Nb_3Sn 11 T Dipole for the High Luminosity LHC (FNAL)). The architectures considered were baseline RRP108/127, as well as alternative layouts, mainly the RRP150/169 and RRP132/

Table 9.1 Nb_3Sn wire geometry and specifications

Description	Value
Strand diameter (mm)	0.70
J_c (12 T, 4.2 K) (kA/mm^2)	>2.45
Effective filament size D_{eff} (μm)	<41
Twist pitch (mm)	14
Cu RRR (virgin state)	>150
Cu fraction (%)	53.5 (±2)

RRR residual resistivity ratio

Table 9.2 Cable geometrical parameters

Parameter	Un-reacted	Reacted
Mid-thickness (mm)	1.25	1.30
Thin edge (mm)	1.15	1.19
Thick edge (mm)	1.35	1.40
Width (mm)	14.70	15.08
Keystone angle (°)	0.79	0.81

169 layouts, including a low RRR RRP108/127. A cable baseline was generated, with the parameters reported in Table 9.2.

Cabling trials at CERN considered both OST-RRP strands, as well as powder-in-tube (PIT) wires from Bruker European Advanced Superconductors (B-EAS). Most of this work was based on two architectures: PIT 114 and PIT 120. A direct result of this R&D was the observation that PIT is much more sensitive to cable compaction than RRP. The baseline cable geometry resulted in an average degradation of 6%, with cases largely in excess of the 10% acceptance limit. To remain compatible with PIT it was found necessary to reduce the keystone angle from the baseline 0.79° to the range 0.4–0.5°. This corresponds to an increase of the cable's thin edge by 40 μm, which was found to reduce both J_c and RRR degradation due to cabling. It should be noted that this change could not be absorbed in the rather strict cross-section, and would require an iteration of the magnetic design.

At this stage of the project a decision on baseline strand and cable geometry was taken, based on the following observations.

- PIT technology at this geometry (small strands) and field level (12 T) has a critical current density up to 20% lower than that of RRP wires (below specification), is more sensitive to cabling compaction, and has a higher cost. While in the initial stage of the project the volume of magnets was sufficient to justify this alternative route, following re-scoping in 2016 the PIT variant was discarded.
- Among the various RRP architectures, the 132/169 and 150/169 layouts have the advantage of a smaller filament size, which is of importance to reduce persistent current effects. The 108/127 layout, on the other hand, had a lower cost owing to better industrial production maturity. For this reason the RRP108/127 was taken as reference, and the alternative RRP strands were discarded.

The final geometrical parameters and performance considered for the magnet design are, then, those reported in Table 9.1 (strand) and Table 9.2 (cable). During the reaction, Nb_3Sn strands expand due to the phase transformation (Andreev et al. 2002). The parameters of the cable cross-section are presented for un-reacted and reacted cables. The geometrical parameters for the coil winding and curing tooling were determined from the un-reacted cable cross-section, whereas the geometrical parameters for the coil reaction and impregnation tooling were based on the reacted cable dimensions. The latter were also used for the magnet electromagnetic and structural optimization.

Cable samples for qualification and model winding, with rectangular and keystoned cross-sections, with and without a stainless-steel core, were fabricated and initially tested at FNAL, and later also at CERN. Based on the experimental data, the measured I_c degradation from cabling after optimization was less than 4%.

9.2.4 Magnetic Design

The main electromagnetic design challenges of the two-in-one-aperture 11 T dipole are to match the MB ITF, to control the magnetic cross-talk between apertures, and to minimize the magnitude and variation of non-allowed multipoles (Auchmann et al. 2012).

The coil cross-section was optimized using the reacted cable parameters and a 100 μm insulation layer around the cable. The early-stage preliminary design used the iron yoke shown in Fig. 9.2a, leaving a radial space of about 30 mm for the collars. The optimal configuration, delivering 11.21 T at 11.85 kA in a two-in-one configuration, was found with a six-block layout of 56 turns with 22 turns in the inner layer and 34 turns in the outer layer, as shown in Fig. 9.2b.

The coil ends were optimized first to find the optimal mechanical configuration based on easy- and hard-way strain in the cable, as well as the amount of torsion over the unit length. The lead-end optimization also included the layer jump and the transitions between the winding blocks. The relative axial positions of the end blocks were then optimized to minimize the integrated harmonics.

It was decided to use this coil design for the short models and then, before scaling up to the full length, re-optimize the coil cross-section with the experimental data from the magnetic measurements of the short model magnets along with the feedback from coil fabrication.

The nature of the magnetic flux pattern in the two-in-one yoke configuration requires additional features to minimize the cross-talk between the apertures, in particular for the b_2 component. The two holes in the yoke insert reduce the b_2 variation from 16 units to 13 units. Such large cross-talk indicates that the distance between the apertures and the overall size of the yoke would need to be increased for such a high field magnet. The magnetic design was optimized for the two-in-one configuration for the accelerator. Most models were assembled and tested as a single aperture structure, without any attempt to adjust the field quality for such a configuration.

In dipole magnets the iron saturation typically gives rise to a variation of the b_3 and b_5 components. The shape of the cut-out on top of the aperture was optimized to reduce the b_3 variation by 4.7 units when compared to a circular shape. In addition to the holes in the yoke insert, an array of three smaller holes and the position of the hole for the tie-rods were used to further reduce the b_3 variation by 2.4 units, such that the iron saturation induced a variation in b_3 and b_5 of 0.51 and 0.35 units, respectively, between injection and nominal current.

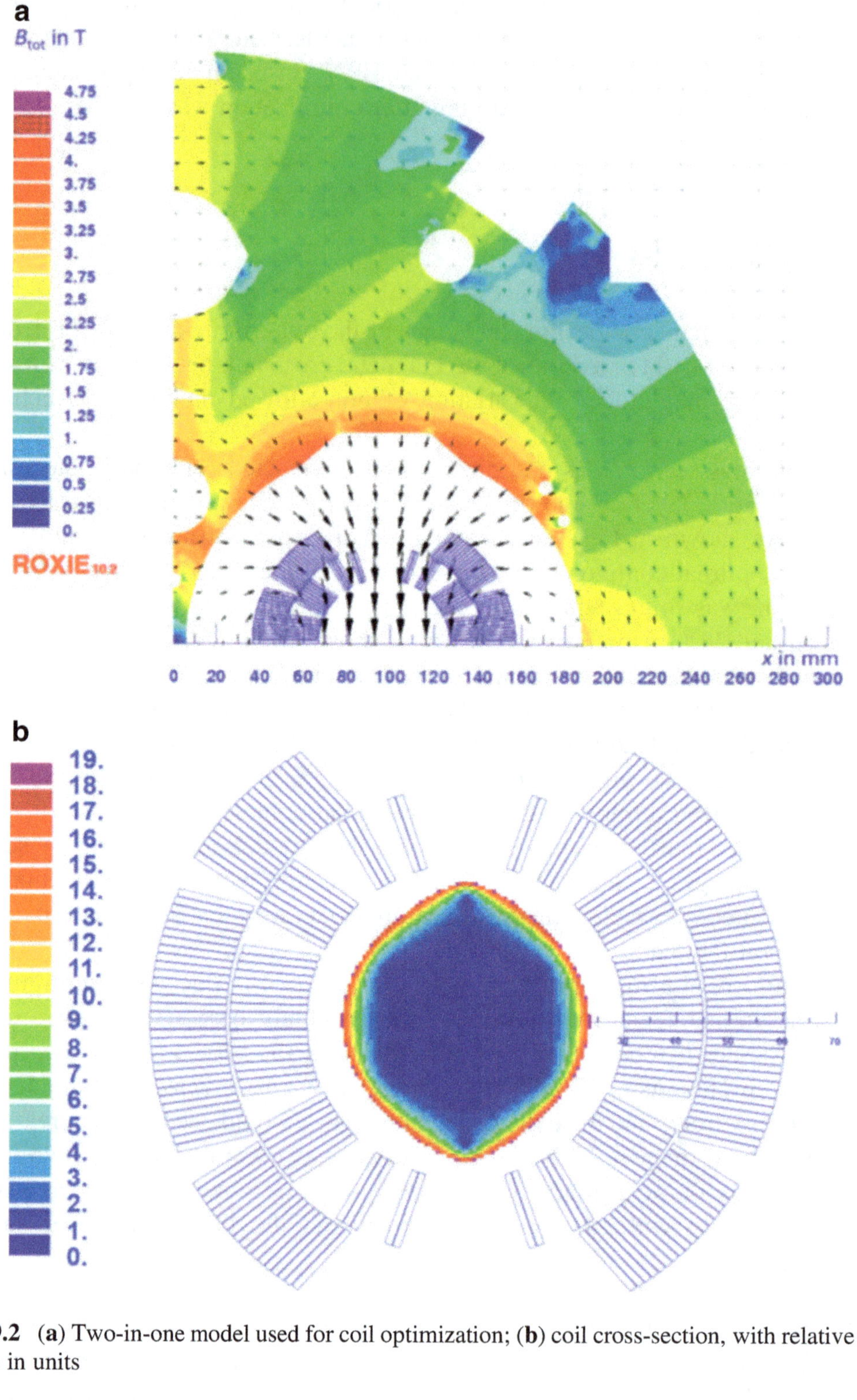

Fig. 9.2 (**a**) Two-in-one model used for coil optimization; (**b**) coil cross-section, with relative field errors in units

Due to the stronger iron saturation effect, the Nb_3Sn dipoles will be stronger than the MB at intermediate excitation levels, the peak difference being 2.4 Tm at 6.7 kA. This difference can be compensated with the foreseen 250 A bi-polar trim power converters to be installed across the 11 T dipoles.

Owing to the larger filament size in Nb_3Sn strands when compared to Nb-Ti strands, the scaled b_3 component due to the persistent currents in the 11 T dipole is about 44 units at the LHC injection current. The magnetization effect strongly depends on the current pre-cycle and the lowest current reached during the cycle, the so-called reset current $I_{res.}$ Using $I_{res} = 100$ A the sextupole component can be reduced to stay within ±20 units between I_{inj} and I_{nom}, which is acceptable for the LHC (Holzer 2014). In addition, as shown by the measurements quoted below, flux jumps strongly affect the filament magnetization below 1 T and reduce the projected b_3 effect, which is a beneficial effect at the expense of reproducibility.

9.2.5 *Mechanical Design*

The goal of the mechanical design is to provide a rigid clamping of the superconducting coil with minimum distortion of the conductor positioning, whilst maintaining at all times the stresses at an acceptable level for the fragile Nb_3Sn. A detailed structural analysis was carried out to explore the optimal parameter space for the magnet assembly.

The design concept for the magnet was inspired by the 1 m long LHC MBFISC model magnet (Ahlbäck et al. 1994), and the collar thickness was chosen for maximum rigidity in the available space to minimize the spring-back effect after the collaring process.

The Nb_3Sn cos-theta accelerator magnets developed at FNAL typically feature the pole being heat-treated and potted in the coil. For a quadrupole it is still possible to apply some pre-stress at the pole through the mid-plane, but for a dipole the angular difference between the pole turns and the mid-plane makes this much more complicated. It either requires deforming the coil elliptically in the mid-plane or applying a uniform radial pressure on the coil's outer surface. The first, most commonly applied, method requires a very delicate balance between overloading the mid-plane turns and not unloading the pole turns at full field. For this reason, an integrated pole is not optimal for a collared dipole structure. To overcome such a shortcoming a new concept based on a removable pole was developed for this project (Karppinen et al. 2012). It allows well-controlled adjustment of the coil pre-compression at the poles without overloading the mid-plane. An additional Cu-alloy filler wedge, which is potted together with the coil, is added to the outer layer to match the azimuthal size of the inner layers to simplify the pole wedge geometry as shown in Fig. 9.3. To protect the fragile Nb_3Sn coils during the collaring process, 2 mm thick stainless-steel loading plates are added at the pole, and a separate impregnation pole provides the accurate size for the pole region of the impregnated coil.

In the collared coil the stress pattern can be tuned by varying the thickness of the shims at the pole. The pole shim opens the possibility of achieving an optimal stress distribution when the magnet is powered by limiting the mid-plane stress while keeping the poles under compression. The separate pole wedge reduces the peak

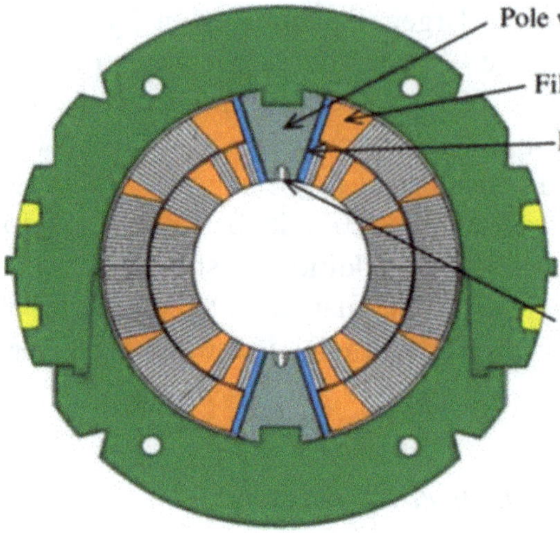

Fig. 9.3 The collared coil concept for the CERN magnet bending, 11 T dipole (MBH)

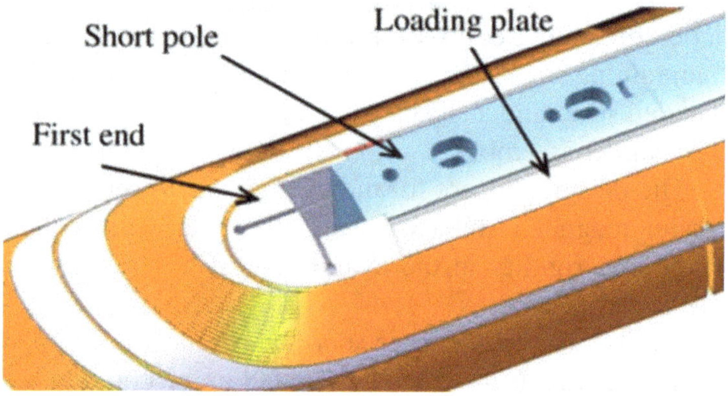

Fig. 9.4 Coil return end showing transition region

stress around the stress-relieving notch by a factor of two as compared to collars comprising the pole area. This reduction is achieved by having every second pole collar carry the load, interleaved with "filler collars" that have a slight clearance from the coil. During the development phase it is also possible to use a shim between the pole wedge and the collar nose to fine-tune the coil pre-stress without taking the coil assembly apart.

The thickness of the collar nose shim is reduced from 0.2 mm in the straight part to 0.1 mm over 55 mm towards the ends, thus linearly reducing the coil stress. This measure has been taken to avoid a discontinuous stress pattern in this delicate area, minimizing any shear stress in the reacted Nb_3Sn coils. The first end spacer has a longitudinal slit to make it more flexible. The other end spacers feature swivel joints in the spacer legs (not shown in Fig. 9.4) to minimize the risk of insulation damage during the fabrication process. Resin-rich volumes, which may have compromised the structural integrity of the coils, have been excluded.

Stainless-steel end saddles are used during the reaction, but they are then replaced with five-axis-machined epoxy-glass (EPGC 3) saddles prior to impregnation, which better matches the transverse rigidity of the coil and reduces the risk of damaging the

Table 9.3 Maximum azimuthal coil pre-stress in pole and mid-plane regions

	Azimuthal coil stress (MPa)				
Position in coil	Under press	Collared coil	Cold mass	Cool-down	$B = 12$ T (min/max)
Inner pole	−126	−92	−143	−115	−27/−5
Outer pole	−87	−52	−65	−61	−37/−5
Inner mid-plane	−115	−55	−65	−58	−134
Outer mid-plane	−91	−66	−95	−94	−127

leads coming out in the mid-plane. They also eliminate the risk of electrical insulation defects in this region.

The wedges separating the winding blocks are usually made of Cu or Cu-alloys. Wedges made from such materials would lose mechanical strength during the coil reaction process. New wedges were developed in collaboration with industry using oxide dispersion strengthened (ODS) Cu-alloy, which retains superior mechanical properties (elastic limit over 240 MPa) after the reaction, and hence minimizes the distortion of the coil geometry due to plastic deformation under stress during assembly and operation. ODS is also thermally and electrically very similar to the "traditional" Cu-alloys.

The collared coils are assembled between the yoke halves and the central yoke insert, and the two 15 mm thick stainless-steel outer shells. The shells are welded together in a welding press to form the shrinking cylinder. During this operation a tensile stress of about 300 MPa is developed in the shrinking cylinder. The yoke inner radius is 0.115 mm larger than the collar outer radius, and there is a horizontal offset of 0.135 mm between the centers of the yoke and the collar radii. This offset increases the radial interference between the somewhat deformed collared coil assembly and the yoke by 0.020 mm, thus minimizing the elliptical deformation of the coils. The reduction in elliptical deformation reduces the sensitivity for assembly tolerances and improves the field quality. This interference is also important for maintaining contact between the yoke and the collared coil at the operating temperature. The yoke gap is closed at room temperature and remains closed up to 12 T field in the bore. Table 9.3 lists the evolution of the coil stress during the fabrication process, at the operating temperature, and during excitation. The poles remain under compression at all times, and the maximum coil stress is less than 145 MPa.

9.3 11 T Dipole Development at CERN

The joint 11 T dipole R&D program at CERN started in 2011 with transfer of the technology developed at FNAL, and with adjustments to the LHC MB specific design features (Savary et al. 2015). At CERN, development was undertaken having in mind the technology suitable for scaling up to the long magnets for installation in the accelerators. In parallel to the R&D, the large tooling for the manufacture of the long magnets was designed, procured, and installed in the CERN Large Magnet Facility.

9.3.1 CERN Model Design and Fabrication Procedure

Twenty-three 2 m long coils were fabricated during 2013–2018, and assembled in seven single-aperture and two double-aperture short models. The main variants explored are given in Table 9.4, including the strand type, the impregnation procedure followed, and the use of the coil. The details of the model design and fabrication procedures are presented in this subsection.

9.3.1.1 Conductor for the R&D Phase and for Series Production

For the R&D phase CERN has procured nearly 400 km of 0.7 mm diameter Nb_3Sn wire of various layouts. Following successful model results, to cover the series construction phase, an additional 1100 km of Nb_3Sn wire of the 108/127 layout are under procurement according to the specifications reported in Table 9.2. About 800 km out of the 1100 km have already been delivered and fully qualified. Table 9.5 gives a summary of the R&D and production conductors in terms of cable unit lengths produced or under procurement and production.

Overall, only a few billets are below I_c specification, but still above 95% of the specified critical current value of 440 A at 12 T and 4.22 K. The baseline cable

Table 9.4 CERN conductors and coil parameters

Coil	Strand type	Impregnation S/P[a]–A/B[b]	Comment
101, 102	Copper	S–A	Practice coil
103	WST[c]	S–A	Practice coil
104	RRP54/61	S–A	Practice coil
105, 106, 107	RRP108/127	S–A	
108, 109	RRP132/169	S–A	
110	RRP132/169	P–B	Practice coil
111, 112, 113	RRP132/169	S–B	
114, 115	RRP150/169	P–B	
116, 117	RRP150/169	P–B	
118	RRP108/127	P–B	Practice coil
119, 120, 121, 122, 123	RRP108/127	P–B	

[a]"S" stands for gelling and curing under vacuum; "P" stands for gelling and curing under pressure, typically 3–3.5 bar

[b]"A" stands for single resin entry point for impregnation; "B" stands for uniformly distributed injection channels along the coil in the mid-plane

[c]Western Superconducting Technologies Company, Ltd. (WST), China

Table 9.5 Summary of all unit lengths of 0.7 mm Nb_3Sn cables, by different type or layout, produced for the 11 T project by CERN

Wire layout	Unit lengths produced	Cable length (m)	Coil
WST (Nb_3Sn low grade)	2	235	Model
RRP54/61 (low RRR)	1	235	Model
RRP108/127 (low RRR)	4	235	Model
RRP169	10	235	Model
PIT114 or PIT120	6	235	Model
RRP108/127, RRP132/169 (low grade)	2	655	Long dummy
RRP169	3	655	Prototype
Final RRP108/127	2	655	Prototype
Final RRP108/127	19 out of 30	655	Series

produced at CERN has an average J_c degradation of 2% with a maximum of 5% for the RRP strands.

The RRR >150 specification for virgin wire has been respected during production, and the RRR is also larger than 100 after a 15% deformation reduction (rolling), which is representative of the deformation undergone by the strand during the cabling process (note that this last test is not a contractual obligation). It is important to remark that, despite strong initial concerns because of the relatively small wire diameter for such a large current density, the large-scale production of Nb_3Sn for the 11 T dipole is turning out to be a great success.

9.3.1.2 Insulation

Despite the tight schedule of the project, a considerable effort was made at CERN to introduce some innovative technology into coil fabrication.

The first example is the re-introduction of mica in the cable insulation, as illustrated in Fig. 9.5. Similar mica-based insulation was used in the early stages of R&D for the KEK race-track (see Chap. 3, Nb_3Sn Accelerator Magnets: The Early Days (1960s–1980s), and then for the CERN-Elin (see Chap. 4, CERN–ELIN Nb_3Sn Dipole Model) and University of Twente/CERN (see Chap. 5, The UT-CERN Cos-theta LHC-type Nb_3Sn Dipole Magnet) Nb_3Sn dipole projects in the 1980s and 1990s, but this was not pursued further in later realizations.

The insulation of the cables used in CERN models (until SP106), and the first prototype, is made of a 25 mm wide, 0.08 mm thick mica tape shaped around the cable, with a 0.07 mm thick layer of S2-glass braided over the mica layer. The mica layer provides a continuous and solid dielectric barrier around the cable, while the S2-glass layer braided over the mica layer provides an additional turn separation and gluing after impregnation with epoxy. Room-temperature tests confirmed the excellent electrical properties of this insulation. The cable insulation thickness at

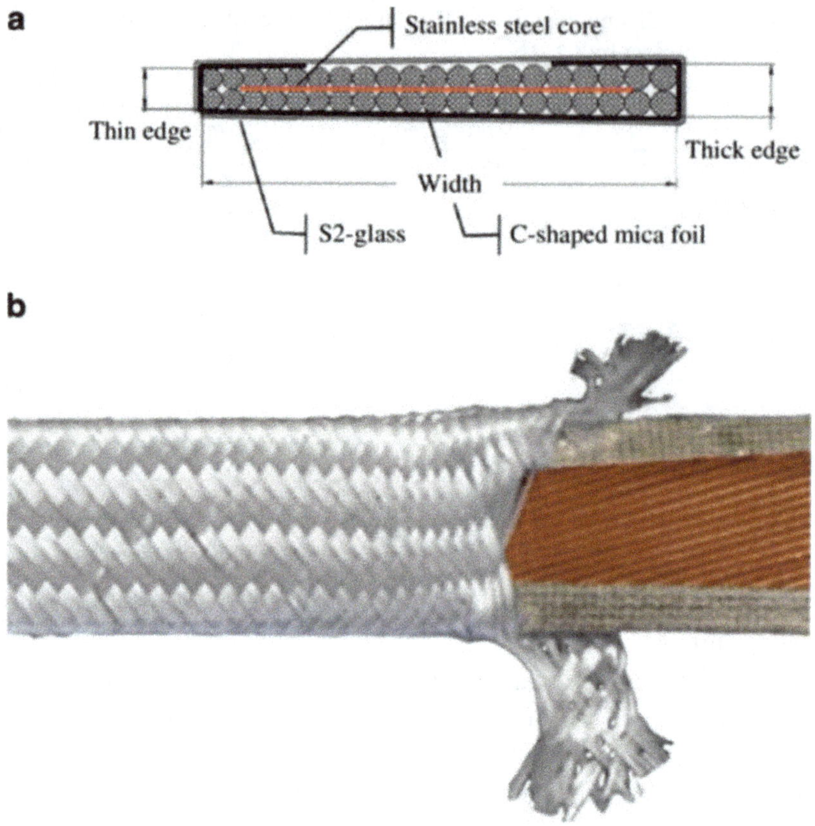

Fig. 9.5 Cable insulation based on S2-glass braided on mica tape (CERN insulation): (**a**) schematic; and (**b**) photograph

production is 0.15 mm. It is reduced to about 0.11–0.14 mm when compressed at about 30 MPa.

The mica tape (see Fig. 9.5) is folded over the cable and covered by a layer of S2-glass braided through a continuous process. At the beginning there was much concern about the possibility that the mica layer would have been a barrier to easy penetration of the resin inside the Rutherford cable. Experience proved that it was not an issue. To facilitate resin flow inside the cable, the mica layer only wraps the two smaller edges and one wide face of the cable, and was designed to cover only about 40% of the other wide face. The layout turned out to be non-optimal for stress distribution over the cable and has recently been modified, see Sect. 9.3.4 below.

9.3.1.3 Coil Technology

The two layers of the coil are wound from one cable unit length without an inter-layer splice. In addition to the re-introduction of mica, these are the main technological innovations from CERN R&D:

a

b

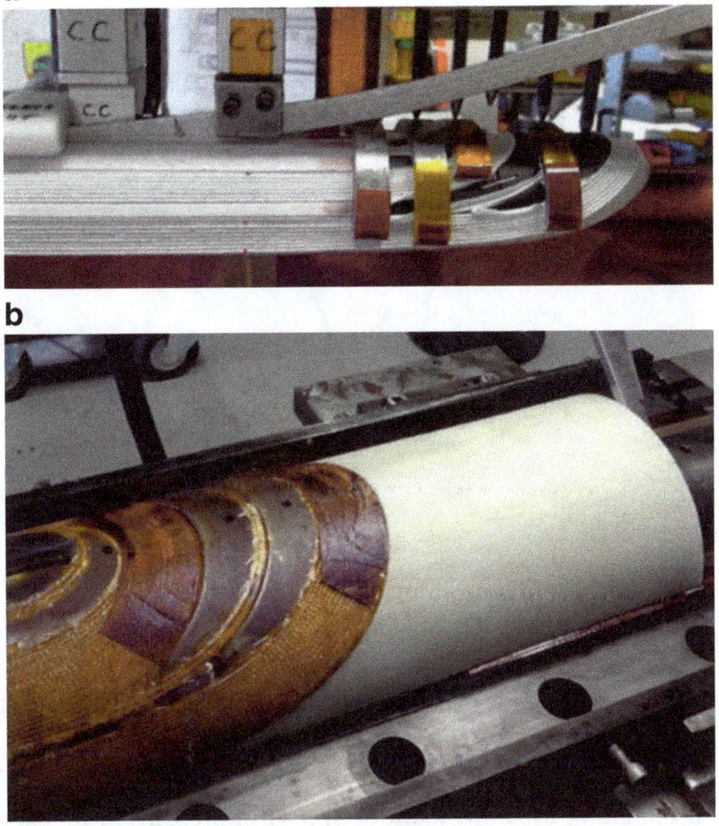

Fig. 9.6 (**a**) Inner-layer lead end with flexible-leg spacers and stainless-steel technology saddle; and (**b**) reacted coil equipped with EPGC22 saddle before impregnation

1. Wedges made from high-performance copper alloy, DISCUP C3/30 ODS copper. ODS wedges minimize coil plastic deformation, while having the elastic, electrical, and thermal properties of Cu even after the heat treatment.
2. Coil end spacers made of AISI 316 L stainless steel, manufactured by using the selective laser sintering process, which provides a short turnaround time if shape modifications are required.
3. Coil ends, some with longitudinal slits and others with swivel joints in the spacer legs. Besides avoiding the resin not reaching the required zone and possible damage to the insulation, it also facilitates winding accuracy, as shown in Fig. 9.6a.
4. Replacement of the metallic end saddle with EPGC-22 non-metallic material.

The first coils were fabricated using the same process and tooling design that was developed at FNAL (see Chap. 8, The Nb_3Sn 11 T Dipole for the High Luminosity LHC (FNAL)). Modifications were then introduced to address the specific features of the CERN coil design. Coil winding mostly follows the process set up at FNAL, with ceramic binder (ceramic matrix CTD-1202-X) but cured to the nominal cross-section. The coil is reacted in a reaction fixture using a three-step cycle, similar to HT1 used at FNAL: 210 °C for 48 h, 400 °C for 48 h, and finally 640 °C for 50 h in an argon atmosphere.

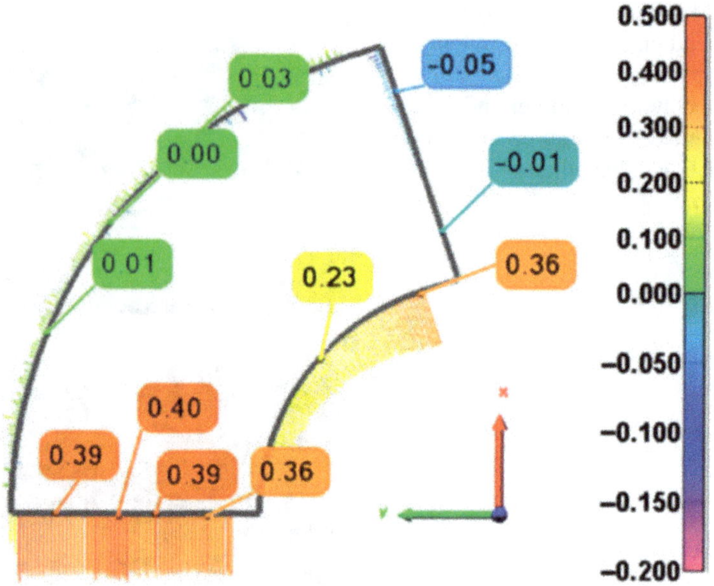

Fig. 9.7 Coil 105 size measurements

After reaction, the coil is transferred into the impregnation mold, the coil leads are spliced to flexible Nb-Ti cables, and the metallic end saddles are replaced by the non-metallic saddles. The usual FNAL flushing, thermal treatment, and final impregnation with CTD-101K epoxy is carried out.

The impregnated coils are measured with a portable 3D measurement arm every 15 cm along the coil, without applying pressure. Each set of measured points in each cross-section is fitted to the CAD model to define the deviations from the nominal sizes. The outer diameter and the loading plate are used as datum surfaces. An example of the measurement data is shown in Fig. 9.7. The variations for the coil outer radius and the pole surface from the design values are small, less than 0.05 mm. The coil mid-plane position is off by 0.4 mm (compared with 0.1 mm in FNAL coils). The deviation of the inner radius is also quite large, ~0.4 mm, but this is not an important parameter for magnet assembly. The measurements also show a considerable longitudinal variation (sometimes called waviness), typically ±0.2 mm, i.e., three to four times that for Nb-Ti. The issue related to the coil size is discussed below (see Sect. 9.3.4).

9.3.1.4 Ground Insulation and Quench Protection Heaters

Robust ground insulation around the coils consists of four layers of 0.125 mm thick polyimide film, which provide the required dielectric strength of 5 kV at room temperature.

The quench protection heaters are large, flexible copper-coated austenitic steel strips glued onto a 0.050 mm thick polyimide film that withstands high voltages, low temperatures, high compression forces, and ionizing radiation. The 0.025 mm thick

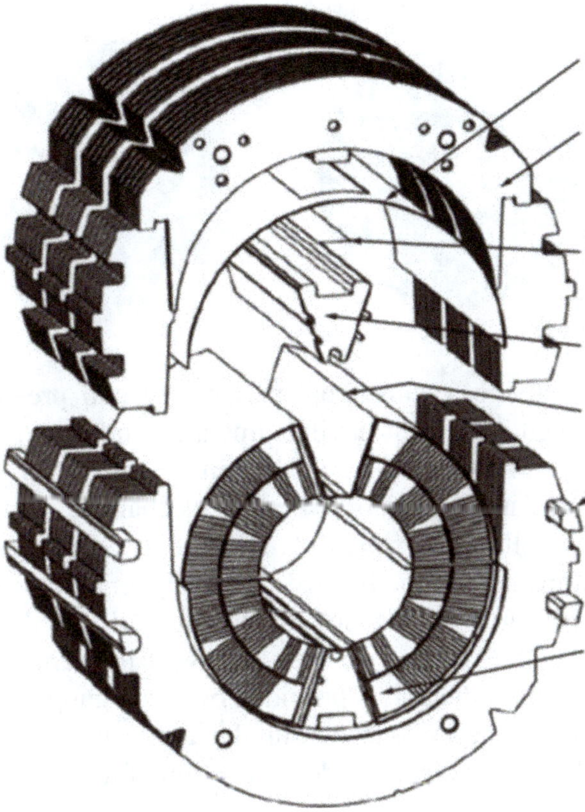

Fig. 9.8 Collared coil assembly

304 L stainless-steel strips are laminated onto the polyimide film with a 0.015 mm thickness of epoxy adhesive. A 0.005–0.015 mm thick Cu layer is coated between the heating stations to lower the total electrical resistance of the heater strips. The quench heaters for coils 101–109 and 111–113 were assembled with the coils after they had been impregnated; whereas those for coils 110 and 114–123 were impregnated with the coils to minimize the thermal resistance between them and the coil, and thus improve their efficiency. Recently, during standard re-commissioning electrical tests (test voltage 2.1 kV), a breakdown in the insulation system occurred. Based on this test result, it was decided to assemble the quench heaters with the coils after they had been impregnated, as previously done.

9.3.1.5 Collared Coil

The collared coil assembly is shown in Fig. 9.8. A rigid clamping and minimum coil distortion is attained with collars made of YUS130S austenitic steel (X8CrMnNiN 19-11-6 grade), pole wedges made of Ti-6Al-4 V titanium alloy, and a loading plate made of AISI 316 L stainless steel, covering the two layers on each side of the removable pole. Shims are placed between the pole wedges and the loading plate, and their size can be altered to regulate the azimuthal coil pre-stress, as needed. Extra shimming is also possible between the pole wedge and the collar nose.

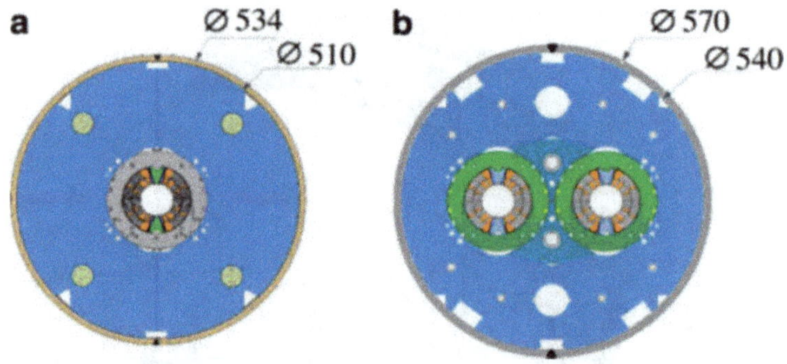

Fig. 9.9 CERN 11 T dipole cross-sections: (**a**) single-aperture; and (**b**) two-in-one-aperture. Dimensions in mm

The collaring procedure delivers a substantial part of the coil pre-stress. The collars are alternately assembled with their cavities upwards and downwards, and clamped around the coils in a press applying a force in the range 16–32 MN/m during collaring. To avoid coil damage during collaring, mechanical stops are used between the top and bottom collaring fixtures. To insert the collaring keys, the collars and the mechanical stops have to be pushed further by an extra 0.1 mm.

The stress in the coils during collaring should not exceed 150 MPa in the mid-plane, whereas it is substantially lower in the vicinity of the removable pole. Indirect control of the stress level in the coil is provided by strain gauges glued onto the noses of the collars. The use of the stresses measured in the collars in conjunction with the finite element model (FEM) allows the determination of the stresses created in the coil during collaring.

9.3.1.6 Short Dipole Models

The single- and two-in-one-aperture structures developed at CERN for the 11 T dipole (Savary et al. 2017) are shown in Fig. 9.9.

Single-Aperture Models

The collared coils are assembled with the half-yokes, the end plates, and the stainless-steel shells. The half-yokes have a vertically tapered gap with a resulting opening of 0.1 mm on the outer diameter, which is closed during shell welding and remains closed up to 12 T. A nominal 0.4 mm collar–yoke shim in the region of the collaring keys ensures the rigidity of the assembly, when most of the pre-stress is given by the collars in the single-aperture assembly. The axial forces, of the order of about 215 kN per pole, are taken up by the 55 mm thick end plates and by the shell, both made of 316LN stainless steel.

For MBHSP101 and the other models, the welding process consisted of a double root pass made with the tungsten inert gas (TIG) procedure. The end plates were then welded to the shell, and the bullet gauges (two per pole at each end) tightened with a torque of 25 Nm, corresponding to an axial pre-load of 8–9 kN per bullet.

Two-In-One-Aperture Dipole

The yoke of the two-in-one-aperture dipole is made of three parts. The central block, featuring a tapered surface at the interface with the yoke halves, is used for assembly purposes, and for locking the collared coils in the structure symmetrically during the assembly and welding of the skin. To tolerate the higher stresses in the two-in-one-aperture structure, the skin thickness was increased from 12 mm to 15 mm and the number of welding passes was increased.

9.3.2 *Magnet Test*

The main parameters and model tests in single and two-in-one-aperture dipole configurations are summarized in Table 9.6.

9.3.3 *Quench Performance*

9.3.3.1 Model Magnet Limits

The maximum current reached in single-aperture model coils varies from 11.9–13.5 kA at 1.9 K, see also Table 9.6 (Willering et al. 2017, 2018). The maximum I/I_{ss} reached was 94% in the two-in-one-aperture coil DP101.

Two coils, SP102 and SP104, were limited at the layer jump at 1.9 K with 10 A/s ramp rate, while the other coils that were run to their limit were all limited in the mid-plane turn, see Fig. 9.10. Tests at 4.5 K confirmed the coil limits in mid-plane and layer jump, and in conductor block 3 of the inner layer. V–I measurements in DP102 show that all four coils have a reduction in I_c and n-value in the mid-plane inner layer turn (not localized, rather, distributed over a large fraction of it). Model

Table 9.6 Maximum current, short sample limit, and holding current tests for each model

Model	Coils	I_{max} (kA)	I_{ss} (1.9 K/4.3 K) (kA)
SP101	106, 107	11.9	14.5/13.1
SP102	106, 108	12.8[a]	14.5/13.1
SP103	109, 111	12.8[a]	14.2/12.8
DP101	106, 108, 109, 111	13.3	14.2/12.8
SP104	112, 113	12.3	14.4/13.0
SP105	114, 115	12.4	14.6/13.3
DP102	109, 112, 114, 115	11.4	14.4/13.0
SP106	116, 117	13.47	14.8/13.6
SP107	120, 121	12.85[a]	14.4/13.1

The nominal operating current is 11.85 kA, while 12.8 kA is the ultimate current

[a]Test target: no attempts to train higher

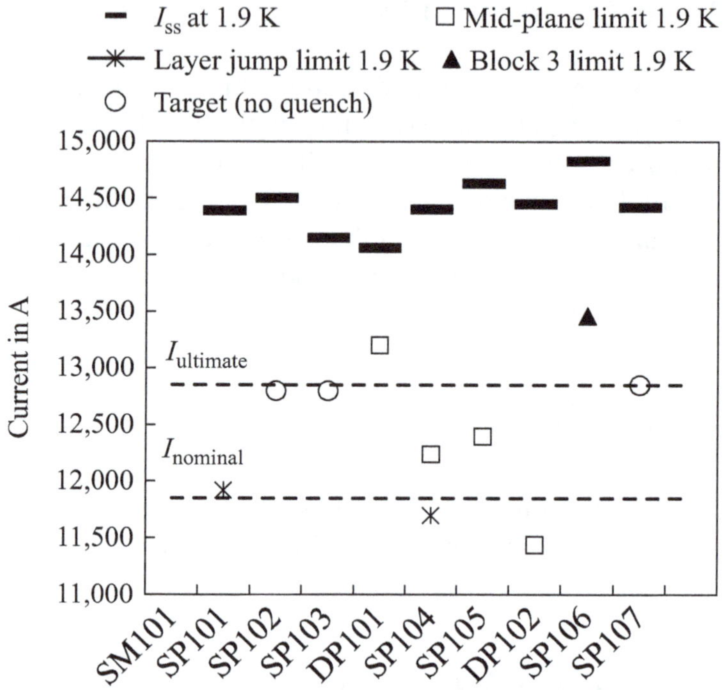

Fig. 9.10 Short sample limits and training limits at 1.9 K for the single-aperture and two-on-one-aperture models

magnets DP101 (coils 109 and 111), SP104, and SP105 also had mid-plane inner layer turn limitations. V–I measurements were not performed, however, and, especially for DP101, the limitation was at a much higher current. DP102 was recombined from coils that had previously been tested, but it was limited at a much lower current, showing that the conductor had been damaged during one of the production steps between the dismantling of the single-aperture model and the recombination into a two-in-one-aperture model. With the data available for these models it was suspected that the mid-plane suffered from irreversible degradation due to high levels of stress during the magnet's production. This behavior might also have been present in the first magnets; it was, however was only clearly detected from assembled coil 109–111 in DP101. From then it was clearly visible in all magnet assemblies. An extensive study was performed to investigate the maximum local stress level in the coils during all of the magnet construction steps, like collaring and yoking, and the damage levels due to stress on the cables, see Sect. 9.3.4 below. Meanwhile, SP106 was tested, showing a strongly reduced I_c and n-value, localized this time in the inner layer coil block 3 of coil 116. After magnet training, however, it reached a very high current value, see the following section.

9.3.3.2 Model Magnet Training

The single-aperture magnets were trained at nominal conditions at 1.9 K with a ramp rate of 10 A/s. Some of the models included coils that were already trained in a previous assembly. The training curves are shown as training per virgin coil, see Fig. 9.11.

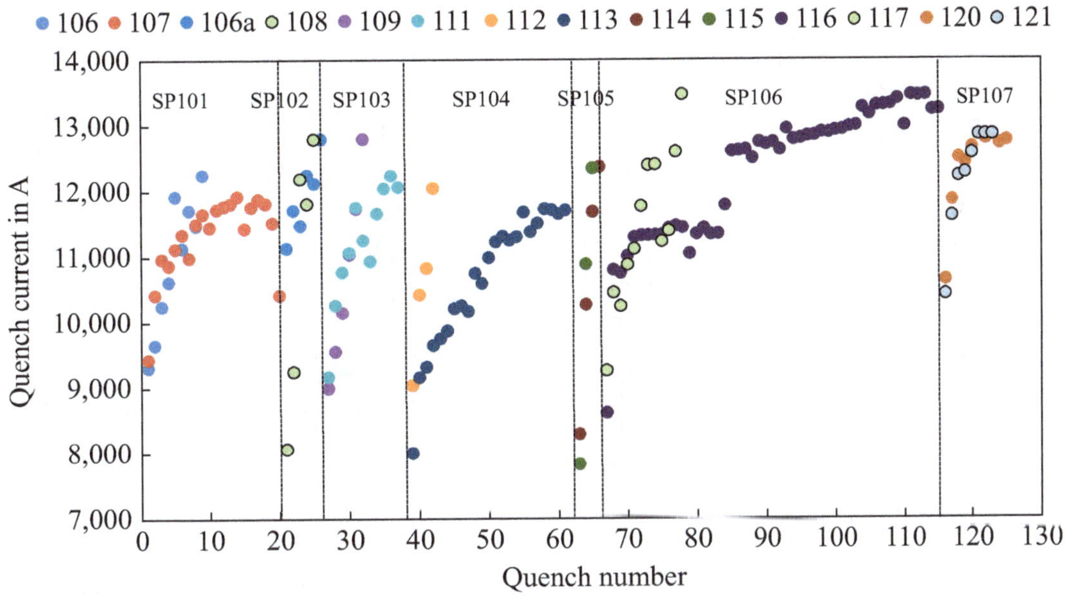

Fig. 9.11 Training quenches for all coils tested

In particular, the coils with layer jump issues, coils 107 and 113, show slow training. An excessive number of quenches are located in the pole turn and layer jump. These coils also show a low quench current limit in the layer jump region and, in addition, there is a strong suspicion of a local non-homogeneous defect in or close to the layer jump, which may have influenced the training rate (Willering et al. 2017). The particularly slow training of coil 116 may be linked with the I_c reduction in coil block 3 of the inner layer. Following quench 19, as shown in Fig. 9.11, the high MIITs (1 MIIT = 10^6 A^2 s) studies (part of the regular test program) started and the protection was delayed to increase the hot-spot temperature after a quench. While conductor degradation driven by the hot-spot was expected, the performance increased step by step, possibly due to a change of stress distribution in the conductor. This magnet went to 13.5 kA, with a central field of nearly 13 T.

Finally, note the significantly better behavior of coils 120 and 121, assembled in the single-aperture model SP107. They are based on the modifications introduced by the Task Force, see Sect. 9.3.4 below.

9.3.3.3 Training after Thermal Cycle

The number of retraining quenches needed to reach the same current as before the thermal cycle are shown in Table 9.7. In some cases there was only a thermal cycle, in other cases the collared coils were reassembled in a new yoke, while in others the coils were recombined and re-collared. In DP102 the coil limit (after conductor degradation during recombining of the coils) was reached without quench, but since it was at a much lower current it is left out of Table 9.7. In general, retraining after a thermal cycle is very fast, and the memory of the coils is very good. In particular, a

Table 9.7 Number of quenches needed to reach the same current as before the thermal cycle

Coil	Action performed during thermal cycle	Magnet	I_{max} (kA)	Number of quenches after thermal cycle
106	Re-collared	SP101 → SP102	11.8	3
106	Thermal cycle	SP102	12.5	2
106	Re-yoked	SP102 → DP101	12.8	2
108	Thermal cycle	SP102	12.5	0
108	Re-yoked	SP102 → DP101	12.8	0
109	Re-yoked	SP103 → DP101	12.8	0
111	Re-yoked	SP103 → DP101	12.8	0
116	Thermal cycle	SP106	13.0	0
117	Thermal cycle	SP106	13.0	0
120	Thermal cycle	SP107	12.85	2
121	Thermal cycle	SP107	12.85	0

good memory after re-yoking the collared coils is important. Coil 107 data for SP101 is not shown, since it already suffered strongly from erratic quenches before the thermal cycle, and is to be assumed irrelevant for memory after a thermal cycle.

9.3.3.4 Erratic Quenches

In model SP102 multiple erratic quenches occurred between 12–12.8 kA, with the quench origin mainly around the pole turn in the heads of the magnet and also around the layer jump. It is therefore remarkable that the same collared coil reached 13.3 kA while being part of two-in-one-aperture DP101, without showing the same behavior. It is generally thought that this behavior can be attributed to the increased and more uniform pre-stress and loading forces in the two-in-one-aperture dipole model compared to single-aperture models. Coils 107 and 113, having layer jump issues, showed erratic quench behavior.

9.3.3.5 Detraining/Degradation During Power

Of all the coils, only coil 107 in model SP102 had detraining or degradation observed during powering. The quenches occurred in the layer jump, a part of the coil for which a non-homogeneous degradation had already been identified. This area is mechanically sensitive due to the discontinuity between the two layers, however no direct evidence of the cause of detraining/degradation was found during cold tests or after coil cutting after the test.

9.3.3.6 Holding Current Tests

A holding current test was performed for all models, either at the ultimate current level of 12.8 kA, or at a current just below the training limit of the magnets. The duration varied between 1 h and 10 h, depending on the time available for the test (Willering et al. 2017, 2018). It should be noticed that, except for MBHSP104, which quenched after 1 h 5 min in the known defective layer jump of coil 113, the magnets showed stable behavior.

9.3.3.7 Temperature Dependence

In Fig. 9.12 the maximum currents reached at 1.9 K and 4.5 K are compared for six of the models. The expected reduction in quench current is 5–15%, in line with the expected 10% due to I_c reduction from 1.9 K to 4.5 K. SP106 has relatively lower performance at 4.5 K, and SP101 has relatively better performance at 4.5 K.

9.3.3.8 Ramp Rate Dependence

For some models, ramp rate studies have been critical to assess that limitations were not due to mechanical causes (like movements). This showed that quench limits were mainly due to the conductor, mainly in the mid-plane or pole turn. In SP104 the highest quench current was reached at an elevated ramp rate of 50 A/s, pointing at non-homogeneous degradation in the pole turn. The results of ramp rate studies of SP106 and SP107 are shown in Fig. 9.13. The measurements at 1.9 K and 4.5 K give consistent results for I/I_c for both magnets. The slope for both coils is similar, but

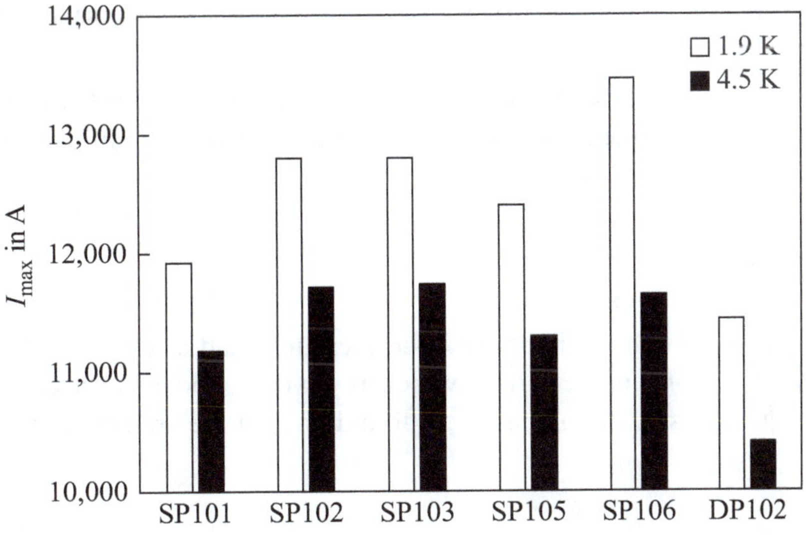

Fig. 9.12 Quench performance comparison between 1.9 K and 4.5 K temperatures for six magnets

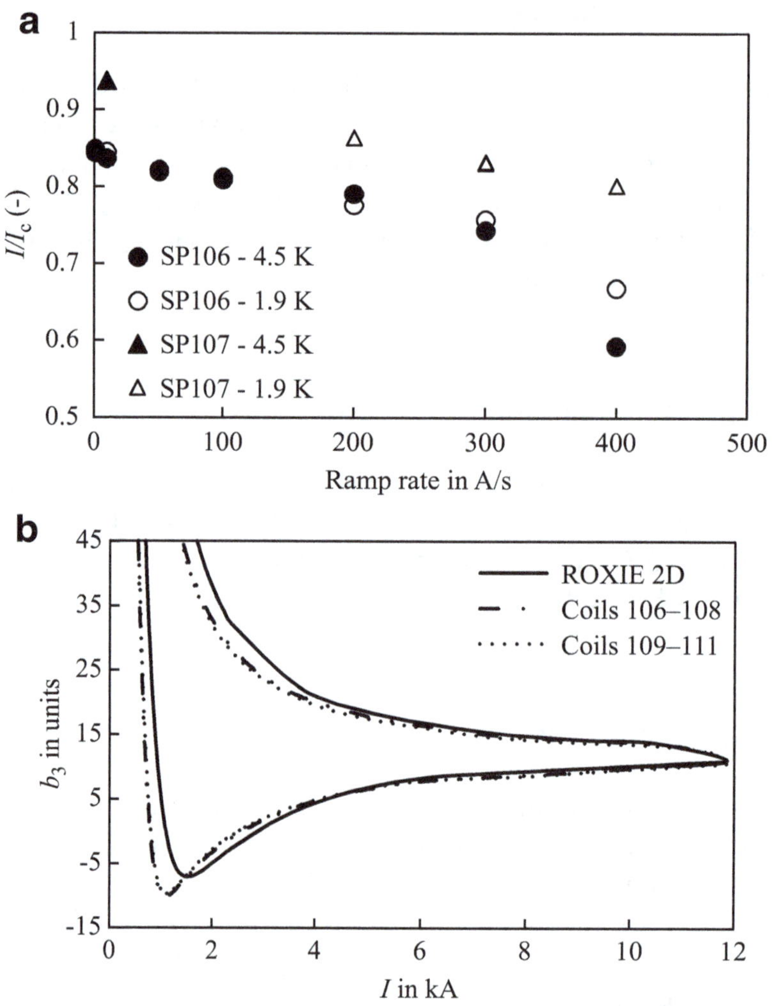

Fig. 9.13 (**a**) Ramp rate quench dependence for SP106 and SP107 models (the latter is the high performance model following the Task Force design revision, see Sect. 9.3.4 below). The current is normalized to the short sample limit. (**b**) Measured and calculated b_3 vs. current for DP101

with SP106 reaches a lower current level throughout. In SP106 most quenches occurred in block 3, while in SP107 the quenches started in the mid-plane turn.

9.3.3.9 Splices

All Nb_3Sn to Nb-Ti splices that have been measured had a resistance between 0.1 and 0.3 nΩ (Willering et al. 2018), which is a very good level according to the design. No quenches or anomalies were found around the splices in all models of magnets.

9.3.3.10 Magnetic Measurements

Comprehensive magnetic measurements were carried out to determine the MBH transfer function (TF) and the field harmonics produced by the coil geometry, persistent currents, iron saturation, and magnetic cross-talk between apertures (Fiscarelli et al. 2016, 2017). The measurements were performed using the CERN Fast Measurement Equipment (FAME) system. It is suitable for tests in a horizontal position at ambient temperature as well as for a vertical cryostat at helium temperature. Field harmonic coefficients are presented at the reference radius $R_{\text{ref}} = 17$ mm.

The magnet TF was measured at 1.9 K by powering the magnet with a stair-step cycle. Measured geometrical and saturation components agree well with the 2D data calculated using the Routine for the Optimization of magnet X-sections, Inverse field calculation and coil End design (ROXIE) at currents above 2 kA. At low currents the persistent current effect is visible, as expected.

Figure 9.13b shows the measured and calculated b_3 vs. the magnet current for both apertures. Measurements and calculations agree well for both apertures. Below 2 kA, the measured persistent current effect in b_3 diverges from calculations. A similar discrepancy was also observed in the 11 T dipole models tested at FNAL. Although the persistent current effect in b_3 is large, the b_3 variation between injection and nominal field is about 22 units, within the acceptable level of 27 units. It should be noted that calculations are performed with a model that limits the magnetic moment of the wires at low field, in the region where the magnetization collapses due to flux jumps. This choice reduces considerably the effect of persistent currents at low field (Izquierdo Bermudez et al. 2016).

The saturation of the iron separating the two apertures results in magnetic cross-talk, with effects mostly visible for b_2 harmonics, see Fig. 9.14. The compensation is not perfect; it is, however, good enough to be acceptable for operation.

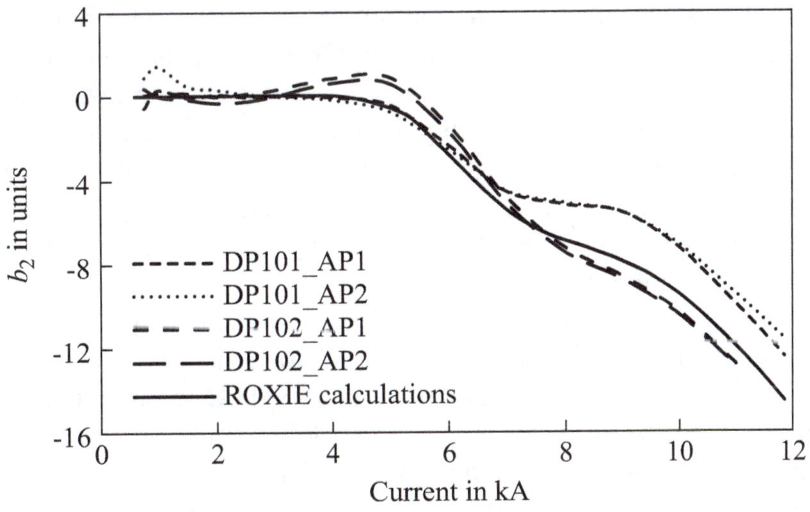

Fig. 9.14 Measured b_2 harmonic over the whole dynamic range for DP101 and DP102, compared with calculation. (Courtesy of L. Fiscarelli, CERN)

Table 9.8 Field harmonics at $I = 3.5$ kA for DP101

	2D calculations		Coils 106, 108		Coils 109, 111	
n	b_n	a_n	b_n	a_n	b_n	a_n
2	−0.07	0.00	−1.80	4.17	−0.09	6.30
3	7.46	0.00	11.85	−1.12	11.54	0.13
4	0.00	0.00	−0.31	0.15	−0.65	0.99
5	−0.01	0.00	1.25	−0.01	1.58	−0.65
6	0.00	0.00	−0.13	−0.11	−0.17	0.31
7	−0.09	0.00	0.15	−0.11	0.21	−0.18
8	0.00	0.00	−0.04	0.04	−0.05	0.02
9	0.91	0.00	0.65	0.06	0.70	−0.31

The ramp-rate effect on normal and skew multipoles was measured in current cycles with ramp rates of 40 and 80 A/s. As expected, and similar to the FNAL models, the effect is small due to the use of a cored cable.

At 1.9 K, the geometrical harmonics were defined as the average measured values on the current up and down ramp branches of the stair-step cycle at the magnet current of 5 kA. Table 9.8 presents the measured harmonics in the central section of the two apertures of DP101 and the calculated values for the nominal 2D coil geometry. The main differences are in a_2 and b_3, which can be explained by errors in the coil's actual position and coil deformations during magnet assembly, and should be corrected with a more rigorous procedure for the long magnets. A very recent measurement on the first prototype confirms this expectation.

9.3.4 Design Revisions by the 11 T Task Force

Following the test of single-aperture models SP104 and SP105, and the test of the same coils in the two-in-one-aperture configuration DP102, it became increasingly clear that the conductor performance in the CERN models was well below the expected level. As anticipated in the discussion of the powering test results, training reached a plateau at high field, at which point quenches appeared to be reproducible, with no precursors, an identical location, and a very similar voltage "signature." A good example of this behavior is the plateau reached after the eighth training quench in the single-aperture model SP105, followed by a sequence of seven identical quenches (one de-training during the sequence).

The reasons for this performance limitation became clear by taking direct high-resolution measurements of the voltage-current characteristics of the limiting coil, in the portion where the quench origin was identified. One such measurement is reported in Fig. 9.15, taken in the worst 1.3 m long straight cable segment in the limiting coil 109 of two-in-one-aperture model DP102. We see that there is a clear voltage rise as the current is ramped, well in advance of the quench, and well below the projected critical current at the field and temperature operating conditions. Based on the early onset, the magnitude of the measured voltage, its reproducibility, and

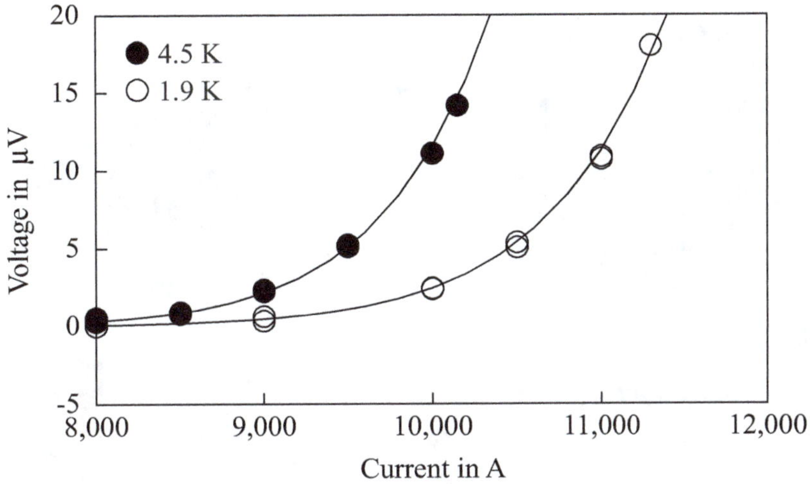

Fig. 9.15 Voltage–current characteristics in model DP102 (mid-plane segment 109 II-I1), for the worst coil

stable magnet behavior under the quench current, we have attributed this early V–I appearance to extensive irreversible degradation of the conductor in the mid-plane of the coil. In fact, measurements of voltage during ramping in all coils of two-in-one-aperture model DP102 have shown that this is not a feature of a single coil location but, rather, a systematic behavior.

The voltage in the other coils is, however, not sufficient to yield to a quench, and can be sustained in stable conditions. The final crucial information is that repeated cycling to high current, as well as repeated quenching, did not cause an observable change in this behavior, indicating that the irreversible degradation is not associated with electromagnetic loads, but must have occurred during manufacturing or cool-down. This observation was complemented by the independent finding that the C-shaped mica sheet wrapped around the cable, with a gap of approximately 7 mm in the middle, is a potential stress intensifier at the cable edges, which has been verified in several instances. An example is shown in Fig. 9.16, where we report the pressure measurement film (PMF) imprints (Fuji prescale film) obtained from a measurement of the mid-plane stress during a pre-collaring operation taken in single-aperture model SP105 at full load (key insertion), corresponding to a target average stress of 120 MPa. As shown, the local stress in the cable edges can be much higher than the average pressure exerted on the whole cable surface, with a distribution that has a range at least as wide as 120 MPa. The actual range may be even larger, but it is difficult to make an accurate assessment because of saturation of the pressure measurement film. The stress peaks are in correspondence with the wrapped edge of the mica foil, and a reduction of stress is observed in the center of the cable where the mica foil wrap has a gap.

The above findings triggered an analysis of coil production and assembly, with the aim of finding the step where the irreversible degradation took place. Measurement of stress in the collar nose, PMF imprints taken during simulated collaring

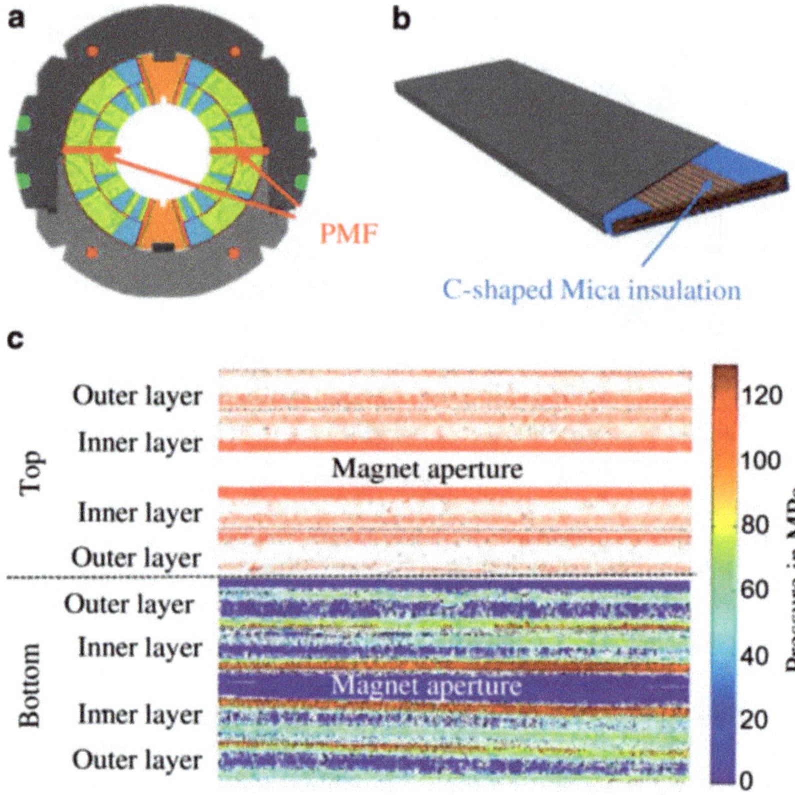

Fig. 9.16 (**a**, **b**) Schematic of the original insulation scheme with incomplete "C" wrap (blue); and (**c**) pressure measurement film (PMF) imprints obtained during collaring tests in single-aperture model SP105. The PMF results have been scanned and translated to values of pressure in MPa. (Courtesy of S. Izquierdo Bermudez and F. Wolf (CERN))

steps, and FEM analysis, point to the collaring step and, in particular, under the press at the insertion of the keys, as the most stressed situation for the coil mid-plane. In particular, combining the direct stress measurements during collaring of the magnet models and the expected stress intensification associated with the mica wrap explained above, it is expected that the local stress exerted on the thin cable edge of the mid-plane of the coils in model magnets SP104 and SP105 (and DP102) reached values well above the 150 MPa target, which at the time was the limit set to avoid irreversible conductor degradation.

A final element in this analysis is the actual conductor insulation thickness. The original insulation system of the 11 T cable consists of a partial wrap with a mica foil of 80 μm thickness and 25 mm width, and 11 tex S2-glass fiber braided on top. The nominal thickness, defined under a pressure of 30 MPa, was taken as 100 μm during the coil design phase. In practice, the thickness obtained in production was quite different from the nominal value. Specifically, the measured insulation thickness was found to be in the range 117 μm (at 30 MPa) to 135 μm (at 5 MPa). This additional thickness accumulates in the coil, resulting in a denser and stiffer coil, which can exacerbate the stress peaking effect described above.

Based on the observations above, some of the design aspects were reconsidered and an iterative process was started, consisting mainly of the following.

Review and Modification of the Insulation Scheme The modifications made to the insulation system consisted firstly in a more complete coverage of the cable perimeter, using a wider mica foil. A nominal width of the mica foil of 31 mm results in a gap reduction from the original 6.9 mm to 0.9 mm. In addition, the number of picks per cm of the glass fibers was increased (from seven to nine) with a reduction of the number of plies per strand (from nine to four). This change to the braiding yields a reduction of the nominal thickness of insulation to the desired 100 μm, measured under a transverse pressure of 5 MPa, i.e., leaving sufficient room for the dimensional changes to the conductor during heat treatment.

Characterization of the Transverse Stress Limit of Insulated Cables The main aim was to determine quantitatively the degradation limits of the 11 T Nb_3Sn cable in its final configuration (strand, cable, and insulation system). The quantity sought is the pressure at which the insulated cable is irreversibly degraded, most likely due to the development of cracks in the filaments. This limit is a critical input for the determination of a safe collaring procedure. The results collected to date point to a critical stress of 175 MPa as the maximum allowable local value, beyond which the conductor is irreversibly degraded. Extensive studies, including micrographs, have shown that beyond this stress the filaments start cracking. These findings confirm the limit applied for the collaring, i.e., 150 MPa peak stress at any location.

Testing of the Mechanical Properties of the 11 T Winding The nominal dimensions of the coil cavity defined by the collars must be adapted by acting on the pole, mid-plane, and radial shims to induce mechanical interference, and thus produce the required azimuthal and radial pre-stress. Accurate knowledge of the stiffness of the coil in relation to the applied stress and coil deformation is hence a critical input for the selection of these shims. In fact, the coil stiffness for an Nb_3Sn epoxy impregnated coil of 20–40 GPa is typically a factor of three to four higher than that for a Nb-Ti polyimide insulated coil in the range 8–15 GPa. Tolerances in the assembly, and uncertainties in the coil dimensions and stiffness, translate therefore into a much greater variation of pre-stress than is customary in Nb-Ti, which, in combination with a hard limit on the maximum allowable stress described above, makes the collaring step a balancing act. For these reasons we have performed extensive mechanical measurements of impregnated stacks of cables and cut-out portions of windings.

The relationship between stress and deformation is highly non-linear, exhibiting significant hysteresis and ratcheting. The equivalent modulus in virgin loading conditions derived from the above measurements is around 10–20 GPa, while further loading gives values in the range 25–35 GPa. This value is to be compared to measurements on cable stacks, which in comparable conditions give a virgin modulus of 15 GPa and successive loading values of 40 GPa. It is clear from these results that a single value for coil stiffness is not an appropriate parameter: rather, the complete behavior of loading and unloading must be taken into account when analyzing the stress and strain state of the coil under the various manufacturing and loading steps. Still, the mechanical properties are affected by a large uncertainty, typically 30% of the equivalent coil stiffness. Given this complexity, the initial working assumption was to use FEM to select the appropriate shim size, on a case-by-case basis. The burden of such analyses, and the inherent uncertainty in the value of the mechanical characteristics, has, however, motivated a different approach to coil shimming, as described below.

Experimental Analysis of the Collaring The last major element of R&D was the experimental analysis of the collaring procedure, with the aim of establishing a direct correlation between the coil–collar interference and the resulting pre-stress. A number of collaring trials were performed using short segments cut out of representative production coils and highly instrumented collar packs. Pressure measurement systems were used for additional diagnostics in the collared coil assembly to supply a direct measurement of the stress at the coil mid-plane and at the collar–pole interface. Among the systems considered, scanned and processed PMF imprints were the most practical and effective. Several tests were performed on different coils, changing the azimuthal shims so as to modify the coil–collar interference. We define the coil size *excess* as the sum of the interference of the right and left coil halves, taking into account that each different coil segment may have different dimensions. The results are very interesting in that they show that, to limit the maximum mid-plane stress to less than 150 MPa, the coil size excess must be kept below 300 μm, which provides a very simple recipe for the collaring operation. Note that with this choice of interference, and given the specific collaring mechanics and collar dimensions (influencing spring-back), the residual stress at the pole is around 50 MPa. This stress is in principle not sufficient to maintain contact at full powering.

The changes above were implemented as part of a fast-track effort in two single-aperture short models, MBHSP107 and MBHSP108, for which a total of six coils were built, coil 118 to coil 123, before the construction of the first long magnet for the HL-LHC. Care was taken that the short coils were all built with final components, and, in particular, strand, cable, and insulation. Also, the models were produced using as much as possible the same methods and procedures applied to the construction of the long magnets, so as to provide early and valuable feedback.

To date, the results for single-aperture model MBHSP107, achieving ultimate performance with the fastest training among all short models, and showing no limitations within the operating range, have confirmed the soundness of the modifications introduced. While definitive conclusions can only be drawn after the testing of SP108, in order to verify the reproducibility of these results, the design changes described in this section were introduced into the series production magnets (it was too late for their introduction in the prototype).

This result also indicates that the pre-load required for operation at high field may not be as high as the computed average electromagnetic load, which has relevance both for the 11 T dipole as well as for R&D for higher fields.

9.3.5 Full-Scale Prototype

A full-length prototype has been fabricated at CERN (Savary et al. 2018). This prototype has all the features of the final magnets, except for the modifications introduced by the Task Force's revised design and procedures, see above. It has been a key learning and debugging exercise for such a complex object: indeed, this 5.5 m long dipole is the first long Nb_3Sn magnet built at CERN, and the longest Nb_3Sn accelerator magnet built so far. Besides the importance of the 11 T dipole itself, we consider it a strategic step paving the way for the 7.2 m long coils of MQXF, the inner triplet quadrupole for HL-LHC. The prototype is targeting full accelerator quality in terms of quench performance, field errors, electrical robustness, protection, and geometry. Although most of the major tooling was available at CERN, recovered from the production of the Nb-Ti magnets for the LHC, significant modifications and upgrades were necessary. The largest additional tooling required to fabricate the coils were an argon oven for the reaction of the Nb_3Sn conductor, and a vacuum pressure impregnation system. The design and procurement of the contact tooling started in the middle of 2013 with the winding mandrel and curing mold, followed by the reaction fixture and impregnation mold, to finish later with the collaring tooling. Because of a tight schedule resulting from the need to install the magnets during the accelerator's LS2 during the years 2019–2020, an important overlap between the different phases of the project was inevitable. The assembly work profited from the large amount of experience in the assembling and collaring of long coils and the assembly of long cold masses at CERN for the LHC magnets.

To optimize the overall duration of development and production activities, it was decided to involve industry from the beginning of the prototyping activities at CERN. Contracts were placed with four companies, namely Alstom Power Systems in France, ASG Superconductors in Italy, BNG in Germany, and Oxford Instruments in the UK. These companies contributed to the design of the tooling, the construction activities, and to a certain extent also to the analysis of the magnet models' performance. To our knowledge, it is the first time that European Union (EU) industry was asked to contribute at this stage of an accelerator magnet project. This also allowed technology transfer to be carried out *in the shadow* of the development activities, and the anticipation of the industrialization of magnet production.

Three practice coils were made, one with copper conductor and two with low-grade (meaning not fully qualified) superconducting wire, to develop the procedures, train the technical staff, and commission the machines and tooling. Five full performance coils were then fabricated: one was lost in an incident and four have been assembled in the prototype. Special attention was required for the fabrication of the long coils, especially for the reaction and impregnation processes, which are specific to Nb_3Sn. Dealing with the thermal contraction of such long coils, however, proved to be more difficult than expected, and large gaps appeared after reaction, especially in the heads region. Although the gaps were filled with glass fiber this is

Fig. 9.17 The first 5.5 m long 11 T MBH dipole prototype in its cryostat at CERN ready to be transported to the test bench

most likely to be a weak point in the fabrication. The revised design suggested by the Task Force, discussed above, is supposed to address this issue.

The two practice coils (CR01 and CR02), made of low-grade wire, were assembled with the collars to validate the collaring process and tooling. The first collared coil assembly for the prototype was made with coils CR04 and CR05, and the second collared coil assembly with coils CR06 and CR07.

The prototype with its cryostat was finished in May 2018, see Fig. 9.17, and swiftly tested. The first test campaign showed that coil CR07 was damaged at both ends, limiting the performance to a value below nominal of 10 kA. Damage to one coil is suggested by the fact that both ends of that coil have a quench current limit well below the nominal value. The same coil shows resistive behavior (4.5 nΩ) at low current. Although at present the exact nature and location of the damage are not clear, production records show that the ends of coil CR07 had large gaps after heat treatment, nearly 10 mm that needed to be filled and required some manipulation at the level of the pole and end spacers. Ramp rate studies showed typical "inverse" behavior, i.e., higher quench values at higher ramp rates, which is a sign of an issue with the conductor (either localized or distributed, it is difficult to disentangle). Investigations about a possible correlation and cause are ongoing, especially trying to understand whether there is any scale-up effect, from the models to the long prototype coils. The fact that the first long prototype contains at least one coil that is damaged shows the importance of the adjustments devised by the Task Force in order to have more uniform and predictable results.

9.3.6 Series Production

Series production of the 11 T models comprises four magnets for installation in the LHC machine, and two spare magnets. For this, 12 collared coils are to be produced; the plan foresees the production of 30 single coils, of which six coils may be used to cover for possible incidents during production. The design and construction procedures integrate all recommendations issued by the Task Force.

In order to cope with the tight schedule requirements, following competitive tendering a service contract was placed with Alstom Power Services, Belfort, France, now part of General Electric. The work is to be carried out at CERN in the Large Magnet Facility, where all of the necessary machines and tooling are available from the prototyping phase. All of the components and consumables are to be provided by CERN. Some of the tooling has been duplicated in order to allow the achievement of the scheduling requirements. The contract comprises a start-up phase during which the personnel of the contractor can learn the manufacturing and inspection procedures that have been made available by CERN. The total duration of the contract will be of the order of 30 months. The other construction activities, like the cold mass assembly, the cryostating, the cold tests, and the final preparation prior to installation in the accelerator, will be carried out by a mixed team of CERN personnel and contract labor under direct CERN technical responsibility.

9.4 Conclusions

The 11 T dipole project for the HL-LHC is very demanding from both the technical and scheduling points of view. The initial technology transfer from FNAL to CERN, especially in cable design and coil technology, has been very important to curb the work and meet the schedule. The design contains both some revisited as well as novel features, e.g., the insulation system and other coil components, featuring a new original removable pole collaring concept. A total of 23 coils (including practice coils of copper or low-grade superconductor cables), all 2 m long, have been manufactured to date, and most of them have been assembled in seven magnet models with single apertures at CERN. Two two-in-one-aperture dipole short models were also assembled and tested. Models MBHSP102, MBHSP103, and MBHDP101 have shown satisfactory training performance and memory, and acceptable field quality, compatible with the installation requirements for the LHC machine. Models SP101, SP104, and SP105 showed limitations that were usually understood and traced to damage to the layer jump region or to the conductor due to excessive local pre-stress in the mid-plane. One magnet with slow training behavior, SP106, was pushed to a very high current quench value by gradually increasing the hot-spot temperature (delaying protection). The second two-in-one-aperture assembly, DP102, was limited to below the ultimate current due to damage to all of the coils that were good in the single aperture test. Altogether, these results indicated that the conductor and the coil design are adequate to reach the ultimate current, however the design with the assembly procedures used in the model R&D phase is very vulnerable to damage due to an excess of pre-stress, especially in the mid-plane.

The very good results, especially in terms of memory (always very good), have justified proceeding towards long magnets. The issue of a possible excess of stress was thoroughly investigated by a Task Force and a few changes in the design and in the collaring procedure were introduced and validated with SP107, which showed the best quench behavior among all models (very few quenches to ultimate current and almost perfect memory after a thermal cycle). While the prototype has been essential to debug all of the chain of construction for long magnets, only the first series dipoles will contain all elements of the basic and revised design, and we are very confident that we can meet the HL-LHC objective of installing the successful 11 T long cryo-assembly dipole in 2020, the first magnets wound with Nb_3Sn to operate in an accelerator.

The operation in a very demanding and precise collider like the LHC should show definitively the suitability of high field magnet technology for proton colliders, probably the most demanding application of superconductivity at present.

References

Ahlbäck J, Ikäheimo J, Järvi J et al (1994) Electromagnetic and mechanical design of a 56 mm aperture model dipole for the LHC. IEEE Trans Magn 30(4):1746–1749. https://doi.org/10.1109/20.305594

Andreev N, Barzi E, Chichili DR et al (2002) Volume expansion of Nb-Sn strands and cables during heat treatment. Adv Cryog Eng 48(614):941. https://doi.org/10.1063/1.1472635

Apollinari G, Béjar Alonso I, Brüning O et al (eds) (2017) High luminosity large hadron collider (HL-LHC): technical design report V.0.1, CERN-2017-007-M. CERN, Geneva

Auchmann B, Karppinen M, Kashikhin VV et al (2012) Magnetic analysis of a single-aperture 11 T Nb_3Sn demonstrator dipole for LHC upgrades. In: Proceedings of IPAC2012: international particle accelerator conference, New Orleans, May 2012, p 3596

Bottura L, de Rijk G, Rossi L et al (2012) Advanced accelerator magnets for upgrading the LHC. IEEE Trans Appl Supercond 22(3):4002008. https://doi.org/10.1109/tasc.2012.2186109

Brüning O, Rossi L (eds) (2015) The high luminosity large hadron collider – the new machine for illuminating the mysteries of universe. Advanced series on direction in high energy physics, vol 24. World Scientific, Singapore. https://doi.org/10.1142/9581

Carli C (ed) (2010) Proceedings of the Chamonix 2010 workshop on LHC performance, Chamonix, 25–29 Jan 2010, CERN-ATS-2010-026. CERN, Geneva

Damerau H, Funken A, Garoby R et al (eds) (2014) LHC injectors upgrade—technical design report, vol 1, protons, CERN ACC.2014-0337, 15 Dec 2014. CERN, Geneva

de Rijk G, Milanese A, Todesco E (2010) 11 Tesla Nb_3Sn dipoles for phase II collimation in the Large Hadron Collider, sLHC Project Note 0019. CERN, Geneva

Fiscarelli L, Auchmann B, Izquierdo Bermudez S et al (2016) Magnetic measurements and analysis of the first 11-T Nb_3Sn dipole models developed at CERN for HL-LHC. IEEE Trans Appl Supercond 26(4):1–5. https://doi.org/10.1109/tasc.2016.2530743

Fiscarelli L, Izquierdo Bermudez S, Dunkel O et al (2017) Magnetic measurements and analysis of the first 11-T Nb_3Sn 2-in-1 model for HL-LHC. IEEE Trans Appl Supercond 27(4):1. https://doi.org/10.1109/tasc.2016.2639285

Holzer B (2014) Impact of Nb_3Sn dipoles on the LHC lattice and beam optics, CERN ACC-NOTE-2014-0063. CERN, Geneva

Izquierdo Bermudez S, Bottura L, Todesco E (2016) Persistent-current magnetization effects in high-field superconducting accelerator magnets. IEEE Trans Appl Supercond 26(4):1–5. https://doi.org/10.1109/tasc.2016.2519006

Karppinen M, Andreev N, Apollinari G et al (2012) Design of 11 T twin-aperture Nb_3Sn dipole demonstrator magnet for LHC upgrades. IEEE Trans Appl Supercond 22(3):4901504. https://doi.org/10.1109/tasc.2011.2177625

Rossi L (2004) Experience with LHC magnets from prototyping to large scale industrial production and integration. In: Proceedings of EPAC2004. 9th European particle accelerator conference, Lucerne, July 2004, LHC Project Report 730. CERN, Geneva, p 118

Savary F, Apollinari G, Auchmann B et al (2015) Design, assembly, and test of the CERN 2-m long 11 T dipole in single coil configuration. IEEE Trans Appl Supercond 25(3):1–5. https://doi.org/10.1109/tasc.2015.2395381

Savary F, Bajko M, Bordini B et al (2017) Progress on the development of the Nb_3Sn 11 T dipole for the high luminosity upgrade of LHC. IEEE Trans Appl Supercond 27(4):4003505. https://doi.org/10.1109/tasc.2017.2666142

Savary F, Bordini B, Fiscarelli L et al (2018) Design and construction of the full-length prototype of the 11-T dipole magnet for the high luminosity LHC project at CERN. IEEE Trans Appl Supercond 28(3):1–6. https://doi.org/10.1109/tasc.2018.2800713

Willering GP, Bajko M, Bajas H et al (2017) Cold powering performance of the first 2 m Nb_3Sn DS11T twin-aperture model magnet at CERN. IEEE Trans Appl Supercond 27(4):1–5. https://doi.org/10.1109/tasc.2016.2633421

Willering G, Bajko M, Bajas H et al (2018) Comparison of cold powering performance of 2-m-long Nb_3Sn 11 T model magnets. IEEE Trans Appl Supercond 28(3):1–5. https://doi.org/10.1109/tasc.2018.2804356

Zlobin AV (2010) Status of Nb_3Sn accelerator magnet R&D at Fermilab. In: 'EuCARD – HE-LHC'10. AccNet mini-workshop on a "high-energy LHC", Villa Bighi, 14–16 October 2010, CERN Yellow Report CERN-2011-003 (arXiv:1108.1869). CERN, Geneva, p 50

Zlobin AV, Andreev N, Apollinari G et al (2012) Design and fabrication of a single-aperture 11 T Nb_3Sn dipole model for LHC upgrades. IEEE Trans Appl Supercond 22(3):4001705. https://doi.org/10.1109/tasc.2011.2177619

Part III
Block-Type Dipole Magnets

Chapter 10
Block-Type Nb_3Sn Dipole R&D at Texas A&M University

Peter McIntyre and Akhdiyor Sattarov

Abstract A succession of Nb_3Sn block-type dipoles has been developed at the Texas Agricultural and Mechanical (A&M) University and tested, in a progression of stages aimed toward 16 T operation field. This chapter describes the details of magnet design, fabrication procedures, and test results.

10.1 Introduction

The Accelerator Research Laboratory (ARL) at the Texas Agricultural and Mechanical (A&M) University has spent the past 16 years developing block type dipole technology, with fields towards the 16 T range, as a cost-effective basis for the superconducting storage rings of a future hadron collider. The motivation for the work has been to develop coils for a dipole magnet that are easy to manufacture, and to address several aspects of dipole technology that become challenging for thick coils. At the beginning of development, stress management within the windings and correction for persistent current (Kashikhin and Zlobin 2001) due to the large filaments in high-performing Nb_3Sn strands had been seen by the accelerator magnet community as the main challenges. To address these challenges the rectangular winding geometry of a block type geometry was selected. A cross-section and 3D sketch of the dipole coils of a block-type magnet are shown in Fig. 10.1.

The major challenge in this design option is the so-called flared-end geometry to allow for a continuous aperture and to enable the beam tube to pass through the magnet. The transition from the rectangular geometry of the body winding to the flared end is susceptible to de-registration within cables, and poses a complex challenge for stress management. So far, in this program three model dipoles (TAMU1 to TAMU3) were built using Rutherford cable, wound in flat double-pancake windings without flared ends (Diacaenko et al. 1997) to gain experience with Nb_3Sn dipole magnets. Two new model dipoles currently being developed (TAMU4 and TAMU5) utilize a novel cable-in-conduit (CIC) superconductor,

P. McIntyre (✉) · A. Sattarov
Department of Physics and Astronomy, Texas A&M University, College Station, TX, USA
e-mail: mcintyre@physics.tamu.edu; a-sattarov@physics.tamu.edu

D. Schoerling, A. V. Zlobin (eds.), *Nb_3Sn Accelerator Magnets*, Particle Acceleration and Detection, https://doi.org/10.1007/978-3-030-16118-7_10

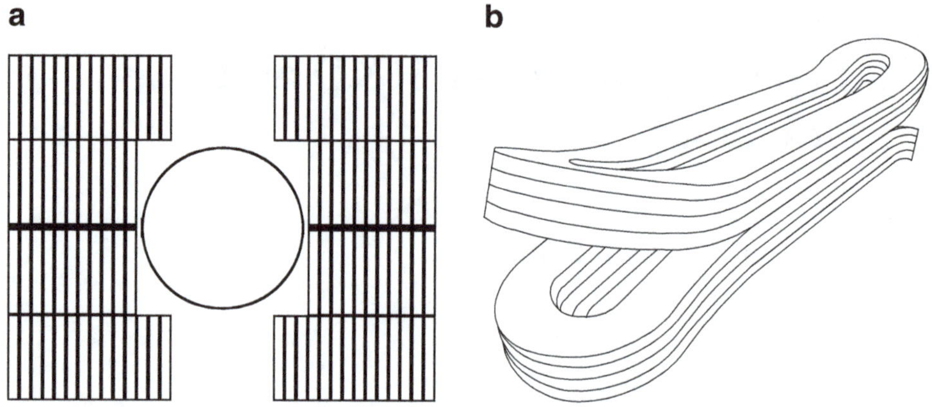

Fig. 10.1 (**a**) Coil cross-section; and (**b**) a 3D sketch of a block-type dipole magnet

Table 10.1 Summary of model dipoles built at ARL

Dipole model	Timeline	Superconductor	Field/current (T/kA)	
			Design	Achieved
TAMU1 Battle et al. (2001)	1999–2001	Nb-Ti	6.6 T/8.1 kA	6.5 T/8.0 kA
TAMU2 Noyes et al. (2006) and McInturff et al. (2007)	2002–2006	Nb_3Sn (ITER)[a]	6.9 T/9.4 kA	6.6 T/8.9 kA
TAMU3 Blackburn et al. (2008a), Holik et al. (2011, 2014) and Elliott et al. (2016)	2007–2013	Nb_3Sn (RRP)[b]	13 T/13.6 kA	6.6 T/7.6 kA
TAMU4 Assadi et al. (2015)	2014–2017	Nb-Ti	4 T/15 kA	NA
TAMU5 Assadi et al. (2017)	2016–	Nb_3Sn/Bi2212	17 T/25 kA	NA

[a]International Thermonuclear Experimental Reactor
[b]Restacked rod process

wound in a barrel wind with flared ends. A summary of the main parameters of TAMU1 to TAMU5 is presented in Table 10.1. Due to the early stage of this work, TAMU4 and TAMU5 are not discussed in this chapter.

10.2 R&D Program and Approach

The block-type dipoles were designed, built, and tested following a strategy that focused upon one challenge or open question at a time. The main questions formulated at the beginning of the project are listed below.

1. How to manage the accumulation of Lorentz stress so that it does not produce strain in the windings beyond the limit for degradation of critical current density J_c?
2. How to suppress persistent-current multipoles due to magnetization in the superconducting filaments, which are much larger in Nb_3Sn than in Nb-Ti strands?
3. How to provide heat transfer in the inner regions of a thick winding to provide cooling and stability against micro-quenches?
4. How to support Lorentz stress in the flared ends?
5. How to form the flared ends of the winding without damaging the internal registration of the cable?
6. How to make a hybrid winding incorporating Nb-Ti windings that are not heat-treated with wind-and-react (W&R) windings of Nb_3Sn and Bi-2212?

10.2.1 Stress Management

A common theme among the first three model dipoles (TAMU1 to TAMU3) developed at ARL is the strategy of stress management, illustrated in Fig. 10.2.

Stress management entails the integration of a high-strength support matrix within the coil, so that Lorentz forces can be intercepted and bypassed. The winding is divided into inner and outer winding blocks, in which the Rutherford cable and the wire used in each winding are graded in their composition and diameter so that the inner and outer elements operate at approximately the same fraction of the critical current.

To enforce a decoupling of forces between the inner and outer windings, a laminar spring is introduced. The laminar spring features a low compressive

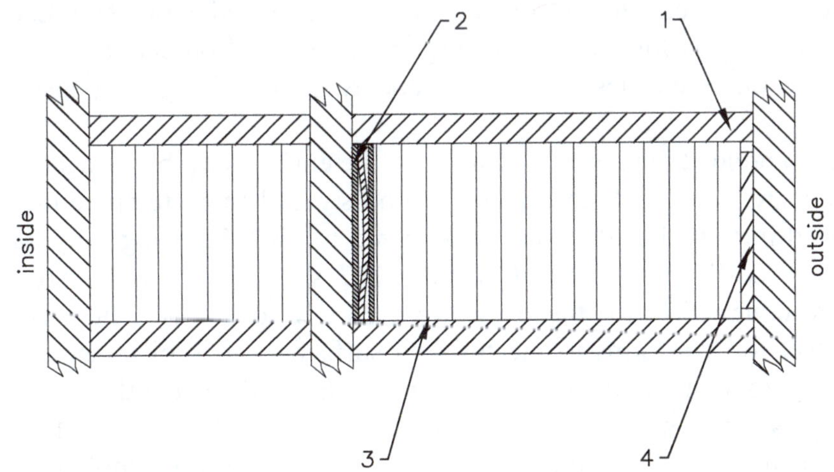

Fig. 10.2 Stress management provisions in one horizontal layer of a block-type dipole winding. *1* – Inconel piers; *2* – laminar spring; *3* – S-glass insulation and mica paper shear release; *4* – laminar strain gauge

modulus, so that very little force can be transmitted from the inner to the outer winding. Almost all the Lorentz force developed in the inner winding is intercepted on a structural beam and transferred through the flanking piers to the support matrix. The laminar spring also presents a uniform face load that maintains compression of the outer winding. The inner winding is impregnated within its compartment of the structure so that no further pre-stress is applied.

Each winding was surrounded by mica paper to provide shear release and low-friction sliding of the winding with respect to the surrounding elements of the support matrix. The mica paper does not absorb epoxy during coil impregnation, so no bonding between the coil blocks and the support matrix is established. The mica paper used, 132P from Cogebi Inc., Dover, NH, USA has a very small amount of binder (<2%), produces no ash during the Nb_3Sn heat treatment, and releases shear at the interface between piers and the coil. Shear release was verified and confirmed through considerable testing.

Because the Lorentz force acting upon the inner winding is bypassed past the outer winding, the maximum stress in any winding is limited to around 100 MPa, which is well below the limit for critical current degradation of Nb_3Sn coils.

The support matrix is made from Inconel 718. The parts are produced from material with mill finish (no surface machining), and are cut from sheet using electric discharge machining.

10.2.2 Pre-stressing the Structure

The support matrix must be loaded to compress the stress management structure so that it is preloaded and cannot move as Lorentz forces are applied within the windings. To achieve the required pre-stress the structure is preloaded with bladders. Each bladder consists of two thin stainless-steel foils, welded together around their edges to form a hermetic enclosure with a fill tube attached at one end. Each bladder is fabricated using hydraulic foil-forming and laser-weld procedures, and has been tested to five times the design pressure and twice the design expansion without failure.

In order to preload the assembled magnet, the entire assembly is heated to 80 °C, the bladders are evacuated, and molten Wood's metal (melt temperature 65 °C) is then pumped into the bladders and pressurized to 14 MPa using a hydraulic hand pump. The Wood's metal alloy Cerrolow 147 selected has near-zero aggregate temperature contraction for the cycle from 343 K to 4.2 K. The bladder's thin stainless-steel foil conforms to the finished surface of the coil assembly and flux return cavity to deliver a uniform pre-stress. By independently controlling the hydraulic pressure applied to the sets of flat and curved bladders, the preload is controlled on the entire assembly. The magnet assembly is then cooled to freeze the metal while holding the pressure.

The dipole can be disassembled by heating the magnet to 80 °C and pumping out the Wood's metal, so that the coil modules could be re-used in subsequent model dipoles. The operation of over-compression that is used in bladder-and-key coil structures is not necessary (Caspi et al. 2001; Hafalia et al. 2003). Moreover, preload is delivered to the support structure, not to the windings themselves. The aim of applying pre-stress in this design is to make the structure stiff without applying pre-stress to the coil, so that any coil movement is blocked by the stiff structure. Relative movement between the structure and the coil is allowed, and sliding planes have been introduced to avoid any slip–slick motion and reduce training. In conventional dipole structures, preload is applied to the windings themselves, and might result in excessive strain at room temperature that could irreversibly damage Nb_3Sn filaments.

10.2.3 Flux Plate Suppression of Persistent Magnetization Fields

When a dipole is ramped from the injection field to the nominal operation field either as an accelerator or as a collider and then back to low field, eddy currents are induced at the cable, strand, and filament levels. Induced currents between strands in the cable and between filaments within a strand must pass through the copper matrix and are resistively damped. Currents induced in the individual filaments circulate entirely within the superconductor, and so can persist indefinitely (Green 1971). These persistent currents produce serious challenges for accelerator magnets. First, the pattern of induced currents produces multipole fields in the magnet aperture, which produce harmful effects upon beam dynamics. The amplitude of the multipoles induced by persistent currents depends typically on the coil geometry, the filament diameter, and the critical current density in the filaments. Second, the persistent current magnetization produces a hysteresis loop. When the magnet current is ramped down to injection, it sets the induced magnetization on the discharge trajectory. After new beam is injected and the current is ramped up as the beam is accelerated, the pattern of persistent currents re-distributes over a small range of magnetic field to that of the charging trajectory. The coil re-magnetization causes a change in the sextupole component b_3 and must be carefully compensated to avert the risk of disrupting the beam just as acceleration begins.

The block-type coil configuration uniquely makes possible a method to suppress the persistent current effect. A pair of horizontal steel flux plates are incorporated within the dipole, as shown in Fig. 10.3. We have simulated the pattern of multipoles that are induced due to persistent current effects in a block-type dipole winding, and compared the multipoles for a filament size of 50 μm and a critical current density of 2500 A/mm^2 at 4.2 K and 12 T, for the cases with and without a flux plate (Blackburn et al. 2003). The flux plate suppresses the persistent-current multipoles at 1 T injection field by a factor of 5 compared to the multipoles without a flux plate. At injection field, the flux plates are unsaturated and present a strong dipole boundary condition. Lines of force re-distribute within the flux plate to cancel multipole components.

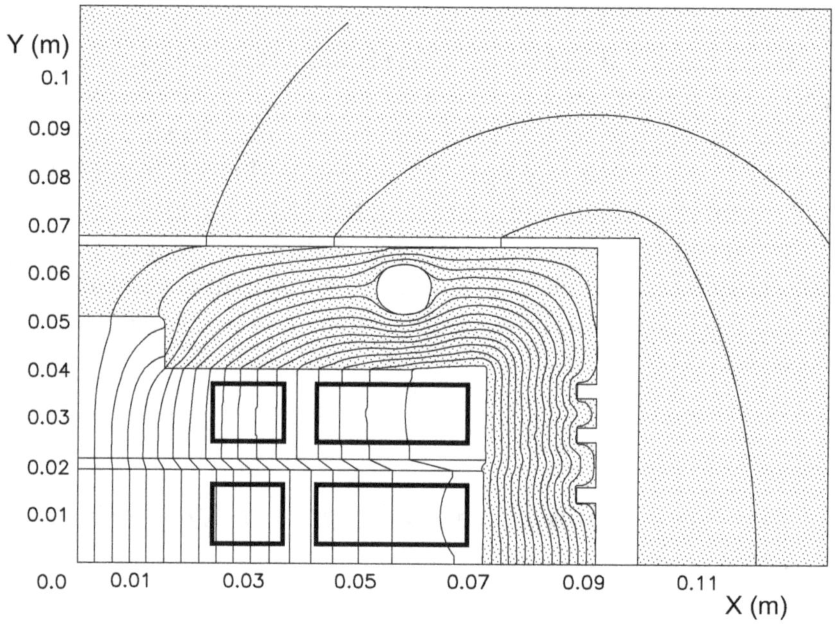

Fig. 10.3 Magnetic field distribution at injection field (1 T) for a 14 T block-type dipole. The steel flux plate (placed between the two coils) re-distributes the field lines and minimizes the sextupole component

Suppression of persistent-current magnetization is important for dipoles that will utilize Nb_3Sn, because the sub-element size in those conductors is about 50 μm, 10 times larger than the Nb-Ti filaments in the wire used for the LHC.

10.3 TAMU1: Single-Shell Nb-Ti Model Dipole

TAMU1 was designed to serve as a learning model for the construction techniques that are required for stress management: integration of the windings in the support matrix; mica paper slip surfaces; epoxy impregnation of the windings; and rectangular steel flux return. The inner and outer sub-windings are assembled within the support matrix. The windings are composed of Nb-Ti Rutherford cable and are arranged in three double-pancake winding layers, as shown in Fig. 10.4. The main parameters of TAMU1 are summarized in Table 10.2.

10.3.1 Magnet Manufacture

Each double-pancake winding was wound with an S-transition at the inner boundary. Successive double-pancake windings were connected by a splice. Before impregnation, the windings were instrumented with voltage taps, quench heaters, and spot heaters. During vacuum impregnation, the coil attained a horizontal bow of

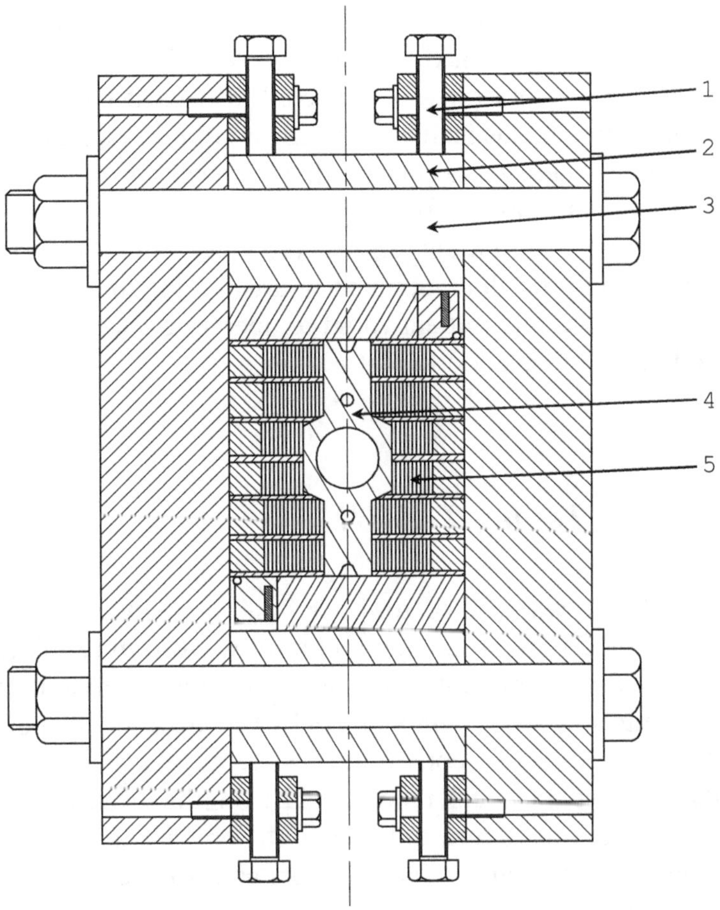

Fig. 10.4 TAMU1: first block-type prototype dipole using Nb-Ti. *1* – side compression press; *2* – compression spacer bar; *3* – main stress bolts; *4* – laminated mandrel; *5* – coil

Table 10.2 Main parameters of TAMU1

Parameter	Value
Max. field in winding (short sample) (T)	7.92
Short sample current (A)	8550
Stored energy (MJ/m)	0.10
Max. Lorentz force (MN/m)	0.47
Superconducting cable	
Number of strands	30
Strand diameter (mm)	0.808
J_c(5 T, 4.2 K) (A/mm^2)	2846
J_{Cu} at quench and short sample current (A/mm^2)	785
Cu/non-Cu ratio	1.5
Number of turns in windings	10/15/15

about 2 mm arising from a stress relieve in the fixture. This bow was corrected using smart shims to provide a uniform contact along the sides of the coil within the flux return. The coil was then installed in the structure and preloaded by using large bolts, as shown in Fig. 10.4.

Incorporation of the smart shim made it impossible to close the structure that was designed to provide stress management, so that the preload of the flux return was delivered directly to the coil. The coil was preloaded to 500 kN, which represents about 30% of the maximum calculated Lorentz force. As the coil is excited to full field, the Lorentz force is expected to exceed the preload, and the retaining bolts should stretch elastically—the maximum bolt elongation is calculated to be around 0.20 mm. There was a concern that the resulting coil motion might produce quenches at high field; however, it did not.

Several intermittent shorts (coil to ground) were encountered during the preloading procedure, apparently arising due to a displaced insulation layer during preload. The intermittent room-temperature shorts were repaired by locating them and adding insulation in that region.

10.3.2 Testing TAMU1

The dipole was cooled to 4.5 K, and all windings were checked for turn-to-turn shorts by driving a saw tooth current waveform (20 A maximum) through the coil. The current ramp rate up was twice the current ramp rate down, so that the inductance of each winding could be checked independently at two frequencies. The voltages across coils 1 and 2 were mismatched by a ratio ~1.7, much more than the resistive mismatch ratio of 1.22 from the locations of the voltage taps, indicating that there was a turn-to-turn short, which was not found at previous testing at ambient temperature. Such a short could pose a serious hazard in a high-current quench.

A short removal scenario was attempted, in which the coil was exercised with a continuous saw tooth ramp to 300 A peak current, and the current ramp rate was increased in a succession of steps until the coil quenched from alternating current heating. The rationale was that a substantial but limited amount of energy, dumped through the short during the quench, would be sufficient to burn out the short without damaging the coil.

Each succeeding plateau corresponded to a continuous ramp to 300 A, at a series of increasing ramp rates: 50 A/s up to 600 A/s. At 600 A/s, the coil temperature reached the critical temperature $T_c = 9.3$ K, and the coil quenched. After recovery, the coil did not quench again during a repeated 600 A/s ramp sequence, but did with a 700 A/s ramp sequence. Something in the coil had changed to produce this change in behavior.

The voltages across the windings were then measured during a ramp to 300 A with 400 A/s up and 200 A/s down. All winding inductances were consistent with their calculated values L_i, indicating that the short had been successfully removed.

The quench performance of TAMU1 is shown in Fig. 10.5. TAMU1 performed close to its short sample field of 6.6 T on almost every quench showing virtually no training. There was no evidence apparent for the 20% lower current of quench #6; the location inferred from voltage taps was comparable to that of other quenches at short sample current.

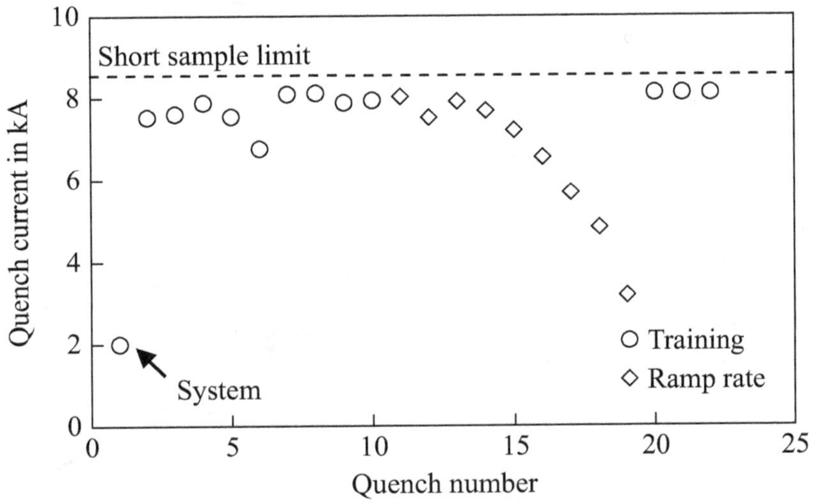

Fig. 10.5 TAMU1 quench history

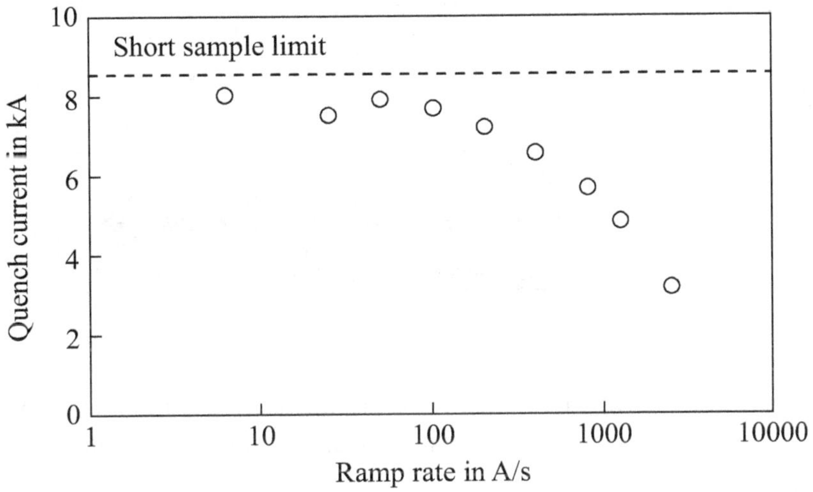

Fig. 10.6 TAMU1 ramp rate dependence

TAMU1 supported rapid ramping up to 1000 A/s (Fig. 10.6). The decrease in quench current at a ramp rate of 1000 A/s was around 50%.

10.4 TAMU2: Mirror-Geometry Nb_3Sn Dipole

TAMU2 is a two-layer Nb_3Sn dipole that contains all of the elements for stress management, and provisions for W&R heat treatment of the windings, vacuum impregnation, and preload of the reacted windings. It was designed as a mirror-magnet, so that it could be built with a single layer.

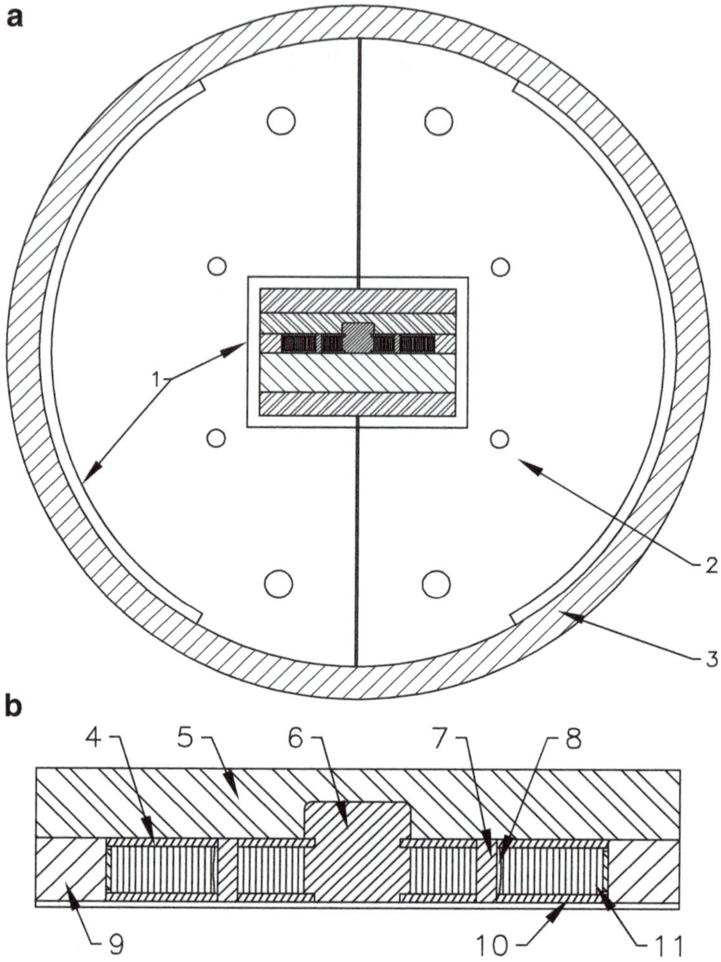

Fig. 10.7 Cross-sections of (**a**) the magnet; and (**b**) the winding module of TAMU2. *1* – spacer for inserting bladders; *2* – Fe flux return; *3* – Al stress tube; *4* – Inconel piers; *5* – Fe thick skin; *6* – Ti mandrel; *7* – Ti beam; *8* – laminar spring; *9* – Fe side bar; *10* – Fe flux plate; and *11* – laminar strain gauge

The Rutherford cable was made using International Thermonuclear Experimental Reactor (ITER)-type Nb_3Sn chrome-plated strand, which was available to ARL at no cost.

Figure 10.7 shows a cross-section of the mirror configuration. The different elements are described in detail in the following sub-sections. The main magnet parameters are summarized in Table 10.3.

10.4.1 Laminar Springs

TAMU2 was the first magnet in which laminar springs were incorporated to enforce force transfer between the inner and outer sub-windings (#8 in Fig. 10.7), as described in Sect. 10.2.1. The laminar spring is formed from annealed 0.2 mm thick Inconel X750 sheets, die-stamped to form the interior spring components, and hermetically sealed within a laser-welded Inconel shell. Hermetic sealing of the spring is vital, so that the spring enclosure does not fill with epoxy during the vacuum impregnation of the coil assembly. The springs are 13 mm wide and

Table 10.3 Main parameters of TAMU2

Parameter	Value
Max. field in coil (short sample) (T)	7.6
Short sample current (A)	10,625
Stored energy at B_{max} (MJ/m)	0.062
Max. Lorentz force at B_{max} (MN/m)	0.93
Superconducting cable	
Number of strands	30
Strand diameter (mm)	0.808
J_c(10 T, 4.5 K) (A/mm^2)	1050
J_{Cu} at quench and short sample current (A/mm^2)	1146
Cu/non-Cu ratio	0.67
Number of turns in windings	
Inner	11
Outer	16

Fig. 10.8 Laminar spring cross-section

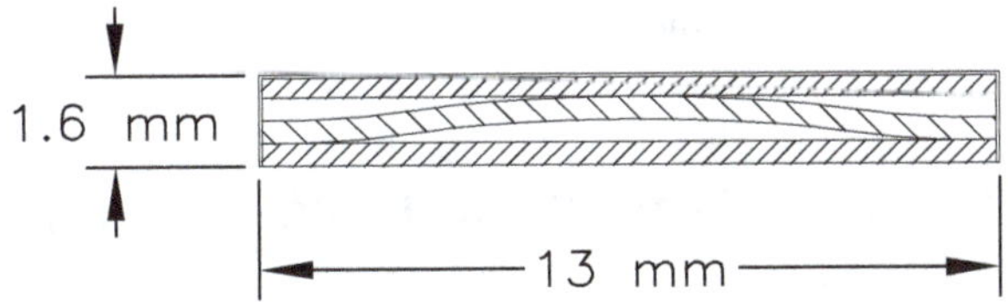

1.6 mm thick (Fig. 10.8). They have an elastic working range of 0.28 mm, the elastic range is 15% of the elongation. It is designed such that 75% of the laminar spring's deflection is used to deliver a pre-stress of about 20 MPa to the outer winding.

The laminar springs have not shown hysteresis or fatigue over 1000 cycles, which is found to be a good performance for a test magnet: The usability of this system for an accelerator-quality magnet, which is designed for tens of thousands cycles (the design number of cycles for the LHC magnets is 20,000) remains an open question.

10.4.2 *Mica Paper Shear Release*

Shear release was evaluated by placing the coil block in a two-axis dynamometer, applying a side load (simulating the vertical stress compressing the coil against the pier (#4 in Fig. 10.7)), and then applying a shear load to the pier until it delaminates.

The shear stress S_x required to delaminate the mica paper is a combination of the intrinsic shear S_0 within the mica and the static friction μS_y produced by the face loading stress S_y pressing the coil block against the face of the neighboring pier: $S_x = S_0 + \mu S_y$. We impregnated a number of model coil blocks and measured S_x at a succession of levels of normal (face-loading) stress S_y. The results for the shear stress S_x are given in Fig. 10.9. Shear release occurs for $S_x < 10$ MPa so long as $S_y < 10$ MPa. The distribution of S_y has been calculated everywhere in the coil assembly, for the preloaded dipole at room temperature. Face-loading stress never exceeds 10 MPa.

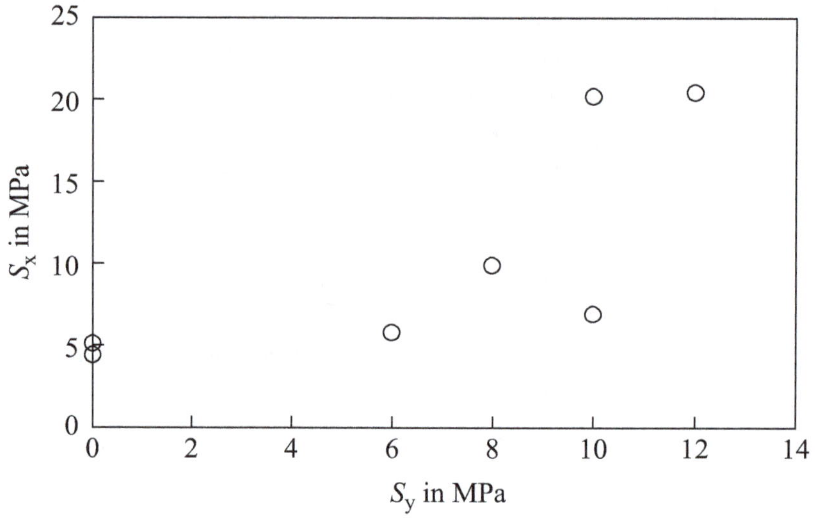

Fig. 10.9 Measured shear release stress as a function of transverse loading stress for mica paper interface in coil blocks

10.4.3 Thermal Contraction Compensation Using Ti Blocks

The axial forces on the ends of each winding are intercepted by an end support structure. This support structure is friction locked into the body of the flux return yoke. Friction locking is achieved through sufficient pre-stress in the structure. To preserve this pre-stress from ambient temperature to cool-down, small blocks of titanium are implemented which are selected such that the thermal contraction of the coil and iron in the end structure is fully compensated by the Ti blocks. The same approach was used within the coil subassemblies to maintain sufficient pre-stress in the structure.

10.4.4 Stress Transducers

Laminar pressure (stress) gauges were placed along the outer boundary of the outer winding, one along each side and one around one end (#11 in Fig. 10.7). Each transducer was prepared as a multi-layer sandwich of 316 stainless-steel foils and polyimide foils, coated with epoxy and cured under pressure to form a fully dense sandwich. Some of the stress gauges were built with five layers, some with seven layers (the capacitance response is proportional to the number of layers). The fabrication followed similar procedures developed at the European Organization for Nuclear Research (CERN), Fermi National Accelerator Laboratory (FNAL), and Lawrence Berkeley National Laboratory (LBNL), but with significant improvements that eliminated glue delamination and response creep that had plagued earlier transducers (Benson et al. 2012). Figure 10.10 shows a fabricated five-layer stress

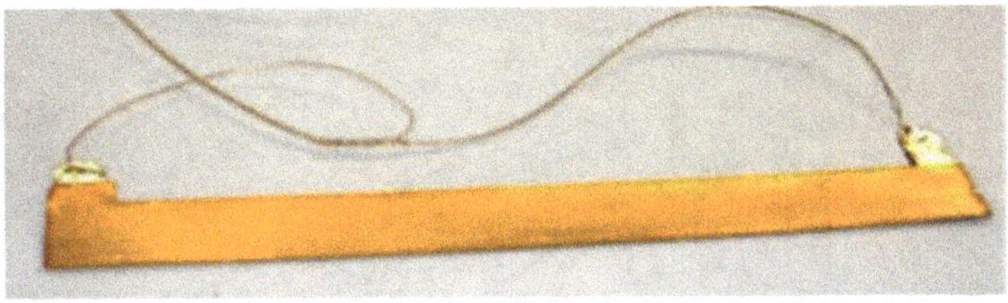

Fig. 10.10 Stress transducer

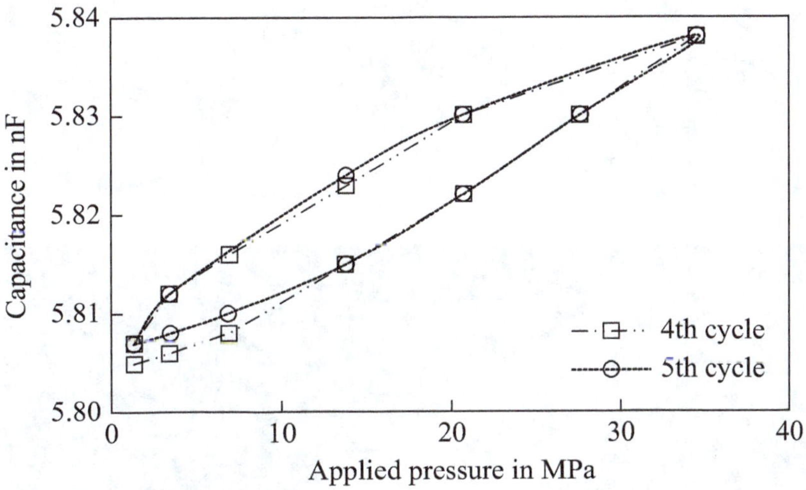

Fig. 10.11 Calibration of a stress transducer through multiple cycles

transducer. Figure 10.11 shows the calibration of a transducer over the entire range of stress from preload to full field for the fourth and fifth cycles. The calibration curve converged to a closed hysteresis curve.

10.4.5 Coil Fabrication

A double-layer coil was wound onto the laminated titanium mandrel (Fig. 10.12). The lead end of the mandrel was anchored to the support plate, and the other end of the mandrel was left unanchored, so that the mandrel could expand and contract with the winding during heat treatment. A gap of 2 mm was opened in the center of the laminated mandrel so that the net shrinkage of the winding during reaction could pull the gap closed and the winding would not be put into a strain state after heat treatment.

10.4.5.1 Heat Treatment

For the heat treatment, the coil was mounted with its side rails and end shoes in a reaction fixture that held all coil dimensions through reaction heat treatment. A removable post segment was located midway along the laminated center mandrel. It

a

b

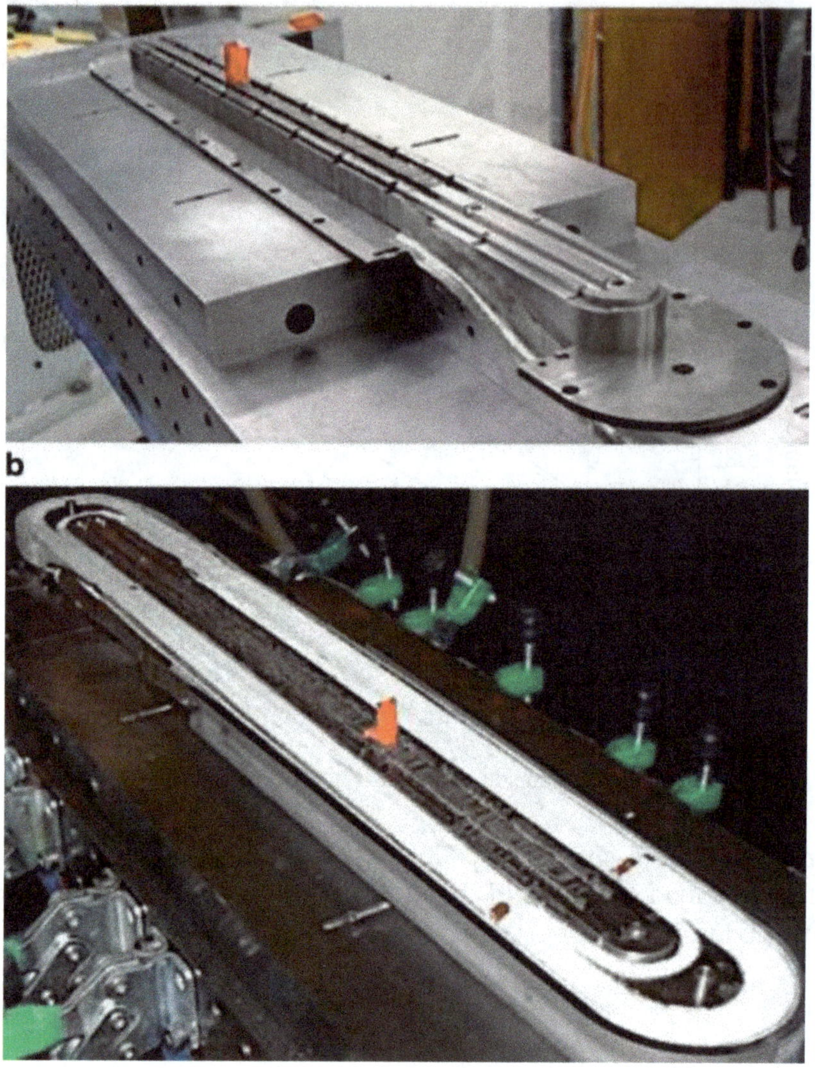

Fig. 10.12 (**a**) Winding mandrel ready for first turn; and (**b**) inner winding

was removed after coil winding before reaction, to provide room for coil axial expansion. The coil module was heat-treated: 210 °C for 100 h, 340 °C for 48 h, and 650 °C for 180 h. A purge of argon gas was maintained through the gas manifold within the reaction fixture to remove volatilized hydrocarbons from the sizing of the coil.

After heat treatment, it was observed that the pier that spans the outer layer piers was stretched. It was concluded that the deformation resulted from net elongation during heat treatment of the titanium pier (#7 in Fig. 10.7) separating the inner and outer windings and net shrinkage of the Inconel piers (#4 in Fig. 10.7), both caused by release of internal stress from the rolling of the original sheet materials. The coil was not visibly damaged and the assembly of the module and of the dipole was continued. In subsequent tests, rolling stresses were released by performing a stress relief heat treatment before machining.

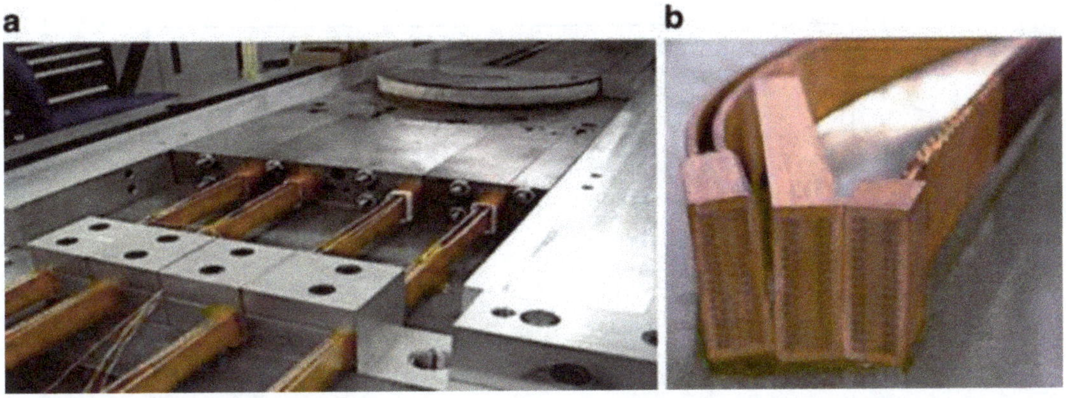

Fig. 10.13 (**a**) Splicing of Nb-Ti lead cables onto Nb_3Sn windings; (**b**) cross-sections of test splices

10.4.5.2 Nb_3Sn to Nb-Ti Cable Splicing

Each Nb_3Sn lead was spliced to a pair of flexible Nb-Ti cables. During heat treatment, each lead segment of Nb_3Sn cable is supported within a removable channel, so that it cannot deform. After heat treatment, the segments of the channel are removed and a pair of Nb-Ti cables is assembled to sandwich the Nb_3Sn lead. The sandwich contains flat strips of fluxed low-melt SnAg solder between successive layers. The support channel is then re-assembled and heater cartridges are energized to heat the joint to the flow temperature of the solder. The process was developed to produce a fully soldered joint from each face of the Nb_3Sn lead to an Nb-Ti cable. Measured splice resistance was 0.28 nΩ. The splicing and the joints are shown in Fig. 10.13.

10.4.6 Structure Assembly

The flux return yoke is divided into two halves in the vertical mid-plane surrounded by a super-alloy aluminum stress shell. The coil assembly surrounded by flat bladders is inserted into the rectangular cavity of the flux return. The flux return itself is preloaded by the pair of curved bladders that are inserted into the aluminum stress shell.

To achieve the required pre-stress, a set of four flat bladders supports the coil module in the box space within the two halves of the steel flux return (#1 in Fig. 10.7). Another pair of curved bladders are located in a space between the outside surface of the flux return steel and the super-alloy aluminum stress shell. This shell provides the overall containment of all forces and additional pre-stress during cool-down thanks to its larger thermal contraction factor.

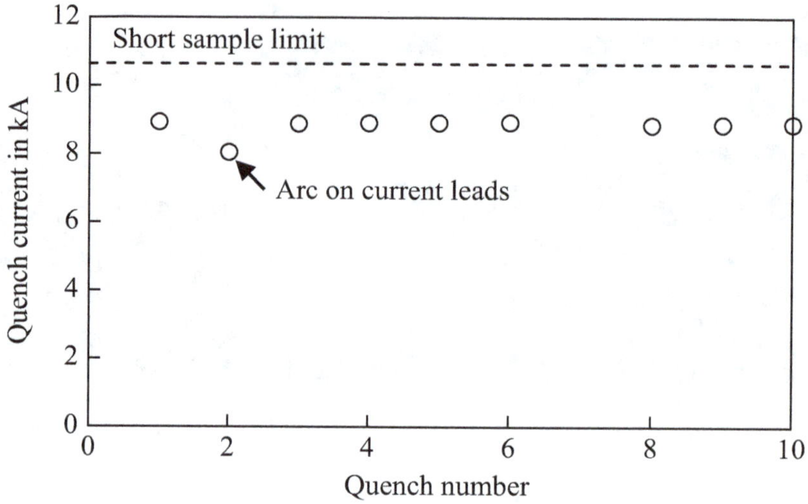

Fig. 10.14 Quench history of TAMU2 (data for quench #7 was lost due to a defect in the data acquisition)

10.4.7 *Testing of TAMU2*

The magnet was tested in boiling LHe at 4.4 K at LBNL. The chronological quench history of TAMU2 is given in Fig. 10.14. The ramp rate for training quenches was 10 A/s. The cable and strand data available at the time of construction indicated that the short sample current for TAMU2 is 8914 A. Short samples of the same wire were wound on an ITER barrel, heat-treated by the same schedule, and tested for short sample current $I_c(B)$. The dipole quenches in Fig. 10.14 correspond to 93.4% of witness strand short sample currents. There was no training: within the small spread in quench current values observed: the magnet went to the same value every quench. The best recent performance for cable made from ITER strand would have resulted in a short sample current of 9633 A in the windings of TAMU2.

Some excitement came with quench #2. As current was ramping up towards the previous quench value, a peculiar clanking sound was heard from the region of the Cu bus bar channels that carry current to the test cryostat. Just as we were endeavoring to find its origin, a flash of light and an explosive bang came from the same location. A large bolt had been left on top of one of the Cu bus bars during preparations for the test. Although made from stainless steel the bolt had no small permeability. When the bus current reached ~8 kA, the bolt magnetically levitated, and by the second cycle happened to shift in position to make a short circuit between the side-by-side bus bar leads. After repair of the test station the magnet training was continued, and magnet training ramp rate studies were continued to be performed. For the ramp rate studies, the coil current was ramped successively faster and the quench current was measured. The ramp rate dependence is plotted in Fig. 10.15. It was anticipated that coupling currents would be suppressed by the orientation of the cable parallel to the magnetic field in the windings. As previously mentioned, the conductor was chrome-plated, which increased the contact resistance between strands and therefore aided in the suppression of coupling currents. Nevertheless, the extreme robustness of TAMU2 for fast ramp rates was

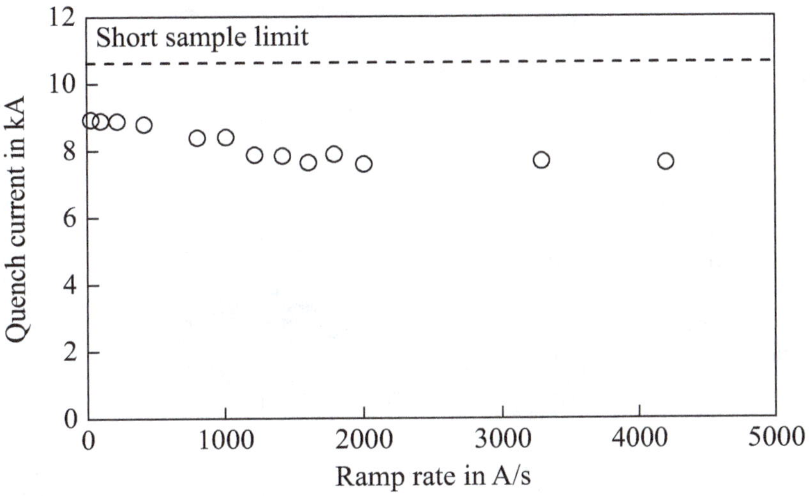

Fig. 10.15 Ramp rate studies of TAMU2

surprising. The only limitation that was encountered was the power supply's inability to regulate with the low inductance of TAMU2 at very high ramp rates (>2000 A/s). For these points, the stated ramp rate value was estimated from the current read-back data. The ramp rate data indicates that an Nb_3Sn block-coil dipole could be suitable for cycling at a few T/s in the 6 T field range, a topic of interest for future fast-cycling accelerator projects.

10.5 TAMU3: 13 T Racetrack Dipole

TAMU3 is an Nb_3Sn block-type dipole built to embody all the above provisions for stress management. Its flat racetrack coils were built as two identical modules, each containing an inner and outer winding, as shown in Fig. 10.16. The inner and outer windings were graded—the strand size in each was chosen so that the inner and outer windings would operate at the same fraction of short sample limit. The strands intended for both windings were 54/61-filament restacked rod process (RRP) Nb_3Sn produced by Oxford Instruments Technology, USA. Table 10.4 summarizes the main parameters of TAMU3.

10.5.1 Fine-Filament S-Glass Insulation

An improved S2-glass fabric insulation was developed for use in the Nb_3Sn windings of TAMU3. Following some lead development at the Commissariat à l'énergie atomique (CEA) in Saclay, France (Canfer et al. 2008), the company AGY, Aiken, SC (www.agy.com), produced a fine-filament S2-glass yarn containing 204 filaments of 5.5 μm diameter with a linear mass density of 11 tex (g/km). The yarn was treated

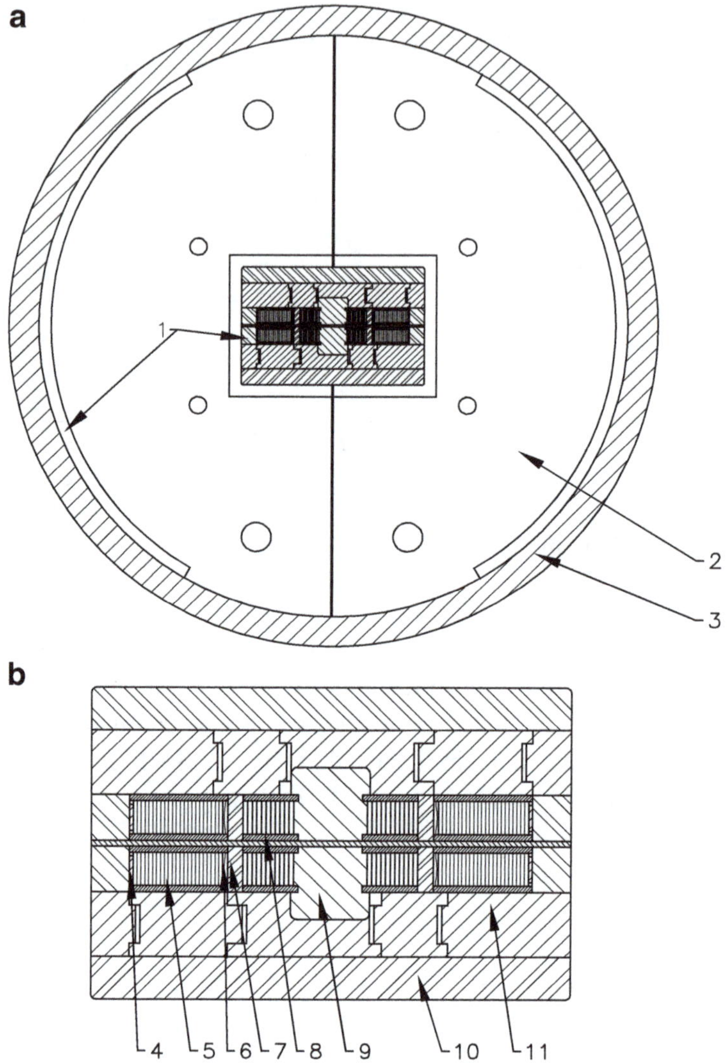

Fig. 10.16 Stress management elements in TAMU3. (**a**) The magnet: *1* – bladders; *2* – Fe flux return; *3* – Al stress tube; (**b**) the coil: *4* – stress transducers; *5* – Inconel piers; *6* – laminar spring; *7* – Ti beams; *8* – Mica shear release; *9* – Ti central mandrel; *10* – Fe filler; *11* – Fe side bar

with a silane-based sizing. Silane sizing is stable under the 650 °C heat treatment required for Nb_3Sn. Silane provides the benefit of promoting enhanced surface adhesion between epoxy and the S2-glass.

To braid this yarn onto the Rutherford cable the company A&P Technology, Cincinnati, OH developed in collaboration with ARL a forming guide and stabilizing technique to secure the Rutherford cable during processing (Blackburn et al. 2008b). A uniform, tight-weave fabric with a compressed thickness of 55 μm/side was produced. Figure 10.17a shows the fabric direct-woven onto the cable for TAMU3; Fig. 10.17b shows a micrograph of the yarn in the fabric.

10.5.2 Mechanical and Electrical Characterization

To perform mechanical and electrical testing, 10-stack assemblies of cable segments were stacked, compressed, and vacuum-impregnated. The 10-stack samples were

Table 10.4 Main parameters of TAMU3

Parameter	Value
Max. field in coil (short sample) (T)	14.1
Short sample current (A)	13,750
Stored energy at B_{max} (MJ/m)	0.425
Max. Lorentz force at B_{max} (MN/m)	1.76
Superconducting cable inner layer	
Number of strands	30
Strand diameter (mm)	0.808
J_c(10 T, 4.5 K) (A/mm^2)	2750
J_{Cu} at quench and short sample current (A/mm^2)	2047
Cu/non-Cu ratio	0.67
Superconducting cable outer layer	
Number of strands	34
Strand diameter (mm)	0.70
J_c(10 T, 4.5 K) (A/mm^2)	2750
J_{Cu} at quench and short sample current (A/mm^2)	2097
Cu/non-Cu ratio	1.0
Number of turns in windings	
Inner	11
Outer	23

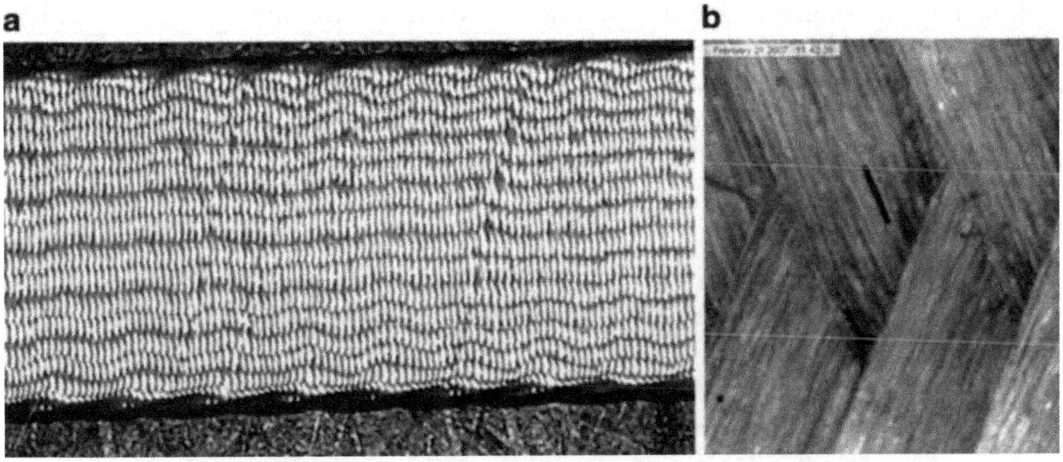

Fig. 10.17 (**a**) Thin-filament silane-sized S2-glass fabric direct-wound on Nb_3Sn Rutherford cable for TAMU3); (**b**) micrograph of yarn in the fabric

tested for mechanical strength against delamination and for electrical strength of turn-to-turn insulation. The interfaces were delaminated at a pressure of 33 MPa, corresponding to a shear stress of about 5 MPa. Similar tests were made using segments of windings from the TAMU2 dipole, which were prepared using conventional S2-glass insulation technology. The shear strength of the new fabric was about 20% higher than that of the segments from TAMU2. The improvement was attributed to the silane, retaining its adhesion-promoting benefits even after 650 °C heat treatment.

The 10-stack samples were instrumented to measure turn-to-turn dielectric strength within an impregnated coil. First, voltage was applied between the two outermost turns of the 10-stack. Up to an applied voltage of 3 kV, corresponding to a local electric field of about 3 MV/m, the resistance was greater than 130 GΩ. Second, voltage was applied between adjacent turns throughout the stack. The resistance was larger than 130 GΩ up to >300 V/turn. These results are acceptable for this magnet. The volume fraction of insulation within the coil is about 7%, compared with 12–20% for conventional insulation. The thinner insulating layer should improve heat transfer for quench protection of high-stored-energy magnets.

10.5.3 Heat Treatment

A heat treatment of 210 °C for 48 h, 340 °C for 48 h, and 670 °C for 70 h was employed to maximize the critical current density in the windings. Tin leakage was observed on the exposed face of the inner winding of the first module, at locations between the third turn and the ninth turn of the inner conductor. One tin leak is shown in Fig. 10.18. It was 1.5 cm long and occurred 15 cm away from the inner lead end.

Tin leakage was not expected, so it was puzzling that multiple tin leaks had occurred. Extracted samples of wire from a reacted witness sample were sent to the National High Magnetic Field Laboratory (NHMFL) for measuring the critical current I_c. The results showed a resistive onset at about 400 A, at around half the critical current expected in the RRP wire. We investigated the source of the wire, and found that a spool of long-obsolete modified jellyroll (MJR) wire had been shipped to us by mistake for use in the inner winding. This mistake resulted in a

Fig. 10.18 Tin leakage at turn 3 of the inner winding of one module

compounding of errors, with the effect that the short sample current and field was almost halved; from an expected 13.75 kA to 7.8 kA.

This dreadful pair of mistakes were discovered at a time when the modules had been completed, and the funding and schedule could not accommodate re-fabricating both modules. We had no choice but to complete the magnet assembly and test it to its severely compromised short sample limit.

10.5.4 Testing TAMU3

TAMU3 was tested at the Brookhaven National Laboratory (BNL) magnet test facility in a liquid He bath at 4.4 K. The first quench occurred at 5695 A (73% of the expected short sample limit on load line). On subsequent quenches a modest degree of training was observed—quench current improved to a plateau of around 6600 A (85% on load line), corresponding to 7.57 T in the center bore of the dipole.

The locations of the quenches were estimated from the timing of quench arrival at each of the voltage taps on the inner winding. The quench initiated at the location where the outer lead of the inner winding is hard-way bent and then guided along a lead channel in the strong back structure of each module. That location is indicated by the blue arrow in Fig. 10.19.

We concluded that the conductor was damaged in that location in one of the modules. Indeed, the MJR conductor was severely embrittled in the high-temperature heat treatment that was used for the TAMU3 windings, so the extra

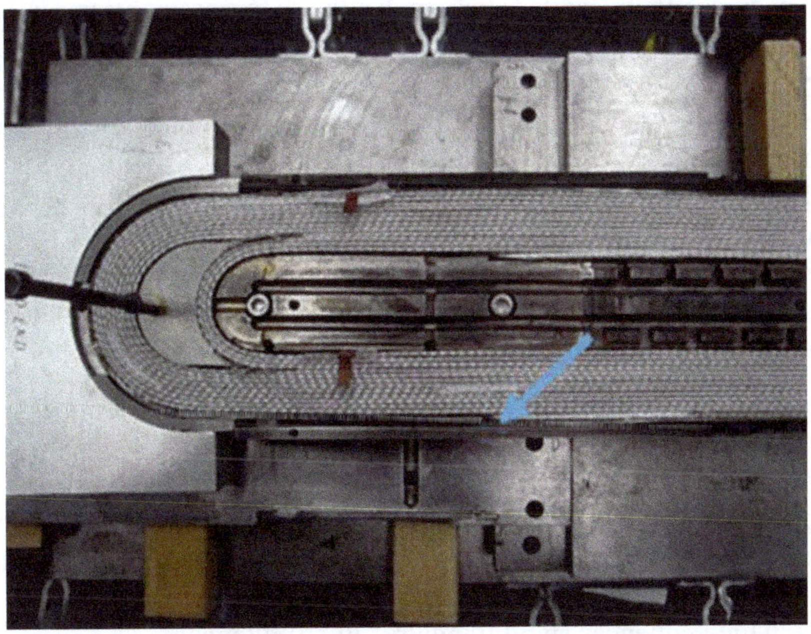

Fig. 10.19 TAMU3 winding showing the location where quenches originated

stress of the hard bend for the lead may have produced a weak point location for damage.

We learned a bitter lesson from the experience of TAMU3: it is vital to perform short sample measurements on samples of all wire that will be used in a dipole, in order to avoid mistakes of the kind that cost the performance of this dipole. Another important lesson was that enough space for coil expansion during reaction needs to be provided to avoid high stresses of the coil during the heat treatment in the reaction mold.

10.6 Conclusion

A succession of dipoles has been built using block-type coil geometry and stress management elements. TAMU1 first introduced stress management. TAMU2 introduced stress management in an Nb_3Sn dipole for the first time, and achieved remarkable ramp rate performance. TAMU3 introduced a fine-filament S2-glass insulation that enhances mechanical and electrical properties of an impregnated winding. Unfortunately, a mistake in the conductor used in one winding severely compromised the magnet's performance.

Block-type geometry naturally accommodates stress management within windings in the body of the dipole, and provides just as efficient use of superconductor as do other coil geometries. It is the only coil geometry that can accommodate a flux plate to suppress persistent-current multipoles from magnetization of filaments within the cables.

The difference in amount of superconductor for the block-type and shell-type (also known as cos-theta) coil design under the same assumptions is small. The choice of cable and coil geometry for any application should be based upon the field homogeneity, cryogenic stability, robustness against strain degradation, support of flared ends, manufacturability, compatibility with hybrid-coil strategies, and overall collider cost per TeV.

Recently, ARL has developed a coil-forming technology that enables rapid, precise forming of compact flared ends for windings based on CIC conductor. ARL believes that this conductor type offers a promising basis for high-field collider dipole development, in that it addresses each of the challenges that pace such development:

- Stress is managed at the cable level and is intercepted by sheath tubes and beams;
- Beams precisely position each cable turn;
- A simple winding procedure provides stability against de-registration when forming flared ends;
- Strands are bathed in LHe within the cable, the mechanical stability is provided by the metallic support structure, and no impregnation around the strand is required, which enhances stability against micro quenches;

- CIC provides the possibility of argon buffer gas flow during Nb_3Sn heat treatment;
- CIC can be made with sheath and center tubes from Al-bronze to suppress cation migration during Bi-2212 heat treatment;
- CIC provides the possibility of O_2 gas purging within the sheath tube at high pressure to support Bi-2212 over-pressure processing without exposure to the overall furnace and dipole components.

In a recent paper (McIntyre et al. 2018) ARL presented two designs for CIC-based hybrid-coil dipoles suitable for operation at 16 T. The quantity of superconducting wire in each is similar than that in high-field designs currently being developed using Rutherford cable. The work on this concept continues at ARL.

References

Assadi S, Chavez D, Gerity J et al (2015) Magnet design and synchrotron damping considerations for a 100 TeV hadron collider. In: Henderson S et al (eds) Proceedings of the international particle accelerator conference IPAC2015, Richmond, 5 May 2015, pp 4034–4037

Assadi S, Breitschopf J, Chavez D et al (2017) Cable-in-conduit dipoles to enable a future hadron collider. IEEE Trans Appl Supercond 27(4):1–5. https://doi.org/10.1109/tasc.2017.2656157

Battle C, Blackburn R, Diaczenko N et al (2001) Testing of TAMU1: a single-aperture block-coil dipole. In: Lucas P, Webber S (eds) Proceedings of the 2001 particle accelerator conference (Cat. No.01CH37268), Chicago, 18–22 June 2001, vol 5, pp 3642–3644

Benson CP, Holik EF III, Jaisle A et al (2012) Improved capacitive stress transducers for high field superconducting magnets. Advances in cryogenic engineering: transactions of the cryogenic engineering conference—CEC, vol 57. AIP Conf Proc 1434(1):1337–1344. https://doi.org/10.1063/1.4707059

Blackburn R, Elliott T, Henchel W et al (2003) Construction of block-coil high-field model dipoles for future hadron colliders. IEEE Trans Appl Supercond 13(2):1355–1357. https://doi.org/10.1109/tasc.2003.812667

Blackburn R, Diaczenko N, Elliott T et al (2008a) Fabrication of TAMU3, a wind/react stress-managed 14 T Nb_3Sn block coil dipole. J Phys Conf Ser 97(012128). https://doi.org/10.1088/1742-6596/97/1/012128

Blackburn R, Fecko D, Jaisle A et al (2008b) Improved S-2 glass fabric insulation for Nb_3Sn Rutherford cable. IEEE Trans Appl Supercond 18(2):1391–1393. https://doi.org/10.1109/tasc.2008.920762

Canfer S, Ellwood G, Baynham DE et al (2008) Insulation development for the next European dipole. IEEE Trans Appl Supercond 18(2):1387–1390. https://doi.org/10.1109/tasc.2008.921862

Caspi S, Gourlay S, Hafalia R et al (2001) The use of pressurized bladders for stress control of superconducting magnets. IEEE Trans Appl Supercond 11(1):2272–2275. https://doi.org/10.1109/77.920313

Diacaenko N, Elliott T, Jaisle A et al (1997) Stress management in high-field dipoles. In: Comyn M et al (eds) Proceedings of the 1997 particle accelerator conference, Vancouver, 12–16 May 1997. IEEE, Piscataway, pp 3443–3445

Elliott T, Garrison R, Holik T et al (2016) Testing of TAMU3: an Nb_3Sn block-coil dipole with stress management. U.S. Department of Energy, Office of Scientific and Technical Information report, Oak Ridge, TN, DOE-TAMU-11685-5

Green MA (1971) Residual fields in superconducting dipole and quadrupole magnets. IEEE Trans Nucl Sci 18(3):664–668. https://doi.org/10.1109/tns.1971.4326146

Hafalia R, Caspi S, Chiesa L et al (2003) An approach for faster high field magnet technology development. IEEE Trans Appl Supercond 13(2):1258–1261. https://doi.org/10.1109/tasc.2003.812632

Holik EF, McInturff AD, Benson CP et al (2011) Current progress of TAMU3: a block-coil stress-managed high field (>12 T) Nb_3Sn dipole. In: Satogata T, Brown K (eds) Proceedings of the particle accelerator conference, New York, 26 Mar–1 Apr 2011. IEEE, Piscataway, pp 1163–1165

Holik EF, Garrison R, Diaczenko N et al (2014) Construction challenges and solutions in TAMU3, a 14 T stress-managed Nb_3Sn dipole. In: Adv Cryo Eng: AIP conference proceedings vol 1573, pp 1535–1542

Kashikhin VV, Zlobin AV (2001) Correction of the persistent current effect in Nb_3Sn dipole magnets. IEEE Trans Appl Supercond 11(1):2058–2061. https://doi.org/10.1109/77.920260

McInturff A, Bish P, Blackburn R et al (2007) Test results of an Nb_3Sn wind/react "stress-managed" block dipole. IEEE Trans Appl Supercond 17(2):1157–1160. https://doi.org/10.1109/tasc.2007.898437

McIntyre P, Breitschopf J, Chavez D et al (2018) Block-coil high-field dipoles using superconducting cable-in-conduit. IEEE Trans Appl Supercond 28(3):1–7. https://doi.org/10.1109/tasc.2018.2797915

Noyes P, Blackburn R, Diaczenko N et al (2006) Construction of a mirror-configuration stress-managed Nb_3Sn block-coil dipole. IEEE Trans Appl Supercond 16(2):391–394. https://doi.org/10.1109/tasc.2006.871326

Chapter 11 The HD Block-Coil Dipole Program at LBNL

Gianluca Sabbi

Abstract This chapter reports on the HD program, a series of high field models exploring the block-coil layout for application to the arc dipoles for future high-energy colliders.

11.1 Introduction

The HD program is a series of block-coil model dipoles developed at the Lawrence Berkeley National Laboratory (LBNL) to explore Nb_3Sn technology at the highest possible field, for application in future energy frontier accelerators (Sabbi 2003). Table 11.1 shows a summary of the series. The first model (HD1) used a flat racetrack configuration to probe the fundamental characteristics of this approach. Two assemblies were tested in 2003–2004 using a shell-based support structure that, for the first time, included provisions for axial preload (Lietzke et al. 2004, 2005; Ferracin et al. 2005). The magnet achieved a coil peak field of 15.4 T, demonstrating that block-coils could effectively utilize the latest generation of Nb_3Sn wires to operate above 15 T, while coping with coil stresses approaching 200 MPa. The HD2 model (Fig. 11.1) focused on incorporating the main design features required for high-energy collider applications: a magnetically efficient layout; a clear aperture in the 40 mm range; a cost-effective fabrication process; and high field quality over the full operating range from injection to high energy (Sabbi et al. 2005; Ferracin et al. 2006, 2008). Five assemblies were tested in 2008–2009 (Ferracin et al. 2009, 2010). The HD2c test achieved 13.8 T in the bore, which, at the time of writing, is still the highest field recorded in a model dipole with accelerator-relevant bore size and field quality. Further progress was limited by repeated quenches at the boundary between the coil straight section and ends. The HD3 model attempted to address these issues through more accurate conductor positioning at the critical locations, but showed similar quench patterns, indicating the need for additional analysis and optimization (Cheng et al. 2013, Felice et al. 2013; Marchevsky et al. 2014).

G. Sabbi (✉)
Lawrence Berkeley National Laboratory (LBNL), Berkeley, CA, USA
e-mail: glsabbi@lbl.gov

D. Schoerling, A. V. Zlobin (eds.), *Nb_3Sn Accelerator Magnets*, Particle Acceleration and Detection, https://doi.org/10.1007/978-3-030-16118-7_11

Table 11.1 HD model timeline, clear bore, and highest field at 4.5 K

Model	Year tested	Bore (mm)	B_{max} coil (T)	B_{max} bore (T)
HD1a	2003	8	15.3	16.0
HD1b	2004	8	15.3	16.0
HD2a	2008	36	14.0	13.3
HD2b	2008	36	14.0	13.3
HD2c	2008	36	14.5	13.8
HD2d/d2	2009	43	14.1	13.4
HD2e	2009	43	13.2	12.5
HD3a	2011	43	10.9	10.3
HD3b	2013	43	14.2	13.4

Fig. 11.1 HD2 magnet. This model (HD2c) achieved 13.8 T at 4.5 K, the highest value recorded to date in a dipole with accelerator-relevant bore and field quality

11.2 Block-Coil Dipole Design Features

The maximum field in superconducting accelerator dipoles depends primarily on the available current density and the coil width. When the coil width is relatively small compared to the required aperture, the traditional shell-type (cos-theta) layout offers the most efficient solution (Mess et al. 1996). The arc dipoles of future colliders, however, enter a new regime of field, aperture, and forces that, coupled with the special properties of high-field superconductors, has prompted researchers at LBNL and elsewhere to explore alternative approaches (Sabbi 2002). Among these options is the block-coil layout, where a rectangular Rutherford cable is wound with its wide side parallel to the main dipole field.

A fundamental characteristic of the block configuration, compared to cos-theta, is that the coil width is controlled by the number of turns rather than the number of layers. In the aperture range of interest for high-energy colliders, this feature can be exploited to access the highest field levels allowed by Nb_3Sn properties using only two layers, while meeting the most critical cable design, field quality, and coil fabrication requirements, and without introducing any internal spacers (Sabbi et al.

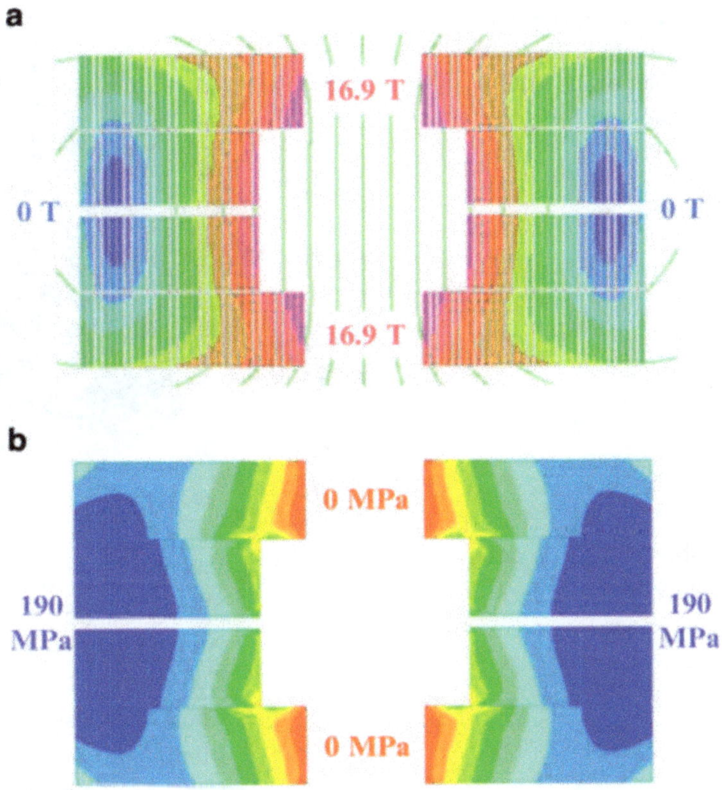

Fig. 11.2 HD2 coil cross-section showing the coil field map with: (**a**) flux lines; and (**b**) coil stress calculated assuming a 16 T target dipole field

2005). This solution, shown in Fig. 11.2, results in high efficiency and compactness, and minimizes the number of coil parts.

Another important feature of the block-coil winding orientation is the separation between the high field and high stress locations in the energized coil. The superposition of preload and Lorentz forces results in low stress in areas next to the bore, where the field is highest, while the highest stresses are located in the low field regions of the coil (Fig. 11.2). This characteristic can be exploited to minimize the impact of stress degradation on magnet performance.

Among the main technical challenges for the block layout is a reduction of the available aperture to provide structural support to the magnet bore. This is not the case in the cos-theta design, which exploits a self-supporting Roman arch layout using keystone Rutherford cables and wedges. In addition, a flared coil design needs to be developed to clear the beam path in the end regions while achieving high magnetic efficiency and field quality in the straight section.

11.3 HD1 Model

11.3.1 Magnet Design

HD1 was designed to explore the maximum dipole field that could be achieved using the best-performing Nb_3Sn wires available at the time. The magnet cross-section is

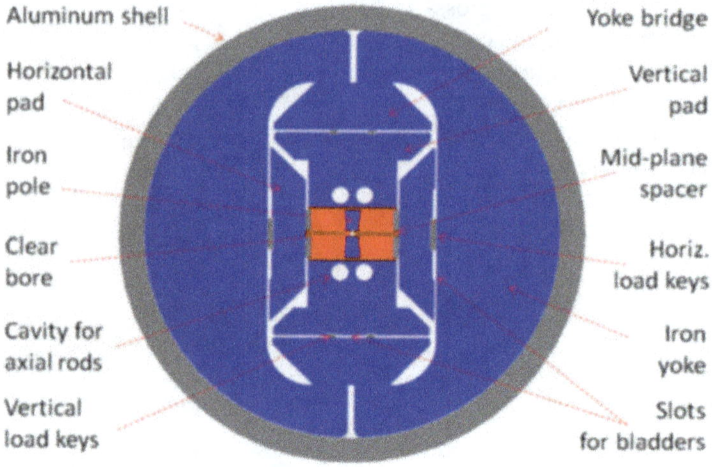

Fig. 11.3 HD1 magnet cross-section. The outer diameter of the shell is 74 cm

Table 11.2 HD1 design parameters

Parameter	Value
Strand diameter (mm)	0.8
Number of strands	36
Cable width (bare) (mm)	15.75
Cable thickness (bare) (mm)	1.40
Minimum bending radius (mm)	10
Cable insulation thickness (mm)	0.11
Number of turns (layer 1)	35
Number of turns (layer 2)	34
Mid-plane gap (total) (mm)	10
Clear bore diameter (mm)	8

shown in Fig. 11.3. Each pole is composed of a double-layer racetrack coil module wound from a continuous cable length around a 20 mm wide iron pole. A 10 mm thick G10 plate separates the coils and includes a small bore (8 mm in diameter) where the magnetic field can be measured using a Hall probe.

The conductor and coil parameters are listed in Table 11.2. The strand was produced by Oxford Superconducting Technology (OST) using the recently developed restacked rod process (RRP). The design was based on the 54/61 layout, with 54 hexagonal sub-elements composed of superconducting filaments embedded in a copper matrix, and seven sub-elements made of pure copper (Parrell et al. 2003). This conductor achieved a critical current density at the level of 3 kA/mm^2 at 12 T and 4.2 K, a significant improvement over previous generations. Cabling was performed at LBNL using a two-step process aimed at improving mechanical stability while minimizing damage to the sub-elements. The cable is initially fabricated with a thickness slightly larger than the final target. It is then annealed at 200 °C for 6 h and re-rolled to the final dimensions.

HD1 used a simplified coil design that severely limited the clear bore size and field quality, but retained the most fundamental technological challenges of high field accelerator dipoles: high coil stresses, approaching 200 MPa; mechanical and magnetic forces acting primarily on the wide face of the cable and accumulating over many turns; and high axial forces in the end regions. In order to confront these challenges, a full 3D preload system was implemented. The support structure, shown in Fig. 11.3, is based on a thick aluminum shell surrounding the iron yoke, which is vertically split in two halves with an open gap. Inside the yoke, horizontal and vertical iron pads support the coils. The transverse preload is achieved using water-pressurized bladders inserted between pads and yoke to compress the coil pack and tension the shell. Interference keys are inserted to lock the pre-stress and allow for bladder deflation and removal. The thermal contraction differentials between yoke and shell are exploited to generate a large increase of the pre-stress during cool-down, preventing over-stress and possible conductor damage during the warm assembly step.

Axial preload is also provided to minimize longitudinal displacements under excitation. Aluminum rods connect stainless-steel plates pushing on the coil ends. The axial rods are pre-tensioned during assembly. Similar to the transverse preload system, thermal contraction differentials between the aluminum rods and the iron winding pole generate a substantial increase of the axial force during cool-down in order to reach the target preload.

11.3.2 Fabrication and Test Results

Lorentz force accumulation during magnet excitation is not a major concern in the block-coil geometry, since the highest stress occurs in a region where the field is low and a margin is available. A high coil preload is required, however, to prevent conductor motion under Lorentz forces. The mechanical stress associated with the preload can cause permanent degradation of the conductor properties, and limit the magnet performance. In addition, the stress variation during cool-down needs to be accurately predicted and monitored to avoid overshoot. For these reasons, a conservative preload target was selected in the first assembly of HD1 (HD1a). The shell azimuthal tension was 25 MPa at assembly and increased to 120 MPa after cool-down. The corresponding coil pre-stress was calculated to be 25 MPa at assembly and 155 MPa after cool-down (Ferracin et al. 2005). The HD1 shell and yoke were previously used for the RD series of common-coil dipoles, which generated a much higher horizontal Lorentz force. Therefore the shell thickness was larger than optimal, and resulted in most of the preload being generated by the cool-down differentials.

Based on the finite element analysis (FEA) results, the transverse preload of HD1a was expected to maintain contact at the coil–pole interface at up to 95% of the short sample limit. Magnet training (Fig. 11.4) started at a coil peak field of 12.8 T, and progressed to 15.2 T in 12 quenches (Lietzke et al. 2004). In this phase, most quenches were located in the return end, at the outer tip of a spacer introduced to reduce the peak field. Afterwards, training became more erratic and the quench

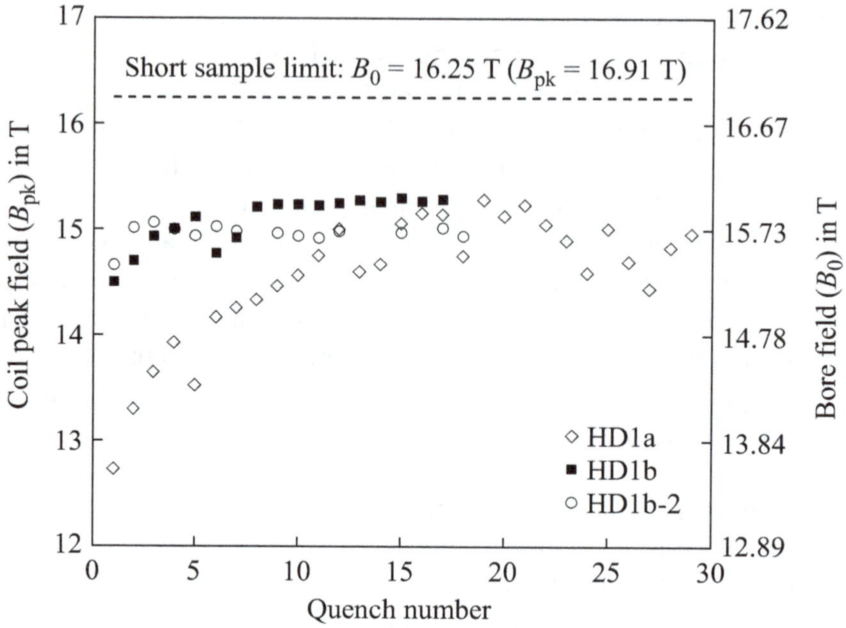

Fig. 11.4 HD1a, HD1b, and HD1b-2 training quenches at 4.5 K

locations shifted to the straight section at the high-field pole turn. The maximum coil field oscillated between 14.5 T and 15.3 T. The estimated short sample is 16.25 T, based on critical current measurements of extracted strands reacted with the coils.

A detailed analysis was performed following the HD1a test in order to understand and correct the observed limitations (Ferracin et al. 2005). The 3D FEA model showed a 15 μm gap developing during cool-down at the outer tip of the return end spacer, and increasing to 85 μm at the highest current achieved during the test. This behavior was attributed to a stainless-steel rail surrounding the coil and intercepting the axial preload. A similar but smaller gap (65 μm) appeared in the lead end, which featured an end shoe sliding between side rails. Increasing the axial force did not result in significant reduction of the computed gap. A reduction of the gap size by a factor of about 2 was obtained, however, by slightly reducing the rail thickness in a short segment at the end of the straight section. This relief allowed some bending in the side rail and facilitated the transfer of axial preload to the coil. A visual inspection of the coils performed after the HD1b test showed epoxy tearing at the coil-to-end-spacer interface, confirming the mechanical analysis results (Fig. 11.5).

The straight section quenches were attributed to non-uniformities in the coil transverse size, resulting in areas of low horizontal preload. In order to address this issue, shimming was introduced to improve the coil size uniformity. In addition, the coil preload was increased by 30 MPa, reaching a peak of 185 MPa. HD1b had faster training to a stable plateau with a field level close to the maximum reached in HD1a (Fig. 11.4). The initial quenches were again located in the end regions, while plateau quenches occurred in the highest field region and had the characteristics of a conductor-limited, rather than motion-triggered, origin. Following a thermal cycle, the magnet exhibited a stable plateau but at a 2% lower level than previously

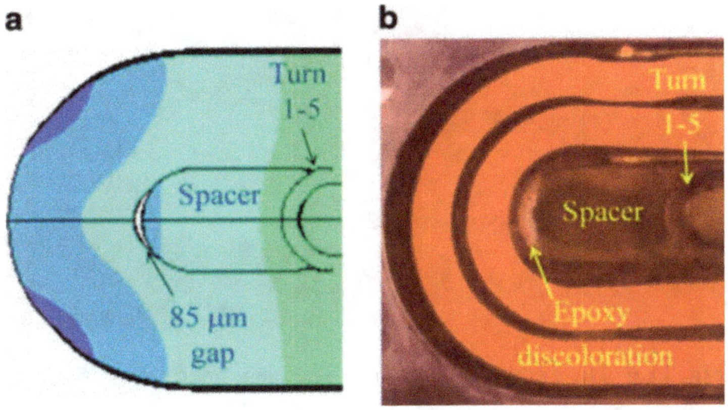

Fig. 11.5 (**a**) HD1a coil displacements calculated for the actual preload conditions and maximum achieved current, showing a gap of 85 μm between coil and spacer. (**b**) Picture of HD1 coil 1 showing epoxy discoloration and tearing at the coil–spacer interface. (Ferracin et al. 2005)

achieved. A temperature dependence study performed at the end of this test provided further confirmation of a conductor-limited plateau (Lietzke et al. 2005). These results indicate that the HD1b preload is at the limit for permanent critical current degradation. However, the HD1 demonstration of consistent performance at the 15 T level, achieved with fast training and no retraining, opened the way to the development of a new generation of high-field magnets enabling future colliders such as the High Luminosity Large Hadron Collider (Sabbi 2013) and the 100 TeV Future Circular Collider (Tommasini et al. 2017).

11.4 HD2 Models with 36 mm Bore

11.4.1 Magnet Design

The main goal of HD2 was further developing the block-coil approach for high-field dipoles by introducing a clear bore and field quality in the range required for high-energy physics colliders. The coil cross-section is shown in Fig. 11.6. Each pole is composed of a double-layer winding using a continuous cable length. In order to satisfy the design objectives with only two layers in each pole, a wide cable is required. In fact, the cable width should be comparable with the vertical aperture, so that the second layer can protrude toward the vertical axis to adjust the field quality, while the first layer can provide a high conductor packing factor in the region closer to the mid-plane. Due to the limited experience with the new RRP process, the strand diameter and layout was kept the same as in HD1, and the number of strands was increased up to the mechanical stability limit for coil winding. Following a series of cabling and winding tests, a 51-strand design was selected. The 22 mm cable width could support a coil aperture of about 45 mm on each side, and was compatible with a bending radius of 12.5 mm as required for layer 2. In recent years, the available diameter of high-performance RRP wires has increased to about 1.2 mm, with a corresponding expansion of the aperture range accessible with this coil layout (Sabbi et al. 2016).

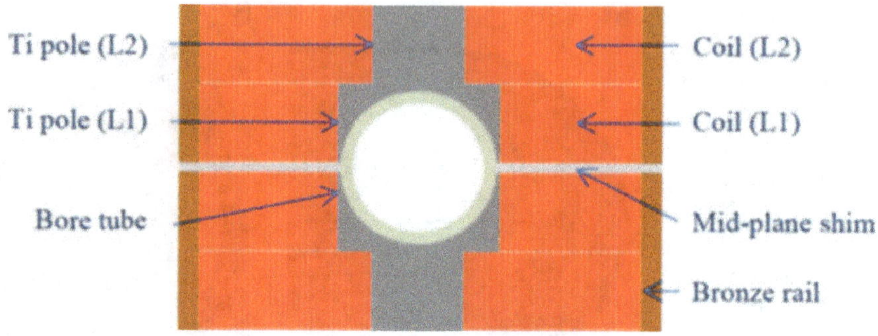

Fig. 11.6 HD2 coil module detail. The clear aperture is 36 mm in diameter

Fig. 11.7 HD2 cross-section showing the main coil and structural elements

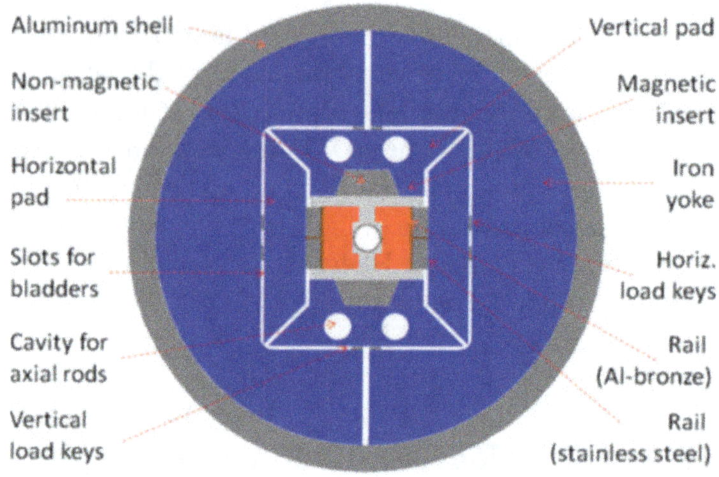

The magnet cross-section is shown in Fig. 11.7. The iron yoke has an outer diameter of 623 mm. The coil field quality is optimized by adjusting the number of turns and positions of the two conductor blocks, without using internal spacers. This requires a mid-plane gap of 1.37 mm. In order to control the saturation harmonics, a spacer is introduced between the coil and horizontal pads, and inserts are placed between the coil and the vertical pads to create a trapezoidal iron profile.

A reference current of 16 kA, corresponding to a dipole field of 14.1 T, was chosen for optimizing the geometric field quality. The harmonic components, expressed in units of 10^{-4} relative to the dipole field at a reference radius of 13 mm, are: $b_3 = -0.13$, $b_5 = -0.20$, $b_7 = -0.09$, $b_9 = -0.89$. At a current of 2 kA, $b_3 = 4.06$, $b_5 = -1.15$. The harmonic variation due to the iron saturation effect from 2 kA to 16 kA is therefore $\Delta b_3 = -4.2$, $\Delta b_5 = +0.9$. This variation is monotonic and occurs in the first half of the current range, between 2 kA and 8 kA (Ferracin et al. 2008).

The coil peak field is located at the pole turn of layer 2, and the bore field to peak field ratio is 0.95. The main design and performance parameters are summarized in Tables 11.3 and 11.4. The coil field at short sample is about the same in HD2 as in HD1, while the bore field is lower by 1.4 T in HD2 due to its larger aperture and the use of a non-magnetic winding pole, as required to avoid a strong saturation effect

Table 11.3 HD2 design parameters

Parameter	Value
Strand diameter (mm)	0.8
No. strands	51
Cable width (bare) (mm)	22.000
Cable thickness (bare) (mm)	1.400
Insulation thickness (h/v) (mm)	0.110/0.110
No. turns/quadrant (layer 1)	24
No. turns/quadrant (layer 2)	30
Clear bore diameter (HD2a, b, c) (mm)	36.0
Clear bore diameter (HD2d, e) (mm)	43.3

Table 11.4 HD2 performance parameters

Parameter	HD2a, b	HD2c
Short sample current I_{ss} at 4.3/1.9 K (kA)	17.3/19.2	18.1/20.0
Bore field at I_{ss} (4.3/1.9 K) (T)	15.1/16.5	15.6/17.1
Coil peak field at I_{ss} (4.3/1.9 K) (T)	15.9/17.4	16.5/18.1
F_x/F_y layer 1 (quadrant) at 17.3 kA (MN/m)	+2.3/−0.4	
F_z layer 1 (quadrant)at 17.3 kA (kN)	90	
F_x/F_y layer 2 (quadrant) at 17.3 kA (MN/m)	+3.3/−2.2	
F_z layer 2 (quadrant) at 17.3 kA (kN)	126	
Stored energy at 17.3 kA (MJ/m)	0.84	
Inductance (mH/m)	5.6	

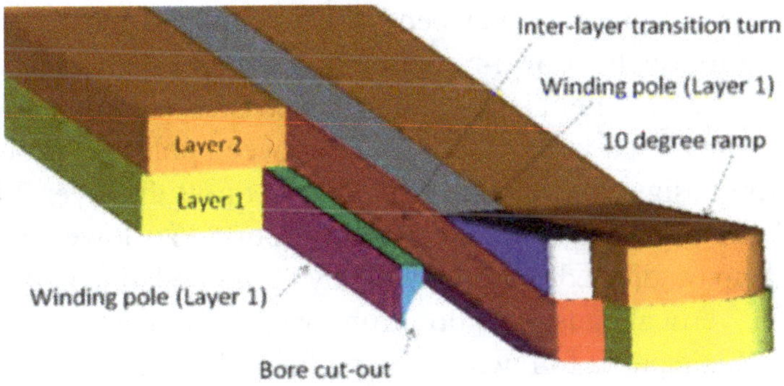

Fig. 11.8 HD2 end design concept. The pole turn of layer 2 proceeds parallel to the magnet axis and meets layer 1 at the end of the 10° ramp

during the ramp to high field. Further details on the cross-section design and performance parameters are provided by Sabbi et al. (2005) and Ferracin et al. (2006, 2008).

The main features of the coil ends are shown in Fig. 11.8. The design generally follows the D10 block-dipole (Fig. 11.9), which was the first Nb_3Sn magnet fabricated at LBNL and reached a bore field of 8 T (Taylor et al. 1985). A hard-way bend is introduced to ramp the conductor blocks away from the bore with a 10° angle. After the hard-way bend, a flat racetrack end configuration is recovered on the inclined plane. The transition from layer 1 to layer 2 is accomplished by continuing the pole turn of layer 2 parallel to the magnet axis. As the transition turn approaches

Fig. 11.9 Roy Hannaford winding the D10 coils (ca. 1983). The flared end and inter-layer transition of HD2 closely follows the concept developed for D10

the layer 1 elevation, a hard-way bend is included to match the angle of the ramp. The subsequent easy-way bend has two sections with different radii, the first using the layer 2 radius (12.5 mm), and the second using the layer 1 radius (22.6 mm). At the end of the easy-way bend, the transition turn has become the layer 1 pole turn and can follow the down-ramp into the straight section.

The peak field in the coil ends is easier to control in a flared configuration, due to the increased vertical separation between poles. Without optimization of the coil and iron design, however, the field at the tip of the layer 2 innermost turn would still be 1.5% higher than in the straight section. To lower the end field, the length of the ramp is adjusted to shift layer 2 by 20 mm relative to layer 1 (Fig. 11.10). In addition, the vertical pads transition from iron to stainless steel, and the iron insert above the coil is tapered following the end flare. As a result of these modifications, a 7% field margin is obtained in the end regions, without requiring internal coil spacers.

Figure 11.11 shows the dipole field profile on the magnet axis. The field is stable within 1 unit up to $z = 142$ mm, and is within 1% of the central field up to $z = 217$ mm. The coil geometric straight section ends at $z = 237.4$ mm.

A critical new element of HD2 is the bore structure required to react the preload applied to the coil. This component needs to be compatible with the coil fabrication process, maximize the aperture available to the beam, and minimize deflections and stress during assembly, cool-down, and excitation. After investigating alternative options (Sabbi et al. 2005) it was decided to rely on the combination of two main structural elements: winding poles, made of a titanium alloy, and a stainless-steel tube (Nitronic 40), inserted in a round cavity with 21.65 mm radius carved in the layer 1 winding pole. The winding poles are an integral part of each double-layer coil

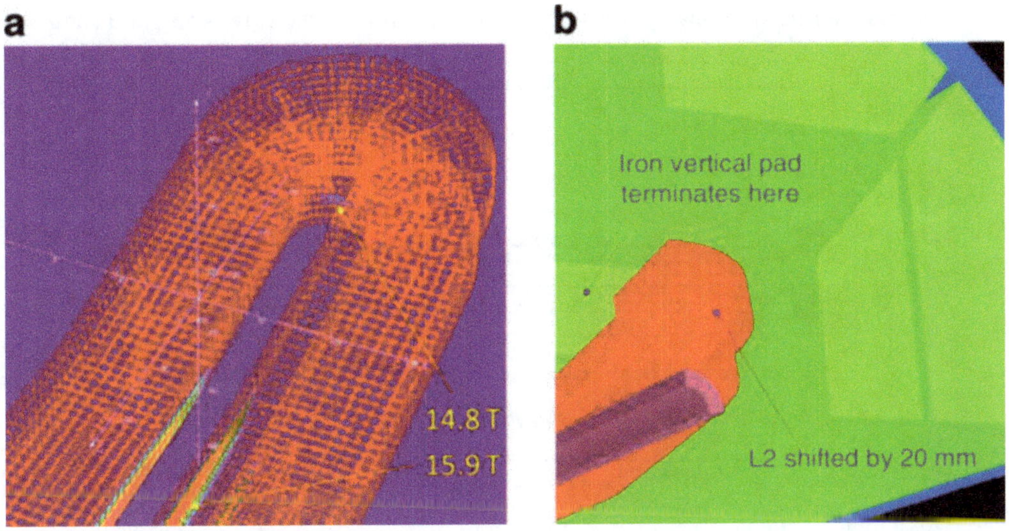

Fig. 11.10 Magnetic configuration of the end regions to reduce the peak field. (**a**) Peak field in the straight part and the end region; (**b**) iron configuration in the end region

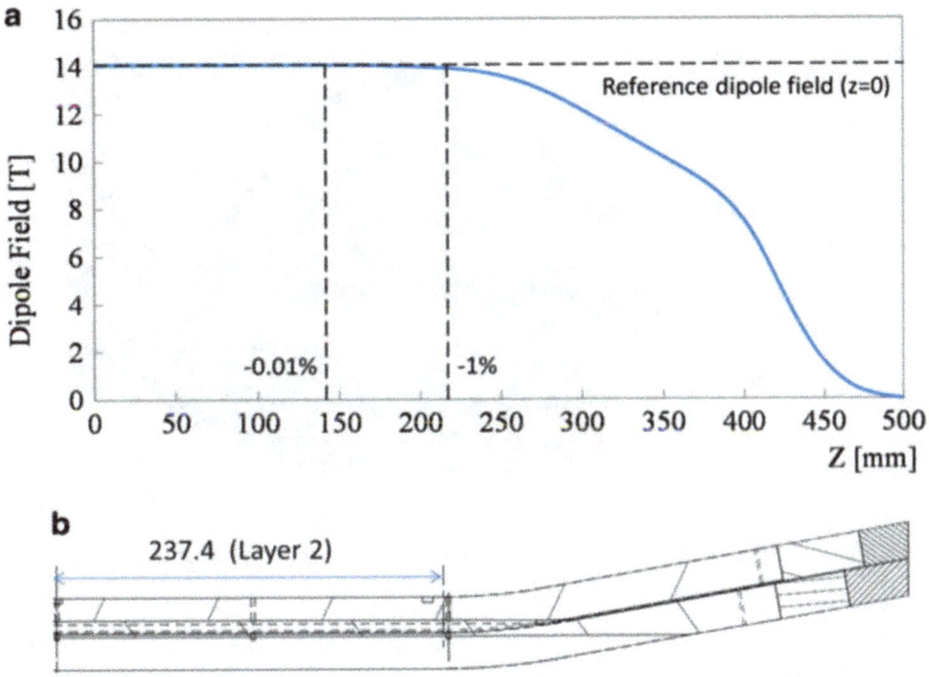

Fig. 11.11 (**a**) Longitudinal profile of the dipole field along the axis at 16 kA. The vertical lines indicate the longitudinal coordinates at which the dipole field decays by 0.01% (142 mm) and 1% (217 mm) relative to the central field; (**b**) a longitudinal section of the coil is also included for reference

and define its geometry. The central tube provides additional support in particular at the mid-plane, where the winding poles are thin and discontinuous. A conservative 3.65 mm wall thickness was selected for the first three assemblies of HD2, resulting in a clear aperture of 36 mm.

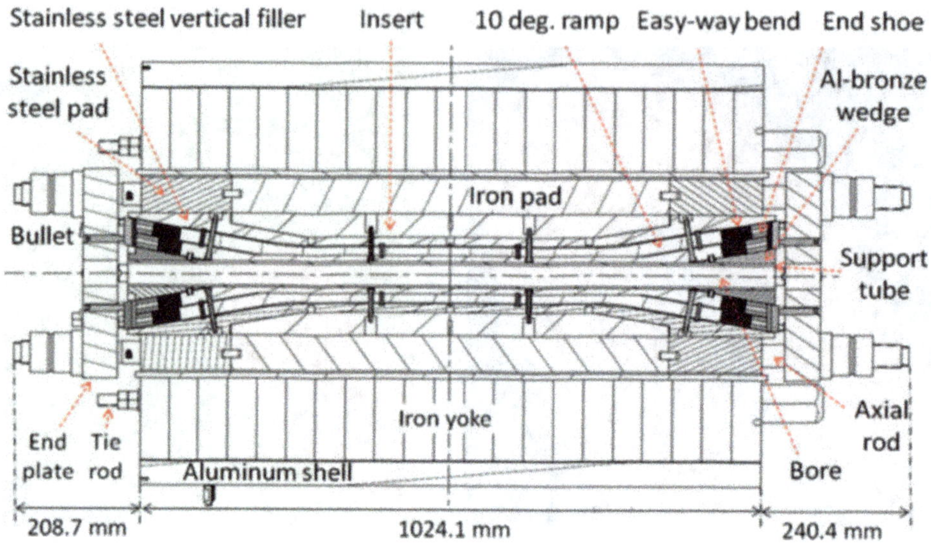

Fig. 11.12 HD2 longitudinal section showing the coil and structure elements

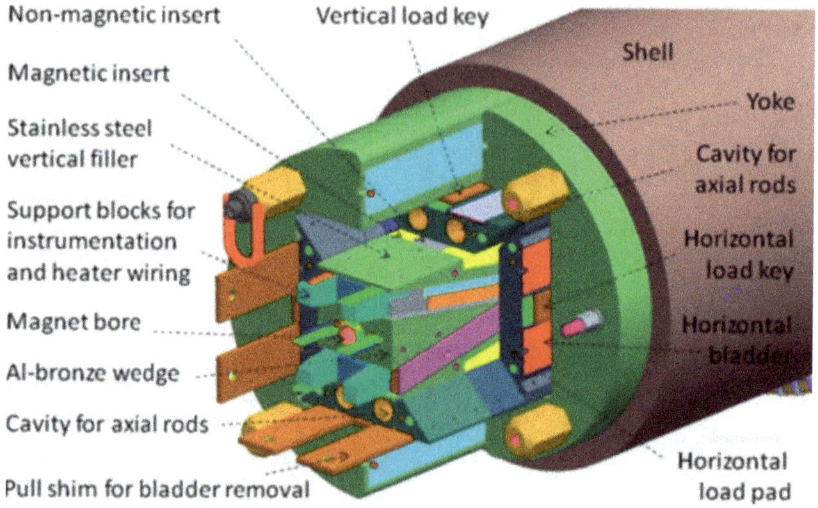

Fig. 11.13 HD2 CAD model with an exploded view of the main components

The external support structure is conceptually similar to HD1, but optimized for HD2 with all new components. The aluminum shell has an outer diameter of 705 mm and a thickness of 41 mm. Axial support and preload is provided by 50 mm thick stainless-steel endplates (Nitronic 40), connected by four 18.5 mm diameter aluminum rods. Details of the coil and structure design are shown in Figs. 11.12 and 11.13. Wedge-shaped elements surround the bore tube and support the flared ends on the bore side. The top surface of the coil is covered with a stainless-steel plate that also accommodates iron and stainless-steel inserts filling the cavity between end flares. These elements, labeled as "magnetic insert" and "non-magnetic insert" in Figs. 11.7 and 11.13, create a trapezoidal iron profile in the magnet cross-section, which is used to control the saturation harmonics.

Table 11.5 HD2 cable production data

Parameter	Coil 1	Coil 2	Coil 3
Cable ID	956R	966R	970R
Billet ID	7419, 7424	9770	9562, 9731
Strand diameter (μm)	802, 802	801	801, 804
Non-Cu fraction (%)	51.2	54.3	55
Reaction temperature (C)	668.9	679.5	678.4
Reaction time (h)	48	75	75
J_c(15 T, 4.2 K) (kA/mm^2)	1.72	1.90	1.88
Cabling degradation (%)	4.4	1.92	2.6
Production length (m)	130	135	115
Twist pitch (mm)	130	132.7	127
Thickness (first pass) (mm)	1.48	1.48	1.48
Thickness (final) (mm)	1.40	1.40	1.40

11.4.2 Magnet Fabrication

One practice coil and three production coils were fabricated for HD2. The first two assemblies (HD2a/b) used coils 1 and 2, while coils 2 and 3 were used in all subsequent assemblies.

Table 11.5 shows a summary of the cable production parameters, and Fig. 11.14 shows images of the fabricated cables. Coils 2 and 3 had significantly higher critical current density due to a change in the high-temperature stage of the heat treatment, from a nominal 665 C/48 h to 680 °C/75 h. The first two stages were the same for all coils: 210 C/72 h, 400 °C/48 h. The cabling degradation is about 4% in coil 1 and 2% in coils 2 and 3. These low values are consistent with scanning electron microscopy images showing minimal deformations of the sub-elements at the cable edges, a primary goal of the cable design and process optimization. Other parameters are similar for all cables. The finished cable lengths are in the 115–130 m range. The length of cable in one coil is 103.4 m.

The cable is insulated with a fiberglass sleeve of 0.11 mm thickness. Winding is performed starting from layer 2, using the stainless-steel filler as a base plate (Fig. 11.15). A 0.25 mm thick fiberglass sheet covers the base plate for electrical insulation. The winding spool is split in two portions, one for each layer. The winding pole is mounted on the base plate and the inter-layer transition is clamped in place. Layer 2 is wound, and the end shoes and side rails are installed. The same sequence is then repeated for layer 1, without an intermediate curing step. The coil pack is completed with end wedges and enclosed by reaction plates. End pushers are installed to maintain the end parts in position during reaction.

No axial pole gap is incorporated to release the winding tension and allow the coil to contract during the reaction process. With a flared end design, the tooling and parts should be designed to allow sliding in the straight section, while keeping the end components anchored to the coil. In order to simplify the design, it was decided

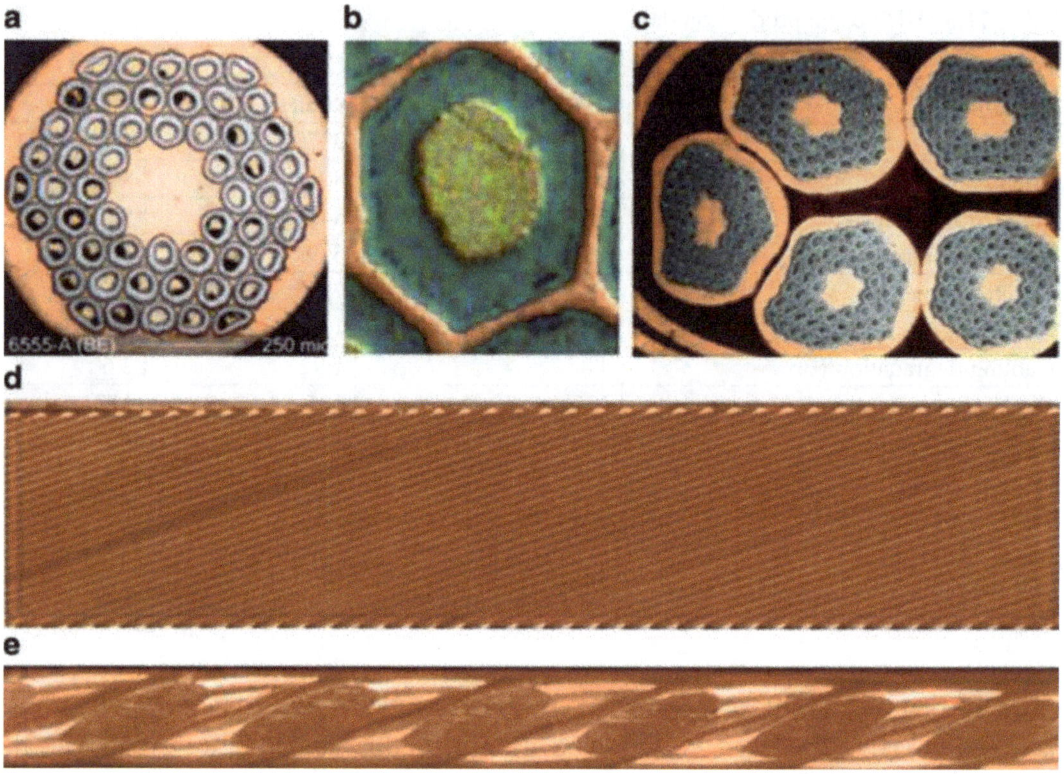

Fig. 11.14 HD conductor: (**a**) RRP54/61 strand used in HD1 and HD2; (**b**) sub-element detail; (**c**) cross-section of HD2 cable edge showing strand and sub-element deformations; (**d**) top view of HD2 cable; (**e**) side view of HD2 cable (not in scale with (**d**)) showing the flat surfaces (facets) at the layer transition

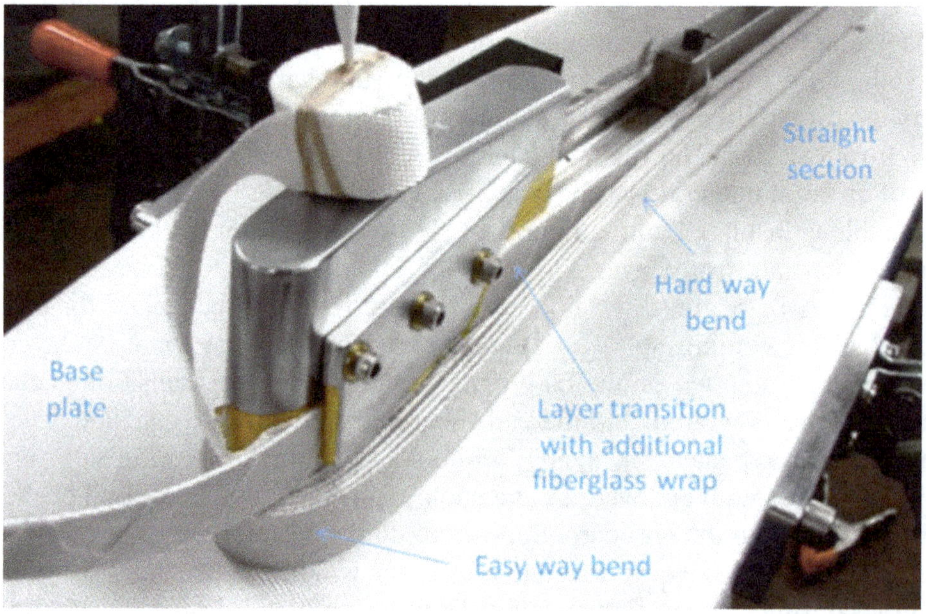

Fig. 11.15 Detail of HD2 coil winding, showing the base plate covered with a fiberglass sheet, the layer transition, and the first turns of layer 2

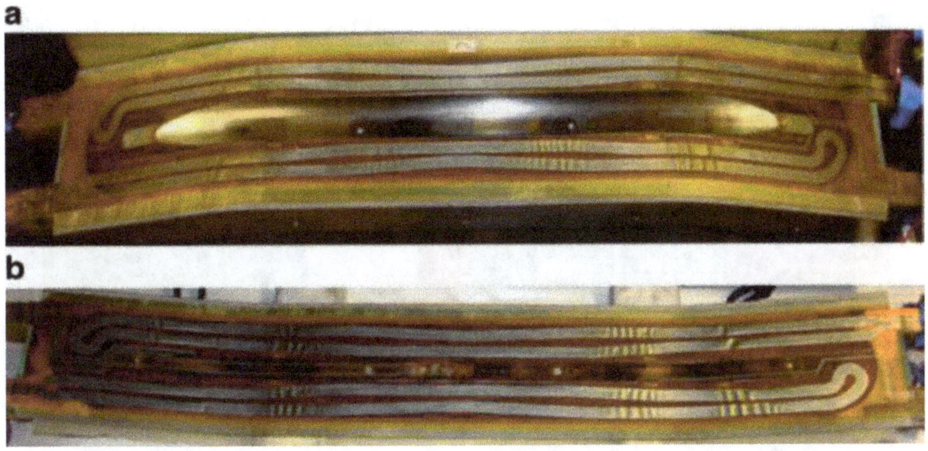

Fig. 11.16 HD2 coil: (**a**) top; and (**b**) bottom views. Two independent protection heaters are integrated in the instrumentation trace on each side of the coil

to delay implementing these features until a later phase in the program. Apart from this, the reaction process follows established procedures for Nb_3Sn magnets.

After reaction the instrumentation trace, integrating quench heaters with voltage tap and strain gauge wiring, are placed on the coil surface. The impregnation fixture is then assembled, sealed, and installed in a vertical position in a vacuum chamber. Epoxy filling takes place over a period of about 5 h, and is followed by a curing step. Figure 11.16 shows a picture of the completed coil.

In preparation for loading, the two coils are assembled around the supporting tube and mid-plane shim. The structure components up to the horizontal and vertical pads are added and bolted to form the coil pack. In a separate operation, the yokes halves are installed in the shell, pre-tensioned, and locked in place by temporary interference keys placed in the vertical gap. The coil pack is then inserted in the shell–yoke sub-assembly and bladders are installed in the horizontal and vertical gaps. An intermediate loading step is performed to compress the coil pack, tension the shell, and extract the yoke gap keys, replacing them with interference keys placed between pads and yokes. As a next step, the axial support system is installed, and the rods are tensioned using a hydraulic system. A final transverse loading step takes place to shim the interference keys and reach the design preload based on the strain gauge measurements.

11.4.3 Assembly and Test

The first HD2 assembly was initially cooled to liquid nitrogen to verify the FEA calculations. Tension increased from 49 MPa to 135 MPa in the shell, and from 34 MPa to 83 MPa in the rods. These results were consistent with the design targets, and no further modifications were required before the HD2a magnet cold test. At 4.5 K, the shell reached 144 MPa and the rods reached 90 MPa, corresponding to a

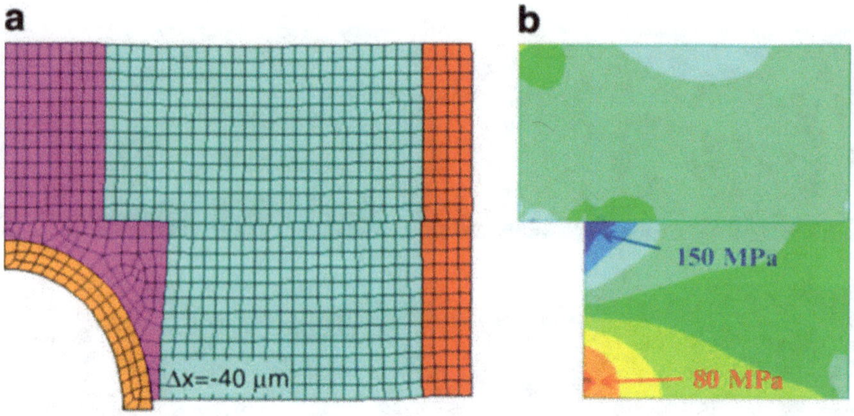

Fig. 11.17 (**a**) Coil deformation; and (**b**) stress at cool-down for a 14 T preload level

total force of 560 kN on the magnet ends. The calculated peak stress in the coil is about 150 MPa. Under these conditions, the coil–pole interface is expected to remain in contact up to a 14 T bore field. A minor adjustment of the coil end support was performed before the second test (HD2b) without unloading the shell. Details of the mechanical measurements and analysis are provided in Ferracin et al. (2009, 2010).

Figure 11.17 shows the calculated coil deformations and stress at cold. Due to the large thickness variation of the layer 1 winding pole, and the associated bending under the preload force, the peak coil stress is in the first turn of layer 1, at the corner away from the mid-plane where the winding pole deflections are at the minimum.

The training histories for HD2a and HD2b are shown in Fig. 11.18. HD2a had a first quench at 11.4 T (73% of I_{ss}). After 16 quenches it reached a maximum bore field of 13.3 T (87% of I_{ss}), with an estimated coil peak field of 14.0 T. In the second test (HD2b), the magnet did not exhibit memory of the previous training, and after a first quench at 11.0 T (71% of I_{ss}) it reached the same maximum field in 12 quenches.

Despite a significantly higher margin in coil 2, the HD2a quenches were evenly distributed between the coils. All quenches originated in the pole turn of layer 1, which has a 4% lower field with respect to the pole turn in layer 2, but significantly higher stress at cold (Fig. 11.17). The longitudinal locations, shown in Fig. 11.19, are concentrated at the end of the straight section, close to but before the start of the hard-way bend.

In the initial phase of the HD2b training, quench patterns were similar to HD2a (Lizarazo et al. 2009). In quench 12, however, a failure of the extraction system caused an increase of the quench integral released to coil 1 from 16 to 23 $(kA)^2s$ (million ampere squared seconds or MIITS). After this event, all quenches took place in coil 1, and three out of four were in the pole turn of layer 2 (peak field region). Higher ramp-rate sensitivity was also observed, and a reduction of the ramp rate from 20 A/s to 10 A/s was required to achieve the previous field level. These results indicate a degradation of the conductor properties of coil 1 due to the high temperature and stress experienced in quench 12.

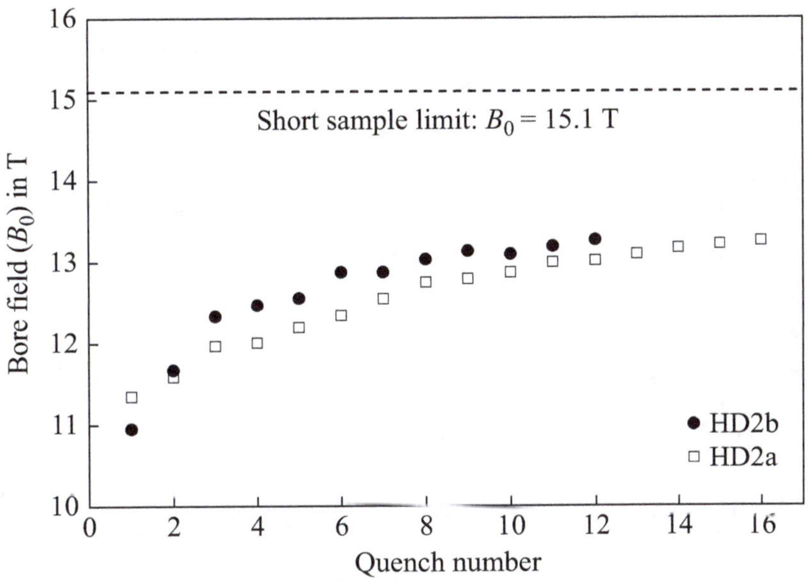

Fig. 11.18 Training quenches in HD2a and HD2b

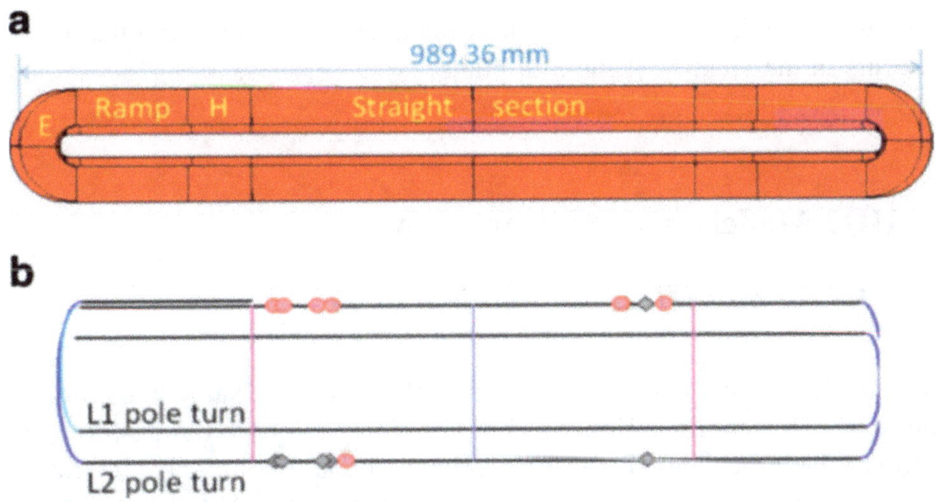

Fig. 11.19 HD2a longitudinal quench locations. (**a**) Schematic of coil: E: easy-way bend; H: hard-way bend; and (**b**) location of quenches. Coil 1 quenches are shown in grey; coil 2 quenches are in red

In HD2c, coil 1 was replaced by coil 3, matching the coil 2 performance and increasing the short sample field by 0.5 T. The shell tension was increased by about 13 MPa. A total of 28 training quenches were performed (Fig. 11.20). All 12 quenches in coil 2 started in the layer 1 pole turn, but were concentrated in a single straight section interval at the lead end transition side, within 100 mm from the start of the hard-way bend. In coil 3, two quenches started in the pole turn of layer 2, and 14 started in the layer 1 pole turn, at the lead end non-transition side, in a straight section interval of 50–100 mm from the hard-way bend. Quenches occurring in the layer 1 pole turn of one coil were often closely followed (within 1 ms) by an

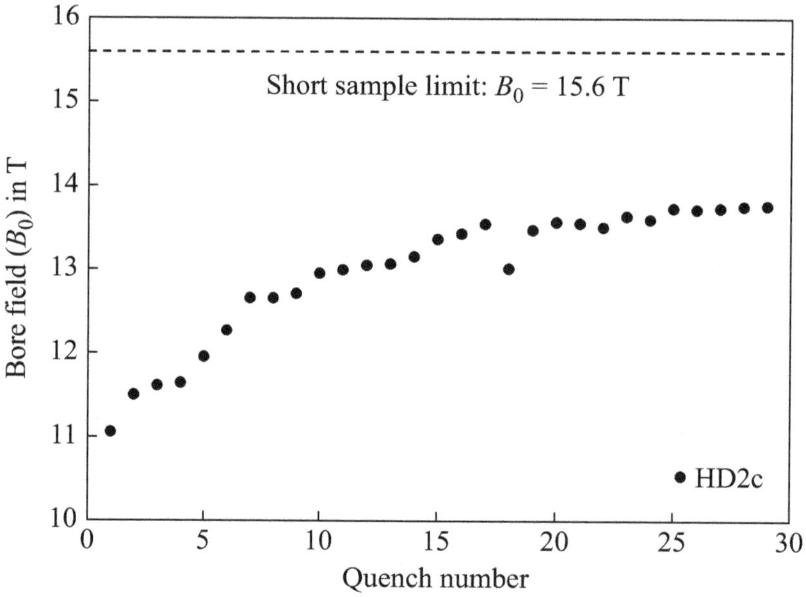

Fig. 11.20 HD2c training history and calculated short sample limit

adjacent quench in the layer 1 pole turn of the other coil, suggesting that the trigger event occurs in the vicinity of the mid-plane.

11.5 HD2 Models with 43.3 mm Bore

11.5.1 HD2d and HD2e Assembly and Test

After the HD2c test, the magnet was reassembled using the same coils, but without the bore tube. This configuration was not chosen to improve the performance, but rather to provide a benchmark for the mechanical analysis and characterization. In particular, it removed the uncertainty related to the contact between the tube outer diameter and the pole inner diameter, which is highly sensitive to fabrication tolerances, and made it possible to install strain gauges in the round cutout of the titanium pole.

Analysis showed that the layer 1 titanium pole could withstand the preload force, although deflections at the mid-plane increased by a factor of about 2, and the coil peak stress for a given shell tension also increased by about 15%. In order to maintain the peak stress at the level of previous tests, the shell tension was initially reduced to 130 MPa (HD2d) with a corresponding reduction of the field at which the coil would separate from the pole. In the next test (HD2e) a complementary case was studied by increasing the preload to the level required to maintain coil-to-pole contact up to 15 T, but at the risk of degrading the conductor under a calculated peak stress at cool-down of 185 MPa.

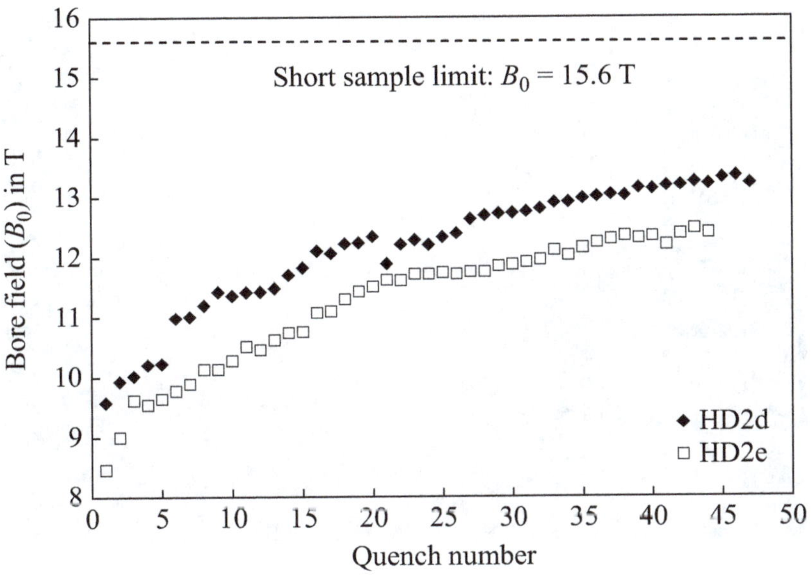

Fig. 11.21 Bore field (T) as a function of training quenches for HD2d and HD2e. The short sample bore field of 15.6 T corresponds to a coil peak field of 16.5 T

The quench histories are shown in Fig. 11.21. In the HD2d test, the magnet started training at a bore field of 9.6 T (59% of I_{ss}) and reached a maximum bore field of 13.4 T (84% of I_{ss}), corresponding to an estimated conductor peak field of 14.1 T, in 46 quenches. After increasing the coil preload in HD2e, the magnet trained from 8.5 T (52% of I_{ss}) to 12.5 T (74% of I_{ss}) in 43 quenches.

The locations identified in the HD2d and HD2e tests were at the pole turn of layer 1, in a segment of approximately 100 mm length at the end of the straight section, before the hard-way bend. While this is consistent with previous tests, training became significantly slower, and quenches were more widely distributed at the corresponding locations of both coils, both sides and both ends. In addition, the HD2e performance was about 1 T lower than HD2d, indicating further conductor performance degradation under the higher preload.

11.6 Analysis of HD2 Quench Performance

Most training quenches of the five HD2 assemblies originated in the layer 1 pole turn, within a short longitudinal interval at the end of the straight section, and at symmetric locations in both coils and both ends.

Quench evolution following the initial transition generally follows one of the following scenarios. In one case, a quench start is detected at an adjacent location in the layer 1 pole of the other coil. In the other case, a quench is detected in layer 2 of the same coil, at a location facing layer 1, without involving the highest field turns of layer 2 next to the pole (Ferracin et al. 2010).

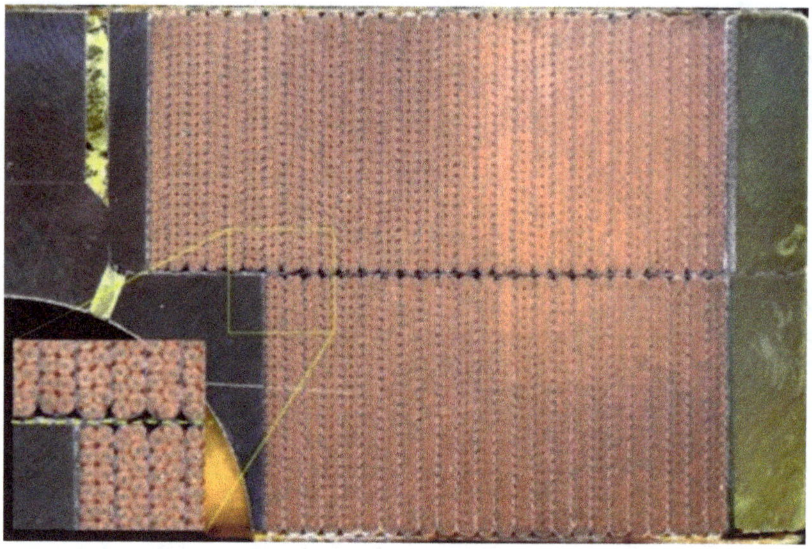

Fig. 11.22 Cross-section cuts of HD2 coil #1 in the center of straight section (Ferracin et al. 2010; Cheng et al. 2013). The two layers are close to the design positions

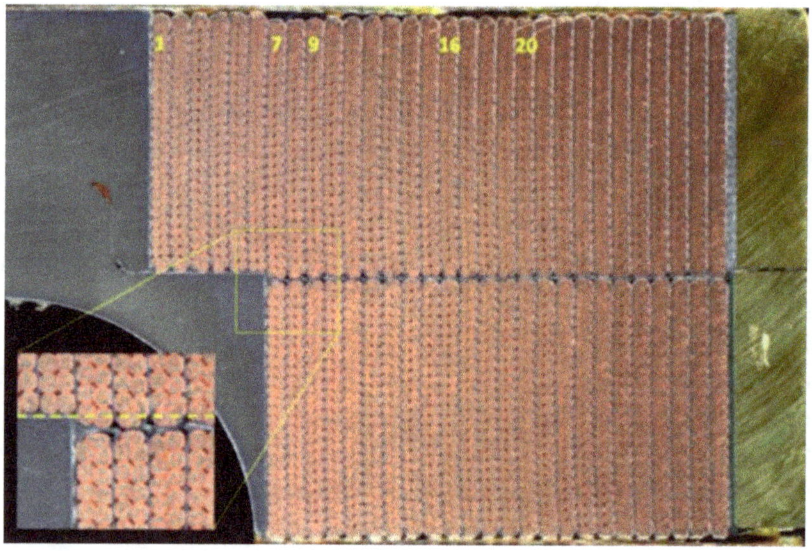

Fig. 11.23 Cross-section cuts of HD2 coil #1 at the end of the straight section, near the beginning of the end flare (Ferracin et al. 2010; Cheng et al. 2013)

Additional information is provided by images of coil 1 sections taken after it was replaced with coil 3. Figures 11.22 and 11.23 show a comparison of the cut at the magnet center, where no quenches were recorded, with the cut at the end of the straight section, where most of the quenches were located. In the latter, a shift toward the mid-plane can be noticed in the layer 2 turns after turn 6. The shift determines a compression of the layer 1 turns, and leaves a gap at the top of layer 2. The gap is filled with epoxy, indicating that the turn had already moved during the winding/reaction stages, before impregnation. Turn 7 of layer 2 moves below the corner of the layer 2 island, creating the potential for stress concentration and insulation failure.

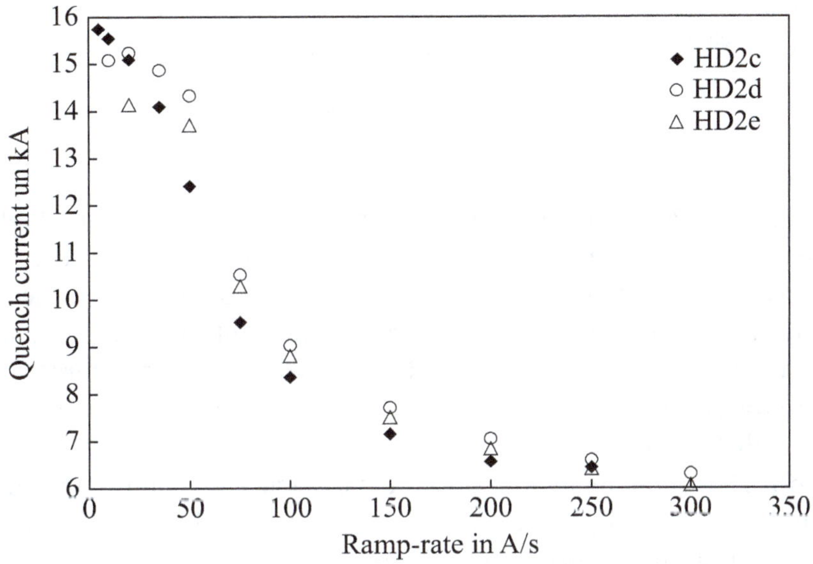

Fig. 11.24 Quench current as a function of ramp rate in HD2c, HD2d, and HD2e

Figure 11.24 shows the ramp rate dependence of HD2c, HD2d, and HD2e. In HD2c, the absence of a plateau at low ramp rate indicates that the highest quenches are conductor-limited. In HD2d, the maximum current is reduced from 16 kA to 15 kA, which may be attributed to the lower preload. Although fewer ramp rate measurements are available for HD2e, a comparison of the 20 A/s and 50 A/s points with the other two magnets appears to confirm the additional degradation of the conductor performance under the increased preload.

Summarizing the above findings, one can infer that two mechanisms are limiting HD2 performance: (a) stick–slip motions at the interface of the layer 1 first turn with the mid-plane spacer, where pole deflections and coil displacements are highest as the Lorentz forces acting on the turns balance the initial preload; (b) conductor degradation in the layer 1 first turn at the interface with layer 2, where the preload stress reaches its peak, and the conductor positions deviate from the design geometry. The lack of provisions to accommodate longitudinal coil expansion and contraction during reaction may be considered as an additional factor contributing to the HD2 stress degradation limit, which appears to be significantly lower than in HD1 (150 MPa vs. 185 MPa).

11.7 HD3 Models

11.7.1 Design and Fabrication

HD3 implemented several modifications aimed at better understanding and correcting the performance issues observed in HD2 (Cheng et al. 2013). The nominal coil envelope was unchanged, but the number of turns was reduced by one in each

Table 11.6 HD3 performance parameters

Parameter	Value
Short sample current I_{ss} at 4.4 K (kA)	18.67
Bore field at I_{ss} (T)	15.37
Peak field at I_{ss} (T)	16.26

layer. This change was implemented following the experience of the high-gradient quadrupole (HQ) models developed by the US LHC Accelerator Research Program, where it was found that previous criteria used to estimate the cable transverse expansion during reaction were not sufficient to prevent excessive strain and conductor degradation (Felice et al. 2012). Reducing the number of turns in the same coil envelope was the simplest and fastest route to provide additional transverse space, although the maximum field and field quality were negatively impacted (Table 11.6). The implementation of axial pole gaps to control the longitudinal strain during reaction was planned for the next set of coils, since it required more extensive modifications of coil parts and tooling.

In HD2, the hard-way bend connecting the straight section with the 10° end flare had a minimum bend radius of 349 mm (layer 2). Despite being more than 15 times larger than the cable width, this radius proved difficult to follow precisely during the winding process. As a consequence, the bend region extended into the adjacent straight section. The reaction and potting tooling and process helped to restore the desired geometry, but positioning errors were observed in the transition area after cutting HD2 coil 1 (Fig. 11.23). In order to avoid this effect, the hard-way bend radius of the HD3 coils was significantly increased to 873 mm. A curing step was also introduced after winding of each layer to help achieve and maintain the design geometry throughout the subsequent operations. The thickness of the fiberglass sheets used for inter-layer insulation and coil–pole insulation was increased for improved electrical robustness and to mitigate the effect of any residual positioning errors of the conductor. In order to ensure the correct axial positioning of the wedge-shaped spacers supporting the end flare, a mating surface was incorporated in the pole-island and the wedge.

Three coils were fabricated. One of them (coil 2) had to be discarded due to carbon contamination during reaction, leading to electrical shorts at several locations. Therefore, coils 1 and 3 were used in the assembly. The wire architecture was RRP 60/61 for coil 1 and RRP 54/61 for coil 3. The 60/61 wire had a copper fraction of 0.67, compared to 0.83 in the 54/61 wire. The critical current of the 54/61 wire was similar to previous HD2 coils. For the 60/61 wire, higher critical current density was found in virgin wires, but cabling degradation was also higher so that the critical current density of extracted strand measurements showed similar results as the 54/61. Nevertheless, the critical current (Fig. 11.25) was still considerably higher due to the larger area of superconductor relative to the stabilizer.

HD3 was assembled without a bore tube, similar to the last two tests of HD2. This approach facilitates the mechanical instrumentation of the pole, and the interpretation of the strain gauge data. It also provides a route towards larger clear apertures for given coil apertures, which is critical for collider applications. The inner radius of the

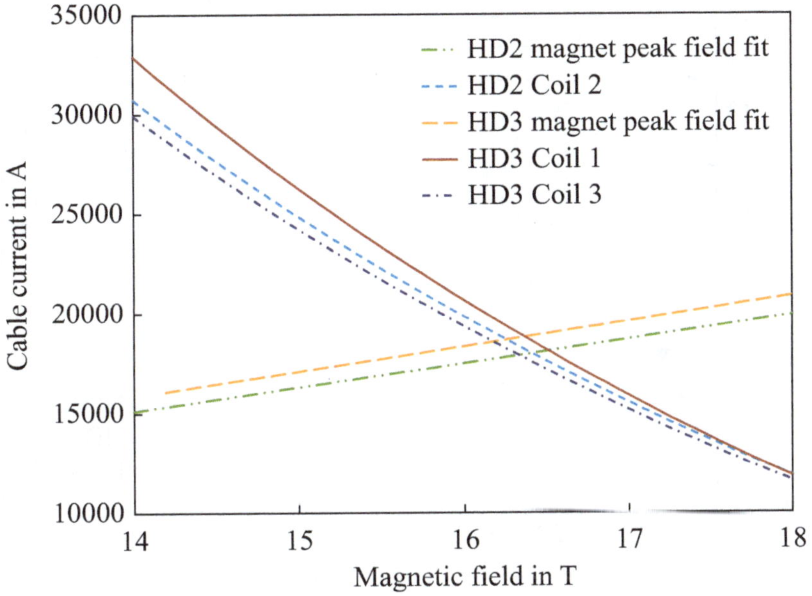

Fig. 11.25 Load line and critical current parametrizations of HD3 coils 1 and 3, and HD2 coil 2, at 4.5 K, based on measurements of extracted strands

round cutout for the pole-island of layer 1 was slightly reduced in HD3 from 22.3 mm to 21.5 mm, however, in order to reinforce the area close to the mid-plane. The main structure components were identical to HD2, and the preload targets were similar.

11.7.2 Test Results

The first test of HD3 (HD3a) had to be interrupted due to a short in the NbTi current leads. While the origin of this short was in the cryostat, the resulting arcing caused damage up to the NbTi–Nb_3Sn joints at the coil ends, requiring a complete magnet disassembly to perform a repair. In the second test (HD3b) two thermal cycles were performed. The training curves for the three tests are shown in Fig. 11.26. HD3a had a first quench at 6.5 T and progressed to 11.5 T in 14 quenches. HD3b started training just below 10 T and reached 13 T in about 45 quenches, after which progress became very slow. The maximum dipole field recorded was 13.4 T. After the thermal cycle, training started about 2 T higher and progressed to 13 T in five quenches. After that, progress became slower and seven additional quenches were required to reach 13.3 T. Therefore the magnet demonstrated some memory of the previous training, but the first quench was still substantially lower than the maximum level previously achieved, and recovering this level required significant retraining. Quenches were at the same locations as in the previous HD2 assemblies, indicating that improving the conductor positioning at the hard-way bend did not remove the HD2 performance limitations.

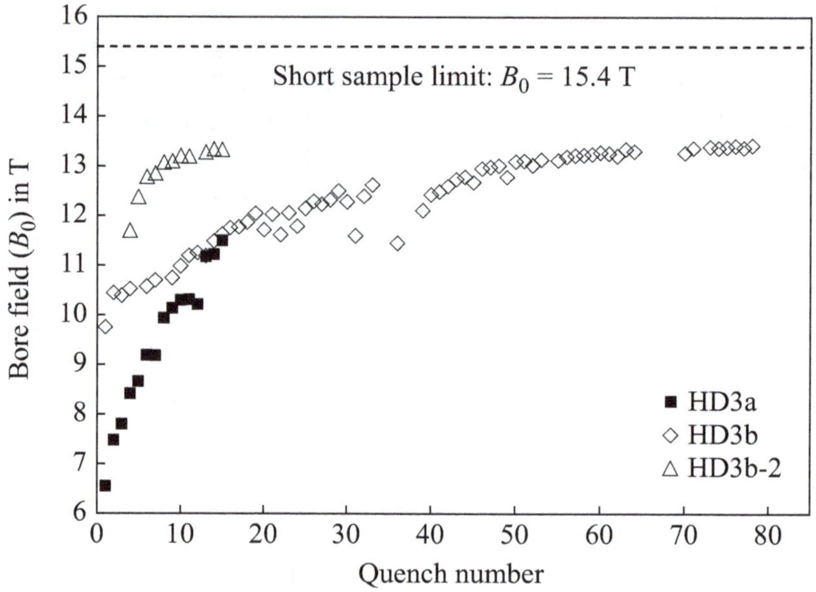

Fig. 11.26 Training histories of HD3a, HD3b, and HD3b-2 (second thermal cycle). (Marchevsky et al. 2014)

11.8 Summary and Outlook

The HD models represented an important step in the development of Nb_3Sn accelerator magnets for future high-energy colliders. Record fields were achieved in both technological (HD1a/HD1b) and accelerator (HD2c) configuration, coping with stress levels in the range 150–200 MPa. The coil design is magnetically efficient and requires only two layers per pole, with no spacers. Field quality features were implemented and a complete process for magnet fabrication was developed. Several critical questions, however, remain to be addressed.

First training was quite satisfactory for the first two assemblies of HD1 and HD2 using virgin coils. Both magnets achieved 80% of the short sample limit after two quenches, and progressed to 87% in 10–15 quenches.

After a preload increase, HD1b started training at 88% of the short sample limit and reached a stable plateau at the 95% level. Following a thermal cycle, the plateau level decreased by about 2%, but was achieved from the first quench.

Contrary to HD1, HD2 did not demonstrate a significant improvement of the first quench level in subsequent tests. The configurations including a bore tube had fast training to about 87% of the short sample limit, while those without a bore tube had significantly slower training. Based on these results, the bore tube or a thicker winding pole appears to be required, but the optimal configuration and thickness need to be further explored in order to effectively support the coil while preserving the largest possible aperture.

Analysis of the highest quenches in HD2 indicates that the magnet is ultimately limited by degradation of the conductor properties, rather than mechanical motion. The absence of a gap in the HD2 pole to control the axial strain during coil reaction, as typically implemented in Nb_3Sn magnets including HD1, is a possible factor contributing to this degradation. A new set of tooling and coil parts was designed and procured to test this hypothesis (Cheng et al. 2013) but has not yet been used.

Achieving high dipole field in accelerator configuration was the main focus of the HD2 and HD3 models. Additional development will be required to fully satisfy the aperture and field quality requirements of future colliders. In particular, a 0.8 mm strand of the RRP 54/61 or 60/61 design was chosen for the HD models since it provided the best critical current performance. This layout, however, has a large effective filament diameter, which results in limited stability margins, and significant field distortions due to conductor magnetization (Wang et al. 2015). New wire architectures with a higher number of sub-elements should be implemented to increase the strand diameter and extend the design to a larger aperture while reducing the effective filament size (Sabbi et al. 2016). Compensation of the persistent current harmonics with thin magnetic inserts was also explored in the HD2 design (Ferracin et al. 2008). A detailed program of field quality characterization and correction was not carried out, however, and remains one of the main priorities for future development of the block-coil concept.

The extension of HD2 to a twin-aperture configuration, and the potential to reach higher field using grading, were explored by Sabbi et al. (2015). Detailed quench protection studies in accelerator conditions were also performed in recent years, with particular emphasis on the incorporation and optimization of the Coupling Loss Induced Quench (CLIQ) system to achieve a safe discharge in a broad range of design parameters and operating conditions (Ravaioli et al. 2016).

References

Cheng DW, Caspi S, Dietderich DR et al (2013) Design and fabrication experience with Nb_3Sn block-type coils for high field accelerator dipoles. IEEE Trans Appl Supercond 23(3):4002504. https://doi.org/10.1109/tasc.2013.2246811

Felice H, Ambrosio G, Anerella MD et al (2012) Impact of coil compaction on Nb_3Sn LARP HQ magnet. IEEE Trans Appl Supercond 22(3):4001904. https://doi.org/10.1109/tasc.2012.2183843

Felice H, Borgnolutti F, Caspi S et al (2013) Challenges in the support structure design and assembly of HD3, a Nb_3Sn block-type dipole magnet. IEEE Trans Appl Supercond 23 (3):4001705. https://doi.org/10.1109/tasc.2013.2243794

Ferracin P, Bartlett SE, Caspi S et al (2005) Mechanical analysis of the Nb_3Sn dipole magnet HD1. IEEE Trans Appl Supercond 15(2):1119–1122. https://doi.org/10.1109/tasc.2005.849508

Ferracin P, Bartlett SE, Caspi S et al (2006) Mechanical design of HD2, a 15 T Nb_3Sn dipole magnet with a 35 mm bore. IEEE Trans Appl Supercond 16(2):378–381. https://doi.org/10.1109/tasc.2006.871323

Ferracin P, Caspi S, Cheng DW et al (2008) Development of the 15 T Nb_3Sn dipole HD2. IEEE Trans Appl Supercond 18(2):277–280. https://doi.org/10.1109/tasc.2008.922303

Ferracin P, Bingham B, Caspi S et al (2009) Assembly and test of HD2, a 36 mm bore high field Nb_3Sn dipole magnet. IEEE Trans Appl Supercond 19(3):1240–1243. https://doi.org/10.1109/tasc.2009.2019248

Ferracin P, Bingham B, Caspi S et al (2010) Recent test results of the high Nb_3Sn dipole magnet HD2. IEEE Trans Appl Supercond 20(3):292–295. https://doi.org/10.1109/tasc.2010.2042046

Lietzke AF, Bartlett S, Bish P et al (2004) Test results of HD1, a 16 T Nb_3Sn dipole magnet. IEEE Trans Appl Supercond 14(2):345–348. https://doi.org/10.1109/tasc.2004.829122

Lietzke AF, Bartlett SE, Bish P et al (2005) Test results of HD1b, an upgraded 16 Tesla Nb_3Sn dipole magnet. IEEE Trans Appl Supercond 15(2):1123–1127. https://doi.org/10.1109/tasc.2005.849509

Lizarazo J, Doering D, Doolittle L et al (2009) Use of high resolution DAQ system to aid diagnosis of HD2b, a high performance Nb_3Sn dipole. IEEE Trans Appl Supercond 19(3):2345–2349. https://doi.org/10.1109/tasc.2009.2019036

Marchevsky M, Caspi S, Cheng DW et al (2014) Test of the high-field Nb_3Sn dipole magnet HD3b. IEEE Trans Appl Supercond 24(3):1–6. https://doi.org/10.1109/tasc.2013.2285881

Mess KH, Schmüser P, Wolff S (1996) Superconducting accelerator magnets. World Scientific, Singapore

Parrell JA, Zhang Y, Field MB et al (2003) High field Nb_3Sn conductor development at Oxford Superconducting Technology. IEEE Trans Appl Supercond 13(2):3470–3473. https://doi.org/10.1109/tasc.2003.812360

Ravaioli E, Ghini JB, Datskov VI et al (2016) Quench protection of a 16-T block-coil dipole magnet for a 100-TeV hadron collider using CLIQ. IEEE Trans Appl Supercond 26(4):1–7. https://doi.org/10.1109/tasc.2016.2524527

Sabbi G (2002) Status of Nb_3Sn accelerator magnet R&D. IEEE Trans Appl Supercond 12 (1):236–241. https://doi.org/10.1109/tasc.2002.1018390

Sabbi G (2003) Nb_3Sn Magnet development at LBNL. In: Advanced accelerator magnet workshop, Archamps, 17–18 March 2003

Sabbi G (2013) Nb_3Sn IR quadrupoles for the high luminosity LHC. IEEE Trans Appl Supercond 23(3):4000707. https://doi.org/10.1109/tasc.2012.2233844

Sabbi G, Bartlett SE, Caspi S et al (2005) Design of HD2: a 15 Tesla Nb_3Sn dipole with a 35 mm bore. IEEE Trans Appl Supercond 15(2):1128–1131. https://doi.org/10.1109/tasc.2005.849510

Sabbi G, Bottura L, Cheng DW et al (2015) Performance characteristics of Nb_3Sn block-coil dipoles for a 100 TeV hadron collider. IEEE Trans Appl Supercond 25(3):1–7. https://doi.org/10.1109/tasc.2014.2365471

Sabbi G, Ghini JB, Gourlay SA et al (2016) Design study of a 16-T block dipole for FCC. IEEE Trans Appl Supercond 26(3):1–5. https://doi.org/10.1109/tasc.2016.2537538

Taylor C, Scanlan R, Peters C et al (1985) A Nb_3Sn dipole magnet reacted after winding. IEEE Trans Magn 21(2):967–970. https://doi.org/10.1109/tmag.1985.1063680

Tommasini D, Auchmann B, Bajas H et al (2017) The 16 T dipole development program for FCC. IEEE Trans Appl Supercond 27(4):1–5. https://doi.org/10.1109/TASC.2016.2634600

Wang X, Ambrosio G, Chlachidze G et al (2015) Validation of finite-element models of persistent current effect in Nb_3Sn accelerator magnets. IEEE Trans Appl Supercond 25(3):1–6. https://doi.org/10.1109/tasc.2014.2385932

Chapter 12
CEA–CERN Block-Type Dipole Magnet for Cable Testing: FRESCA2

Etienne Rochepault and Paolo Ferracin

Abstract The FRESCA2 dipole magnet, with a nominal field of 13 T in a 100 mm aperture, has been developed within a collaboration between the Commissariat à l'Energie Atomique et aux Energies Alternatives (CEA) and the European Organization for Nuclear Research (CERN), in the framework of the European Coordination for Accelerator Research and Development (EuCARD) project. This chapter summarizes magnet design, technology development, fabrication, and test results.

12.1 Introduction

In 2009, the European Coordination for Accelerator Research and Development (EuCARD) project was initiated, with the goal of carrying out research on new concepts and technologies for future upgrades of European accelerators (de Rijk 2012). Work Package 7 was dedicated to superconducting high-field magnets towards higher luminosities and energies. Key partners were the Commissariat à l'Energie Atomique et aux Energies Alternatives (CEA), Saclay, and the European Organization for Nuclear Research (CERN). The main objective of the work package was the design, fabrication, and test of the Facility for the REception of Superconducting CAbles (FRESCA2), a 100 mm aperture dipole generating a nominal bore field of 13 T and an ultimate field of 15 T, aiming at providing a new facility to test superconducting cables at CERN in high magnetic fields.

E. Rochepault (✉)
CEA Paris-Saclay (Commissariat à l'énergie atomique et aux énergies alternatives Paris-Saclay), Saclay, France
e-mail: Etienne.Rochepault@cea.fr

P. Ferracin
CERN (European Organization for Nuclear Research) Meyrin, Genève, Switzerland
e-mail: Paolo.Ferracin@cern.ch

D. Schoerling, A. V. Zlobin (eds.), *Nb_3Sn Accelerator Magnets*, Particle Acceleration and Detection, https://doi.org/10.1007/978-3-030-16118-7_12

Table 12.1 Comparison of FRESCA and FRESCA2

Parameter	FRESCA	FRESCA2
Superconductor	Nb-Ti	Nb_3Sn
Nominal bore field (T)	9.5	13
Short sample bore field at 1.9K (T)	10	17.5
Free bore aperture (mm)	88	100
Short sample current at 1.9 K (kA)	13.6	14.9
Stored energy at short sample (MJ/m)	0.7	5.5
Lorentz force Fx/quadrant at I_{nom} (MN/m)	3.7	7.7
Good field length at 1% (mm)	600	540
Total length (mm)	1700	2200
Outer diameter (mm)	740	1030
Mass (t)	7.8	8.8

From 2009–2013 the work was performed in the framework of EuCARD; after 2014 the project continued in the framework of a collaborative contract between CEA and CERN. The fabrication of the Nb_3Sn coils started in the spring of 2015 at CEA. CEA wound the coils and CERN carried out their heat treatment and impregnation. After finalizing the first four coils, magnet assembly was carried out at CERN in early 2017. The subsequent test campaigns took place at CERN. During the first test run a faulty coil was detected. It was exchanged, and the magnet was retested. This second assembly, called FRESCA2b, exceeded the nominal target and reached a bore field of 13.3 T. A third assembly, FRESCA2c, with increased pre-stress, was tested in the spring of 2018 and reached a maximum bore field of 14.6 T at 1.9 K, setting a new field record for dipole magnets with a usable aperture. FRESCA2 will therefore increase the available field of the dipole test station at CERN from around 10 T for FRESCA (Leroy et al. 2000) to around 14 T. FRESCA2, with high temperature superconductor (HTS) inserts generating up to 5 T (Durante et al. 2018; Lorin et al. 2016; van Nugteren et al. 2018), will make accessible a field range approaching 20 T. Table 12.1 compares the main parameters of the FRESCA and FRESCA2 magnets.

The first ideas towards such a Nb_3Sn dipole magnet accessing the 14–15 T field range with an aperture in the order of 100 mm date back to the Next European Dipole (NED) program, which was launched in 2004 within the Coordinated Accelerator Research in Europe (CARE) project (Devred et al. 2006). In this program the focus was put on Nb_3Sn conductor development, aiming at a critical current density (J_c) of 1500 A/mm^2 at 4.2 K and 15 T, but also initially included the detailed design and fabrication of a Nb_3Sn model magnet. Due to a lack of resources in this program the scope of the program was realigned to concentrate on conductor development, and the magnet design was stopped. The powder-in-tube (PIT) conductor developed in this program was used for the fabrication of some coils for FRESCA2.

12.2 Magnet Design

12.2.1 Initial Choices

The design relies on rectangular block coils with flared ends (see Figs. 12.1 and 12.2), which was inspired by HD2 and HD3 (see Chap. 11 of this book), following a concept proposed in the framework of the Superconducting Super Collider (SSC) studies (Taylor et al. 1985). After a comparison between cos-theta and block-coil layouts at the beginning of the project (Devaux-Bruchon et al. 2010), the block-coil has been chosen for the following reasons:

1. Less cable degradation due to keystoning.
2. Easier winding: the cable only has to withstand a hard-way bend in the flared section, and no twist. The coil ends do not require end-spacers.
3. Easier assembly: the block-coils provide flat faces that are easier to align and place into contact during coil-pack assembly.
4. Easier stress management: the cables are arranged perpendicular to the main dipole force (x direction), and the structure provides the pre-stress in the same direction. The accumulation of Lorentz forces is in the outer turns, where the field is lower.

The coils are made of double-pancakes (two layers) with a layer jump in the central post. Each of the four double-pancakes is individually wound, reacted, and instrumented. The coils are impregnated with epoxy resin, together with their respective components. A magnet assembly is made up of a total of four coils: two coils 1–2 (layers 1 and 2, coil ID starting with '12',), and two coils 3–4 (layers 3 and 4, coil ID starting with '34'). The posts of coils 1–2 are made of titanium for mechanical strength; the posts of coils 3–4 are made of iron to reinforce the magnetic field in the aperture. The coil layout is optimized to obtain a field homogeneity of <1% within a diameter of two-thirds of the 100 mm aperture.

The support structure follows a shell-based concept. The lateral preload is provided by an external aluminum shell, transferred via two yoke halves and vertical and horizontal pads surrounding the coils, as shown in Fig. 12.2a. The lateral preload is tuned at room temperature using bladders and keys, and then additional preload is applied by the aluminum shell thanks to differential contraction during cool-down. The gap between the two yoke halves is left free during the cool-down to allow for the contraction of the shell.

The longitudinal preload is applied first at room temperature by pre-tensioning the Al rods connecting the two end plates, then during cool-down by differential contraction of the Al rods. A conceptual and detailed magnet design is provided by Manil (2013).

12.2.2 Strand and Cable Parameters

For the magnetic design a Nb_3Sn strand with a non-Cu critical current density of up to 2500 A/mm^2 at 12 T and 4.2 K, and 1500 A/mm^2 at 15 T and 4.2 K (Bordini et al.

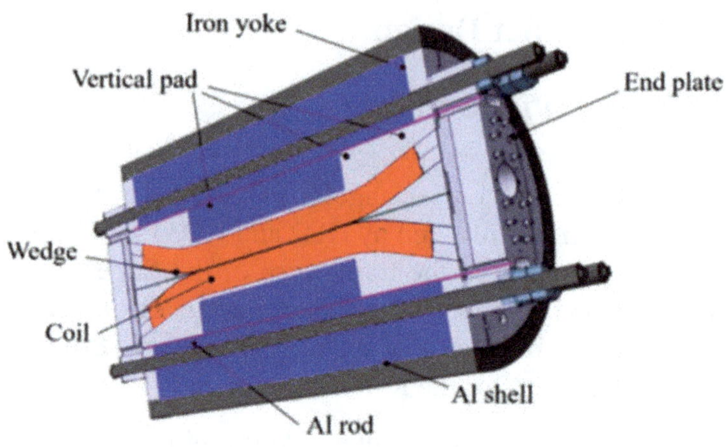

Fig. 12.1 FRESCA2 magnet structure (longitudinal section)

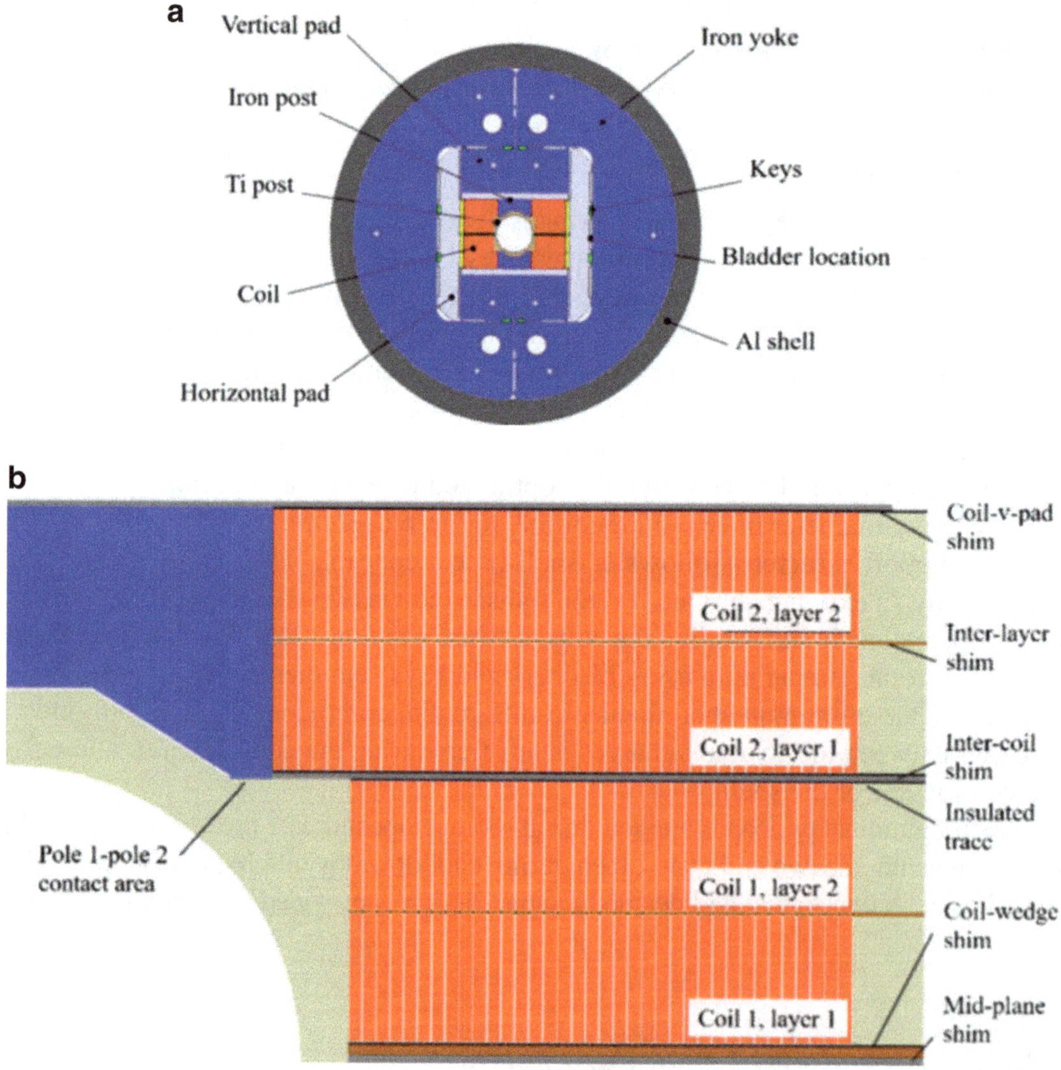

Fig. 12.2 FRESCA2 magnet. (**a**) Cross-section; (**b**) close-up showing the coil cross-section and insulation scheme (one quadrant at the center of the straight section)

Table 12.2 Final strand and cable parameters

Parameter	PIT	RRP
Strand diameter (mm)	1	1
Number of filaments	192	132/169
Effective filament diameter (μm)	48	58
RRR (extracted)	120	84–125
Cu/non-Cu ratio	1.3	1.3
J_c(12 T, 4.2 K)[a] (A/mm^2)	2300	3000
J_c(15 T, 4.2 K)[a] (A/mm^2)	1300	1600
Number of strands	40	40
Transposition pitch (mm)	120	120

[a]With self-field correction and cabling degradation

2012) was assumed. Two different strands have been procured: a PIT strand produced by Bruker-EAS GmbH, Hanau, Germany, featuring 192 filaments; and a Restacked Rod Process (RRP) strand produced by Oxford Superconductor Technology (OST), NJ, USA, featuring a 132/169 stack. The filament sizes are 48 μm for the PIT strands and 58 μm for the RRP strands. A Cu/non-Cu volume ratio of 1.25 was chosen to ensure protection and a large engineering current density. Table 12.2 summarizes the final target parameters of the Nb_3Sn wires.

For FRESCA2 a large rectangular Rutherford cable without a core was chosen to minimize the inductance (and increase the current) of the magnet. Due to the resulting cable's large aspect ratio, it was a challenge to make the FRESCA2 cable mechanically flexible enough for winding ('windable') and to limit its cable critical current degradation to below 5% (Oberli 2013).

To limit the number of strands in the cable to 40 (the limit of CERN's cabling machine) a 1.25 mm diameter strand initially developed for the NED program (Devred et al. 2007) was envisioned. After cabling tests, however, the cable was considered to be too large to be windable. A strand diameter of 1.0 mm was finally selected. The initial cable transposition pitch of 160 mm was reduced to 120 mm to improve the windability of the cable. The width compaction factor of the cable was optimized to limit the strand critical current degradation during cabling to 4–5%, for both PIT and RRP strands. The average residual resistivity ratio (RRR) was measured between 84 and 125 for extracted strands. Despite this relatively low RRR, measurements on short samples have shown that the stability current in both PIT and RRP extracted strands was at least twice the operating current, which was deemed sufficient not to affect the magnet's performance (Bordini et al. 2012; Oberli 2013). Cable cross-sections are shown in Fig. 12.3. The cable is insulated with braided S2 glass yarn with a thickness of 0.160 mm with 636 sizing and 11 tex.

The thicknesses of each unit length of cable produced were measured by sampling two cable stacks. The fabrication process for these stacks is considered to be representative of the coil manufacturing process. The stacks were then measured under an average pressure of 5 MPa before and after heat treatment, with insulation, and after removing the insulation. The cable geometrical parameters are summarized in Table 12.3.

The magnetic design has been initially optimized taking into account early measurements on PIT strands, considering a 10% cabling degradation, an insulation

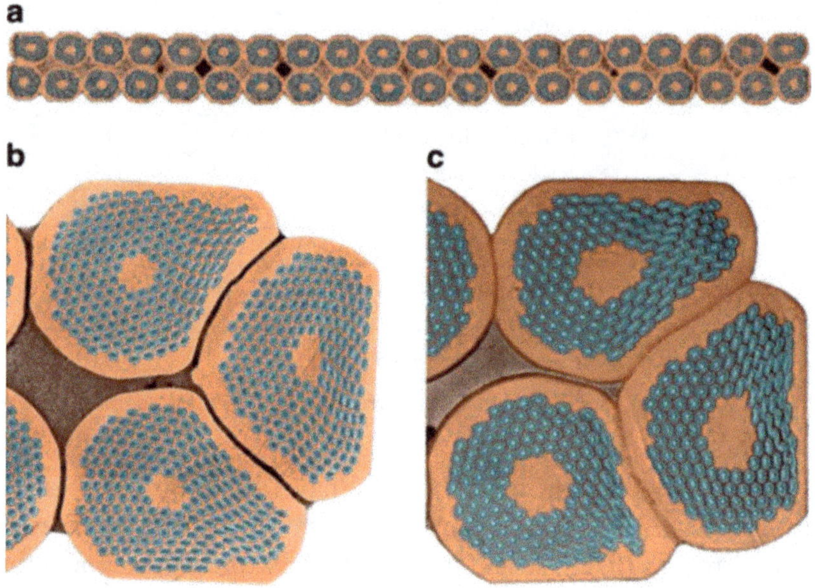

Fig. 12.3 (**a**) Cross-section of a FRESCA2 cable made of 40 PIT strands of 1.0 mm diameter; and cross-section of cable edges with (**b**) PIT strands and (**c**) RRP strands

Table 12.3 Final cable parameters

Parameter	Before reaction	After reaction
Bare thickness (mm)	1.82	1.9
Bare width (mm)	20.9	21.48
Insulation thickness (mm)	0.16	0.16
Insulated thickness (mm)	2.14	2.22
Insulated width (mm)	21.22	21.8

thickness of 0.2 mm, and not taking into account the cable expansion during reaction (Milanese et al. 2012). Later, expansions of 2% in width and 4% in thickness have been considered (Ferracin et al. 2013), corresponding to conservative assumptions with respect to measurements performed for MQXF coils (Rochepault et al. 2016). After the final determination of the expansion of the FRESCA2 cable, the tooling had already been designed and manufactured, so the insulation thickness was decreased from 0.2 mm to 0.16 mm for allowing the insertion of the coil into the tooling (Manil et al. 2014).

12.2.3 Magnetic Design

12.2.3.1 Cross-Section Design

The rectangular block layout de facto limits the magnetic optimization to the following parameters (not considering the geometry of the iron): (a) the total number of turns; (b) the relative number of turns in coils 1–2 vs. coils 3–4; (c) the position of

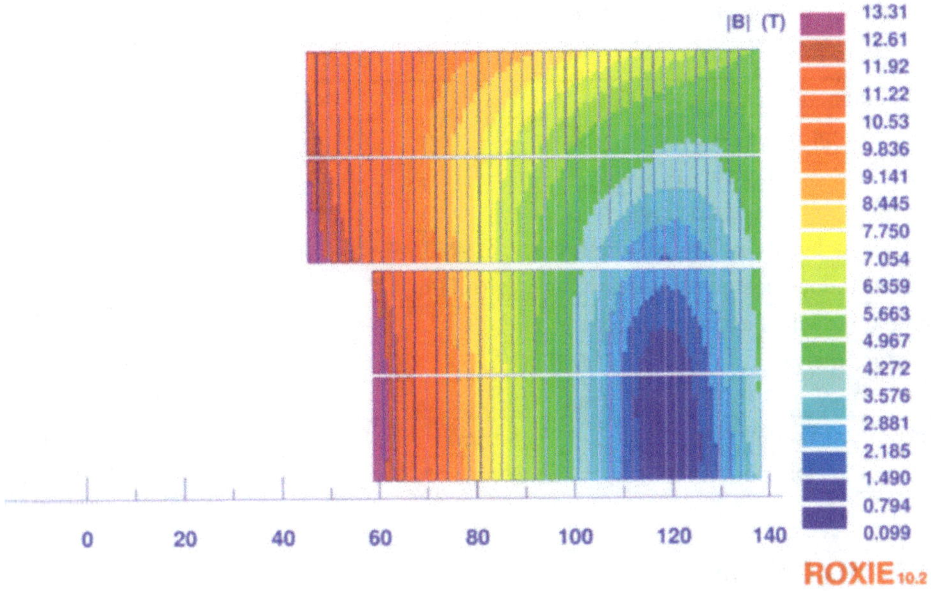

Fig. 12.4 Baseline coil cross-section and baseline magnetic flux density in the coil (T)

the first cable in layer 1; and (d) the thickness of the mid-plane shim. The main objective was to reach 13 T nominal field with sufficient load line margin at 4.2 K (>10%). The required field quality was less stringent (low-order field harmonics of several tens of units) compared to accelerator magnets that are usually of the order of a few units.

The 2D magnetic optimization was performed (Milanese et al. 2012) using the ROXIE (Routine for the Optimization of magnet X-sections, Inverse field calculation and coil End design) program package. The magnetic field map and a baseline coil cross-section is shown in Fig. 12.4. The first conductor of layers 1 and 2 is placed at a distance of 58 mm from the center, whereas the mid-plane shim thickness amounts to $2 \times 3.5 = 7$ mm. The post for coils 3–4 is made of iron, which allows an increase of the central field while reducing the coil peak field and thus increasing the margin. With all the iron parts, the nominal current for 13 T is decreased by 2 kA, and the peak field in the coil is decreased by 0.7 T.

In terms of field homogeneity, the optimization of the position of the windings resulted in a computed field quality characterized by sextupole (b_3) and decapole (b_5) normalized harmonics of 68.5 and −27.8 units, respectively, at a reference radius of 33 mm. The saturation of the iron pole in coil 2 and in the vertical pad (see Fig. 12.2) determined a reduction of the transfer function from 2.4 T/kA at 1 kA to about 1.2 T/kA at a nominal current of 10.6 kA. The effect of magnetization has not been taken into account. A comparison between the computed and measured field quality values will be made in the section Magnetic Measurements below. The main parameters for the magnetic cross-section are summarized in Table 12.4. The nominal current, originally estimated to be 10.82 kA considering the first assumptions (Milanese et al. 2012), was re-calculated to 10.58 kA based on the updated cable parameters, the 3D magnetic finite element model (FEM), and magnetic measurements.

Table 12.4 Main magnetic cross-section parameters

Parameter	Value
Nominal magnetic flux density (T)	13.0
Nominal current I_{nom} (kA)	10.6
Number of turns in layers 1 & 2	36
Number of turns in layers 3 & 4	42
Stored energy density in straight section at I_{nom} (MJ/m)	2.8
Differential inductance per unit length at I_{nom} (mH/m)	41.1
Differential inductance at I_{nom} (mH)	62.5
Temperature margin at I_{nom} (4.2 K/1.9 K) (K)	3.0/5.3
Radius for $\Delta B/B \leq 1\%/2\%$ at I_{nom} (mm)	32/42
Sextupole b_3 at nominal current ($R_{ref} = 33.3$ mm)	68.5×10^{-4}
Decapole b_5 at nominal current ($R_{ref} = 33.3$ mm)	-27.8×10^{-4}

12.2.3.2 3D Coil Design

To clear the aperture, flared coil ends are used (see Fig. 12.1). An inclination angle of 17° has been chosen after fixing the minimum distance from the mid-plane to the ends to 61 mm, to accommodate the mechanical support. This choice of inclination angle is a compromise between long planar transitions (large angles) and long inclined sections (small angles), both resulting in longer ends. The minimum hard-way bending radius of the flared ends is set to 700 mm, based on experience from HD3, for which the bending radius was increased from 349 mm to 873 mm. A short inclined section follows, with a length of 24 mm for layers 1 and 2, and 31 mm for layers 3 and 4. The coil ends follow a circular path, so the easy-way bending radius is imposed by the cross-sectional layout. Winding tests with copper cable have guided the choice of the above parameters for the end geometry. The overall coil end-to-end length is 1500 mm, whereas the minimum straight section is 730 mm. The layer jumps were placed in the ends (Fig. 12.5). The path of the layer jump was chosen to minimize the deformation of the cable.

The nominal cable length needed to wind layers 1 and 2 is 225 m, whereas 255 m are needed for layers 3 and 4. A total of about 40 km of strand (275 kg) for the overall magnet is needed.

Magnetic simulations in 3D confirm that the magnetic flux density in flared coil ends naturally decreases. The peak field can be reduced further by optimizing the geometry of the iron, and no spacers are needed in the coil ends. Contrarily, in flat racetrack coils, spacers are needed in the coil ends to reduce the magnetic flux density (Rochepault et al. 2018b). For this reason, and also considering the manufacturing aspects, the iron yoke covers the full length of the magnet, whereas the iron in the vertical pad only covers the straight section. The central post for layers 3 and 4 is a solid piece of iron. The peak flux density in the coil ends is reduced by 1.0 T with respect to the straight section, at nominal current. The magnetic length is 1.13 m, and the longitudinal 1% uniform field region has a length of 0.54 m. The total stored energy at nominal current is 3.8 MJ.

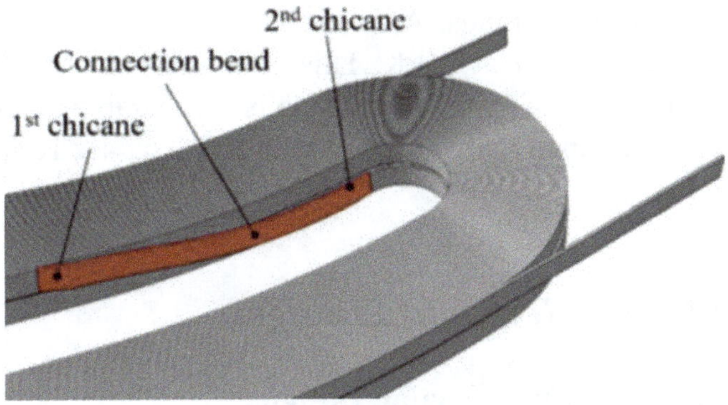

Fig. 12.5 Layer jump (highlighted in red) between layers 3 and 4

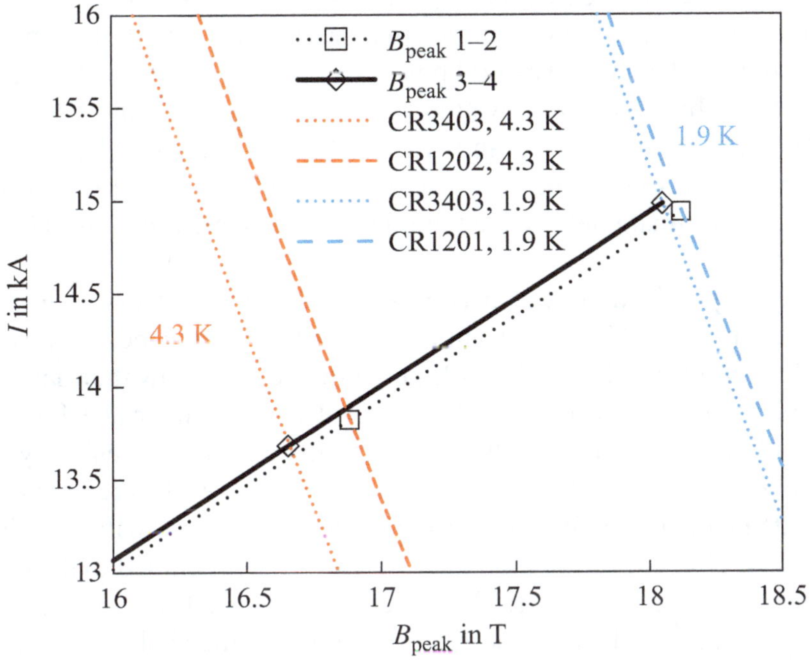

Fig. 12.6 Load lines for coils 1–2 and 3–4 of the magnet and critical current lines at 4.3 K and 1.9 K. Self-field correction and cable degradation are considered. Short sample limits are represented only for the two coils with the lowest short sample current, and are plotted with markers

12.2.4 Short Sample Limits and Load Lines

The critical current is measured on virgin (round) strands and strands extracted from cables using short samples (Rochepault et al. 2017). Measurements have been performed at 4.3 K and the data has been fitted by parameterization curves (Bordini et al. 2012) and extrapolated to 1.9 K. The degradation due to cabling, deducted from the comparison between virgin and extracted strand measurements, has been found to be on average between 2% and 5%.

Figure 12.6 illustrates the estimates for the short sample limit (SSL) current I_{ss} and peak field on the coil B_{peak} based on the computed load lines and the critical line of extracted strands. At a nominal field of 13 T the load line margins are 23% at 4.3 K and 29% at 1.9 K. This large margin was chosen to accept a critical current degradation of up to 15%, which is comparable to HD3. For the ultimate bore field of 15 T, the load margins are 9% and 16% at 4.2 K and 1.9 K, respectively.

12.2.5 Mechanical Design

12.2.5.1 Cross-Section Design

During the first test, for each operating condition, i.e., nominal (13 T central field) and ultimate (15 T central field), the pre-stress was changed with the aim of keeping the cable compressed along the central posts, and to limit the peak stress in the coil to about 150 MPa. This choice follows the approach adapted for the HD3 dipole, and for the SMC (Perez et al. 2015) and RMC (Perez et al. 2016) racetrack models, which reached 86%, 99%, and 97% of their short sample limit, respectively. A pre-stress for the ultimate field will be applied to the magnet that will be installed in the test station.

Mechanical optimization was performed in 2D and 3D using ANSYS finite element software (ANSYS Inc., Canonsburg, PA). For the post for coils 1–2, a titanium alloy has been selected for its high yield stress and low thermal contraction. Early designs considered a removable inner tube as the interface for the magnet inserts, providing support for both the insert and the outsert. A sensitivity analysis of the tube thickness, from 6 to 10 mm, was conducted. It was considered that a value of 8 mm provided a comfortable margin for mechanical support, with a limited impact on the magnetic efficiency. The tube thickness was later included in the central post. For the lateral preload, two lateral keys per side are used in order to balance the preload forces on the coils. A thickness of 70 mm for the shell has been chosen to provide approximately 50% of the preload forces during cool-down for ultimate operation. The horizontal pad is made from stainless steel; the vertical pad is made of iron in the straight section and stainless steel in the ends. The mid-plane shim is in glass-fiber reinforced epoxy (G11).

Bonded contacts between the coils and the central posts were assumed. For all other contacts, between structural components and between the structure and the coils, separation and sliding with a friction coefficient of 0.2 was allowed. The impregnated coils are modeled as blocks with linear properties. The rigidity modulus was initially assumed to be isotropic and equal to 30 GPa and 42 GPa at 293 K and 4.2 K, respectively, with an integrated thermal contraction of 3.36 mm/m from 293 to 4.2 K (Milanese et al. 2012). Based on later mechanical measurements, the modulus was changed to 20 GPa at 293 K and 4.2 K, the friction coefficient to 0.15, and the coil thermal contraction from 293 K to 4.5 K to 3.90 mm/m.

The nominal and ultimate horizontal interference for the lateral preload is 0.6 mm and 1.1 mm, respectively. The pressure required in the bladders to open the corresponding gaps are within reach (in the order of 300–400 bar). The average pressure between a coil and the central post, and between a coil and the horizontal pads, is positive (compression) at all operating steps. According to the model, the corresponding peak stress in the straight section, after cool-down, are 94 MPa and 109 MPa, for the two preload levels, respectively, and 95 MPa and 116 MPa for the nominal and ultimate currents, respectively (see Fig. 12.7). The estimate of the peak stress is 30–40% lower when considering the 20 GPa modulus compared to a 42 GPa modulus. Coils 3–4 are de-bonding (no contact pressure) from their central post during powering (see Fig. 12.8).

12.2.5.2 End Design

The coil ends are supported vertically by two stainless-steel wedges on the mid-plane (see Fig. 12.1), and the vertical pads follow the flared shape, getting progressively thinner. The coil blocks are surrounded laterally with stainless-steel rails and in the ends by stainless-steel end-shoes (reacted and impregnated with the conductor). The length of the coils including the end-shoes is 1.6 m, and is the same as the shell. The design of the end-plate takes into consideration the space needed to insert keys and bladders made of 316 L stainless steel, as well as the instrumentation wires. Its thickness has been set to 150 mm to minimize the bending stress.

12.2.6 Quench Simulations

A 2D electro-thermal model, implemented in CAST3M (a finite element program developed at CEA) has been used to determine the parameters of the quench protection system. Quench-back was not considered. Quench heaters are simulated on the outer surface of each layer, covering 50% of the total allowable surface, and providing 50 W/cm^2 (Felice et al. 2009). According to these computations, the maximum temperature T_{max} is 125 K assuming no detection time (see Fig. 12.9). The highest temperature is in the high-field region. In order to keep T_{max} below 200 K, the detection and circuit operation time (from quench initiation to opening of the switch and activation of the heaters) must be smaller than 100 ms. In the case of failure of two heaters out of four, T_{max} increases by 30 K.

12.3 Technology Development

A series of experimental tests has been carried out prior to coil fabrication, in order to validate the technological choices made during the detailed design phase.

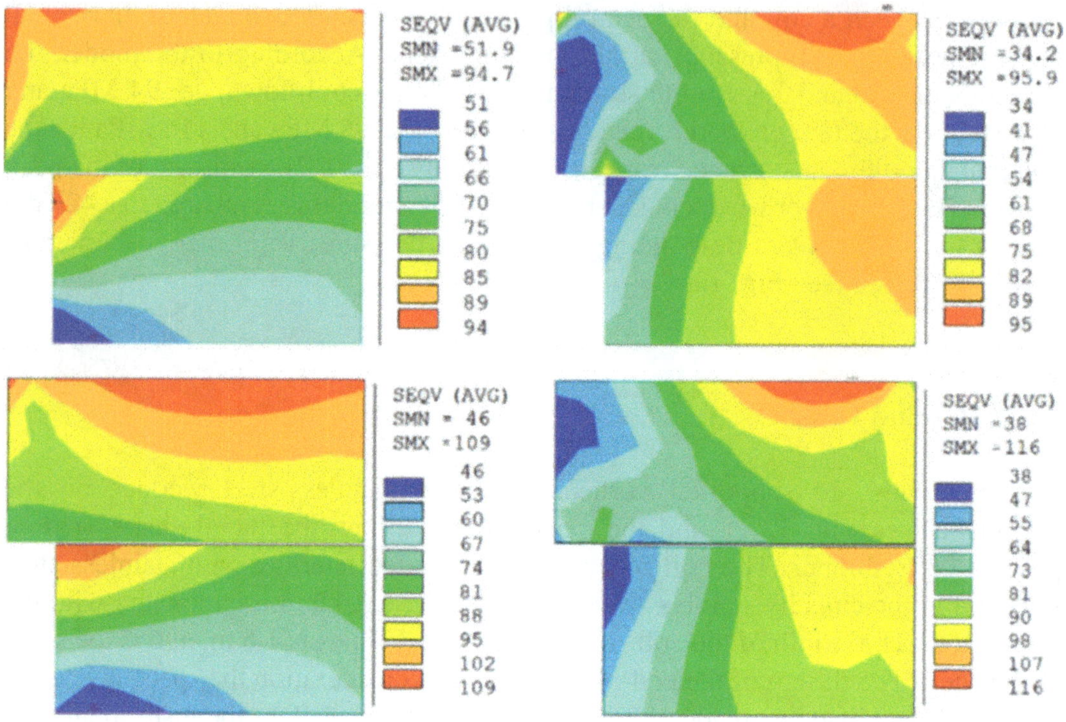

Fig. 12.7 Coil equivalent von Mises stresses for the central sections, generated with the 3D model. Top figures: 0.6 mm preload; after cool-down (left); and powering to nominal 13 T (right). Bottom figures: 1.1 mm preload; after cool-down (left); and powering to ultimate 15 T (right)

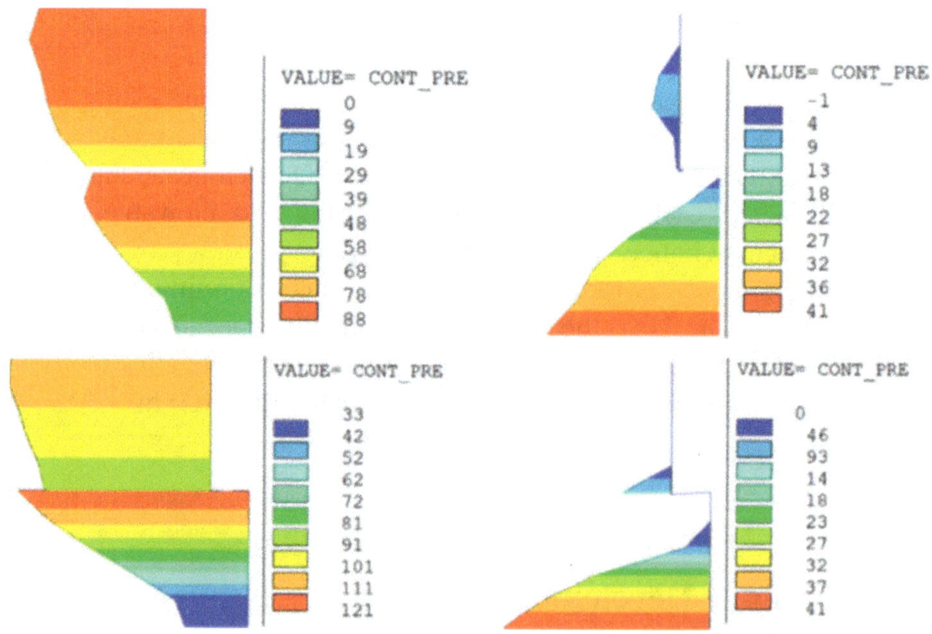

Fig. 12.8 Coil-pole contact pressure, in MPa, for the central sections, generated with the 3D model. Top figures: 0.6 mm preload; after cool-down (left); and powering to nominal 13 T (right). Bottom figures: 1.1 mm preload; after cool-down (left); and powering to ultimate 15 T (right)

Fig. 12.9 Temperature distribution within the dipole coil at the end of the current discharge with zero detection time

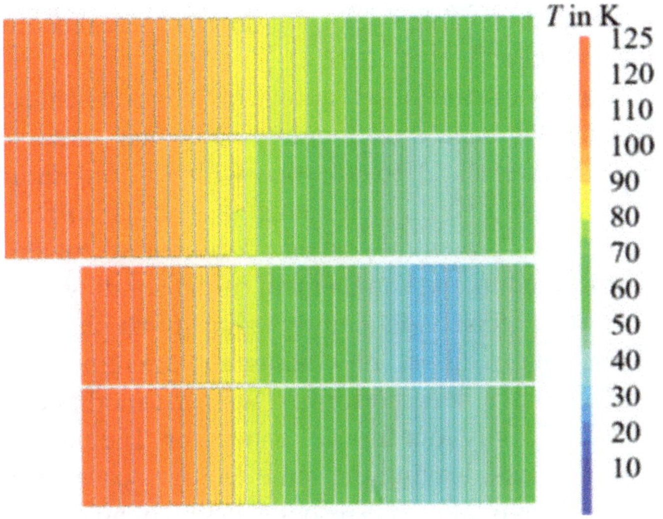

12.3.1 *Sub-scale Winding Tests*

Preliminary winding tests were first performed at CEA on a reduced mock-up (short straight section, only a few turns) using copper cables to explore and choose the winding tension, inclination angle, radius of the hard-way bend, profile of the easy-way bend, and orientation of the winding. Trials were also performed for the layer jump. These tests allowed validation of the windability of the coil ends and confirmed the choice of a block-coil layout.

12.3.2 *Full-Scale Cu Practice Coils*

The coil fabrication process and the tooling design have been tested and validated on two full-scale practice coils, 1–2 and 3–4, using copper cables. The Cu cable had a slightly larger thickness (2.24 mm compared to the nominal 2.22 mm), so only 40 turns have been wound instead of the nominal 42 turns.

12.3.3 *Dimensional Changes During Heat Treatment*

As learned from previous projects, Nb_3Sn cable sections expand and contract longitudinally during heat treatment due to stress release and the phase transition during Nb_3Sn formation. The exact changes in dimensions of the coils is difficult to predict and depends greatly on the cable and coil fabrication process. The tooling has to allow for a contraction in the longitudinal direction and an expansion in the cross-section, in order to reduce residual stress and potential degradation. To determine the exact change of dimension during heat treatment, a series of measurements was performed on cable stacks and on sub-scale racetrack coils (short length, only a few turns) (Durante et al. 2016).

Following the observations on the sub-scale coils, the winding and reaction tooling has been split longitudinally into three sub-elements (Manil et al. 2014; Rondeaux et al. 2016) to cope with the longitudinal contraction. The tooling has been adapted so the sub-elements are able to freely move axially during the heat treatment. The central post and rails are also cut along the straight section. Each post part is fixed to one sub-element of the winding table, and the gaps between the post parts are precisely defined by using shims placed between the sub-elements. These shims are placed before winding and are removed before heat treatment.

Tests have been performed with the modified tooling described above for coils 3–4, using only four turns of conductor. The missing 38 turns have been replaced by filling blocks, leaving the same relative space for the cable to expand transversally during heat treatment. The coils were wound at CEA and heat-treated at CERN. The first test was performed with an RRP cable, assuming a longitudinal contraction of 1% (Manil et al. 2014). The gaps were distributed at the lead end (6 mm) and the return end (10 mm). In that configuration, it was expected that the gaps would not close, which would allow an in situ measurement of the coil contraction. Surprisingly, the gaps reduced by only 3.2 mm (see Table 12.5) instead of the expected 13.2 mm reduction, so a second test was performed introducing a total gap of only 3 mm, expecting that the gaps would fully close. In this configuration, however, a reduction of the gap of only 1 mm was observed. Excessive friction or stuck parts were suspected, and two additional tests were performed (using a PIT cable) in a configuration with reduced friction (removing guiding features and placing additional mica layers). Even with reduced initial gaps, however, the gaps never fully closed.

Following these tests, it was decided to place shims with a thickness of 1 mm between the blocks of the winding table before winding. After winding, the gaps were closed manually by removing the shims and putting the blocks back in contact. This approach provides an extra cable length of 0.13% inside the mold, gives a margin to release the winding tension, and allows for cable contraction during reaction. As a comparison, overall contractions of the order of −0.3 to −0.15% have been reported for MQXF coils (Rochepault et al. 2016).

12.3.4 Quench Protection and Instrumentation

The stored energy density in the coil is comparable to previously built magnets, allowing for similar protection approaches. For FRESCA2, quench protection is provided by using a dump resistor and quench heaters. The quench heater design, shown in Fig. 12.10, is based on 25 μm thick and 12 mm wide stainless-steel strips sandwiched between polyimide films. To cover the 80 mm wide coils two nested families of heaters are installed, going from the connection side to the

Table 12.5 Full-scale coil contractions

Coil	Initial gaps (mm)	Gap contraction (mm)	Coil contraction (%)
DR3401 (RRP)	15.7	−3.3	−0.25
DR3402 (RRP)	3.0	−1.0	−0.08
Sub-scale (insulated, fixed, Fe post, RRP)	25	−2.4 to −0.9	−0.35 to −0.15
DP3401 (PIT)	4.0	−1.6	−0.12
DP3402 (PIT)	4.0	−0.2	−0.01
Sub-scale coils (insulated, fixed, Fe post, PIT)	25	−3.2 to −1.7	−0.45 to −0.25

Fig. 12.10 Quench heater impregnated at the surface of a coil

non-connection side and back, with a wiggling shape and without heating stations. The total integrated length of the heaters is 3.1 m (four turns), with a total surface per circuit of 370 cm^2, and a total resistance of 8 Ω. The required power of 51 kW is provided with a current of 80 A and a voltage of 640 V. This quench heater technology was previously successfully applied to the RMC magnets.

12.4 Magnet Fabrication

12.4.1 Coil Fabrication

The coil fabrication process follows four main stages (Manil et al. 2014; Rondeaux et al. 2016): winding (with un-reacted cable), heat treatment, post-heat treatment operations, and impregnation with epoxy resin. Each single double-pancake is fabricated separately, with two different sets of tooling.

So far, six coils have been produced: five coils using RRP conductor (CR1201, CR1202, CR3401, CR3402, and CR3403, see Fig. 12.11) and one coil using PIT conductor (CP1203). A second PIT coil (CP3404) will be produced: the cable unit length is available and the components are being procured. A detailed description of coil fabrication is provided by Rochepault et al. (2017) and Rondeaux et al. (2016).

Fig. 12.11 Four Nb_3Sn coils of one magnet are ready for the first magnet assembly

12.4.1.1 Winding

The coil winding operation was performed at CEA on a horizontal winding machine. Due to a large spring-back of the cable, a 'swelling' appears in the straight section. Therefore, when compressing the coils laterally with rails, the excess of conductor is transferred to the ends, and gaps open between the coil ends and the central post. Consequently, the coil ends were 1.4–4.0% thicker than nominal, the coils were 0.2–0.5% longer, and the end-shoes were outwardly misaligned by a few mm.

The winding tooling is kept for the heat treatment. All of the following operations were performed at CERN.

12.4.1.2 Heat Treatment

The heat treatment cycle was optimized on short samples for a large critical current and RRR. The following heat treatment cycle for the RRP cable has been used:

- Ramp at 25 °C/h, held for 72 h at 210 °C;
- Ramp at 50 °C/h, held for 48 h at 400 °C;
- Ramp at 50 °C/h, held for 50 h at 650 °C.

Six witness samples for critical current measurements were systematically reacted with the coils.

As explained above, coils usually contract longitudinally during the heat treatment. For coils 1–2, the gaps introduced during winding remained closed, and the coils contracted globally as expected. The remaining additional length was compensated for by varying the insulation thickness between the coil and the end-shoes. Unexpectedly, however, coils 3–4 contracted very little and remained a total of 0.4–0.5% longer. Additional gaps even re-opened between the post pieces, and between the coil ends and the post. This behavior is attributed to the higher thermal expansion of the iron posts, preventing the contraction of coils 3–4 during heat treatment. The extra length has been compensated for by re-machining the end-shoes.

After the heat treatment of CR1201 (coil with open bore), a significant deformation of the Ti post (up to 1.5 mm) was observed. This deformation was attributed to the volume expansion of the coil during reaction. The post was machined after impregnation to allow assembly. For the next coils, filler pieces were inserted into the bore during heat treatment. This technique was first applied in the HD3 coils and allowed the avoidance of any deformation.

12.4.1.3 Instrumentation, Splicing, and Preparation for Impregnation

Each layer is instrumented with traces consisting of the abovementioned quench heater circuits and connections for up to 10 voltage taps. On the outer side of the trace 0.2 mm of glass-fiber cloth was added.

The Nb_3Sn leads are spliced to Nb-Ti leads and the splice is attached to the end-shoes. During the splicing of coil CR3401, some strands were accidentally damaged. The area was then repaired by soldering a copper stabilizer to provide both a mechanical support and stabilization in case of a quench.

12.4.1.4 Impregnation

The coils are impregnated inside a vacuum impregnation tank, using CTD-101K epoxy resin from Composite Technology Development (CTD), Lafayette CO, US. The pressure inside the mold is around 10 mbar before impregnation, with the autoclave being at 10^{-3} mbar. Resin curing is made at atmospheric pressure, with a first step of 6 h at 110 °C, followed by 16 h at 125 °C.

Impregnation of coil CR1201 failed during the first attempt and only two-thirds of the coil volume was filled with epoxy. A second impregnation was performed by machining new injection holes into the impregnation mold.

For CR3401, some air volumes were accidentally trapped in the cavity and released during impregnation, which created some voids at the surface of the coil. The impregnation was repaired by locally re-impregnating each void. This issue is likely to be the source of the damage to the quench heaters. For the subsequent coils, adequate filler pieces were used, the sealing of the mold was improved, and no further issues were observed.

12.4.2 Magnet Assembly

12.4.2.1 Coils' Pairing

Out of the first four coils produced, pairs of coils 1–2 and 3–4 were selected for the best dimensional fit. All possible combinations have been studied based on the dimensional measurements performed after coil fabrication. The most favorable case was to pair CR1201 with CR3402 and CR1202 with CR3401, providing a minimum of 0.1 mm clearance for assembly. In the second magnet, FRESCA2b, CR3401 was replaced by the spare coil CR3403 (Bourcey et al. 2018).

The ground insulation scheme initially proposed relied on coil-to-post insulation, with contact post-to-post. The measured post-to-coil resistances were, however, lower than expected; therefore, the insulation scheme was modified by adding polyimide layers between the posts during magnet assembly.

12.4.2.2 Double-Coil Assembly

In order to compensate for potential shape defects of the coils, a tailor-made shim of about 0.9 mm thickness is impregnated between coils 1–2 and coils 3–4 (Fig. 12.12). Coil 3–4 is placed on top of coil 1–2, making sure that the posts are in good contact in the straight section. The coils are then clamped and the tooling is closed and sealed to allow for vacuum impregnation with Araldite MY750 charged resin to fill the gap between the coils. A mold release agent is applied on of all the surfaces in contact with the resin so as to allow disassembly of the two coils in the case of a replacement being required.

The electrical resistance between the central posts of coils 1–2 and 3–4 was checked before and after the tailored shim impregnation by applying different voltage levels. After the impregnation of pole 2 (CR1202/CR3401), a short was detected. After disassembly, it became evident that the resin did not flow in between the central posts. The polyimide insulation was reinforced for the second attempt and the result was satisfactory (16 GΩ at 1 kV).

12.4.2.3 Pole Assembly

Once the coil assembly is finalized, a vertical pad is placed on coil 3–4, with a 1 mm G10 insulating shim between. Two wheels have been designed to be able to rotate the assembly by 180°, lift, and then position it. The mid-plane insulation and the wedges are then installed to complete the half coil-pack, providing a flat mid-plane for the assembly of the second pole. Two layers of 0.125 mm thick polyimide film were added in the mid-plane and bent in the aperture to ensure electrical insulation between the Ti central posts. The same operations are repeated for the second pole. The first pole is then lifted and positioned carefully on top of the second pole (see Fig. 12.13). To do so, a lifting beam is mounted into the aperture of coil 1–2, and attached at the ends of the vertical pad. The lifting beam is finally slid out of the aperture, taking care to not damage the strain gauges and wires.

Fig. 12.12 Coils 1–2 and 3–4 installed in the tailored shim impregnation mold

Fig. 12.13 Manual assembly process of the two poles together using dedicated tooling

12.4.2.4 Magnet Preloading

Finally, the coil-pack is centered horizontally and vertically in the structure by using the bladders and inserting the first keys (see Fig. 12.14). The axial loading system is assembled before starting the preloading operations (see Fig. 12.15).

The preload is subdivided in three loading sub-steps but, depending on the available shim sizes and the evolution of the stress, some intermediate steps may be added. The stress is monitored using strain gauges at various locations on the poles (CR1201 and CR1202) and on the shell (only the mid-plane locations S1M and S2M will be described). For the three assemblies, the post stress vs. shell stress (transfer function) was linear, as expected from the FE model. The agreement with the measurements are within a few MPa. Both FRESCA2a and FRESCA2b were loaded at room temperature until there was a stress of around 150 MPa in Ti post 1–2 and 30 MPa in the Al shell, corresponding to a peak stress in the coil of approximately 30 MPa. More details on the analysis can be found in Rochepault et al.

Fig. 12.14 Magnet after coil-pack insertion. The picture shows the elements of the assembly

Fig. 12.15 Magnet ready for preloading with axial loading system assembled. The loading structure is shown

(2018a). The coil-ends were preloaded by tensioning the rods to 70 MPa, corresponding to 20% of the Lorentz forces in the coil ends during powering. The total preload provided by the rods at cold corresponds to 60% of the Lorentz forces.

In FRESCA2a, after the first loading step, a relative lateral shift of the coils of about 0.2 mm, caused by a misaligned coil pack, resulted in an unbalanced preload (a difference of 100 MPa in the two Ti posts). To compensate for this effect, an asymmetric loading was performed by removing 0.1 mm in the loading shims on both sides of coil CR1202. For FRESCA2b and FRESCA2c, the alignment procedure was improved using coil measuring machine (CMM) data, resulting in a well-balanced preload.

FRESCA2c was loaded for ultimate operation at 15 T, corresponding to 60 MPa in the shell, and 330 MPa in the post for CR1201 (CR1202 showing artificially higher stress for the reasons explained below). The rods were tensioned to 120 MPa, corresponding to 35% of the coil-end Lorentz forces at warm temperature and 70% at cold.

12.5 Magnet Tests

The FRESCA2 magnet was tested three times in the CERN magnet test facility: FRESCA2a in February 2017, FRESCA2b in August 2017, and FRESCA2c in May 2018. In addition to ramps to quench and the investigation of training performance, the test plans included ramp rate, protection studies, and magnetic measurements. The magnets were tested at both 1.9 K and 4.3 K, and thermal cycles were carried out to check quench memory. Quench locations were determined by analyzing signals recorded by voltage taps installed on the pole turns of each layer (10 per layer for a total of 40 voltage taps). The mechanical behavior was monitored by strain gauges mounted on the aluminum shell, the aluminum axial rods, and the poles of coils CR1201 and CR1202. We provide in this section a summary of the quench performance, the protection studies, and the magnetic and stress measurements.

12.5.1 Quench Performance

12.5.1.1 First Assembly, FRESCA2a

The tests were carried out as follows. The first protection studies were performed with quench heaters and energy extraction tests. Training to 13 T was then started at 1.9 K, followed by training at 4.2 K. Quench 1 of FRESCA2a was a typical pole turn quench, originating in the lower layer pole turn of coil CR1201, in the flared segment close to the head. Quenches 2–6 all originated due to the known cable damage in the splice region of coil CR3401 (described above). These quenches were reproducible, with the same pattern, same location, and same precursor. Therefore, one training quench can be considered for reaching 12.2 T in the first assembly (see Fig. 12.16).

The powering tests ended for FRESCA2a after quench 6, because during the discharge a quench heater of coil CR3401 failed, causing damage to the coil insulation.

12.5.1.2 Second Assembly, FRESCA2b

The second assembly, FRESCA2b, reached its initial target field of 13 T after two quenches in the first cool-down at 1.9 K, see Fig. 12.16. The quenches occurred in

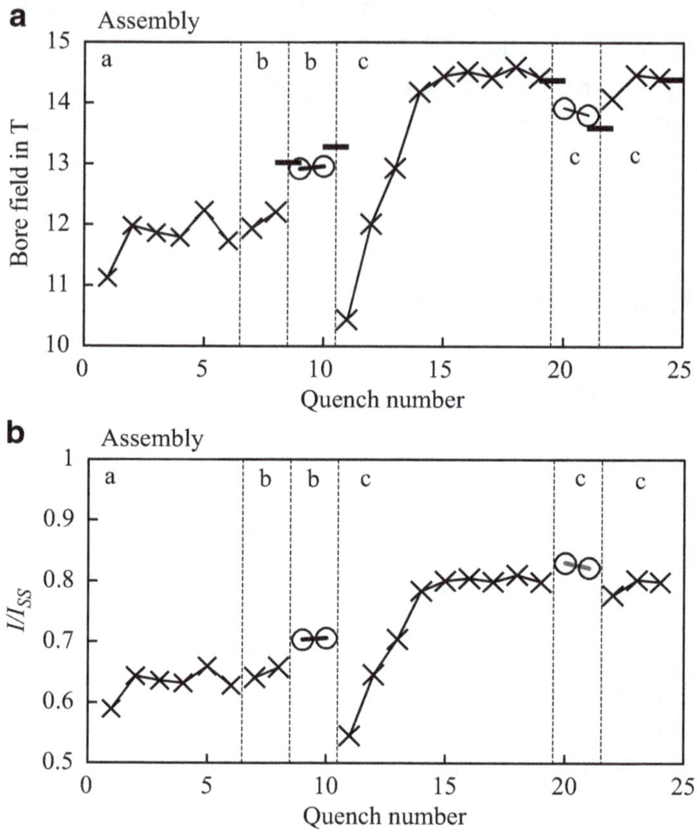

Fig. 12.16 Training curves for FRESCA2a, b, and c: (**a**) bore field at quench current; (**b**) quench current normalized to the short sample. The marker crosses represent quench currents at 1.9 K, the round markers quench currents at 4.5 K, and the flat markers represent stable currents with no quench

the pole turn of coil CR1202 and in the second turn of coil CR1201, all in the high-field region in the straight section of the magnet.

Powering at 4.5 K resulted in two quenches just below 13 T. Both quenches occurred in the high-field region, in the straight segment of the pole turn of coil CR1202 and in the straight segment of the pole turn of coil CR1201. The number of training quenches up to 73% of the short sample limit is small compared to the predecessor magnets and model coils like HD and RMC.

A thermal cycle was performed, with the magnet reaching 280 K before cool-down to 1.9 K. After a thermal cycle the increased target field of 13.3 T was reached without quench.

All five recorded high-field quenches in the pole turns were preceded by rather large precursors in the voltages, indicating conductor movement as the cause of the quench. A detailed analysis of the test results can be found in Willering et al. (2018).

12.5.1.3 Third Assembly, FRESCA2c

A third assembly, FRESCA2c, has been prepared, with an increased preload for reaching the ultimate field of 15 T. The magnet reached a maximum record bore field of 14.6 T after seven additional quenches at 1.9 K. After a thermal cycle the magnet showed a good memory with only one retraining quench. Two stable currents at

14.4 T during 1 h were obtained at 1.9 K, and one stable current at 13.6 T during 1 h at 4.5 K. The magnet did not reach the conductor limit, but was limited by vibrations, in just one coil, indicating the potential for higher currents. The analysis of the tests is ongoing, and will be described in Willering et al. (n.d.).

12.5.2 Mechanical Analysis

12.5.2.1 Cool-Down

The data acquired during the different load-steps provided useful information for model tuning and FEM parameter validation. To better match the measurements with FEM simulations during the cool-down, 2D parametric studies for different material properties, contact conditions, and friction coefficients were performed. The cool-down is modeled with linear thermal contractions and homogeneous temperatures. The material properties are similar to the values proposed for MQXF magnets.

In particular, the contact conditions between the vertical pad, the iron post 3–4, and the Ti post 1–2 have a significant impact on the post stress. For FRESCA2a, the stress in the posts closely followed the simulated values. During the first cool-down of FRESCA2b, however, the contact conditions in coil CR1202 changed and a higher stress appeared in the post (see Fig. 12.17a). This behavior is attributed to a loss of contact at the top faces between Ti post 1–2 and iron post 3–4 (a detailed explanation can be found in Rochepault et al. (2018a)). The consequence is an additional bending stress, increasing the total stress measured in the post. Fortunately, an additional bending of the post has a marginal impact on the coil peak stress. The hypothesis of a lack of contact between the horizontal pad and coil 3–4, translating into a higher preload of coil 1–2, has been ruled out using stress-sensitive films during magnet assembly (Bourcey et al. 2018).

After warm-up, a loss of stress of about 20 MPa is observed in the Al shell. This effect is reproduced in the FEM by allowing a plasticization of the coil pack. According to the FEM, it is not attributed to friction.

12.5.2.2 Powering

As the coil is energized, the posts will be laterally un-loaded and the shell will be azimuthally stretched under the outward Lorentz forces. The post unloads linearly with the square of the current (with the first-order proportional to the Lorentz forces, see Fig. 12.17a). The measured slope is higher than predicted by the model, which could be due to changes in contact conditions, as suggested by Rochepault et al. (2018a).

Figure 12.17b shows the evolution of the shell azimuthal stress with the current squared. The measured data are compared with two 3D FEMs: the nominal model, assuming perfectly bonded contacts between the coils and posts; and a model considering sliding contact between the coils and the posts, with friction and

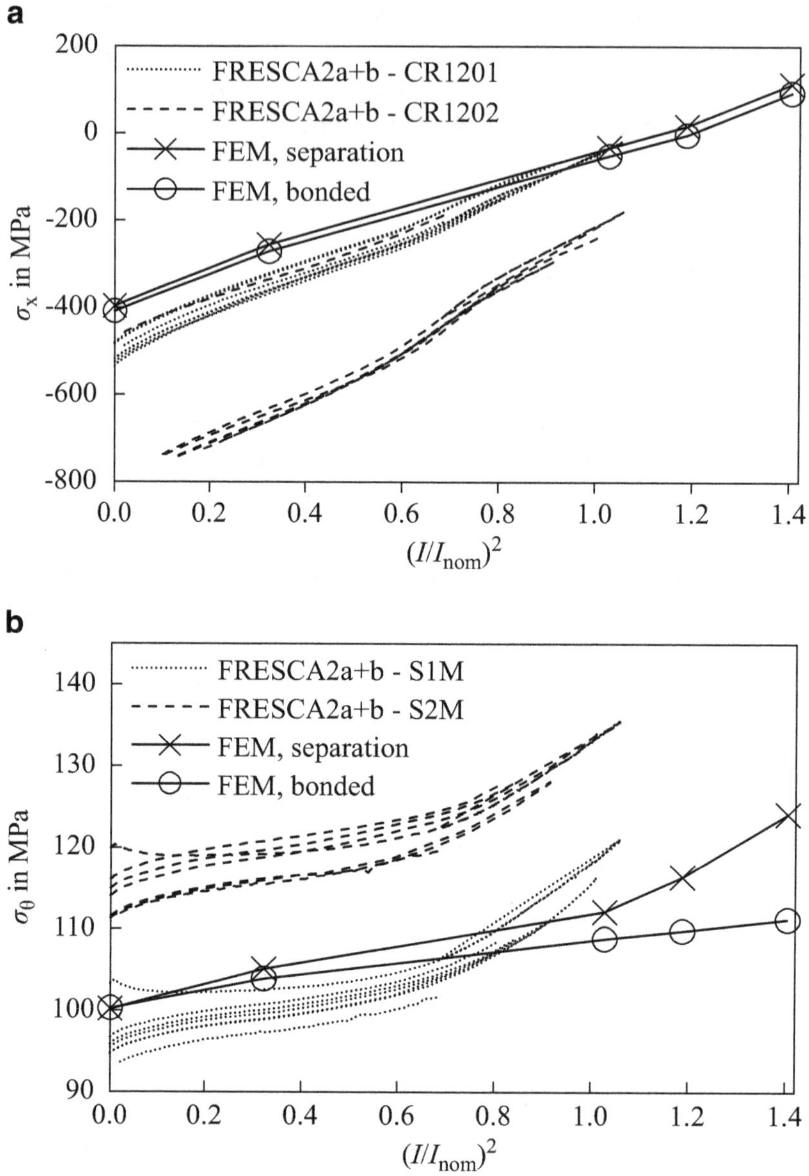

Fig. 12.17 (**a**) Lateral post stress; and (**b**) azimuthal shell stress during powering as a function of the current squared. The data from FRESCA2a (last quench) and FRESCA2b (six current ramps) are compared with 3D finite element models using two different coil–post contact conditions (bonded and separation)

separation allowed. At low current, the azimuthal stress σ_θ in the shell follows the model. The slope changes at $(I/I_{nom})^2 = 0.6$, so $I = 8.2$ kA. This effect can be an indication of a de-bonding, after which the coil is pushing on the shell. This behavior is suggested by the 3D FEM, for which separation is allowed. This de-bonding, however, occurred at a lower current (8.2 kA) than predicted by the model (10.55 kA). This is currently unexplained, and could be attributed either to a loss of preload during cool-down, or additional magnetic forces between the iron parts. The impact of de-bonding on the pole stress is marginal.

Similarly, the axial rod stress is slightly increasing with the current squared, as predicted by the model. For FRESCA2c, the strain gauges show a similar behavior, except that the curves are shifted due to the higher preload. The change of slope occurs at a higher current and the poles remains in compression. The mechanical analysis is described in Willering et al. (n.d.).

12.5.3 Magnetic Measurements

The magnetic field of FRESCA2b has been measured with a cold rotating shaft with five coil segments, each 247 mm long. In Fig. 12.18, the measured data is compared with the bore field profile given by the 3D FEM, and the values averaged on the segments' length. The measurements confirm that the nominal field is reached with 13.04 T at 10.6 kA. The homogeneity of the magnetic field has been measured for the five middle segments of the magnetic measurement shaft. Figure 12.18 shows that the field profile is homogeneous within 1.5% over a length of 0.7 m, which is within the original 2% specification.

Field quality measurements have also been performed, and the field harmonics and transfer functions are within a few percent of the predictions (as shown in Figs. 12.19 and 12.20).

12.5.4 Protection Studies

12.5.4.1 Quench Heaters

In FRESCA2a, one heater strip failed at quench 6, with large damage of the coil-to-heater insulation, and the magnet test was not continued. In FRESCA2b, another strip failed during the first discharge, fortunately with less damage to the insulation, and the powering tests could continue. The failures have been linked to bad impregnation of the heater. It was decided to continue powering with protection from energy extraction alone.

12.5.4.2 Energy Extraction

For energy extraction, the test station is equipped with a thyristor switch that is allowed to switch in an extraction resistor with a delay of less than 4 ms after triggering. The maximum switching voltage of this switch is 1 kV, therefore a maximum extraction resistance of 80 mΩ was chosen (81 mΩ is considered when including the rest of the warm part of the circuit). The magnet protection was assessed by a series of energy extraction tests, where the energy extraction (in the dump resistor) was manually triggered and the current decay was analyzed, see

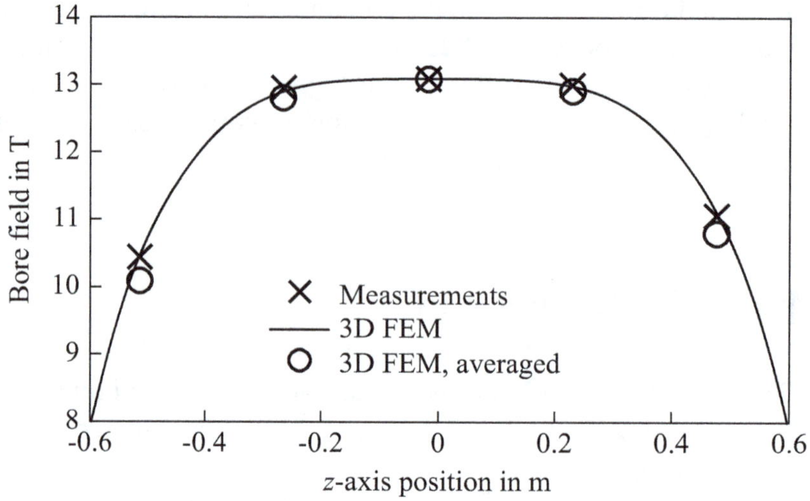

Fig. 12.18 Average magnetic field measured in five 250 mm long segments (shown as crosses). The calculated field profile along the z axis is shown by the line

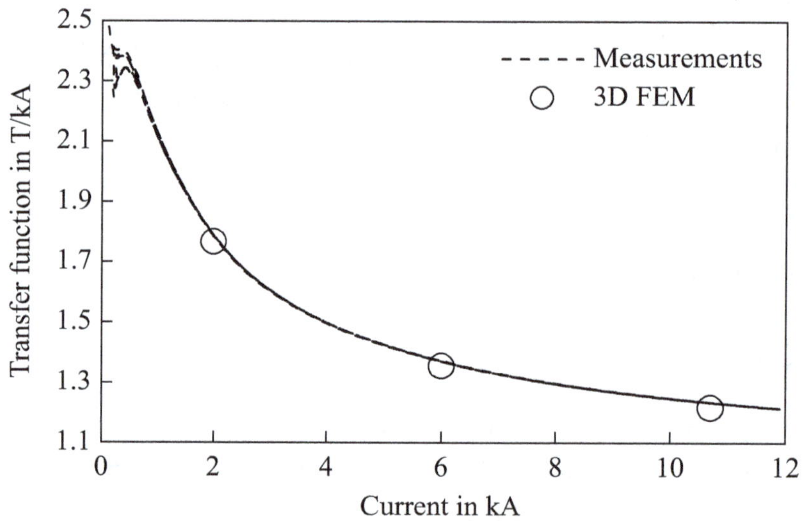

Fig. 12.19 Measured and computed transfer functions at the central section of the magnet as a function of magnet current

Fig. 12.21. The measured quench integral (QI) at 10.85 kA is 33.4 MA^2s, corresponding to a hot-spot temperature of 157 K, which is a reduction of more than 30% compared to the calculated 48.3 MA^2s, corresponding a hot-spot temperature of about 250 K. The difference can be entirely assigned to the effect of quench-back. The tests demonstrate that the contribution of the quench heaters to the reduction in QI was insignificant compared to the energy extraction and quench-back.

An effect of quench-back is a larger energy dissipation in the magnet and helium bath. At 10.85 kA about 30% of the magnet energy is dissipated in the magnet.

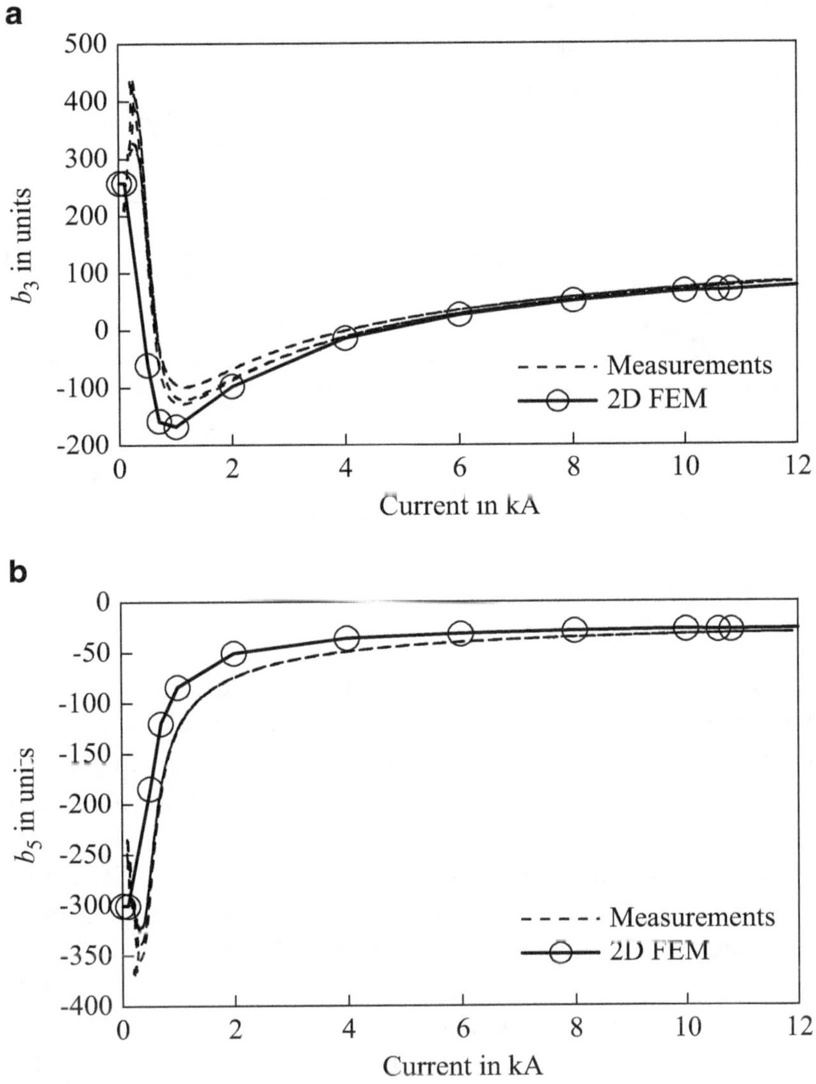

Fig. 12.20 Measured and computed (**a**) sextupole b_3; and (**b**) decapole b_5 normalized harmonics at a reference radius of 33 mm for the central section of the magnet, as a function of magnet current

12.5.4.3 Protection Parameters and Hotspot Temperature

The electrical resistance of several high-field cable segments during current decay was measured, which allowed estimation of the local temperature. This estimate showed that the temperature in the high-field part of the coil did not exceed 100 K.

Based on the analysis of the time to rise to 100 mV for the five quenches in the high-field zone and the five quenches in the low-field zone, a threshold of 100 mV and a validation time of 10 ms are foreseen for operation. A variable threshold protection card should reduce the sensitivity to low-field flux jumps. Using these parameters, the maximum temperature that is expected to be reached in the magnet is about 160 K at 13 T and 180 K at 15 T.

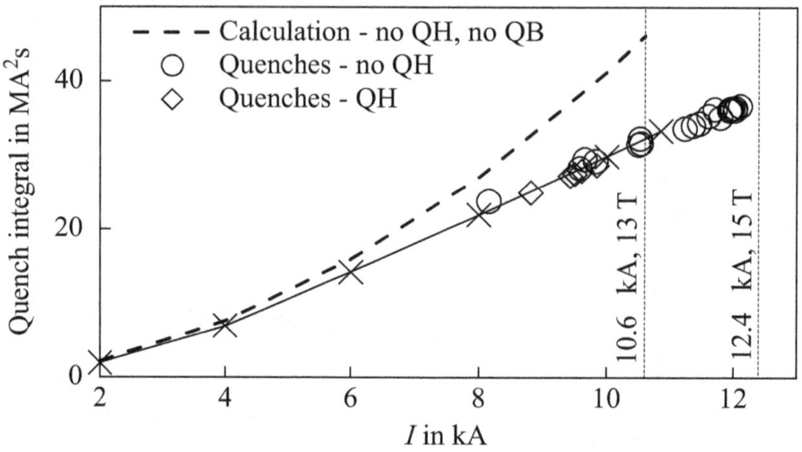

Fig. 12.21 Quench integrals from the triggering of the quench protection at 1.9 K with and without quench heaters (QH) and the consideration of quench-back (QB)

12.6 Conclusions

The FRESCA2 project started in 2009, with the aim to design, fabricate, and test a 13 T Nb_3Sn dipole for CERN's cable test facility. A block-coil layout with two double pancakes and flared ends was chosen, with a design inspired by the HD magnets. The structure is shell-based with a room-temperature preload using bladders and keys.

Six coils have been fabricated following different steps at CEA Saclay and at CERN. So far all tested coils have used RRP wire. One additional coil using PIT conductor is foreseen. The first magnet assembly, FRESCA2a, has been prepared and tested in liquid helium, reaching a maximum central field of 12.2 T. Unfortunately, coil CR3401 showed limitations during the tests because of fabrication issues, and was replaced with coil CR3403 in the second assembly FRESCA2b. The magnet reached a maximum bore field of 13.3 T in only three quenches, slightly above its nominal current. At 13 T the magnet showed stable operation for the 4 h test duration, which validates the powered performance of the magnet. Protection studies were performed, showing an important contribution of quench-back in the reduction of the quench integral at high current levels. During powering, the stress on the posts and the shell showed potential signs of coil–post detachment, at a current lower than the nominal. Even though there was no sign of premature quenches or degradation, it was decided not to power the magnet at higher currents. A final assembly, FRESCA2c, has been prepared, increasing the preload in order to reach a central field of 15 T. The magnet was able to reach a field record of 14.6 T after only seven additional quenches at 1.9 K. All magnet assemblies showed a negligible reduction of quench currents after a thermal cycle.

With this achievement, the FRESCA2 magnet demonstrated that the technology of block-coils with flared ends is able to reach fields close to 15 T in a rather large aperture. It confirms a great potential of this design for future 16 T accelerator dipoles.

In addition to a proof of concept of high-field Nb_3Sn magnet technology, the FRESCA2 magnet, once installed in the test facility, is foreseen to be extensively used to test superconducting cables and HTS inserts in a large background field. Together with HTS inserts, magnetic fields of up to 20 T may be approached in FRESCA2 in a dipole configuration with apertures relevant for accelerators.

References

Bordini B, Bottura L, Mondonico G et al (2012) Extensive characterization of the 1 mm PIT Nb_3Sn strand for the 13-T FRESCA2 magnet. IEEE Trans Appl Supercond 22(3):6000304. https://doi.org/10.1109/tasc.2011.2178217

Bourcey N, Zurita AC, Durante M et al (2018) Assembly of the Nb_3Sn dipole magnet FRESCA2. IEEE Trans Appl Supercond 28(3):4007505. https://doi.org/10.1109/tasc.2018.2809703

de Rijk G (2012) The EuCARD high field magnet project. IEEE Trans Appl Supercond 22 (3):4301204. https://doi.org/10.1109/tasc.2011.2178220

Devaux-Bruchon M, Durante M, Karppinen M et al (2010) EuCARD-HFM dipole model design options. EuCARD-REP-2010-002, Oct. CERN, Geneva

Devred A, Baudouy B, Baynham DE et al (2006) Overview and status of the next European dipole joint research activity. Supercond Sci Technol 19(3):S67–S83. https://doi.org/10.1088/0953-2048/19/3/010

Devred A, Boutboul T, Oberli L (2007) Status of NED conductor development. IEEE/CSC & ESAS European Superconductivity News Forum no. 2, no. ST5, Oct

Durante M, Garcia Fajardo L, Ferracin P et al (2016) Geometrical behavior of Nb_3Sn Rutherford cables during heat treatment. IEEE Trans Appl Supercond 26(4):4802705. https://doi.org/10.1109/TASC.2016.2530166

Durante M, Borgnolutti F, Bouziat D et al (2018) Realization and first tests of the EuCARD 5.4-T REBCO dipole magnet. IEEE Trans Appl Supercond 28(3):4203805. https://doi.org/10.1109/TASC.2018.2796063

Felice H, Ambrosio G, Chlachidze G et al (2009) Instrumentation and quench protection for LARP Nb_3Sn magnets. IEEE Trans Appl Supercond 19(3):2458–2461. https://doi.org/10.1109/TASC.2009.2019062

Ferracin P, Devaux M, Durante M et al (2013) Development of the EuCARD Nb_3Sn dipole magnet FRESCA2. IEEE Trans Appl Supercond 23(3):4002005. https://doi.org/10.1109/TASC.2013.2243799

Leroy D, Spigo G, Verweij AP et al (2000) Design and manufacture of the large-bore 10 T superconducting dipole for the CERN cable test facility. IEEE Trans Appl Supercond 10 (1):178–182. https://doi.org/10.1109/77.828205

Lorin C, Durante M, Fazilleau P et al (2016) Development of a Roebel-cable-based cos-theta dipole: design and windability of magnet ends. IEEE Trans Appl Supercond 26(3):4003105. https://doi.org/10.1109/TASC.2016.2528542

Manil P (2013) Design report for the dipole magnet (Part I). In: Bajas H, Baudouy B, Benda V et al (eds) Dipole model test with one superconduction coil; results analysed. Deliverable: D7.3.1 EuCARD-REP-2010-013, June. CERN, Geneva

Manil P, Baudouy B, Clement S et al (2014) Development and coil fabrication test of the Nb_3Sn dipole magnet FRESCA2. IEEE Trans Appl Supercond 24(3):4001705. https://doi.org/10.1109/TASC.2013.2285879

Milanese A, Devaux M, Durante M et al (2012) Design of the EuCARD high field model dipole magnet FRESCA2. IEEE Trans Appl Supercond 22(3):4002604. https://doi.org/10.1109/TASC.2011.2178980

Oberli L (2013) Development of the Nb_3Sn Rutherford cable for the EuCARD high field dipole magnet FRESCA2. IEEE Trans Appl Supercond 23(3):4800704. https://doi.org/10.1109/TASC.2012.2236602

Perez JC, Bajko M, Bajas H et al (2015) Performance of the short model coils wound with the CERN 11-T Nb_3Sn conductor. IEEE Trans Appl Supercond 25(3):4002805. https://doi.org/10.1109/TASC.2014.2381364

Perez JC, Bajas H, Bajko M et al (2016) 16 T Nb_3Sn racetrack model coil test result. IEEE Trans Appl Supercond 26(4):4004906. https://doi.org/10.1109/TASC.2016.2530684

Rochepault E, Ferracin P, Ambrosio G et al (2016) Dimensional changes of Nb_3Sn Rutherford cables during heat treatment. IEEE Trans Appl Supercond 26(4):4802605. https://doi.org/10.1109/TASC.2016.2539156

Rochepault E, Bourcey N, Ferracin P et al (2017) Fabrication and assembly of the Nb_3Sn dipole magnet FRESCA2. IEEE Trans Appl Supercond 27(4):9500205. https://doi.org/10.1109/TASC.2016.2636145

Rochepault E, Bourcey N, Ferracin P et al (2018a) Mechanical analysis of the FRESCA2 dipole during preload, cool-down and powering. IEEE Trans Appl Supercond 28(3):4002905. https://doi.org/10.1109/TASC.2017.2781704

Rochepault E, Izquierdo Bermudez S, Perez JC et al (2018b) 3D magnetic and mechanical design of coil ends for the racetrack model magnet RMM. IEEE Trans Appl Supercond 28(3):4006105

Rondeaux F, Ferracin P, Durante M et al (2016) "Block-type" coils fabrication procedure for the Nb_3Sn dipole magnet FRESCA2. IEEE Trans Appl Supercond 26(4):4002405. https://doi.org/10.1109/TASC.2016.2528049

Taylor C, Scanlan R, Peters C et al (1985) A Nb_3Sn dipole magnet reacted after winding. IEEE Trans Magn MAG 21(2):967–970. https://doi.org/10.1109/TMAG.1985.1063680

van Nugteren J, Kirby G, Bajas H et al (2018) Powering of an HTS dipole insert-magnet operated standalone in helium gas between 5 and 85 K. Supercond Sci Technol 31(6):065002. https://doi.org/10.1088/1361-6668/aab887

Willering G, Petrone C, Bajko M et al (2018) Cold powering tests and protection studies of the FRESCA2 100 mm bore Nb_3Sn block-coil magnet. IEEE Trans Appl Supercond 28(3):4005105. https://doi.org/10.1109/TASC.2018.2797907

Willering G, Bajko M, Bajas H et al (n.d.) Performance update of the FRESCA2 100mm bore Nb_3Sn block coil magnet. IEEE Trans Appl Supercond submitted

Part IV
Common-Coil Dipole Magnets

Chapter 13
The LBNL Racetrack Dipole and Sub-scale Magnet Program

Steve Gourlay

Abstract After the test of D20, a cos-theta dipole design, the program steered away from resource-intensive magnet development to focus on simple geometries with an emphasis on fundamental technology development. This change in direction stimulated creativity and productivity. During this period 14 magnet tests were completed, a new field record of 14.7 T was achieved, and a new support structure, suitable for high field magnet development was implemented.

13.1 Introduction

The Lawrence Berkeley National Laboratory (LBNL) D20 dipole (see Chap. 6) was successfully tested in early 1997 but, having taken almost six years to complete, was considered a "near-miss" by the US Department of Energy. Taking the lessons learned from the D20 project, the program embarked on a new development path that emphasized simplicity and an incremental approach. The core of this program was based on simple racetrack coils using a double-pancake winding that simplified the lead geometry and avoided internal splices. These coil modules could be powered in a common-coil dipole or quadrupole configuration. Another aspect of the revamped program was the development of a simpler structure better suited to a research and development (R&D) environment. In the course of pursuing a new structure, it was found that small, simply constructed magnets, referred to as "sub-scale" magnets, using racetrack coils, could provide a rapid prototyping platform for dedicated studies. This chapter will describe the design, fabrication, and test results of some examples of the magnets produced by the program from 1998–2003.

S. Gourlay (✉)
Lawrence Berkeley National Laboratory (LBNL), Berkeley, CA, USA
e-mail: sagourlay@lbl.gov

D. Schoerling, A. V. Zlobin (eds.), *Nb_3Sn Accelerator Magnets*, Particle Acceleration and Detection, https://doi.org/10.1007/978-3-030-16118-7_13

13.2 Overview

During this five-year period, the program produced a series of 5 meter-scale magnets and 13 sub-scale magnets with a focus on eventually achieving fields above the range of Nb-Ti. Each of the magnets is briefly described here and will be followed by a more detailed description of design, fabrication, and test results.

RD-2, the first magnet of the new program, was built using 0.808 mm International Thermonuclear Experimental Reactor (ITER) strand manufactured by Teledyne Wah Chang Albany (TWCA). The critical current density (J_c) was 610 A/mm^2 at 12 T and 4.2 K, notably well below the high performance strand now being used. The main purpose was to launch the new program with a quick start while the design of a larger, higher field magnet was underway.

RT-1 was a scaled-up version of RD-2. The two double-pancake coils would become the outer modules for RD-3 when combined with a new, two-layer inner coil module. A similar bolted support structure was used. The main difference was that the horizontal coil spacing was reduced from 40 mm to 9.5 mm. These coils used state-of-the-art (at that time) superconductor with a J_c of 2100 A/mm^2, more than a factor of three greater that the ITER strand used for RD-2. The field level was the absolute limit for a bolted structure.

RD-3 was the first attempt at achieving maximum field and was the first utilization of the "key-and-bladder" concept. The first test (RD 3a) suffered a voltage breakdown during a quench between a heater trace and ground early in the test that destroyed the inner module and one of the outer modules. During the autopsy, it was discovered that a mica slip plane, inserted between the coil modules with the intent of allowing slippage, was actually strongly bonded. It was surmised that if the short had not occurred the magnet would have exhibited extensive training. Upon further investigation, it was found that the wrong type of mica paper was used. Two new coil modules were fabricated and the magnet was reassembled. RD-3b achieved 14.7 T at 4.2 K after some significant training. The next step was to build a new inner module with a bore and move toward accelerator-type field quality (RD-3c). This was the first magnet in this series that used end spacers. Because of the much-increased coil spacing, due to the bore, the field was significantly reduced (but not the complexity).

The sub-scale coil test program utilized small racetrack coils in a simple support structure. The concept was driven by several factors: the need for a less complex, versatile, high field support structure, an ability to perform tests with focused performance goals within a reasonable timeframe, and with a much smaller investment in resources. Performance data derived from these tests were then incorporated into the larger, more demanding model magnets.

Table 13.1 RD-2 cable and design parameters

Parameter	Value
Coil geometry	Two-layer pancakes
Number of turns (turns/coil)	40
Coil inner-radius (mm)	40
Straight section length (mm)	500
Horizontal coil spacing (mm)	40
Vertical bore spacing (mm)	150
Transfer function (T/kA)	0.71 (linear, no iron)
Cable	Rutherford type
Number of strands	30
Cable thickness (mm)	1.45
Cable width (mm)	12.34
Strand diameter (mm)	0.808
Manufacturer	Teledyne Wah Chang Albany
Strand technology	ITER modified jellyroll (MJR)
J_c(12 T, 4.2 K) (A/mm^2)	610 (TWCA)
B_0 (Max, strand) (T)	6.6
B_0 (Max, cable) (T)	5.8
B_0 (Max, achieved) (T)	5.9

13.3 RD-2 Model

As noted above, this magnet was the first step in a program to develop a high field, accelerator quality magnet. The simple geometry, using two flat racetrack coils sandwiched together with a 10 mm bore, was considered compatible with the use of brittle superconductors and necessary for eventually reaching high field levels. In addition, fewer and simpler parts were used in fabricating these coils compared to the more conventional cos-theta cross-section coils.

The designation RD-2 is an anachronism from an earlier naming convention where the number of coil modules, in this case "2," would be followed by a sequence number. Hence, the original name of this magnet was RD-2-01. The nomenclature was abandoned for subsequent models as it was deemed to be too complicated.

The ongoing program for the development and utilization of brittle superconductors for accelerator magnets at LBNL has been focused on coils with a simple racetrack geometry. The ultimate goal of the program was to develop accelerator-quality dipoles with fields up to 15 T. This goal was approached by building a few lower field magnets to demonstrate the feasibility of the design, develop fabrication techniques, and understand relevant performance parameters.

Details of RD-2 have been described previously (Gourlay 1998; Chow et al. 1998a, b; Jackson et al. 2001). An overall summary of the magnet design, fabrication, and test is given here. The main parameters are summarized in Table 13.1.

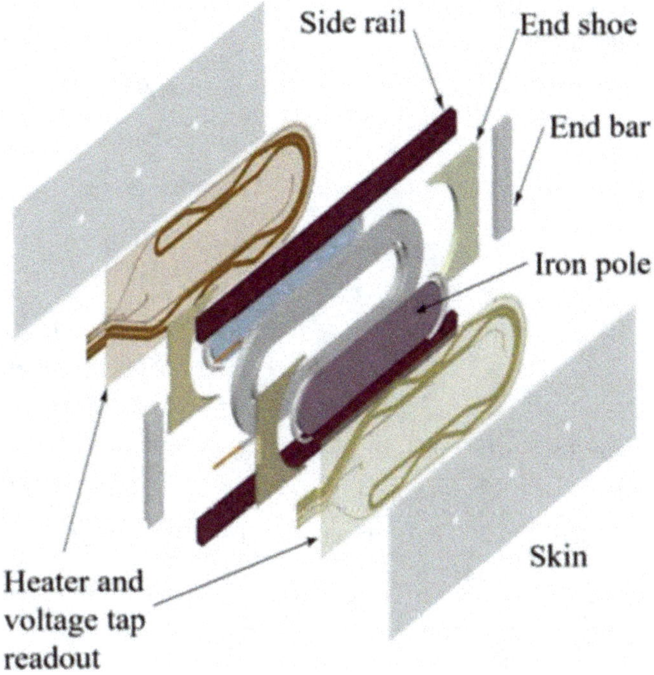

Fig. 13.1 RD-2 coil module

The cable is made from 0.808 mm diameter strand manufactured by TWCA for the ITER project, which had a J_c of about 610 A/mm^2 at 12 T and 4.2 K. Short sample measurements of single strands indicate a bore field of 6.64 T at short sample. A single measurement of a bifilar cable sample gave a lower value of 5.87 T.

13.3.1 Coil Module

The fundamental component of this design is the coil module, which consists of a double-layer coil contained in a support structure. The coil module components are shown in Fig. 13.1.

13.3.2 Support Structure

The magnet structural support was designed for modular coil assembly (Fig. 13.2). End forces and vertical forces (forces in the plane of the racetrack coils) are supported from within the coil module. A vertical pre-stress of 50 MPa is applied through 50 mm thick aluminum-bronze (AlCu) rails running the full magnet length in the coil package, and an end preload of 50 MPa was applied using a series of setscrews loaded against the end shoes. To apply horizontal pre-stress and structural support, the coil packages are sandwiched between stainless-steel clamping bars pulled together by aluminum tension bolts. The horizontal preload is 16 MPa at

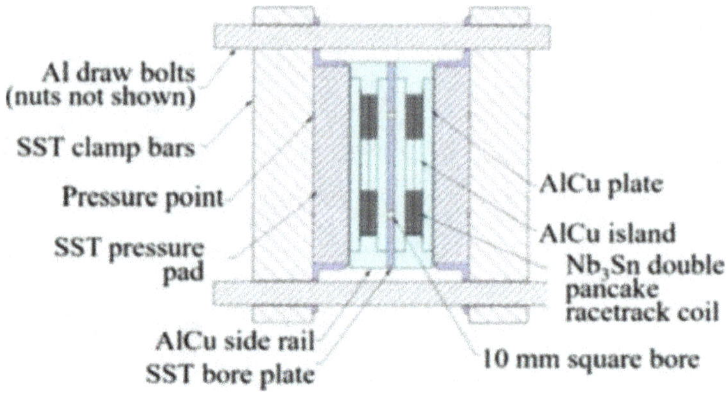

Fig. 13.2 RD-2 structure components

Table 13.2 RD-2 preload combinations

Magnet	Coil preload (MPa)			
	300 K Horizontal	300 K Vertical	4.2 K Horizontal	4.2 K Vertical
RD-2-a	14	50	30	30
RD-2-b	6	50	6	30
RD-2-c	6	21	6	7

room temperature and increases to 30 MPa at liquid helium temperatures. This simple support structure allows easy change-out of coil modules and independent control of vertical and horizontal pre-stress. Several tests of this magnet were done under varying preload conditions.

13.3.3 Fabrication

The double-layer coils were wound around a center island (pole piece) on a flat plate with a ramp between layers to avoid internal splices. All metal parts, which would be in contact with the coil, were made from AlCu in order to survive the high-temperature heat treatment and because of its relatively high heat transfer coefficient compared to other materials such as stainless steel. During the winding process, strips of stainless-steel foil were wrapped around the cable in strategic locations to provide voltage taps. All metal parts were insulated with 0.086 mm thick strips of mica paper to augment the electrical integrity of the coil and provide a parting plane if needed.

13.3.4 RD-2 Test Results

Taking advantage of the flexibility of the RD-2 design, we performed a series of tests with reduced horizontal and vertical preload. Three preload combinations, as summarized in Table 13.2, were tested.

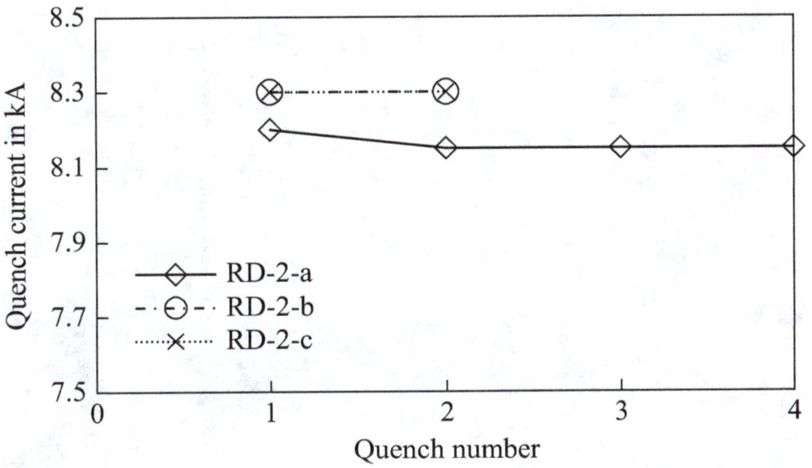

Fig. 13.3 RD-2 training history

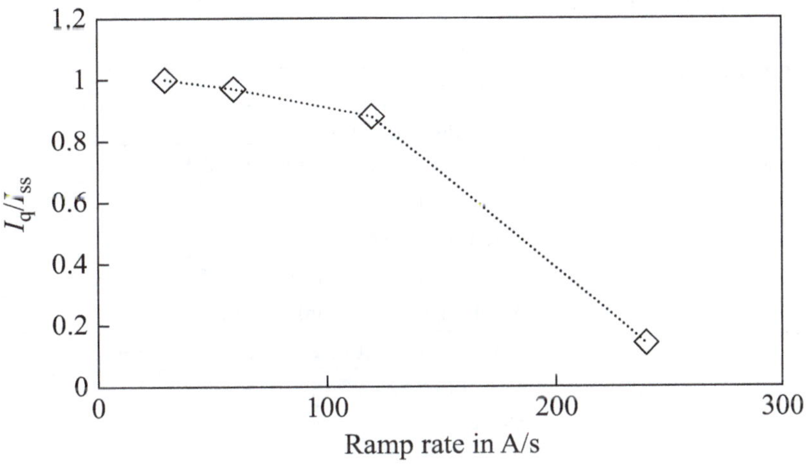

Fig. 13.4 RD-2 ramp rate dependence

The preliminary design was for a 10 mm aperture magnet (40 mm coil spacing) with an emphasis on maintaining the simplicity of the racetrack geometry.

RD-2-a (originally named RD-2-01) required no training and achieved a thermally dependent 4.4 K plateau current of 8.29 kA on the first ramp, slightly above the limit based on the cable measurement. The mechanically modified versions of the magnet (RD-2-b and RD-2-c, originally named RD-2-02 and RD-2-03) performed identically to the original load configuration, despite the sizable differences in loading and loading histories. During the initial cool down, 18 high current quenches were used to establish training, ramp rate, and temperature dependencies. Figure 13.3 shows the spontaneous quench history of the RD-2 series, and in Fig. 13.4 the ramp rate dependence is shown. More details can be found in Gourlay et al. (1999a, b).

Fig. 13.5 RT-1 support structure

Table 13.3 RT-1 design parameters

Parameter	Value
Coil spacing (mm)	9.5
Computed quench field at 4.2 K (T)	12.2
Peak field (T)	12.36
Quench current (kA)	10.5
Straight section length (mm)	500
Number of turns (half magnet)	49
Nominal height of each main coil (mm)	80
Minimum coil bend radius (mm)	70
Vertical bore spacing (mm)	220

13.4 RT-1 Model

With the success of RD-2 demonstrating proof-of-principle, the team at LBNL continued work on a high field, Nb_3Sn racetrack dipole. These next steps systematically led to the construction of a 14 T common-coil dipole.

During the design phase of the 14 T dipole the program took advantage of an opportunity to test the outer coil modules prior to final assembly in RD-3. The incorporation of several new fabrication techniques and the much higher field/stresses that would be encountered in RD-3 encouraged us to take this interim step as a means of substantiating fundamental design assumptions. The highest field, and therefore the highest stresses, were obtained by simply sandwiching the two outer coil modules between a scaled-up version of the bolted support structure successfully used for RD-2 (Fig. 13.5). The RT-1 design parameters are given in Table 13.3.

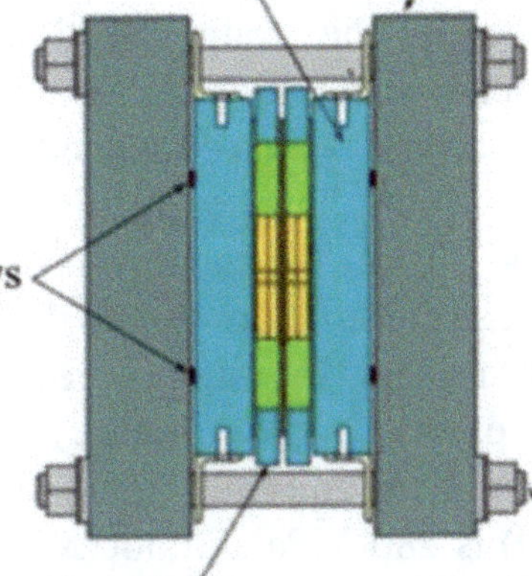

Fig. 13.6 RT-1 coil module support structure. The coil modules are a larger version of the design used by RD-2 shown in Fig. 13.2

13.4.1 Support Structure

Vertical preload and support of the magnet coil windings is completely provided by the coil module structure (Fig. 13.6). To maximize the bore field and provide a more efficient cross-section, the amount of material in the support structure between the bore and the inner conductor and adjacent coil layers must be minimized. This minimization required the use of thin sheets of stainless-steel "skins," replacing the thick AlCu plates used for RD-2 (see Fig. 13.2). The previous method of using fasteners screwed into the AlCu plates was replaced by welding these skins to the module while being loaded hydraulically in the vertical and axial directions. The magnitude of the vertical pre-stress (approximately 43 MPa at 4.2 K) was determined by the requirement that the conductor remain in contact with the support structure interface at all times. The ratios of the vertical and axial preloads were calculated such that the stresses in the end and transition region were as uniform as possible under maximum Lorentz loading.

13.4.2 Assembly

The coil was loaded in steps, alternating between axial and vertical to minimize the development of shear between the coil/skin interfaces. The load was monitored with resistive strain gauges attached to the skins, directly from the press and axial loader cylinder pressures, and by measuring the coil displacement. During cycling the coils were subjected to loads exceeding 100 MPa. Final preload values (after welding) were around 40 MPa in the vertical direction and approximately 20 MPa axially. These values increased by 20–30% on cool-down.

To minimize coil bending, the coil modules were placed between two 60 mm thick stainless-steel pads that were supported by structural beams through a set of "pressure point" keys. The pads and keys ensure minimum bending within the coils regardless of the structural beam deformation. The beams, 114.3 mm thick, were restrained with tie rods across the coils. The beams are longitudinally split into four parts along the straight section of the coils, with two special beams at the ends (compare with Fig. 13.5). Straight section beams were tied with two tie rods on each side, and end beams contained single tie rods (compare with Fig. 13.5). The tie rods are 38.1 mm diameter stainless steel. The bolted support system, while acceptable for the 6 T RD-2 magnet, was marginally adequate for the 12 T coil test. Although in this case the structure served the purpose, it is not considered a viable choice for high field magnets. In order to prevent displacement due to the Lorentz forces, the total force exerted by the structure (pre-stress) would have to exceed the total Lorentz load at maximum current. At full field the maximum stress in the beams and tie rods (cold) is 280 MPa and 300 MPa, respectively. The maximum torque that could be applied to the bolts was 1.02 kN.m, which resulted in a bolt stress of approximately 90 MPa, about a quarter of what is required. To prevent separation during excitation, the required bolt stress would need to be about 360 MPa (warm), exceeding the yield by almost a factor of two. With no pre-stress the coils would separate by approximately 1.9 mm. Based on the applied room temperature pre-stress of the bolts, the measured coil separation was reduced to 1.6 mm.

13.4.3 Test Results

The training history of RT-1 is shown in Fig. 13.7. Two of the first three quenches were associated with runaway ramping of the coils due to noise feeding into the control system. Even though the magnet was in the midst of ramping down at 6.8 A/s during the second quench, voltage tap signals indicated evidence of a motion-induced event. Given that this quench occurred at 79% of short sample, it may be assumed to have had an effect on the subsequent quench behavior. On the fourth quench, the coils reached the limit predicted by the manufacturer's virgin strand

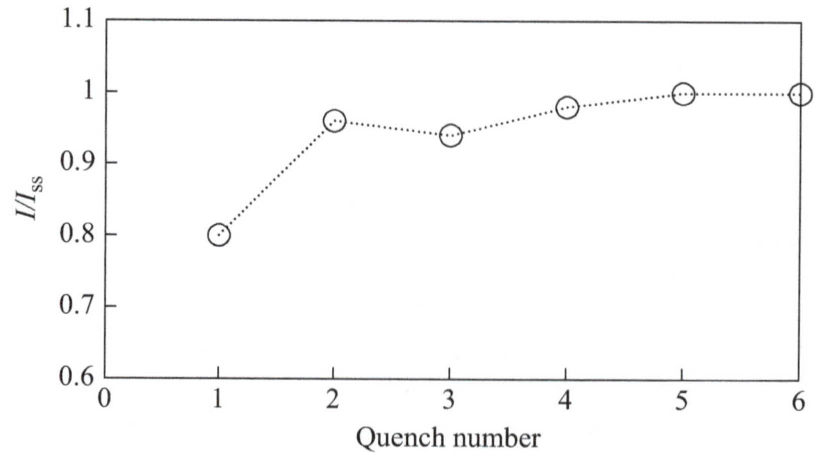

Fig. 13.7 RT-1 training history

Table 13.4 RT-1 strand parameters

Parameter	Value
Cu (%)	59.5
Cu RRR	15
I_c(12 T, 4.2 K) (A)	446
J_c(12 T, 4.2 K) (A/mm^2)	2190
I_c(15 T, 4.2 K) (A)	230
J_c(15 T, 4.2 K) (A/mm^2)	1129
Length (m)	5000
Number of pieces	14
Manufacturer	Oxford Superconducting Technology
Strand technology	MJR

Table 13.5 RT-1 cable parameters

Parameter	Value
Strand diameter (mm)	0.8
Number of strands	26
Thickness (mm)	1.408
Width (mm)	11.338
Length/coil (m)	387

measurement. These values have historically been systematically conservative. Short sample cable measurements predicted a 3.5% higher quench current. Following two more quenches, a plateau corresponding to the cable predicted limit was reached at 10.5 kA, indicating that there was no degradation during cabling or fabrication of the coils. The bore field at this current was 12.2 T (12.36 T peak field on the winding), more than twice that of RD-2. This field level was made possible by the availability of state-of-the-art (at that time) superconductor with a non-Cu J_c, of over 2000 A/mm^2 at 12 T and 4.2 K, a factor of three increase compared to the conductor used for RD-2.

The RT-1 strand and cable parameters are listed in Tables 13.4 and 13.5, respectively. While the conductor non-copper J_c exceeded 2000 A/mm^2, the residual resistivity ratio (RRR) for this wire was approximately 15. This low RRR was caused by nearly total conversion of the Nb barrier and Sn leakage into the copper matrix. At that time, a solution was proposed to increase the RRR closer to a value of 70 by reducing the heat treatment time. One of the possible disadvantages that one might anticipate with such a low RRR is the development of instabilities in the conductor. The coils showed very little training however, and in fact the quench protection was simplified due to the increased quench velocity (about a factor of three over a conductor with an RRR of 50) and subsequently reduced the hotspot temperature in the coil.

13.4.4 Conclusion

The outer coils for RD-3 were tested in a high field/stress configuration and achieved short sample with little training. The new skinning method proved to be an effective way of providing preload, producing very robust coil packages that were able to withstand large displacements without quenching. Within the measurement errors, there was no indication of stress degradation due to either high loading at room temperature (>100 MPa) or during excitation (~50 MPa). With the addition of the inner coil module in the RD-3 configuration, the field in the outer coils would decrease to approximately 10.4 T, but the stress under Lorentz loading will double to approximately 100 MPa. More details on the design, fabrication, and test of RT-1 may be found in Gourlay et al. (2000) and Benjegerdes et al. (2001).

13.5 RD-3 series

The next step toward higher fields was to combine the two coil modules tested in RT-1 with an inner coil module in an attempt to reach fields in the range of 14 T. These coil modules are assembled in RD-3, a common-coil configuration. At the time of the RT-1 test the group still planned to use the wire wrap method to apply the necessary preload. The original design concept is shown in Fig. 13.8.

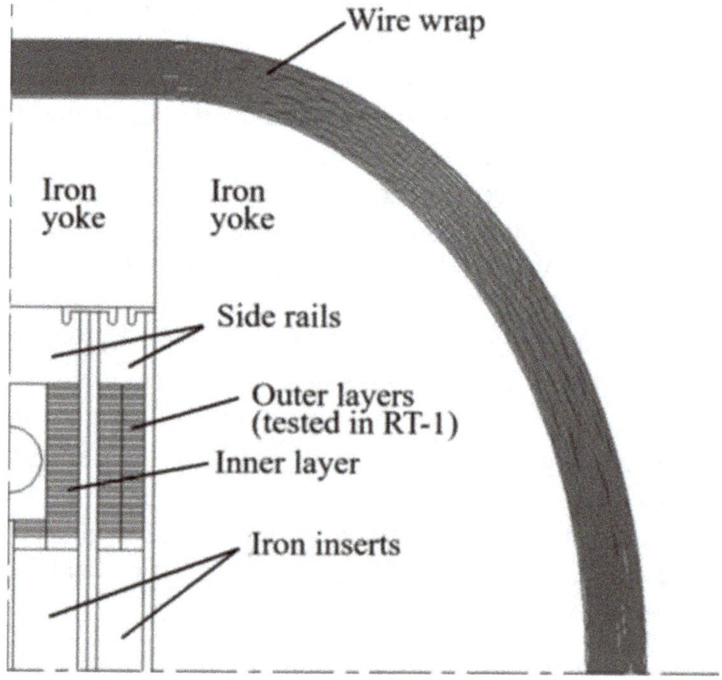

Fig. 13.8 Original concept for RD-3 using wire wrap

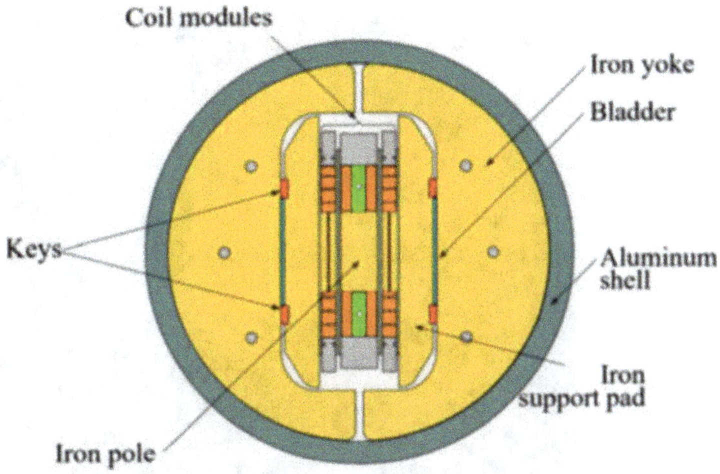

Fig. 13.9 RD-3 cross-section with key-and-bladder support structure

Though the wire wrap method was successfully demonstrated on D20, it was felt by many in the group that the process was much too complicated and time-consuming. So, what were the alternatives? The bolted structure used for RD-2 and RT-1 could not supply the required preload. Coil separation in this case could not be tolerated. In 1997, S. Caspi was exploring the use of epoxy-filled bladders to provide preload, based on a concept successfully used on the Versatile Electron Cyclotron Resonance (ECR) Ion Source for Nuclear Science (VENUS ECR) by Clyde Taylor, former head of the LBNL group. A. Lietzke provided more input on the concept. After more study, S. Caspi presented the idea to S. Gourlay, who suggested using keys to lock in the preload and the use of an aluminum shell instead of stainless steel, which would increase the preload on cool-down as had been successfully demonstrated by the group at the University of Twente for the MSUT magnet (den Ouden et al. 1997). The concept seemed solid but proving it was more complicated. S. Caspi immersed himself in fully developing the idea and quickly realized that a one-third scale model structure would exhibit the same strain levels as full-scale. With the permission of A. Jackson, who was the program head at the time, the team was given two weeks to demonstrate the new concept by building and cold testing an instrumented sub-scale structure model. The new technique (Caspi et al. 2001, Hafalia et al. 2002) was successfully demonstrated and the design for the RD-3 support structure was changed to the so-called key-and-bladder structure shown in Fig. 13.9. Some of the assembly steps are shown in Fig. 13.10(a–c).

13.5.1 Development of the Key-and-Bladder Concept

The large Lorentz forces mean that a cantilever structure is too soft. That effect was demonstrated in RT-1, when the structure allowed the coil halves to separate by more than 1.5 mm at 12 T. The use of a circular shell is more efficient in providing pre-stress that can effectively prevent the coils from separating. The force balance between shell and coils takes place in several steps. Initially the shell pre-stress is set

Fig. 13.10 (**a**) RD-3 coil pack assembly; (**b**) inserting the coil pack into support structure; and (**c**) fully assembled magnet including the LBNL team

Table 13.6 RD-3 magnet parameters

Parameter	Value
Main coil spacing (mm)	25
Computed quench field at 4.2 K (T)	14.6
Peak field, inner layer (T)	14.9
Peak field, outer layer (T)	11.5
Maximum current (predicted) (kA)	10.9
Number of main coil layers	3
Straight section length (mm)	500
Number of turns (half magnet)	50 + 49 + 49
Nominal height of each main coil (mm)	80
Minimum bend radius (mm)	70
Vertical bore spacing (mm)	220
Yoke outer height and width (mm)	300
Cu (%) inner/outer (%)	47.3/57.3
I_c inner 12 T/15 T (A)	540/265
I_c outer 12 T/15 T (A)	382/180

Table 13.7 RD-3 strand parameters

Strand	Inner	Outer
Cu (%)	51.3	59.5
I_c(12 T, 4.2 K) (A)	485	446
J_c(12 T, 4.2 K) (A/mm^2)	1981	2190
I_c(15 T, 4.2 K) (A)	250	230
J_c(15 T, 4.2 K) (A/mm^2)	1021	1129
Length (m) (pieces)	9035 (16)	5000 (14)
Diameter (mm)	0.8	0.8
Manufacturer	Oxford Superconducting Technology	
Strand technology	MJR	

to around 150 MPa by the bladders and keys. During cool-down the stress increases to around 250 MPa due to the relatively large shrinkage of the aluminum shell and remains unchanged during operation. The force on the shell is reacted by the force between the symmetrical halves of the magnet. Most of the reactive force will be carried by the iron pole. The Lorentz force loads the coils and unloads the bore, posts, and side rails. The coil modules will separate only after the Lorentz force overcomes the reactive force, which is not expected to happen below 16 T—well above the short sample field for RD-3.

The integrated double-bore Lorentz force at 16 T is $F_x = 22$ MN/m, $F_y = -3.0$ MN/m, and $F_z = 700$ kN. The ANSYS finite element software (ANSYS Inc., Canonsburg, PA) was used to calculate the magnetic forces and perform the structural analysis. The load case progression followed the steps of assembly, cool-down. The shell pre-stress of RD-3 was designed for an equivalent field of 16 T, providing a sufficient safety margin for the current design field and was later confirmed by strain gauge measurements. Parameters for RD-3 are shown in Tables 13.6, 13.7, and 13.8.

Table 13.8 RD-3 cable parameters

Cable	Inner	Outer
Strand number	40	26
Thickness (mm)	1.418	1.408
Width (mm)	17.159	11.338
Length (m/coil)	210	387

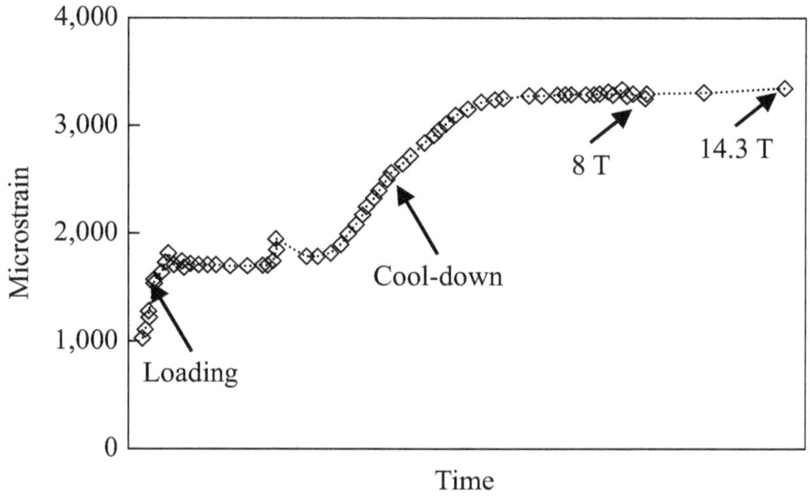

Fig. 13.11 RD-3b pre-stress history, from loading through cool-down and magnet excitation for the first cycle

RD-3 was a Nb_3Sn common-coil dipole designed to exceed 14 T. At that field, the average Lorentz side force is 15.4 MN/m, or a total of 12.0 MN over the 780 mm coil length, acting to push the windings apart. As a means to manage these large forces, a new support structure design was developed for RD-3 that uses inflatable bladders as a temporary internal "press" to load the coil modules inside an aluminum shell. This configuration utilized the best features of the simple cantilevered structure used in the successful test of RT-1, but was intended to eliminate coil displacement.

The bladders, placed between the coil pack and the iron yoke, simultaneously compressed the coil pack to 70 MPa, and tensioned a 40 mm thick structural aluminum shell to 155 MPa. The shell was highly instrumented with strain gauges to record all phases of assembly and testing. Keys were inserted to maintain pre-stress when the bladders were deflated and removed, leaving the shell with 140 MPa of tension. Measurements were compared with finite element analysis using ANSYS to determine the final stress in the shell and coil.

No creep was observed over many days at room temperature. During cool-down, stress in the shell increased to the desired preload of 250 MPa (Fig. 13.11).

This system simplifies magnet assembly substantially, and easily contains the large Lorentz forces without the need for precision parts. The magnet structure reaches maximum stress after cool-down. During operation, reactive forces between the two halves of the magnet are replaced with Lorentz forces within the coils,

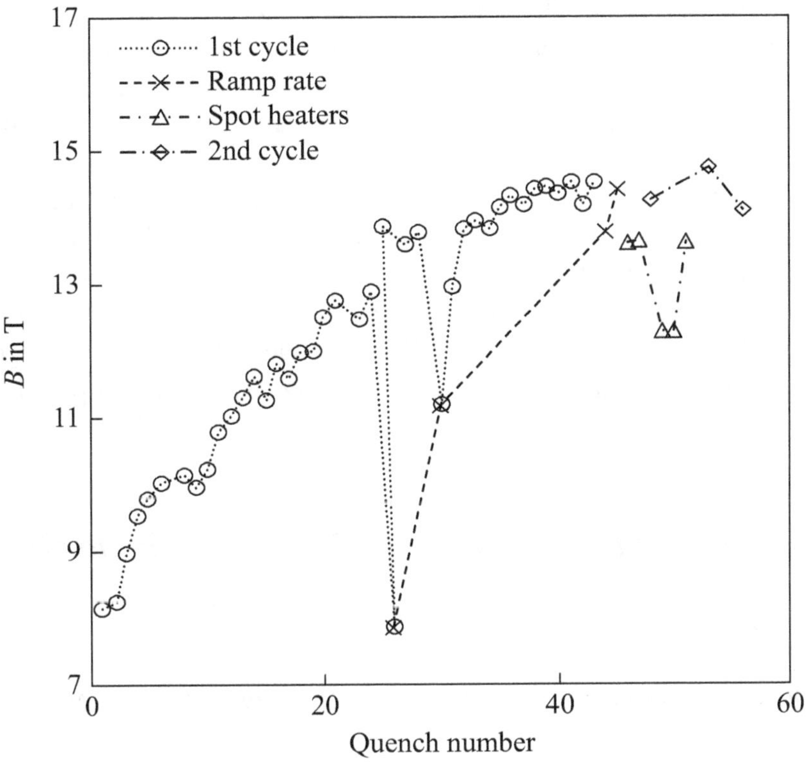

Fig. 13.12 RD-3b quench history

leaving the stress in the structure unchanged. As long as the coil structure is pre-stressed beyond the Lorentz forces, the two coil halves will not separate.

13.5.2 RD-3a, RD-3b

The final assembly of RD-3a was completed in August 2000. During the first ramp, an insulation failure occurred, which resulted in arc damage to two coils. The damaged coils were rebuilt and RD-3b was tested in two test campaigns: one starting in late April 2001, followed by a thermal cycle in early June 2001. All the Nb_3Sn/Nb-Ti joint resistances were measured, and found to be very low ($R < 1$ nΩ). The first training quench occurred at 8.1 T (Fig. 13.12). The magnet trained slowly, exceeding a field of 14 T after 35 quenches. A plateau was reached at an average value of approximately 14.2 T. The quenches at this field level still however, exhibited features typical of motion-induced training quenches, indicating that the magnet may not have been at the short sample limit.

Most of the training quenches below 13.7 T originated in the central (high field) section of the inner coil module. Above this field, most of the quenches originated in the outer coil modules.

After a thermal cycle, the first training quench (Q48) occurred at the previously established plateau of 14.2 T, indicating excellent retention of its previous training. During this test, the magnet reached a bore field of 14.7 T, which is near the short sample critical current limit for both coils. More details on RD-3b can be found in Benjegerdes et al. (2001).

RD-3b showed that Nb_3Sn coils wound in the common-coil (or racetrack dipole) configuration can achieve unprecedented dipole fields (14.7 T) in accelerator magnets. The magnet performed within the range predicted by short sample strand data, and behaved mechanically as predicted by TOSCA (now part of the Opera Simulation Software, Dassault Systèmes) and ANSYS models. There was no degradation in critical current due to cabling or due to the Lorentz loads during operation.

13.5.3 RD-3c: Continuation of the RD Series—Field Quality

13.5.3.1 Magnet Design Features

After developing a coil geometry and support structure that adequately supports the coil, the next step was to explore cross-section designs that provided accelerator-grade field quality. Racetrack cross-sections that have good magnetic field quality are intrinsically more complex, and the simple cross-sections will need to be augmented. Efficient magnetic field generation also requires that the conductor be as close as possible to the bore. A series of designs was planned that progressively implemented field-quality features, starting with the simplest. The first (and as it turned out, the last) in this series, RD-3c, was successfully tested in the spring of 2002.

RD-3c was envisioned as an economical test of the ability to design, build, and measure a reliable, accelerator-quality, common-coil magnet. Accordingly, the following constraints were imposed.

1. Reuse RD-3b's outer coils, yoke, and shell structure.
2. Reduce the associated large field errors to accelerator levels with the simplest harmonic-correction coils.
3. Make no effort to correct either the end, or up-down asymmetric harmonics (best corrected by changes in the iron).
4. Design and fabricate the correction coils from single-layer, flat racetrack coils with only one spacer per layer.
5. Package the correction coils as a double-layer insert coil-module, with no internal splices.
6. Provide a mid-plane "bore-plate" that would allow the insertion of a 25 mm outer diameter (OD) warm, rotating-coil probe.
7. Fabricate and assemble the insert module in a manner that might improve the training rate (compared to that of RD-3b).
8. Apply diagnostics that could better localize conductor motions.

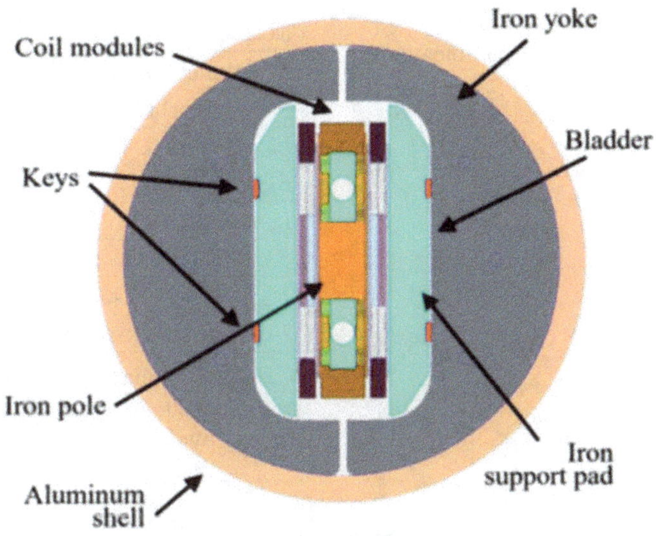

Fig. 13.13 RD-3c cross-section and coil configuration

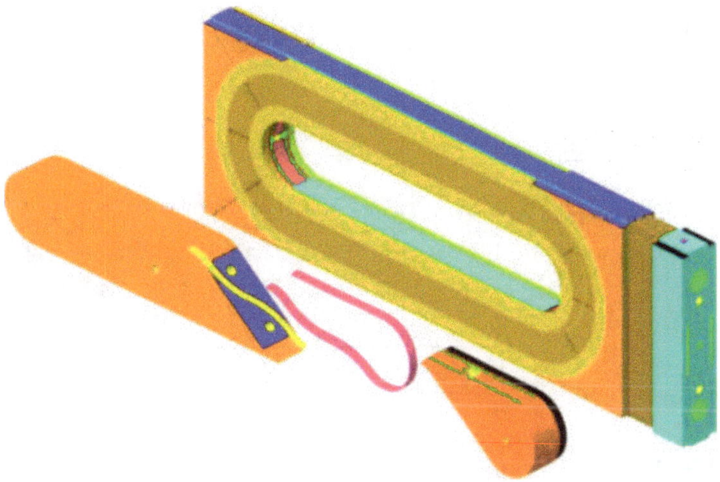

Fig. 13.14 RD-3c inner coil module showing the "s-bend" transition between coils across the bore

Figure 13.13 shows the magnet cross-section that resulted from the above constraints. The harmonic-correction insert module was clamped between RD-3b's outer coil modules.

As in RD-3b, an AlCu bore-plate separated two single-layer insert coils. An internal S-bend ramp supplied the required current reversal across the bore-plate (avoiding an internal splice) (Fig. 13.14).

Each coil used a wide coil-spacer near each bore-hole to counteract the large positive sextupole from the outer coils and each coil layer had 16 turns in two equal blocks. The bore-plate was thick enough (39.5 mm) to avoid excessive deformation of the bore-hole, and allow access for a 35 mm OD anti-cryostat. The insert module was reacted and encapsulated as in previous coil modules. It was not, however, "exercised" in a press before welding. Final pre-stress, applied by bladders between iron pads and the yoke, was maintained via iron keys (as before for RD-3b). Due to the aluminum-iron containment system, the average 300 K coil face pre-stress (~65 MPa, scaled from the measured Al-shell stress) nearly doubled during cool-down to the 4.4 K operating temperature. RD-3c's performance parameters are compared

Table 13.9 RD-3c performance parameters

Magnet	RD-3c[a]	RD-3b
I_{max} (inner) (kA)	13.3	10.8
B_{max} (inner) (T)	13.1	14.8
I_{max} (outer) (kA)	11.9	10.8
B_{max} (outer) (T)	11.3	11.5
I_{ss} (kA)	11.9	10.8
$B_{0\text{-}ss}$ (T)	10.9	14.7
$J_{Cu\text{-}ss}$ (inner) (kA/mm^2)	1.6	1.1
$J_{Cu\text{-}ss}$ (outer) (kA/mm^2)	1.6	1.5

[a]Calculated values. Training was aborted before maximum current was reached, to assess training improvements in a subsequent thermal cycle

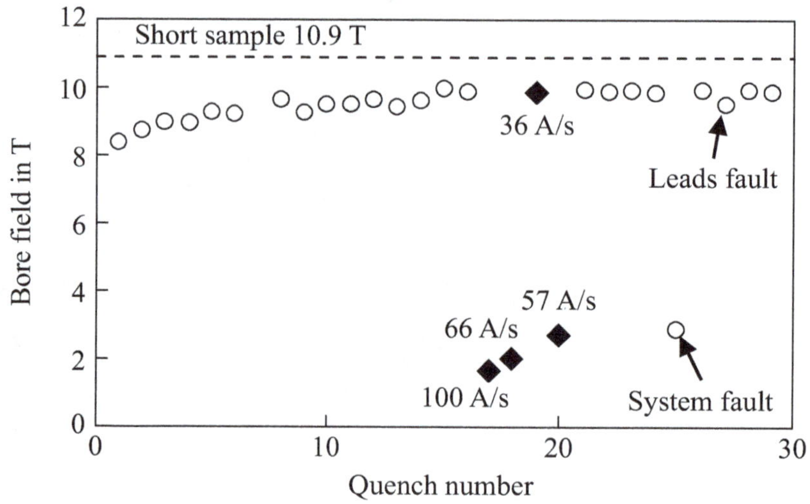

Fig. 13.15 RD-3c training history

with RD-3b in Table 13.9. RD-3c's insert coil used Rutherford cable with Oxford Superconducting Technology strand, similar to that used in RD-3b's high field insert coil (Tables 13.7 and 13.9).

13.5.3.2 RD-3c Test Results

Magnet training began after the magnet protection system was operating reliably at 2 kA. The magnet was ramped to quench with a variety of ramp rates while being cooled by a (4.4 K) two-phase LHe bath. Training quenches (Fig. 13.15) started at 77% of the un-degraded short sample limit but was pursued to only 92%, where the magnet appeared to reach a plateau. These quenches exhibited large voltage spikes indicative of conductor instability. At this point ramp rate studies were initiated as shown in Fig. 13.15. It should be noted that only two of the training quench origins were in the virgin insert-module. The other quenches were equally distributed between the two outer modules, most starting simultaneously in the multi-turn segments of both high and low field layers, near the straight-to-curved end transition. These quenches were preceded by what was interpreted as a flux jump instability approximately 1 ms before resistance growth.

The RD magnets proved to be an excellent vehicle for the development of a number of important aspects of high field magnet design. But for very high field R&D dipoles, a common-coil configuration (two bores in one coil) results in significantly greater challenges than a single-bore configuration for the same field and aperture. Not only are the usage of conductor and the magnetic forces doubled because of the second bore, but the magnet needs to be relatively longer because of the space occupied by the large radius ends. The diameter of the iron yoke also had to be increased in order to shield the opposite fields generated in the two apertures. All these necessities result in increased demand for expensive high-performance conductor and structural materials, and increases the requirements on facilities, for example the size of reaction furnaces, impregnation fixtures, and cryostats, and the cost of testing. For these reasons, the program started considering alternative solutions that would allow them to push the field limit in configurations that are relevant to accelerators and would extend the performance envelope even further.

13.6 Sub-scale Magnet Program

It was quickly realized that the sub-scale mechanical structure used to demonstrate the utility of the key-and-bladder concept would make an excellent R&D vehicle for a large number of parametric studies at a much higher rate and lower cost (Hafalia et al. 2003). In previous common-coil racetrack configurations, the cable was wound around an iron pole-island, and separate side rails or rectangular bars were used to compress the straight sections of the coil layers to a prescribed load. The coil ends were supported by separate end shoes. A coil is compressed at least four times — for reaction, potting, skinning, and final assembly. To further simplify the assembly the sub-scale magnet design incorporated a stainless-steel U-shaped coil-support system (Fig. 13.16) where the return-end shoe and the side rails are integrated into one component. The thickness of the shoe was slightly larger than the cable width to prevent direct contact with the fragile conductor. The small gap was filled with epoxy during impregnation.

A traditional end shoe is used at the lead-end. A push-block is used to axially pull against the horseshoe while simultaneously pushing on the lead-end shoe in a bootstrap fashion. In the past, there were undesirable situations where the feathered ends of the end-shoes deformed the outer turn of the coil. Half of these "pinch-points" were eliminated with this new design. Machined to the same thickness as the dual layers, the horseshoe also provides a continuous sealing surface for epoxy impregnation—improving containment of the epoxy within the coil region during potting. The same horseshoe remains with the assembly from reaction through impregnation, becoming an integral part of the final coil module. A completed sub-scale magnet is shown in Fig. 13.17.

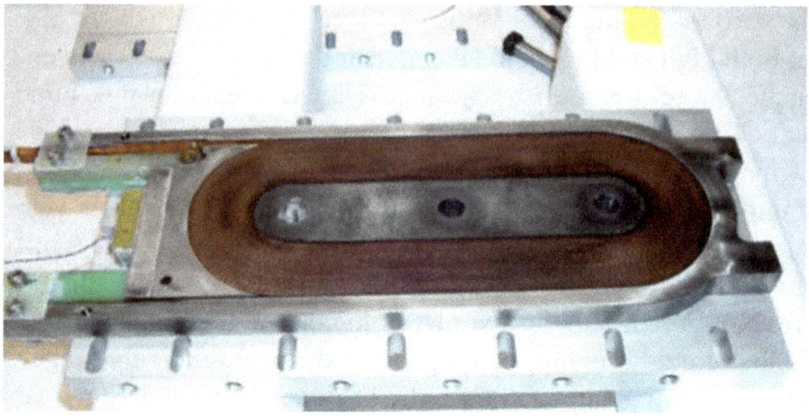

Fig. 13.16 Sub-scale coil module

Fig. 13.17 Sub-scale test module

Since the coils were interchangeable, the philosophy was to use one "base-line" coil and one test coil. The base-line coil was a standard coil that, in most cases, had already been tested. The coils were assembled using the newly developed key-and-bladder technique.

The sub-scale magnet program was governed by the following guidelines:

1. Simplicity, encompassing design and fabrication processes to facilitate iterative designs;
2. Rapid production with maximum information return;
3. Vary parameters to learn how to build low-cost, reliable coils for full-size magnets;
4. Test ideas that are too risky or expensive for full-size magnets;
5. Single-parameter tests;
6. Effectively utilize resources (both personnel and materials).

Table 13.10 General sub-scale coil parameters

Parameter	Value
Number of strands	20
Strand diameter (mm)	0.7
Cable width (mm)	7.8–8.0
Cable thickness (mm)	1.3
Pitch angle (°)	16–17
Packing factor	0.83
Insulation (mm)	0.15

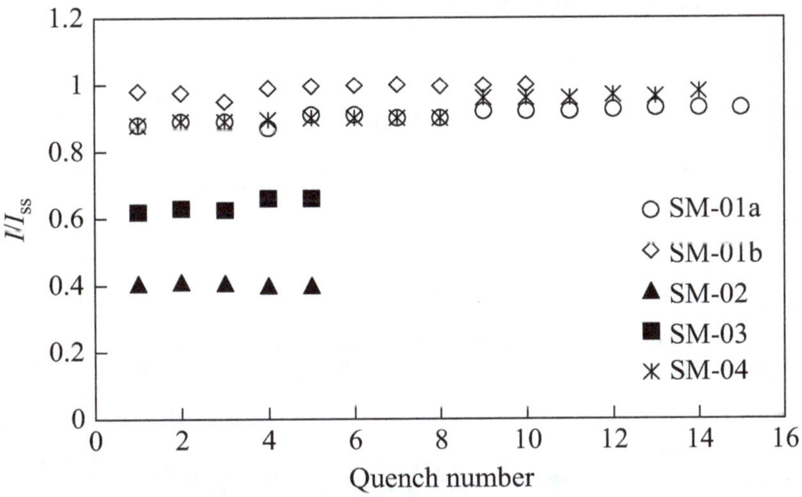

Fig. 13.18 Quench performance of SM-01, 02, 03, and 04

Over the course of the program, 14 coils were produced. The main parameters are shown in Table 13.10.

13.6.1 SM-01

The first pair of coils was tested in December 2001 and exhibited little training. One coil, SC-01, was highly loaded transversely in a press with a stainless-steel skin welded over it to preserve the pre-stress. A second coil, SC-02, was assembled with low pre-stress. The first test, designated SM-01a, was on a highly loaded coil pack with a shell stress of 103 MPa at room temperature. After cool-down, the shell structure generated 248 MPa at 4.2 K. The magnet was warmed to room temperature and the preload was significantly reduced. After cool-down, the shell stress was 90 MPa. SM-01b started training at a higher current and quickly reached 100% of the short sample limit based on the virgin strand, see Fig. 13.18.

A summary of coil/magnet test combinations and performance is shown in Table 13.11. Strand and cable parameters are given in Table 13.12.

Table 13.11 Sub-scale coil performance summary

Magnet	Coils used	Current at first quench (A)	Highest quench current (A)
SM-01a	SC-02, SC-01	8925 (89%)	9391 (94%) Q13
SM-01b	SC-02, SC-01	9701 (97%)	9914 (99%) Q11
SM-02	SC-02, SC-03	3801 (40%)	4005 (42%) Q02
SM-03	SC-02, SC-06	5560 (63%)	6101 (69%) Q29
SM-04	SC-02, SC-08	8776 (88%)	9789 (98%) Q32
SM-05	SC-01, SC-10	8644 (98%)	8757(101%) Q5

Table 13.12 Sub-scale strand and cable parameters

Coils	Number of strands	Cu/non-Cu	I_{ss} (A)	B_{peak} (I_{ss}) (T)
SC-01, SC-02, SC-08	20[a]	46.5/53.5	10,010	11.8
SC-03	14 + 7 (Cu)[†]	41/59 + pure Cu strands	9,429	11.4
SC-06	14 + 7 (Cu)[a]	46.5/53.5 + pure Cu w/ss core	8,851	10.5
SC-10	20[a]	60/40	9,101	10.6

[a]Manufacturer: Oxford Superconducting Technology (OST), Nb_3Sn technology: MJR; [†]Manufacturer: Intermagnetics General Corporation (IGC), Nb_3Sn technology: internal tin

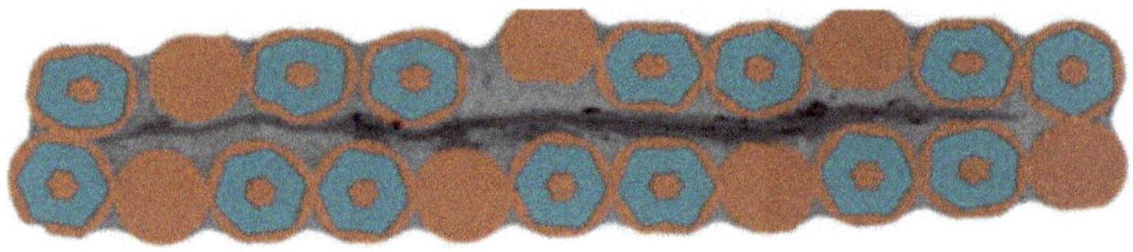

Fig. 13.19 Mixed-strand cable with stainless-steel core

13.6.2 *SM-02 and SM-03—Mixed Strand*

Two coils made of mixed-strand cables consisting of 14 Nb_3Sn strands and seven strands of pure copper were manufactured and tested. One, SC-06, used mixed strand cable with a stainless-steel core, shown in Fig. 13.19.

They were tested as SM-02 and SM-03, respectively. Each of the mixed strand coils was combined with the best-performing coil from SM-01 (the low pre-stress coil SC-02). Because of the difference in elastic modulus, the mixed strand cable was mechanically unstable and proved very difficult to wind, even with a much lower winding tension (89 N compared to the usual 178 N). It was discovered in the post mortem that the mixed strand cable in SM-02 was severely damaged near the lead. SM-03 performed better (Fig. 13.18), though indicating that there might still have been some damage. One should also note that, compared to the coil wound with non-mixed strand, each superconducting wire had to carry a larger current, and the critical current is likewise reduced to the ratio of Nb_3Sn strands in the two cables:

14/20 = 70%. Moreover, the large difference in thermal contraction between the Nb_3Sn and copper strands may lead to damage of the conductor during cool-down.

13.6.3 SM-04—Composite Technology Development (CTD)/ Fermi National Accelerator Laboratory (FNAL) Ceramic Insulation

In SM-04 the standard glass cable insulation was replaced by a sleeve of ceramic fibers. The performance was comparable to the best of the standard versions (Fig. 13.18).

13.6.4 SM-05

Another example of the use of sub-scale coils and magnets was a test designed to investigate possible thermal shock effects on Nb_3Sn performance, which became the thesis topic of a Fermi National Accelerator Laboratory (FNAL) Ph.D. student (Imbasciati et al. 2004).

Sub-scale magnet SM-05 was tested in February–March 2003. The primary purpose of this test was to evaluate the degradation of Nb_3Sn cable under repetitive applied thermal stresses. During a quench in a superconducting magnet, parts of the coils can reach very high temperatures if the proper protection measures are not taken. Nevertheless, even in the case of actively protected magnets, it is necessary to determine the temperature and voltage levels that can be sustained by the magnet parts. An absolute upper temperature limit is given by the melting point of solder (~500 K), since the quench might start near the conductor joints. For impregnated coils, a second limit could be the glass transition point of the insulation, which occurs at about 400 K for epoxy resins. At that temperature, the epoxy becomes soft and, even if the transition is reversible, the changes in electrical properties increase the probability of a short circuit. In the case of magnets using Nb_3Sn superconductor, an additional complexity is introduced via the brittleness of Nb_3Sn, which could suffer permanent degradation under excessive stress.

The goal of this test was to reproduce as realistically as possible the thermo-mechanical conditions in a cable during a magnet quench and determine the effects of high peak temperature on magnet performance.

To simulate a high thermal stress situation in the magnet, one of the coils (SC-10) was instrumented with a spot-heater in the middle of the winding in a high field region with two voltage taps across the spot-heater section. The magnet was trained until a quench plateau was reached.

At a current below the quench current, a quench was initiated with the spot-heater. The quench was left propagating along the cable instead of immediately

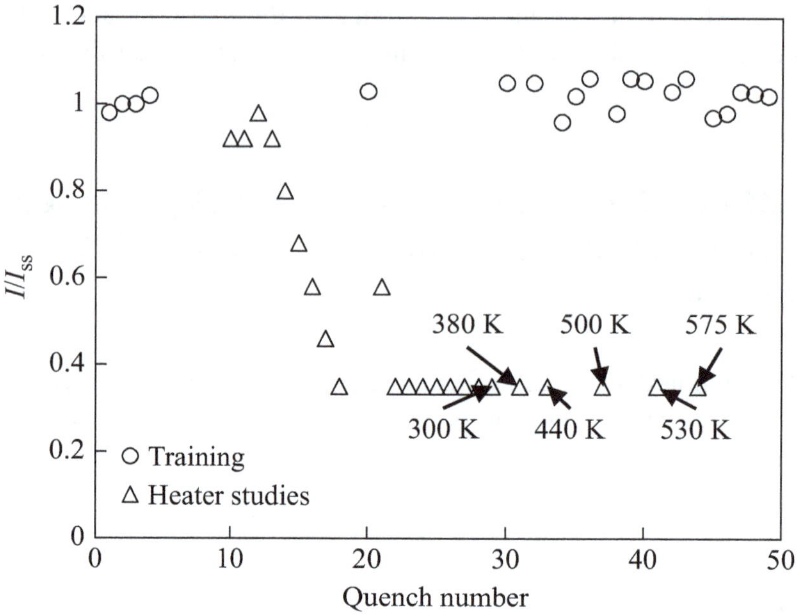

Fig. 13.20 Quench history for SM-05

switching off the current, using a pre-defined delay. The current in the normally conducting matrix allowed a well-defined amount of heating of the cable. The normal conducting zone propagates in the coil with a temperature profile that goes from the peak temperature of the starting point, to the bath temperature in other regions of the coils and in the supporting structure. The temperature gradients that are created in this fast process can induce thermo-mechanical stress. Repeated measurements of the quench current of the coils, after each excursion to high temperature, allow assessment of the critical current degradation as a function of the peak temperature during a quench.

13.6.4.1 Training Summary

The quench history is summarized in Fig. 13.20. The first quench was at 98% of the calculated short sample, which was based on virgin strand measurements. Maximum coil temperatures are indicated for the spot-heater-induced quenches. All training quenches occurred in the virgin coil, SC-10. The ramp rate dependence is shown in Fig. 13.21.

Several spot-heater-induced quenches were performed before reaching the desired temperature. The ramp rate sensitivity of SC-10 is very similar to coils previously tested. The magnet trained quickly and it was possible to induce quenches with the spot-heater after five ramps. The current at which quenches were induced varied from 8 kA down to 3 kA. During the first thermal cycle it was not possible to reach temperatures higher than 300 K due to the 60 V limit of the power supply system. Once the limit is reached, even if the extraction system has not switched off

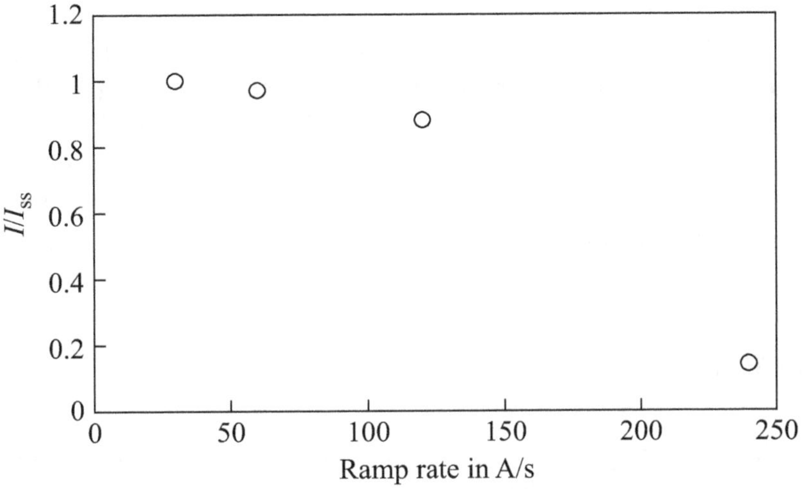

Fig. 13.21 Ramp rate dependence of SM-05

the power supply, the current starts dropping rapidly and heat propagates quickly inside the whole magnet, reducing the peak temperature values. Before starting a second cool-down, the power supply voltage was doubled to 120 V and the time delays were increased up to 3.5 s. With these long time delays it was possible to introduce the extraction system at a very low current level and collect enough heat locally. After reaching temperatures higher than 350 K the magnet was ramped to normal quench in order to observe any degradation effects. Some degradation was observed after the temperature reached close to 400 K. The degradation was not permanent, and after three quenches the current reached previously recorded values. Only after the last spot-heater event with a delay of 3.5 s and a temperature close to 600 K was a permanent degradation observed. The following points can be made.

1. As observed in previous magnets (RT1, SM-01b) the training is significantly reduced if the pre-stress applied to the magnet during assembly is as low as possible.
2. The thermal stresses applied to the magnet during the spot-heater studies did not reduce the performance of the magnet until a final temperature of 590 K was reached. After this event, the current did not recover its previous values, and a degradation of 3% was observed.
3. The thermal stresses applied with peak temperatures below 500 K seemed to improve the performance of the magnet.

13.6.5 SD-01

SD-01 was developed as a platform for training studies as a collaboration between Commissariat à l'Energie Atomique (CEA) in Saclay, France and LBNL. The two coils (SC-01 and SC-02) were powered in a dipole configuration to better mimic the Lorentz conditions of interest. The sub-scale support structure was designed to allow variable vertical, horizontal, and axial preloads, a more complex departure from the simple structure used for the common-coil configurations of the SM series. The first quench was at 98% of the short sample limit, and the magnet trained quickly to a stable plateau. Based on the agreement of strain measurements with values predicted by the model, the test successfully demonstrated the reliability of the new platform as a basis for further training studies. More details on the design can be found in Felice et al. (2007). A comparison of the performance of the LBNL magnets up to this period in the program can be found in Chiesa et al. (2003).

13.7 Summary

With an emphasis on simplicity, this six-year period of the program was essential for developing the tools and confidence needed to continue the quest for higher fields. Along with a new and versatile support structure for high field magnets and the creation of a new program for rapid technology development, the group also implemented full 3D mechanical analysis that has become an essential tool for high field magnet design.

The R&D program, focusing on the common-coil design, was carried to a reasonable end point with RD-3c. These series of tests indicated quite clearly the difficulties incurred by going from a simple flat-coil geometry to a design that included field-quality features (auxiliary coil structures) and a reasonable bore diameter. Continued work to find ways of mitigating these challenges warrants consideration. A few more magnets of this type may have yielded a definitive answer on the viability of this design for future high field accelerator magnets, but due to finite resources and the need for the LBNL program to pursue its primary mission of pursuing accelerator magnet technology to higher fields, the group moved on to exploring block magnet designs.

The sub-scale magnet component of the program proved to be extremely productive in many ways: effective use of resources, both in terms of cost and manpower, the opportunity to perform focused experiments in a short timeframe, and the development of new diagnostics and instrumentation. This R&D paradigm allows engineers and technicians to exercise creativity in an environment that allows them to make the mistakes that are necessary for learning.

At the end of this period, due to dwindling support, the group was forced to abandon this element of the program. Based on our experience, it is quite clear that some form of this approach should be part of every successful program.

References

Benjegerdes R et al (2001) Fabrication and test of Nb_3Sn racetrack coils at high field. IEEE Trans Appl Supercond 11(1):2164–2167. https://doi.org/10.1109/77.920286

Caspi S, Gourlay S, Hafalia R et al (2001) The use of pressurized bladders for stress control of superconducting magnets. IEEE Trans Appl Supercond 11(1):2272–2275. https://doi.org/10.1109/77.920313

Chiesa L, Caspi S, Coccoli M et al (2003) Performance comparison of Nb_3Sn magnets at LBNL. IEEE Trans Appl Supercond 13(2):1254–1257. https://doi.org/10.1109/tasc.2003.812651

Chow KP, Dietderich DR, Gourlay SA et al (1998a) Mechanical engineering and design of magnet RD-2-01. LBNL-42236; SC-MAG-630, DOE contract number AC03-76SF00098, OSTI identifier 5201571, Lawrence Berkeley National Laboratory, Berkeley

Chow K, Dietderich DR, Gourlay SA et al (1998b) Design and fabrication of racetrack coil accelerator magnets. LBNL-42507, SC-MAG-642, Lawrence Berkeley National Laboratory, Berkeley

Den Ouden A, Wessel S, Krooshoop E et al (1997) Application of Nb_3Sn superconductors in high-field accelerator magnets. IEEE Trans Appl Supercond 7(2):733–738. https://doi.org/10.1109/77.614608

Felice H, Caspi S, Dietderich DR et al (2007) Design and test of a Nb_3Sn subscale dipole magnet for training studies. IEEE Trans Appl Supercond 17(2):1144–1148. https://doi.org/10.1109/TASC.2007.898347

Gourlay SA (1998, June) Racetrack coil model fabrication notes, SC-MAG-621

Gourlay SA, Chow K, Dietderich DR et al (1999a) Fabrication and test results of a Nb_3Sn superconducting racetrack dipole magnet. SC-MAG-628, LBNL-41575. https://escholarship.org/uc/item/9qt824d9

Gourlay SA, Chow K, Dietderich DR et al (1999b) Fabrication and test results of a Nb_3Sn superconducting racetrack dipole magnet. In: Luccio A, MacKay W (eds) 1999 particle accelerator conference, New York, 27 March–2 April 1999. IEEE, Piscataway, NJ, pp 171–173

Gourlay SA, Bish P, Caspi S et al (2000) Design and fabrication of a 14 T, Nb_3Sn superconducting racetrack dipole magnet. IEEE Trans Appl Supercond 10(1):294–297. https://doi.org/10.1109/77.828232

Hafalia RR, Bish PA, Caspi S et al (2002) A new support structure for high field magnets. IEEE Trans Appl Supercond 12(1):47–50. https://doi.org/10.1109/tasc.2002.1018349

Hafalia RR, Caspi S, Chiesa L et al (2003) An approach for faster high field magnet technology development. IEEE Trans Appl Supercond 13(2):1258–1261. https://doi.org/10.1109/tasc.2003.812632

Imbasciati L, Bauer P, Ambrosio G et al (2004) Study of the effects of high temperatures during quenches on the performance of a small Nb_3Sn racetrack magnet. Supercond Sci Technol 17(5): S389–S393. https://doi.org/10.1088/0953-2048/17/5/060

Jackson A et al (2001) Supercon's racetrack magnet project. Archived at https://cds.cern.ch/record/2638136?ln=en

Chapter 14
Common-Coil Nb_3Sn Dipole Program at BNL

Ramesh Gupta

Abstract This chapter summarizes the common-coil dipole research and development program at the Brookhaven National Laboratory (BNL). The program goals included: (a) the development of accelerator-quality dipoles based on the common-coil design concept; (b) the demonstration of "react-and-wind" technology for high field collider dipole magnets; and (c) the development and demonstration of a novel background field test facility with a large open space.

14.1 Introduction

The common-coil design concept (Gupta 1997) is a conductor-friendly design, based on a simple coil geometry with large bend radii, and is particularly suitable for brittle superconductors. The common-coil geometry is considered to be technically attractive for high field magnets as it puts lower strain on the conductors with smaller support structures as compared to other designs. The common-coil design is also expected to produce lower cost magnets in large volume in industry since it allows the use of less expensive production techniques (because of the simpler geometry), since the number of coils required is halved (as the same coils are shared between two apertures), and since it has reduced structural requirements.

Several magnets based on the common-coil dipole designs had earlier been proposed for the Very Large Hadron Collider (VLHC) in the USA (Fermilab 2001). The common-coil design has also been used in the present proposal for a Super proton–proton Collider in China (Wang et al. 2016), and is one of the design options under consideration for the proposed Future Circular Collider (Tommasini et al. 2017) at the European Organization for Nuclear Research (CERN). The basic design concept has been extensively studied (Gupta et al. 1999, 2007; Ambrosio et al. 2000; Sabbi et al. 2000) and demonstrated at Brookhaven National Laboratory (BNL), Fermi National Accelerator Laboratory (FNAL) and Lawrence Berkeley National Laboratory (LBNL) with a series of magnets (see this part of this book).

R. Gupta (✉)
BNL (Brookhaven National Laboratory), Brookhaven, NY, USA
e-mail: gupta@bnl.gov

D. Schoerling, A. V. Zlobin (eds.), *Nb_3Sn Accelerator Magnets*, Particle Acceleration and Detection, https://doi.org/10.1007/978-3-030-16118-7_14

A similar configuration had also previously been proposed independently by Danby for low field dipoles (Danby et al. 1983).

BNL has developed, fabricated, and tested a successful common-coil dipole model called DCC017 based on the react-and-wind (R&W) technique. This work is described in detail in this chapter.

The common-coil design facilitates modular geometry, which is good for combining coils made with different types of conductors in both research and development (R&D) magnets and in large-scale production of hybrid magnets. The existing common-coil magnet DCC017 at BNL has created a unique lower cost and fast turnaround magnet test facility. This new type of test facility allows the testing of coils with a variety of parameters in a 10 T background field without disassembling the DCC017 magnet. Given the fast turn-around and lower cost, this R&D can be both "high risk, high reward" and "systematic," and is likely to introduce a new way of doing high field magnet R&D.

14.2 Common-Coil Dipole Design Concept

14.2.1 Design Concept

The common-coil magnet design (Gupta 1997) is a two-in-one (also known as twin) block design that is primarily based on flat racetrack coils that are shared between two apertures. The main coils are common to both apertures, hence the name "common-coil dipole design." A schematic of the common-coil design for the main coils is shown in Fig. 14.1. A set of coils are placed on the left- and right-hand sides of the two vertically arranged apertures to produce magnetic fields in opposite directions. The bending radius in the ends is much larger than that in a conventional dipole design, as it is determined by the separation between the two apertures rather than the size of the aperture itself.

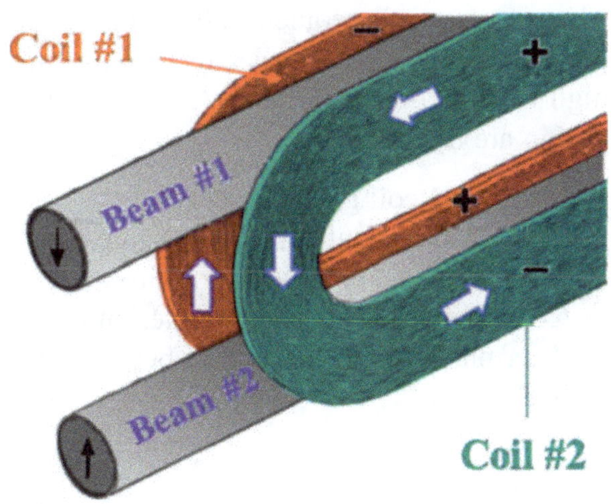

Fig. 14.1 Main coils of the basic common-coil dipole design concept

In addition to the main coils, pole coil blocks (see Sect. 14.3) are also needed to achieve the field quality required in accelerator magnets. These pole coil blocks may or may not have the same simple geometry as the main coils.

14.2.2 Common-Coil Mechanics

For very high field magnets, the magnet mechanics is a key component that determines the technical performance and cost of the magnet. In this respect, the common-coil geometry offers very different mechanics to maintain the large horizontal forces: The individual coils move as a whole, which minimizes the internal strain and stress on the conductor, particularly in the critical end region. Contrarily, in the cos-theta and conventional block designs, large horizontal forces create excessive stress/strain on the conductor in the end region. The BNL common-coil dipole tolerated about 0.2 mm displacement, which is significantly larger than the typical 0.1 mm allowed in cos-theta magnets. This larger allowed displacement, in principle, reduces the need for a large support structure, as long as the negative effects on the field quality are within acceptable limits.

Therefore, lower cost magnets due to a smaller structure and better performance due to less strain in the conductor could be expected. These features could be key for high field magnets where the magnet's structure is a major technical and cost issue. Moreover, the common-coil designs may also offer simpler stress management, if needed, since the layers of coils are stacked horizontally, and therefore planes for intercepting the Lorentz forces can be introduced.

14.2.3 Potential Advantages and Challenges of the Common-Coil Design

The common-coil geometry above described without auxiliary coils is expected to produce a lower cost, easier to manufacture, and technically attractive design for high field two-in-one dipoles. Several anticipated advantages of the common-coil design are listed below:

- Simple 2D coil geometry;
- Fewer coils (about half) as the same coils are common between the two apertures (two-in-one geometry for both iron and coils);
- Conductor-friendly with simpler ends and much larger bending radii, which are determined by the separation between the two apertures rather than the aperture itself;

- Additional technology option of R&W in addition to a wind-and-react (W&R) approach, especially for the main coils;
- Additional material options for insulation and coil material as, in the R&W approach, the coil does not go through a high-temperature reaction;
- More automated manufacturing options may be possible in large-scale production because of the simpler geometry;
- Lower internal strain on the conductor under Lorentz forces as the coils move as a unit;
- Savings from less support structure as much larger deflections are accepted.

The challenges with the common-coil design are listed below:

- Since the common-coil design is only applicable for two-in-one collider dipole geometry, it is a non-suitable design if only one aperture is needed.
- For two apertures, one needs about twice the amount of conductor as compared to that needed in single-aperture cos-theta or block dipole geometries, which could be a significant issue in R&D programs with limited budgets.
- The anticipated advantages of common design, listed above, are yet to be demonstrated.
- Pole coil blocks have to meet the field-quality requirements for accelerators, which require more complicated coil geometries.

Common-coil design offers additional advantages when used as the rapid turn-around magnet R&D test facility:

- A large open space can be incorporated between the apertures, where additional racetrack coils can be inserted and tested as an integral part of the magnet without requiring any disassembly (as in the BNL common-coil magnet DCC017);
- Flexible and modular design is offered with easier segmentation for hybrid high field dipoles using a variety of conductors (Nb_3Sn, Nb-Ti and for very high field high temperature superconductors (HTS));
- Minimum requirements for large, expensive tooling and labor, particularly during the R&D phase, are required;
- Efficient and rapid turn-around magnet R&D due to simpler and modular design might be possible.

14.3 Accelerator-Quality Common-Coil Dipole

Despite the early promise and success of the common-coil dipole design, further development stopped, not for technical reasons, but because of changing US Department of Energy program needs. Subsequent magnets were single-aperture quadrupoles or dipoles, whereas the common-coil design is for two-in-one dipoles. With the growing interest of the high-energy physics community in proton colliders with center-of-mass energies in the order of 100 TeV in a limited tunnel size (due to geographic or other reasons), however, interest in a design option that can produce a lower cost high field magnet has returned (Qingjin Xu et al. 2016; Toral et al. 2017).

14.3.1 Optimization of Field Quality

Field quality in accelerator magnets is expressed in terms of the normal and skew harmonics (b_n and a_n) as defined in the expression

$$B_y + iB_x = B_1 \sum_{n=1}^{\infty} (b_n + ia_n) \left(\frac{x + iy}{R_{\text{ref}}} \right)^{n-1}$$

where B_x and B_y are the components of the field at (x, y), and B_1 is the magnitude of the field due to the most dominant harmonic at a "reference radius" R_{ref}. A reference radius of 10 mm is assumed for 40 mm aperture dipole designs and 17 mm for 50 mm aperture designs.

Field-quality design optimization in superconducting magnets is mostly associated with minimizing geometric and saturation-induced harmonics. The geometric and saturation-induced field harmonics are primarily optimized using the ROXIE software program (Russenschuck 1995).

Persistent-current induced harmonics, another source of errors, are primarily associated with the critical current density and the effective filament diameter of the conductor and the coil geometry. Due to the larger critical current density and typically larger filaments, the persistent-current induced harmonic errors are much larger in current high field conductors (Nb_3Sn, HTS) as compared to those in Nb-Ti composite wires.

All R&D common-coil magnets, except for the FNAL common-coil dipole (Ambrosio et al. 2000; Chap. 15, this book), built to date consisted of the main coil alone (see Fig. 14.1), and achieving a good field quality was not part of the initial considerations. A coil cross-section optimized for geometric harmonics of the level needed in accelerator magnets and minimizing the amount of conductor requires pole coil blocks (Gupta et al. 2000) or field-shaping coils. Type (a), as shown in Fig. 14.2a, uses a smaller amount of conductor but has more complicated ends than type (b), as shown in Fig. 14.2b, where all coils are flat racetrack coils but the conductor of the return side does not contribute to the main field. These additional coils add to the complexity of the magnet. The pole coils have to be designed such that they do not only provide good field quality, but are also well clamped within the structure and have sufficient margin.

Depending on the details of accelerator design, the dipole field created by the return conductors in type (b), which represents a small fraction of the overall conductor, can still be used for an injector (Gupta 1999); and therefore the conductor is efficiently used.

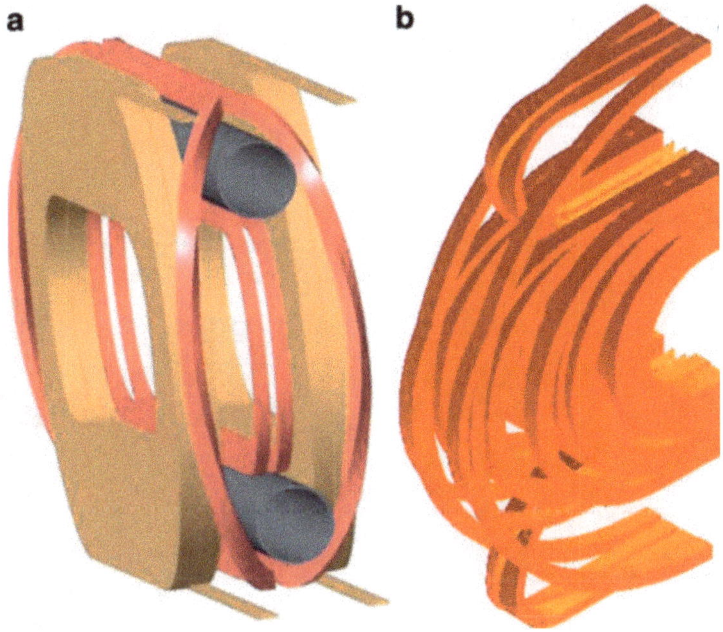

Fig. 14.2 Two possible configurations of coil blocks for good field quality: (**a**) type (a) uses less conductor but requires pole blocks with flared ends to clear the bore tube; contrary to (**b**) type (b), where all of the coils are flat

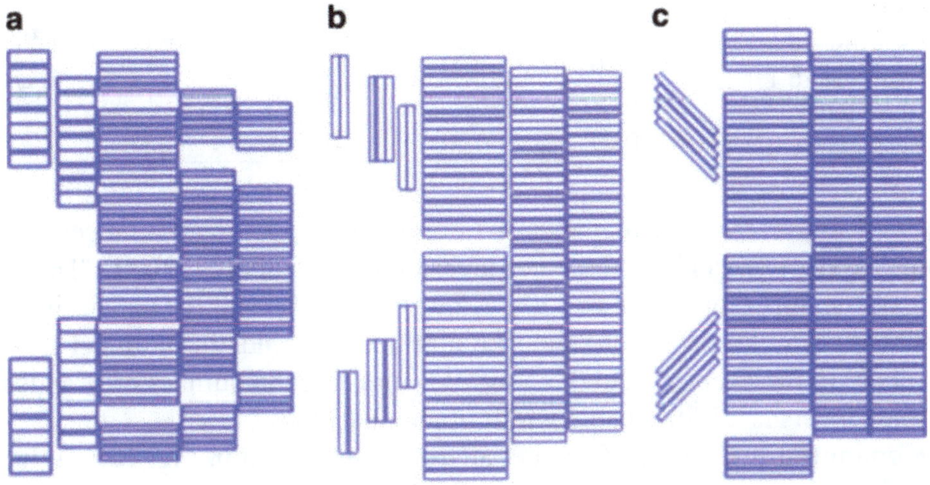

Fig. 14.3 Basic orientation of the pole coils: (**a**) horizontal; (**b**) vertical; and (**c**) inclined

Pole coils can be placed in one of three orientations: horizontal (the same as the main coil, see Fig. 14.3a), vertical (see Fig. 14.3b), and aligned (see Fig. 14.3c), or in combination (Gupta et al. 2000).

14.3.2 Example of an Optimized Common-Coil Dipole

A 50 mm aperture, 16 T common-coil dipole is presented here with the field quality optimized with the help of pole coils. The coil uses rectangular Nb_3Sn Rutherford

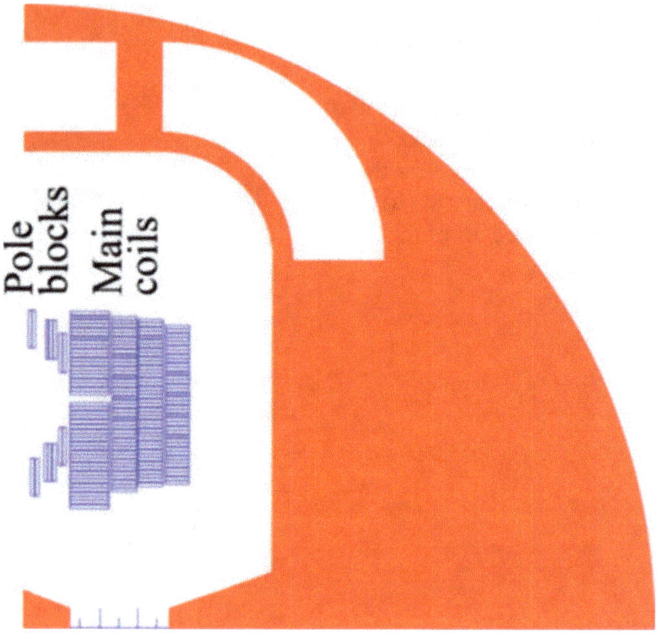

Fig. 14.4 Optimized two-in-one common-coil dipole design

Table 14.1 Skew a_n and normal b_n harmonics at 17 mm radius at 16 T

a_2	a_4	a_6	a_8	a_{10}
0.00	0.00	0.00	0.27	0.21
b_3	b_5	b_7	b_9	b_{11}
0.00	0.00	0.01	−0.16	−0.10

cables with strands having a diameter of 1.1 mm and copper to non-copper ratios for the inner layer of 1.0 and for the outer layer of 1.5 (Toral et al. 2017). The critical current of the superconductor is 1500 A/mm^2 at 4.2 K and 16 T. The insulation thickness is 0.15 mm on either side. The number of strands in the inner layer and pole coils is 36 for a cable width of about 21.3 mm, and the number of strands in the outer three layers is 22 for a cable width of about 13 mm.

The optimized coil design (Gupta et al. 2017a) is shown in Fig. 14.4. It has less than 0.3% peak enhancement (maximum field on the conductor with respect to the field at the center of the bore). Computed harmonics for the design field of 16 T are given in Table 14.1 at a reference radius of 17 mm. Harmonics not listed in the table are zero by symmetry.

Harmonics having significant change as a function of current due to iron saturation are plotted in Fig. 14.5. Small saturation-induced harmonics were achieved ($b_3 <$ 7 units and $a_2 <$ 6 units). Enough space was also left (Gupta et al. 2017a) for the support structure. The fringe field at a radius of 150 mm outside the yoke is about 0.25 T when the yoke outer diameter (OD) is 700 mm (as in the design presented here), ~0.2 T when the yoke OD is 750 mm, and ~0.12 T when the yoke OD is 800 mm. Key parameters of the design are given in Table 14.2.

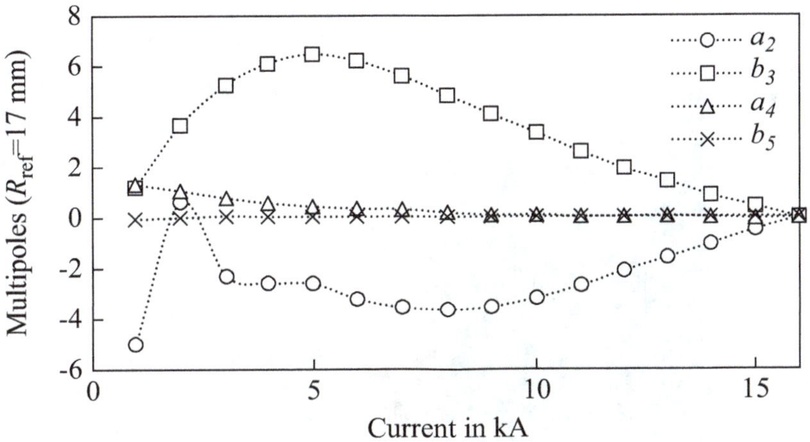

Fig. 14.5 Field harmonics at 17 mm reference radius vs. current

Table 14.2 Key parameters of example design

Parameter	Value
Operating current (kA)	15.96
Field in the aperture (T)	16.0
Margin at 1.9 K (%)	19.3
Intra-beam spacing (mm)	250
Yoke outer diameter (mm)	700
Stored energy per unit length/aperture (MJ/m)	1.7
Inductance/aperture (mH/m)	13
Strand diameter (inner and pole layer) (mm)	1.1
Strands/cable (inner and pole layer)	36
Cu/non-Cu (inner and pole layer)	1.0
Strand diameter (outer layers) (mm)	1.1
Strands/cable (outer layers)	22
Cu/non-Cu (outer layers)	1.5
Total number of turns per aperture	179
Total area of Cu/aperture (mm^2)	5029
Total area of non-Cu/aperture (mm^2)	4026

14.3.3 *Mechanical Analysis*

To examine the basic issues related to the support structure, a simplified one-piece stainless-steel collar is assumed with no joints or connections. The coil modulus of fiberglass/epoxy impregnated Nb_3Sn is assumed to be 20 GPa. Symmetry (frictionless) is assumed at the horizontal split line and at the vertical split line. The thickness of the collar is 37 mm. A frictionless support is assumed on the right-hand edge of the collar.

Figure 14.6 shows the stresses on the coil powered to a field of 16 T at 1.9 K. The maximum stress on the main coil (Fig. 14.6a) is 144 MPa around the mid-plane of

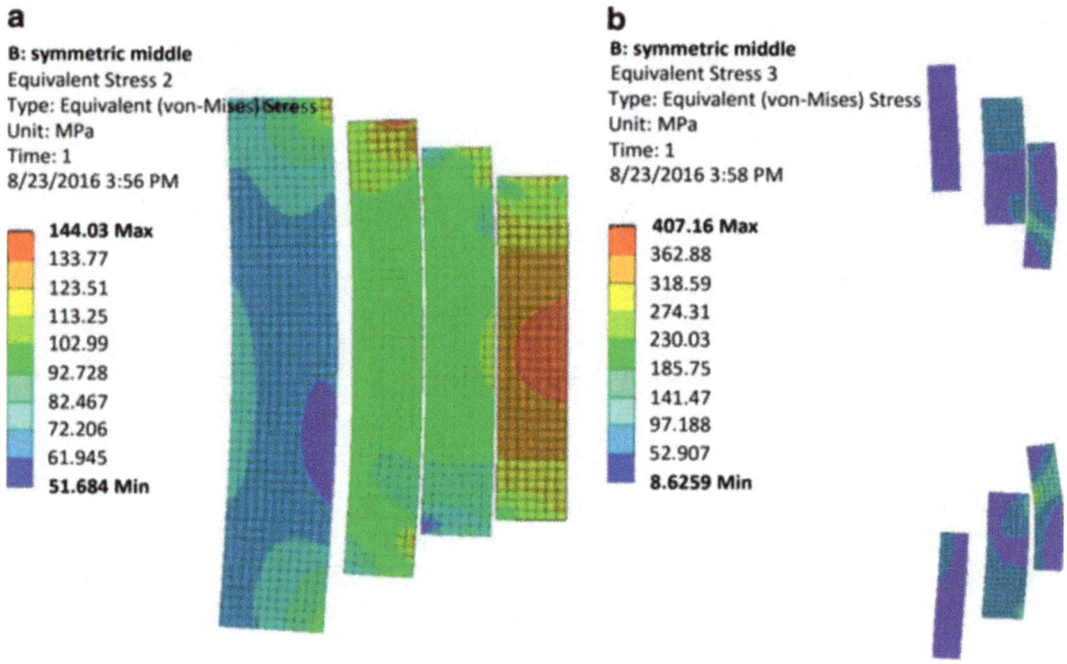

Fig. 14.6 Stresses (**a**) in the main coil; and (**b**) pole coils (MPa)

the outermost coil. This value remained about the same when collars were free to move with no support at the right edge. Stresses on the pole coils (Fig. 14.6b) are also generally below 150 MPa, except at the local area in the right-most pole coil blocks, where the stress is very high, above 400 MPa. This stress is to be reduced in future iterations of the structure.

Figure 14.7 shows the deflections in the main and pole coils under the Lorentz forces at 16 T. We plot the horizontal deflections for the main coil (Fig. 14.7a) and vertical deflections for the pole coil (Fig. 14.7b). The maximum horizontal deflection is about 0.77 mm (in the main coils). This deflection is considered acceptable if the coil moves as a whole (a major benefit of the common-coil design), so long as the relative deflections inside the coil are small to keep the strain within an acceptable limit. The horizontal deflection of the pole coil blocks will be limited by the main coils and the support structure. The goal of future iterations will be to make deflections more uniform. The vertical deflections are less than 0.1 mm, which indicates that the type of support structure considered for the pole coils should be able to hold them against the vertical Lorentz forces.

14.3.4 Influence of Coil Displacements on Field Harmonics

Deflections due to Lorentz forces have an impact on field harmonics, and their change might be significant when the deflections are as large as previously described. The change is expected to be small if deflections are more horizontal

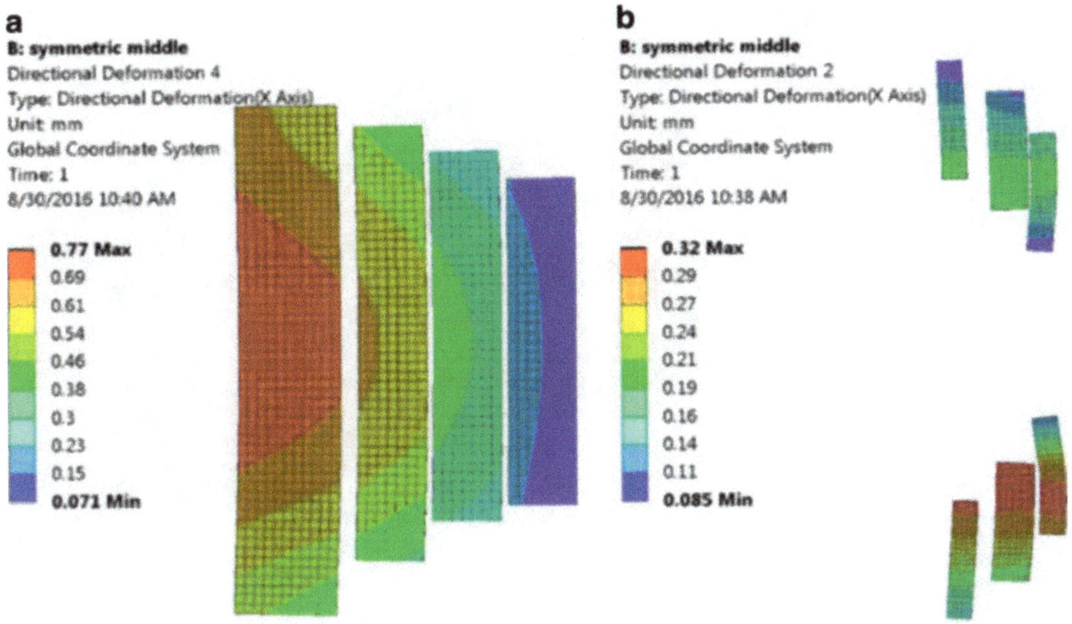

Fig. 14.7 (**a**) Horizontal displacements of the main coils; and (**b**) vertical displacement of the pole coils (mm)

rather than vertical, however, as is the case here. If all blocks are allowed to move horizontally, then a displacement of 1 mm primarily causes a change in the sextupole (b_3) harmonic. The impact is linear as a function of displacement and is computed to be about 9 units/mm. The field harmonics (in particular b_3) also change, however, due to iron saturation. Optimizing b_3 taking these two effects into account shows that its variation can be kept below 10 units from injection to collision energy.

14.4 BNL React-and-Wind Common-Coil Dipole DCC017

All so-far known practical high field superconductors are brittle. Nb_3Sn, however, offers the unique opportunity to wind the magnet with a strand containing the precursors (mainly Nb and Sn) and react the conductor after winding: an approach using W&R. On the other hand, the conductor can be reacted before winding, the R&W approach, which allows a variety of materials (including insulation) to be used since the coil and associated tooling are not subjected to high reaction temperatures.

Moreover, the I_c of Nb_3Sn degrades with strain in various background fields, as reported in Ekin (1980). It was found that I_c bending degradation is a function of strain and magnetic field, with a particularly large degradation at high fields above 10 T. Therefore, R&W high field magnet technology calls for magnetic designs with low bending strain, which has been a major challenge. For this reason, almost all high field Nb_3Sn short R&D accelerator magnets have been built using W&R technology.

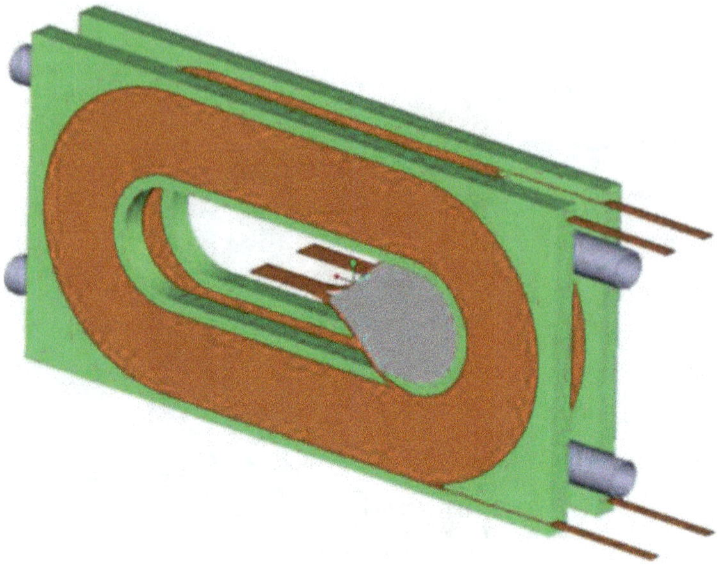

Fig. 14.8 Schematic design of the BNL dipole DCC017 with a pair of racetrack coils

The 10.2 T dipole magnet developed by BNL (Ghosh et al. 1999; Escallier et al. 2001; Gupta et al. 2001, 2002; Cozzolino et al. 2003; Gupta 2015) and described in this chapter, is the highest field R&W Nb_3Sn accelerator R&D magnet ever built. Magnet design, construction, and test results are presented here. The successful construction and test of this magnet opens the possibility of using the R&W approach for longer magnets to be used in accelerators.

14.4.1 Magnet Design

The magnet is based on two pairs of flat racetrack coils (see Fig. 14.8) made with pre-reacted Nb_3Sn cable following the R&W approach. Stainless-steel collars in combination with a stainless-steel shell and an iron yoke contain the Lorentz forces. The coils are subjected to only minimal pre-stress in the horizontal, vertical, and axial directions at cold.

14.4.1.1 Magnetic Design

The magnetic design consisted of two coil layers in a two-in-one common-coil configuration (Gupta et al. 2007) with a minimum bending radius of 70 mm. The main parameters of the design are presented in Table 14.3. One quadrant of the magnet cross-section (one-half of one aperture) is shown in Fig. 14.9.

The bobbin (also known as the central island) on which the coil is wound was made of magnetic steel (the original design had a 5 mm non-magnetic liner). The use of magnetic steel reduces: (a) the peak field on the conductor; and (b) the loss in pre-stress from cool-down because of its lower thermal expansion as compared to

Table 14.3 Main parameters of common-coil dipole DCC017

Parameter	Value
Conductor type	Nb_3Sn
Magnet technology	React and wind
Horizontal coil aperture (clear space) (mm)	31
Vertical coil aperture (clear space) (mm)	338
Separation between apertures (mm)	220
Number of layers	2
Number of turns per quadrant (turns/layer)	45
Coil height (pole-to-pole) (mm)	85
Wire non-Cu J_{sc} (12 T, 4.2 K) (A/mm^2)	1900
Strand diameter (mm)	0.8
Number of strands in cable	30
Cable width (mm)	13.13
Cu/non-Cu ratio	1.53
Computed quench current (kA)	10.8
Computed quench bore field at 4.2 K (T)	10.2
Peak field at quench in inner layer/outer layer (T)	10.7/6.1

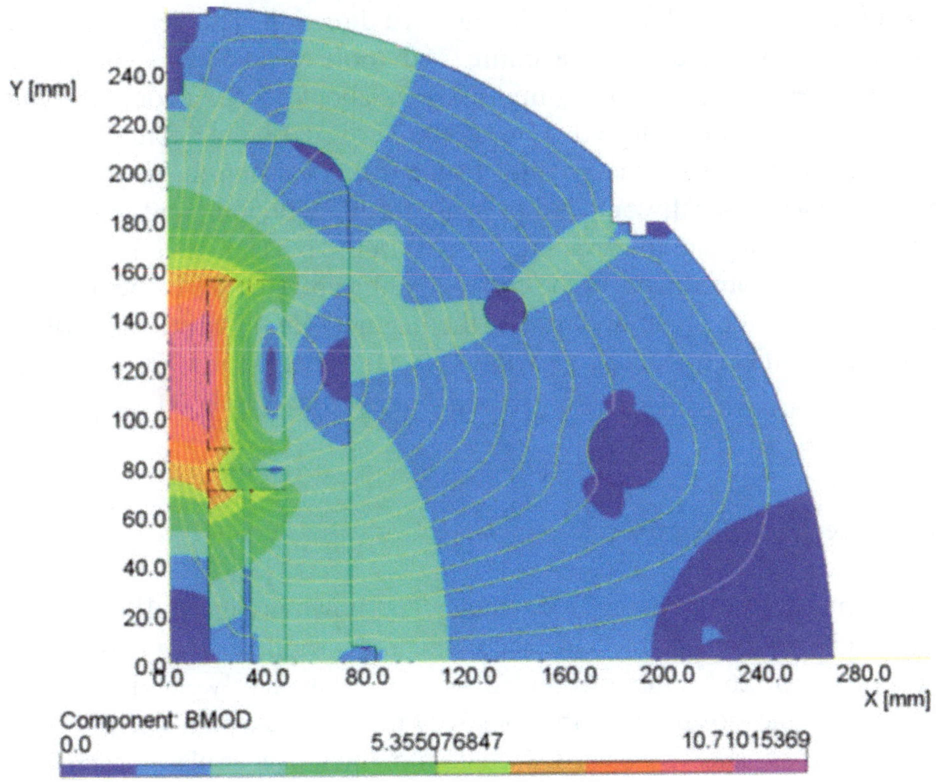

Fig. 14.9 A 2D model of one-quarter of the DCC017 common-coil dipole

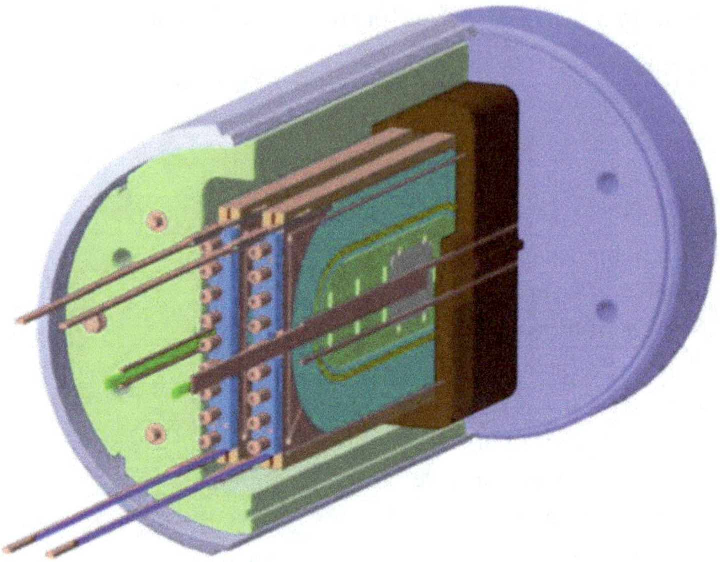

Fig. 14.10 Overall mechanical layout of DCC017

other materials (fiber-reinforced epoxy like G10, coil composite, copper) in the magnet structure.

In the cross-section of the common-coil magnet, the largest component of the Lorentz force is horizontal and is in the outward direction. The net vertical component of the Lorentz force on the entire coil (pole-to-pole) is small, while the conventional vertical stress on the mid-plane is about 35 MPa for one-quarter of the coil in one aperture. The direction of the vertical Lorentz force at low fields depends on the material of the bobbin. In the case of a non-magnetic bobbin the Lorentz forces are vertically outward or away from the bobbin; and in the case of a magnetic bobbin, they are towards the bobbin. In the present design, the computed horizontal/vertical components of the corresponding coil stresses in the first quadrant (right-hand side of the upper aperture of the magnet) are 43 MPa/−1.2 MPa at the computed quench field of 10.2 T, 59 MPa/−0.8 MPa at 12 T, and 77 MPa/−0.3 MPa at 13.8 T.

14.4.1.2 Mechanical Design

The overall mechanical structure of the magnet is shown in Fig. 14.10. The structure consists of 13 mm thick stainless-steel collars, and a rigid vertically split iron yoke having a radius of 267 mm, surrounded by a 25 mm thick welded stainless-steel shell. To restrict the movement of the ends, 127 mm stainless-steel end plates welded to the stainless-steel shell were used. Two 25 mm thick pusher plates, each equipped with nine bolts, were used on each magnet end to transfer coil end forces to the end plates (see Fig. 14.10). No pre-stress in the horizontal or vertical directions was envisaged in the coil's cross-sectional structural design. The screws were closed with minimal torque so as not to apply preload to the coil ends.

The end saddles and sidebars of the coils were made of stainless steel to keep the structure as rigid as possible. The brass spacer was slit at many places, which made it flexible so that its shape could more easily adapt to the layout of the cable in the ends and hence minimize the possibility of pinching the brittle cable. All coils were vacuum-impregnated individually with end saddles and sidebars installed.

To take full advantage of the modular nature of the common-coil design, all coil modules were deliberately made as identical single-layer mechanical cassettes with one splice in the middle so that their relative position in the R&D magnet could be interchanged (see Fig. 14.8). All coils had one 8.5 mm thick end spacer after five turns (counting from the inner radius, not shown in Fig. 14.9) to reduce the peak field in the ends. In addition, there was one wedge in the magnet cross-section that was also 8.5 mm thick and contiguous with the end spacer in the return end.

The guideline for this design was to keep the bending degradation in critical current below 15% to achieve a computed overall degradation of below 5%. Therefore, the bending strain had to be limited to around 0.3%, 0.25%, and 0.2% for 12 T, 14 T, and 15 T peak fields, respectively. Larger margins may, however, be available, as peak field and peak bending strain are not necessarily at the same location. In the design presented here, the peak bending strain is at the coil inner radius, and the peak field is at the coil mid-plane of either aperture.

As the field reduces over the coil width (see Fig. 14.9), the current density can be increased in coil layer 2 with respect to coil layer 1. This so-called grading can be achieved either by varying the cable area or by different currents in coil layers 1 and 2. A variable shunt power supply was introduced, which allowed grading of the current density in layer 1 and in layer 2. The shunt was incorporated in the construction of the magnet, but was never used. Therefore, the short sample field was reduced from the original design value of about 12 T to 10.2 T. The computed short sample limit of the magnet of 10.2 T is based on the actual configuration, cable measurements, and inclusion of the computed bending degradation (see Table 14.3).

The structure was designed to contain Lorentz forces at the original design field of 12 T in a 40 mm aperture. In this design, the horizontal component of the force was 75 MPa. Therefore, there is sufficient margin in the basic support structure, since the outward horizontal force in the present configuration at 10.2 T is only 59 MPa. The present mechanical structure can contain forces for fields up to 13.5 T.

Due to the large forces, the challenge in the common-coil design is to limit the displacement. A structural analysis, using ANSYS finite element software (ANSYS Inc., Canonsburg, PA), is shown in Fig. 14.11, which shows the deflections on the collar (Fig. 14.11a) and end plates (Fig. 14.11b). For a field of 13.5 T, the yoke vertical split remains in contact near the shell but opens 0.05 mm adjacent to the collar. The collars spread apart by 0.28 mm across the 44 mm aperture, but uniformity in the coil region is within 0.08 mm. The end plate deflects 0.41 mm under an axial force of 1.1 MN.

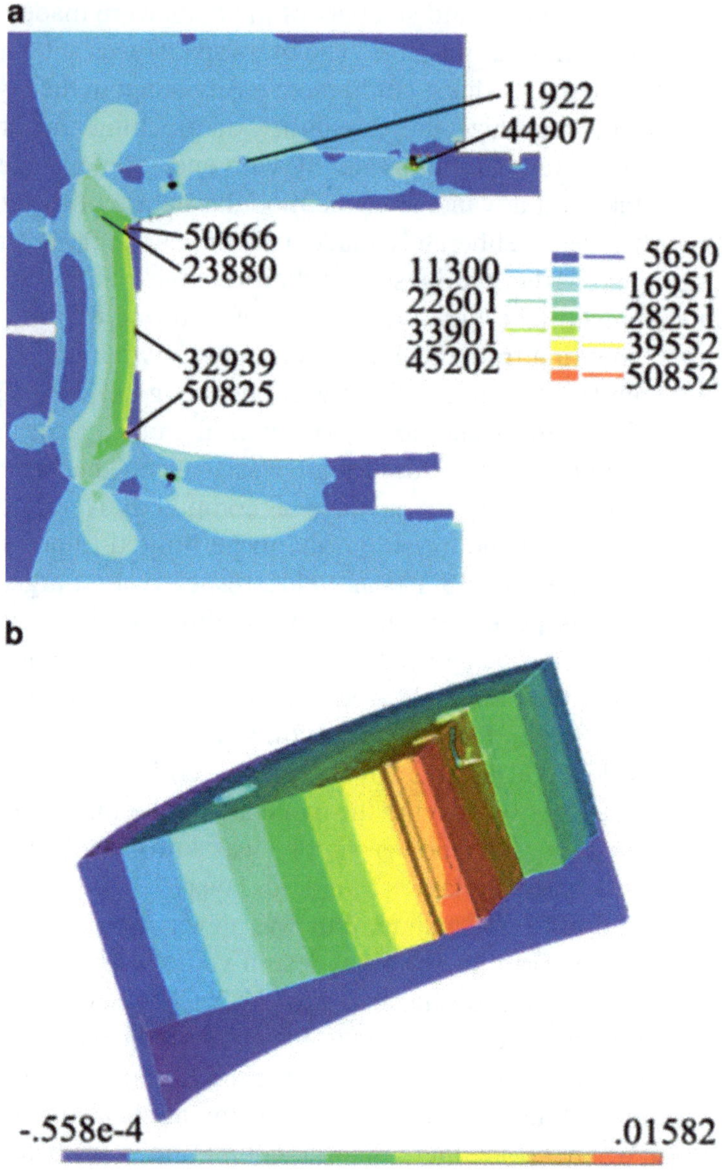

Fig. 14.11 ANSYS simulation showing stresses: (**a**) in the collar; and (**b**) on the end plates in the DCC017 common-coil dipole

14.4.2 Strand and Cable

DCC017 used a 30-strand cable made from 0.8 mm diameter strand using the modified jellyroll process. The strand was manufactured by Oxford Instruments Superconducting Wire LLC ("OST"), which has been acquired by Bruker Energy and Supercon Technologies Inc ("BEST"), a subsidiary of Bruker Corporation (600 Milik Street, Carteret, NJ 07008, USA). The Nb_3Sn wire used came from two billets, ORE-163 and ORE-202. Both billets have the same nominal copper fraction of 60% (measured Cu/non-Cu ratio: ORE-163: 1.54 and ORE-202: 1.6).

Fig. 14.12 Vacuum-impregnation fixture for coating cable with Mobil-1® to avoid sintering

Coils for this magnet were made from two lengths of cable. Cable BNL-N-4-0012 was fabricated by New England Wire Company, 130 North Main Street, Lisbon, NH 03585 USA and cable BNL-6-O-B0899R was fabricated at LBNL. All cable lengths for the four coils were vacuum-impregnated (see fixture in Fig. 14.12) with Mobil-1®, and pre-annealed at 200 °C for 8 h to drive off the volatile constituents in the oil and to remove the strain in the copper. Mobil-1® from Exxon Mobil, 5959 Las Colinas Boulevard, Irving, Texas 75039-2298 USA was used to prevent sintering of the strands after reaction. To avoid sintering is important in a R&W magnet, because the strain degradation of bent sintered cables is larger by a factor of about two, as shown by experiments.

Four sections of cable about 130 m long were reacted in a vacuum furnace using the following heat treatment cycle: 48 h at 200 °C, 48 h at 400 °C, and 72 h at 665 °C. After reaction, the width and mid-thickness of cable BNL-N-4-0012 (used in coil 32) were 12.72 mm and 1.509 mm, respectively, and for cable BNL-6-O-B0899R (used in coils 33, 34, and 35) were 13.17 mm and 1.513 mm, respectively.

Extracted strands from the cable were reacted on stainless-steel barrels using the same reaction schedule as the cable segments. The critical current measurements of the extracted strands were carried out in the range 8–11.5 T, and fitted using Summers' formulation (Summers et al. 1991). Strand data are multiplied by 30 to calculate the cable I_c, as given in Table 14.4.

The cable was reacted on a 280 mm diameter stainless-steel drum. Therefore, after coil winding the bending strains experienced by the strand when: (a) the cable from the drum is straightened; and (b) the cable is bent at a radius of 70 mm, are similar in magnitude but opposite in sign. The effect of the bending is to add tensile strain along the outside of the strand and the same magnitude of compressive strain

Table 14.4 Expected performance of cable used in DCC017

	Cable critical current I_c (A)			
	Cable BNL-N-4-0012 (coil 32)		Cable BNL-6-O-B0899R (coils 33, 34, and 35)	
B (T)	Fitted	With strain	Fitted	With strain
9.0	19,710	16,077	21,348	17,458
9.5	17,970	14,383	19,404	15,687
10.0	16,260	12,838	17,618	14,068
10.5	14,649	11,427	15,975	12,587
11.0	13,188	10,137	14,462	11,230
11.5	11,820	8,959	13,065	9,986
12.0	10,650	7,882	11,776	8,847

along the inside of the strand. The magnitude of the bending strain is equal to r/R, where r is the radius of the filament boundary in the strand (0.58 mm diameter area in the 0.8 mm diameter strand) and R is the bending radius. Since the strand is reacted at a radius of 140 mm, the effective bending radius when the cable is bent at a radius of 70 mm is 140 mm (since the change in $1/R$ is $1/70 - 1/140 = 1/140$). From the as-reacted state, the strand J_c increases with tension and decreases with compression strain (Ekin 1980). Bending increases the compressive strain in the strand from the as-reacted state by $\Delta\varepsilon = 0.21\%$. The minimum I_c of the strand/cable is then calculated using Summers' fit (Summers et al. 1991) and changing the strain by $\Delta\varepsilon = -0.21\%$. In the absence of direct measurements, the calculated I_c sets a lower bound for the effect of bending strain. Since coil 32 is one of the inner coils and has the lowest performance, it will be the limiting coil when the magnet reaches the short sample limit. The calculated critical currents at 4.2 K and the strain-degraded current at 4.5 K are shown in Table 14.4. The category "Fitted" refers to the strand measurements at 4.2 K multiplied by the number of strands, 30, and the category "With strain" refers to the expected performance at 4.5 K in the magnet with a strain of $\varepsilon = -0.21\%$ due to bending, computed using Summers' formulation.

Figure 14.13 shows the magnet load lines and the critical current of the cable vs. magnetic field in coil 32 at a temperature of 4.5 K and a strain $\varepsilon = -0.21\%$. This gives a short sample limit of 10.8 kA using the computed peak field load line based on 2D and 3D models corresponding to a peak field of 10.7 T and a central field of 10.2 T in both apertures.

14.4.3 Tooling

Since pre-reacted Nb_3Sn conductor is brittle and sensitive to local strain, manual handling must be minimized to avoid accidental damage or degradation. The BNL coil winding tooling is shown in Fig. 14.14. Other major pieces of tooling developed for this program are the coil impregnation fixture using vacuum bag technology. One of the coils after impregnation is shown in Fig. 14.15.

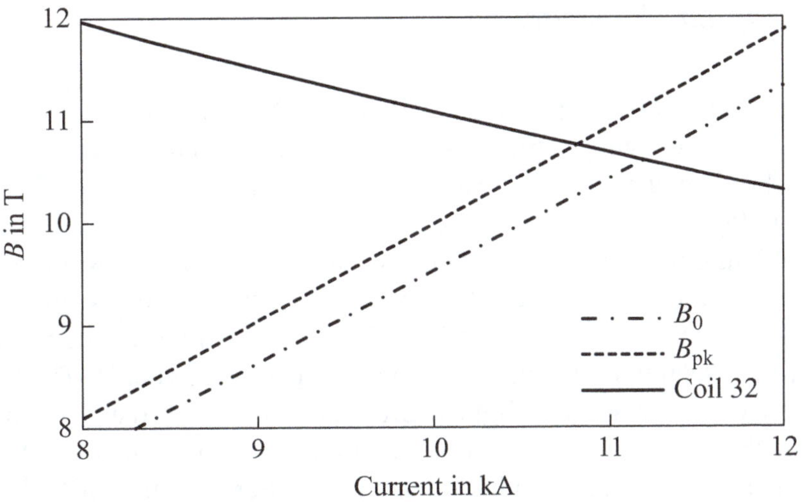

Fig. 14.13 Cable critical surface and DCC017 load lines: B_0 corresponds to the field in the aperture, B_{pk} corresponds to the maximum field in the coil

Fig. 14.14 Coil winding tooling

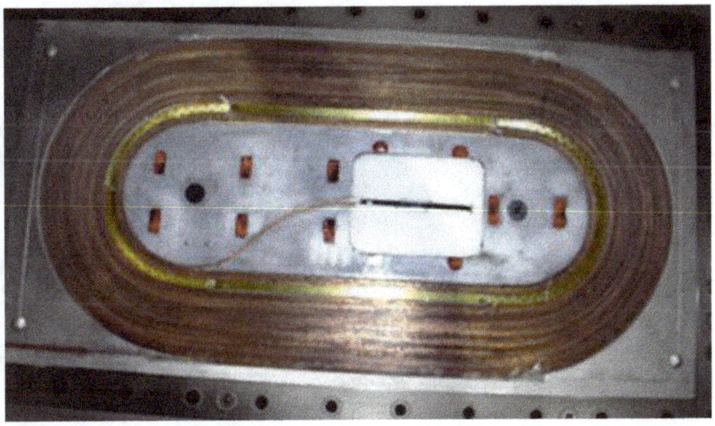

Fig. 14.15 Nb_3Sn coil after impregnation with CTD-101K epoxy (Composite Development Technology, Inc., 2600 Campus Drive, Lafayette, C0 80026, USA)

14.4.4 Magnet Construction

Four coils were wound with pre-reacted 30-strand Nb_3Sn cable on a magnetic steel flat racetrack bobbin having a straight section length of 300 mm and a bending radius of 70 mm. For the turn-to-turn insulation, 0.170 mm thick and 13.2 mm wide Nomex® tape (approximately the same width as the cable) was used. Nomex® is a trademark of DuPont, USA. Utmost care was taken to avoid over-straining the cable. The winding tension did not exceed 53 N for the cable and 67 N for the Nomex® tape. No clamps were used during winding. The cable freely followed its natural path from the straight section to the end, yielding to a negligible buckle. The thickness of the cable from edge to edge was slightly different, which caused a 7° inclination over the width of the 45th turn. No attempt was made to remove this condition due to concerns regarding possible conductor damage. After having finished winding, a minimal 222 N force was applied to straighten the straight section of the coil and to perform shimming against the bobbin. To compensate for the 7° inclination of the cable, tapered shims were used.

The end saddles and sidebars were made of stainless steel and were custom fit to each individual coil. No pre-compression was applied during coil impregnation and curing. All large voids were filled with fiberglass or G10. Each coil was checked for straightness and flatness after curing. The coils were pinned into pairs and shimmed to equal overall size with alumina-filled epoxy. Stainless-steel sheets having a thickness of 1.65 mm were placed between the layers of each coil pair to homogenize the stress distribution at full power. Four 0.025 mm thick stainless-steel strip heaters insulated with polyimide foil were placed on both sides of this sheet.

One of the two coil modules consisting of a pair of coils is shown in Fig. 14.16. This assembly contains an internal splice in a low-field region made with a set of 12 perpendicular Nb-Ti cables. A shunt lead (coming out axially in the middle) can

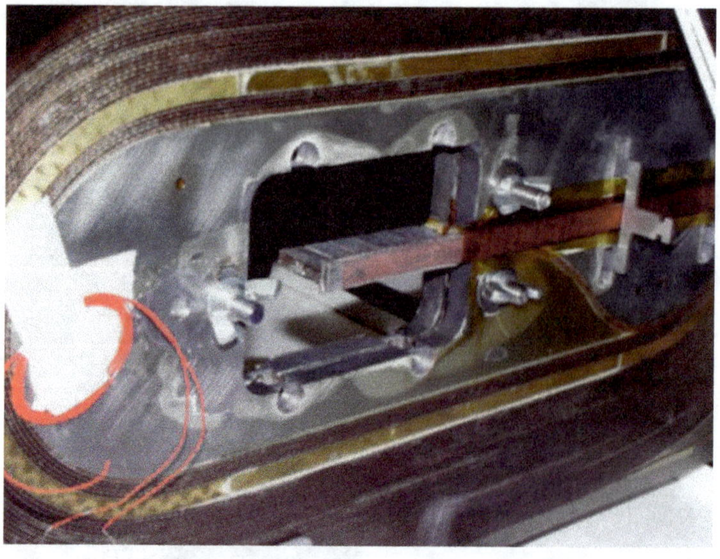

Fig. 14.16 Coil modules with a pair of coils and the shunt lead

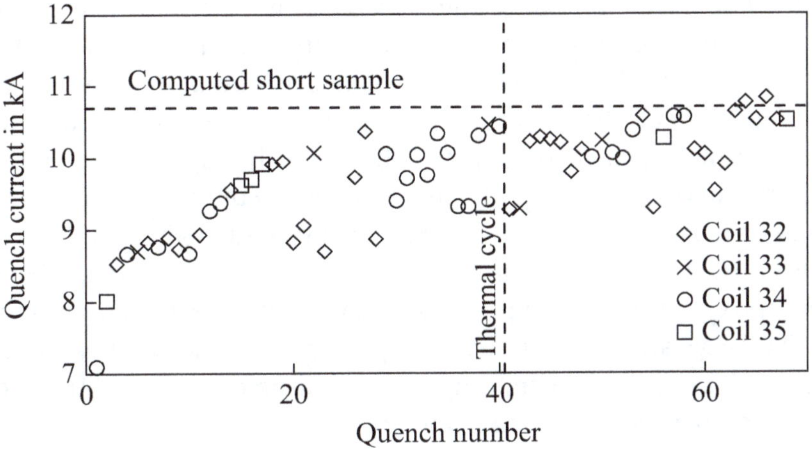

Fig. 14.17 DCC017 quench history. Short sample line at 10.8 kA corresponds to a peak field of 10.7 T and a bore field of 10.2 T

also be seen in Fig. 14.17. This shunt lead contains a Nb_3Sn cable since it passes through a high field region in the ends.

A collaring press was built specifically for this magnet. Coil pre-stress was strictly applied only in the side-to-side direction (cable stack direction) and was relatively low, around 17 MPa. An inflatable bladder was employed to keep the coils hard outward against the collars during collaring. Stainless-steel "keepers" were installed for locking the coils against the collars.

14.4.5 Magnet Test Setup and Test Procedure

The magnet was tested at the Vertical Test Facility at BNL in a liquid helium bath at a nominal temperature of 4.5 K. The magnet was instrumented with voltage taps located between the coils and at the leads to the coils so that the voltages of the four coils could be monitored during testing. Quenches were detected by monitoring the voltage difference between the coil pairs and generating a stop pulse when the voltage exceeded the threshold voltage level. An additional quench detection circuit used the difference between the total coil voltage and the current derivative voltage signal. No voltage taps in the body of the coils were installed.

Quench calculations, using the QUENCH software program (Wilson 1983), showed that coil "hot-spot" temperatures could exceed 400 K. At this temperature level, degradation of Nb_3Sn cable could start to develop due to thermo-mechanical strain effects. Calculations were performed to establish the relationship between the quench temperature and the quench integral quantity $\int I^2\,dt$. It was decided to limit the quench temperature to 300 K where the quench integral limit was 21 $(kA)^2$ s.

Since energy extraction was not available for this test, the coils were instrumented with quench heaters. These consisted of two type 304 stainless-steel strips, each of

0.025 mm thickness and 38.1 mm width separated by 6.35 mm. They were installed between the layers in each coil pair and positioned along each side of the inside surfaces of each layer. The strips were mounted on both sides of the polyimide-wrapped stainless-steel sheet, which separated the layers of each coil pair and essentially prevented quench propagation between the layers. The quench heaters were connected into two circuits. Each circuit was equipped with its own power supply based on a capacitance of 21.7 mF, which provided 450 V and 105 A peak to quench the coils. The total insulation (polyimide with adhesive, fiberglass, and CTD-101K epoxy) was 1.07 mm thick between the heater strips and the bare conductor. There was no copper shunting provided on the stainless steel. Quench delay times at 4 kA after heater firing were measured at 100–200 ms due to the thermal diffusion barrier across the insulation.

Quench tests were performed by powering the magnet in the common-coil electrical configuration using a 30 kA, 15 V power supply. Current ramps were done at rates from 3 A/s to 200 A/s, with most quenches done at 25 A/s or less. On detection of a quench, a stop pulse from the quench detector shut off the power supply, fired the strip heaters, and triggered the fast data logger system to acquire voltage data at a 1 kHz sampling rate.

14.4.6 Test Results

A total of 66 quenches were done, of which 38 were performed before the thermal cycle. The initial training started at 7 kA and continued for 16 quenches up to almost 10 kA. Then, the quench behavior was erratic with quench currents spanning from 8699 A to 10,475 A. After the thermal cycle, the behavior was significantly less erratic, with most quench currents above 10 kA, reaching 10,846 A (10.2 T central field), a value slightly above the calculated short sample limit. Figure 14.17 shows the complete quench history, which included a thermal cycle and a ramp rate study at the end. The magnet was, as expected, mainly limited by coil 32, whose conductor had a slightly lower critical current density and reached its conductor limit before the other coils. The highest quench currents were achieved for 200 A/s ramp rates. No clear evidence for a ramp rate dependence could, however, be found.

The training and erratic quenches occurred in all four coils. Measurements of the initial slope of the voltage increase gave values of dV/dt that varied from 6 V/s to 100 V/s and showed no correlation with the quench current, with different values of dV/dt resulting from similar quench currents, implying that locations within the coils varied. Pre-quench voltage spike detection was limited to a resolution of 60 mV and 1 ms. Most quench signals exhibited voltage spikes, some of which were recognized as flux jump instabilities. The voltage spikes ($\geq$60 mV) were, however, typically not at quench onset. Few quenches exhibited spikes right at quench onset above 60 mV.

Starting with quench number 46, the delay between the quench detector stop pulse and power supply shutoff/strip heater trigger was reduced from 58 ms to 16 ms by inhibiting the power supply firing circuit. The delay was minimized to decrease the amount of heating in the coils after each quench. The quench integral was often above 16 $(kA)^2$ s, with a maximum of 19.1 $(kA)^2$ s. These values corresponded to a quench temperature of about 220 K, which was well below the safe limit of 300 K.

In February 2017, after a 10-year hiatus, DCC017 was recommissioned to test a HTS insert coil (Gupta et al. 2017b). The magnet performed flawlessly, reaching 92% of short sample field without quenching. A limitation of the leads, not the magnet itself, prohibited powering it to higher currents.

14.5 Magnet R&D Approach

The third purpose of building DCC017 was to commission an R&D superconducting coil test facility, which allows racetrack coils to be inserted and tested in the background field magnet of this magnet. This facility will, in principle, allow coils to be tested in a manner in which, previously, only cables have been tested. The DCC017 magnet was specifically designed with a large opening (31 mm wide and 338 mm high) where new R&D coils (HTS or Nb_3Sn) can be inserted and tested together with the existing background field Nb_3Sn coils (see Fig. 14.18). The most appealing part of this powerful magnet facility is that coil testing requires no disassembling of the background field magnet, facilitating a new way of doing "rapid turn-around" and "low-cost" magnet R&D. Another major motivation pursued within this program is to facilitate integration of the field-shaping pole coil in DCC017 to demonstrate a proof-of-principle common-coil dipole.

Fig. 14.18 End view of 10 T Nb_3Sn common-coil dipole DCC017 with a large open space

14.6 Conclusion

The common-coil design offers several inherent technical and cost advantages. It is suitable for high field magnets where the Lorentz forces are large. Simple geometry offers lower cost manufacturing options. Moreover, it also allows the use of R&W technology, as successfully demonstrated in the BNL Nb_3Sn DCC017 magnet.

DCC017 has been commissioned as a unique vehicle for carrying out low-cost, rapid turn-around magnet R&D. The DCC017 magnet has been found to be very robust, as it reached 92% of short sample without quench after a decade of hibernation.

Several designs have been developed, which show that the common-coil design can produce magnets with the field quality needed in accelerator magnets. The proof of the principle of high field magnets of accelerator quality based on this concept is yet awaited.

References

Ambrosio G, Andreev N, Barzi E et al (2000) Study of the react and wind technique for a Nb_3Sn common coil dipole. IEEE Trans Appl Supercond 10(1):338–341. https://doi.org/10.1109/77.828243

Cozzolino J, Anerella M, Escallier J et al (2003) Magnet engineering and test results of the high field magnet R&D program at BNL. IEEE Trans Appl Supercond 13(2):1347–1350. https://doi.org/10.1109/tasc.2003.812665

Danby G, Palmer R, Huson R et al (1983) Panel discussion of magnets for a big machine. In: Cole TF (ed) Proceedings of the 12th international conference on high energy, Fermilab, 11–16 Aug 1983. Fermilab, Batavia, pp 52–62

Ekin JW (1980) Strain scaling law for flux pinning in practical superconductors. Part 1: basic relationship and application to Nb_3Sn conductors. Cryogenics 20(11):611–624. https://doi.org/10.1016/0011-2275(80)90191-5

Escallier J, Anerella M, Cozzolino J et al (2001) Technology development for react and wind common coil magnets. In: Lucas P, Webber S (eds) Proceedings of the 2001 particle accelerator conference PAC2001, Chicago, 18–22 June 2001, vol 1. IEEE, Piscataway, pp 214–216

Fermilab (2001) Design study for a staged Very Large Hadron Collider. Fermilab TM-2149, 4 June, http://lss.fnal.gov/archive/test-tm/2000/fermilab-tm-2149.pdf

Ghosh AK, Cozzolino JP, Harrison MA et al (1999) A common coil magnet for testing high field superconductors. In: Luccio A, MacKay W (eds) Proceedings of the 1999 particle accelerator conference, New York, 27 Mar–2 Apr 1999, vol 5. IEEE, Piscataway, pp 3230–3232

Gupta G (1997) A common coil design for high field 2-in-1 accelerator magnets. In: Comyn M, Craddock MK, Reiser M et al (eds) Proceedings of the 1997 particle accelerator conference, Vancouver, 1997. IEEE, Piscataway, pp 3344–3346

Gupta R (1999) Common coil magnet system for VLHC. In: Luccio A, MacKay W (eds) Proceedings of the 1999 particle accelerator conference, New York, 27 Mar–2 Apr 1999, vol 5. IEEE, Piscataway, pp 3239–3241

Gupta R (2015) Common coil magnet design for high energy colliders (talk). In: CERN MSC seminar, 15 Sep, Todesco E (chair). https://indico.cern.ch/event/442002/

Gupta R, Chow K, Dietderich D et al (1999) A high field magnet design for a future hadron collider. IEEE Trans Appl Supercond 9(2):701–704. https://doi.org/10.1109/77.783392

Gupta R, Ramberger S, Russenschuck S (2000) Field quality optimization in a common coil magnet design. IEEE Trans Appl Supercond 10(1):326–329. https://doi.org/10.1109/77.828240

Gupta R, Anerella M, Cozzolino J et al (2001) Common coil magnet program at BNL. IEEE Trans Appl Supercond 11(1):2168–2171. https://doi.org/10.1109/77.920287

Gupta R, Anerella M, Cozzolino J et al (2002) R & D for accelerator magnets with react and wind high temperature superconductors. IEEE Trans Appl Supercond 12(1):75–80. https://doi.org/10.1109/tasc.2002.1018355

Gupta R, Anerella M, Cozzolino J et al (2007) React and wind Nb_3Sn common coil dipole. IEEE Trans Appl Supercond 17(2):1130–1135. https://doi.org/10.1109/tasc.2007.898139

Gupta R, Anerella M, Cozzolino J et al (2017a) Common coil dipoles for future high energy colliders. IEEE Trans Appl Supercond 27(4):1–5. https://doi.org/10.1109/tasc.2016.2636138

Gupta R, Anerella M, Cozzolino J et al (2017b) Design, construction, and test of HTS/LTS hybrid dipole. IEEE Trans Appl Supercond 28(3):1–5. https://doi.org/10.1109/tasc.2017.2787148

Russenschuck S (1995) A computer program for the design of superconducting accelerator magnets. In: 11th annual review of progress in applied computational electromagnetics, Monterey, CA, 20–24 Mar 1995, CERN AT/95-39, vol 1, pp 366–377

Sabbi G, Ambrosio G, Andreev N et al (2000) Conceptual design of a common coil dipole for VLHC. IEEE Trans Appl Supercond 10(1):330–333. https://doi.org/10.1109/77.828241

Summers LT, Guinan MW, Miller JR et al (1991) A model for the prediction of Nb_3Sn critical current as a function of field, temperature, strain and radiation damage. IEEE Trans Magn 27 (2):2041–2044. https://doi.org/10.1109/20.133608

Tommasini D, Auchmann B, Bajas H et al (2017) The 16 T dipole development program for FCC. IEEE Trans Appl Supercond 27(4):4000405. https://doi.org/10.1109/TASC.2016.2634600

Toral F, García-Tabarés L, Martinez T et al (2017) The EuroCirCol 16 T common-coil dipole option for the FCC. IEEE Trans Appl Supercond 27(4):1–5. https://doi.org/10.1109/tasc.2016.2641483

Wang C, Zhang K, Xu Q (2016) R&D steps of a 12-T common coil dipole magnet for SPPC pre-study. Int J Mod Phys A 31(33):1644018. https://doi.org/10.1142/s0217751x16440188

Wilson M (1983) Superconducting magnets. Clarendon Press, Oxford

Xu Q, Zhang K, Wang C et al (2016) 20-T dipole magnet with common-coil configuration: main characteristics and challenges. IEEE Trans Appl Supercond 26(4):1–4. https://doi.org/10.1109/tasc.2015.2511927

Chapter 15
Common-Coil Dipole for a Very Large Hadron Collider

Alexander V. Zlobin

Abstract A dipole magnet based on the common-coil design and pre-reacted Nb_3Sn cable was developed at the Fermi National Accelerator Laboratory (FNAL) for a Very Large Hadron Collider. Three technological racetrack models and a short dipole model have been fabricated and tested. This chapter summarizes the main results of this program.

15.1 Introduction

In 1998 the Fermi National Accelerator Laboratory (FNAL) launched a new superconducting accelerator magnet research and development (R&D) program with the goal to develop cost-effective and robust high field magnets and technologies for a post-Large Hadron Collider (LHC) machine. The collider concept discussed at that time was called the Very Large Hadron Collider (VLHC) (Fermilab 2001). It was based on a staged approach with ~2 T magnets in low-field Stage 1 and ~10 T magnets in high-field Stage 2. Comprehensive studies of various magnet designs with small coil aperture, various coil cross-sections, cable parameters, etc., were performed to find optimal parameters and sound, affordable Nb_3Sn dipole design suitable for VLHC Stage 2 with a nominal operating field of 10 T.

One of the dipole designs, promoted at that time by Brookhaven National Laboratory (BNL) and Lawrence Berkeley National Laboratory (LBNL), was based on simple block-type racetrack coils arranged in a common-coil dipole configuration (Gupta 1997) (see also Chaps. 13 and 14). Although the common-coil configuration is not the most efficient with respect to the required conductor amount, it has some attractive technological features. For example, the coil radii in this design are defined by the distance between apertures rather than by aperture size, making it suitable for brittle conductors such as Nb_3Sn.

As a part of FNAL high field accelerator magnet R&D, a common-coil accelerator dipole with Nb_3Sn coils was designed and fabricated using the react-and-wind

A. V. Zlobin (✉)
Fermi National Accelerator Laboratory (FNAL), Batavia, IL, USA
e-mail: zlobin@fnal.gov

D. Schoerling, A. V. Zlobin (eds.), *Nb$_3$Sn Accelerator Magnets*, Particle Acceleration and Detection, https://doi.org/10.1007/978-3-030-16118-7_15

(R&W) method. The main results of this work are presented in this chapter. It includes design studies and analyses of the common-coil dipole concept, as well as magnet design, fabrication technology, and specific features and parameters of the first accelerator quality Nb_3Sn common-coil dipole. This magnet has many innovative design and technological features, such as single layer coils, a 60-strand Rutherford cable with large aspect ratio, and stainless-steel collars reinforced by horizontal bridges between coil blocks. The chosen collar structure required simultaneous winding of both left and right coils into the collar structure and impregnating the coil with epoxy inside this structure. Test results of the Nb_3Sn common-coil dipole model are reported and discussed. The broader R&D challenges, including studies and selection of strand and cable as well as fabrication and tests of practice coils and racetrack magnets, are also presented and discussed.

15.2 Magnet Design and Analysis

Several common-coil dipole designs suitable for VLHC were studied in the late 1990s and the early 2000s at FNAL (Ambrosio et al. 2000a; Sabbi et al. 2000). A common feature of all those designs was a multi-layer, multi-block coil based on Rutherford cables. Design analysis revealed serious mechanical difficulties in those designs, in particular large deformations of coils and support structures during magnet excitation, and high stresses in the coil and the support structure. To resolve these problems, a design with a horizontally split yoke was proposed (Ambrosio et al. 2001). Although this design allowed the reduction of stresses in the coils, it complicated the yoke configuration and assembly. This design option ultimately led to complex magnet fabrication process, and eliminated the main advantage of the common-coil design concept—its simplicity.

To avoid these problems, a single-layer common-coil dipole with a strong internal support structure was also developed at FNAL (Kashikhin and Zlobin 2001a). This design maintained all of the advantages of the common-coil approach, including the option of using the R&W method, and also provided aperture size, maximum field, field quality, and coil volume comparable to a corresponding cos-theta dipole design.

15.2.1 Magnetic Design

15.2.1.1 Single-Layer Coil

The cross-section of the single-layer coil developed at FNAL for a common-coil dipole is presented in Fig. 15.1. Each coil contains 56 turns combined into three blocks and divided by 6 mm thick spacers. There are also two 3 mm thick spacers in each central block. The pole blocks are displaced horizontally by 5 mm with respect

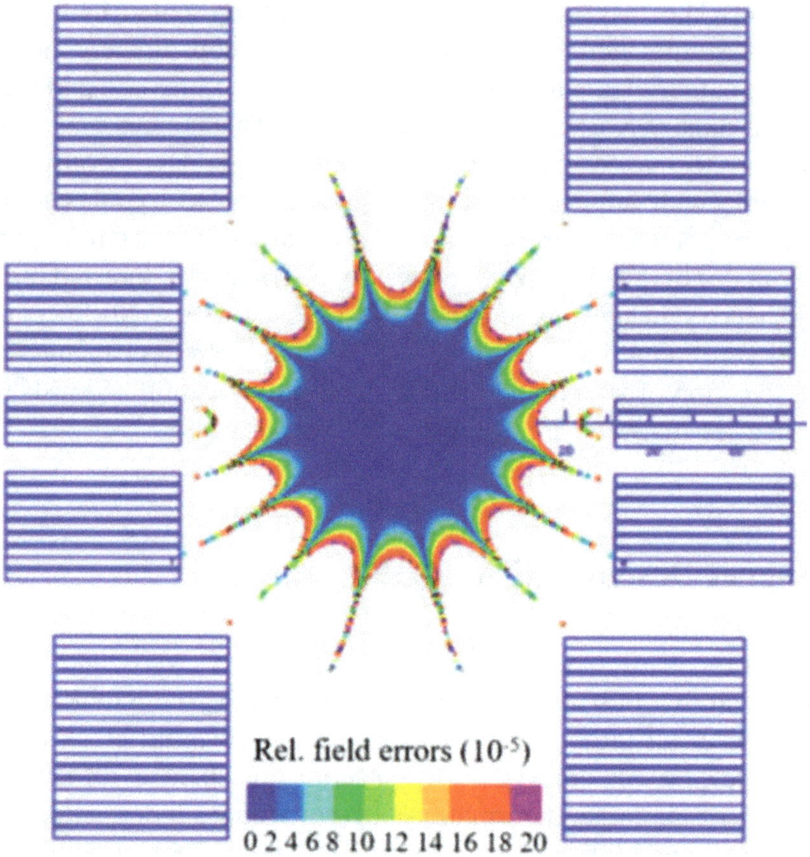

Fig. 15.1 Single-layer coil for a common-coil dipole

to the central blocks. The size and position of blocks and spacers were optimized using the ROXIE computer program (Russenschuck 1995) to achieve the maximum transfer function, minimum coil volume, and small low-order geometrical harmonics. The gap of 40 mm between pole blocks defines the diameter of the coil apertures.

The coil uses 21.09 mm wide and 1.245 mm thick rectangular Rutherford cable made of 60 strands of 0.7 mm diameter. The cable insulation thickness is 0.10 mm. Based on bending degradation studies described below, to use a reacted cable with the chosen strand size, the minimal coil radius must be about 90 mm, which defines the minimal aperture separation in this common-coil dipole.

15.2.1.2 Cold Iron Design

Figure 15.2 shows the initial cross-section of a single-layer common-coil dipole with cold iron. Two coils surrounded by the mechanical support structure are placed inside the round iron separated vertically into two pieces. The gap between the iron pieces is always open to ensure contact of the collared coil with the iron after cooling to operating temperature. Special holes in the iron blocks and magnetic inserts are

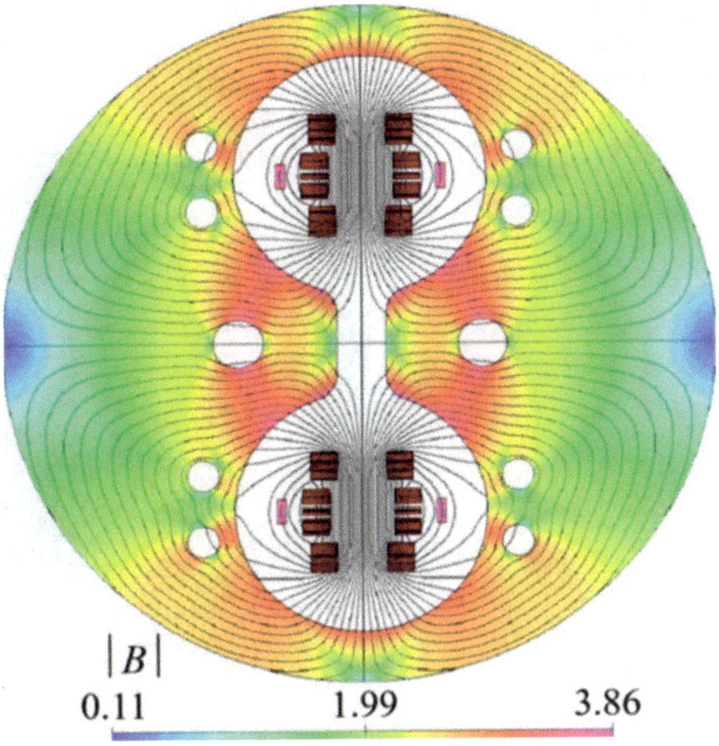

Fig. 15.2 Common-coil dipole concept with cold iron

used to reduce the iron saturation effect. The iron inner surface, outer radius, and the geometry and position of the correction holes and inserts were optimized to achieve good field quality and the appropriate transfer function, and to minimize fringe fields and iron size.

Due to the asymmetry of the iron with respect to the horizontal plane of the magnet apertures, there is a skew quadrupole component a_2 of -7 units (10^{-4} relative to the dipole field at a reference radius of 10 mm). This skew quadrupole component was suppressed by shifting the top and bottom coil blocks by ~0.5 mm with respect to the coil mid planes of each aperture. The optimized iron outer diameter (OD) is 564 mm and the aperture separation is 280 mm.

15.2.1.3 Warm Iron Design

Figure 15.3 shows a cross-section of the single-layer common-coil dipole with warm iron. To compensate for the decrease of the magnet transfer function and the maximum field in this design, the number of turns in the coil was increased from 56 to 61. Magnetic coupling between apertures in this design produces a skew quadrupole component a_2 as large as -35 units. Some small horizontal and vertical displacements of the coil blocks, of less than 0.5 mm, allowed canceling of this component. The effect of iron saturation on the field harmonics was reduced by optimizing the iron inner and outer diameters. The optimal values of the iron OD and thickness in this design were 710 mm and 55 mm, respectively.

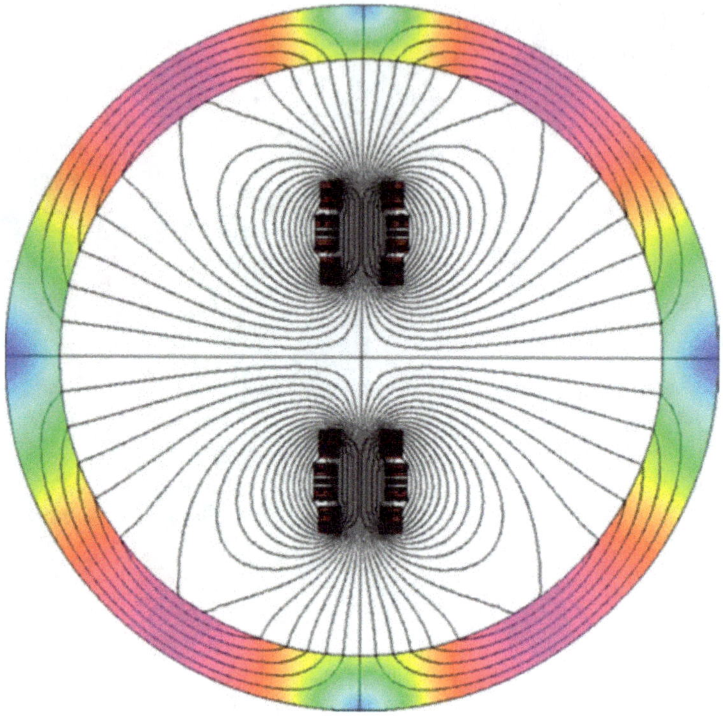

Fig. 15.3 Common-coil dipole concept with warm iron

Table 15.1 Calculated common-coil dipole parameters

	Cold iron	Warm iron
Aperture diameter (mm)	40	40
Aperture separation (mm)	280	280
Iron outer diameter (mm)	564	710
Max. bore field (T)	10.75	10.69
Max. quench current (kA)	24.5	25.4
Stored energy at 11 T (kJ/m)	880	956
Inductance at 11 T (mH/m)	2.78	2.74
Coil area (cm^2)	53.4	58.2

15.2.1.4 Magnet Parameters

The main parameters of the common-coil dipoles with cold and warm iron are shown in Table 15.1. The maximum bore field B_{max} was calculated for a cable packing factor of 0.88, a Cu/non-Cu ratio of 0.85, and a cable critical current density J_c(12 T, 4.2 K) of 2 kA/mm^2.

The maximum field in the magnet aperture at 4.2 K vs. the critical current density of the superconductor in the coil is shown in Fig. 15.4. Assuming 10% J_c degradation during coil fabrication, to provide the nominal operating field of 10 T with 15% margin, the designs described need Nb_3Sn strands with a rather high J_c(12 T, 4.2 K) of 3 kA/mm^2 and a Cu/non-Cu ratio of 1.2.

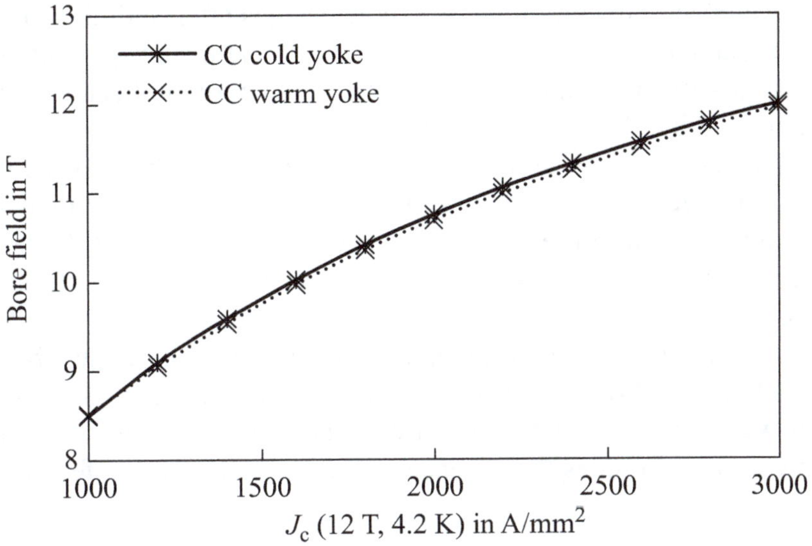

Fig. 15.4 Maximum field in the magnet aperture at 4.2 K vs. the critical current density J_c(12 T, 4.2 K) of superconductor in the coil

Table 15.2 Geometrical harmonics at $R_{ref} = 10$ mm in 10^{-4}

Harmonic number	Cold iron		Warm iron	
	a_n	b_n	a_n	b_n
2	0.005	–	0.000	–
3	–	0.000	–	−0.005
4	−0.001	–	0.002	–
5	–	−0.001	–	−0.001
6	−0.002	–	−0.003	–
7	–	−0.001	–	−0.032
8	0.011	–	−0.128	–
9	–	−0.046	–	−0.059
10	0.003	–	−0.004	–

The field in the aperture of accelerator magnets is represented in terms of harmonic coefficients defined by the formula

$$B_y + iB_x = B_1 \sum_{n=1}^{\infty} (b_n + ia_n) \left(\frac{x + iy}{R_{ref}} \right)^{n-1},$$

where B_x and B_y are horizontal and vertical transverse field components, B_1 is the dipole field strength, and b_n and a_n are the $2n$-pole coefficients at a reference radius R_{ref}. The normal b_n and skew a_n harmonic coefficients are expressed in units of 10^{-4} parts of the main dipole field B_1.

The calculated geometrical harmonics at $R_{ref} = 10$ mm for the common-coil dipoles with cold and warm iron are presented in Table 15.2.

It can be seen that both common-coil dipole designs after optimization have very small geometrical field harmonics. Analysis showed that the design with warm yoke tolerates a coil-to-yoke misalignment of up to 2 mm without noticeable deterioration of field quality and overload of the magnet support system.

It was possible to suppress the effect of iron saturation on the skew quadrupole a_2 and normal sextupole b_3 in both designs up to 12 T. The calculated reduction of the magnet transfer function in the cold iron design reached ~10% at the bore field of 12 T, whereas there was no iron saturation effect on the magnet transfer function for the warm iron design.

Note that contrary to twin-aperture dipole magnets with a horizontal bore arrangement (Kashikhin and Zlobin 2001a) (see also Chap. 7), the common-coil dipole design with warm iron does not have any advantage relative to the common-coil dipole with cold iron, besides the absence of the iron saturation effect.

15.2.2 Mechanical Design

15.2.2.1 Structure Concept

In a common-coil dipole design the two apertures are positioned vertically. This layout requires a rather thick skin (significantly thicker than in designs with a horizontal aperture layout), since the horizontal components of the Lorentz force in each coil are added. The described single-layer coil with current blocks separated by relatively large spacers allows the use of a coil support structure with stress management. This reduces the Lorentz force transferred to the skin (Novitski et al. 2001) and, thus, the skin thickness.

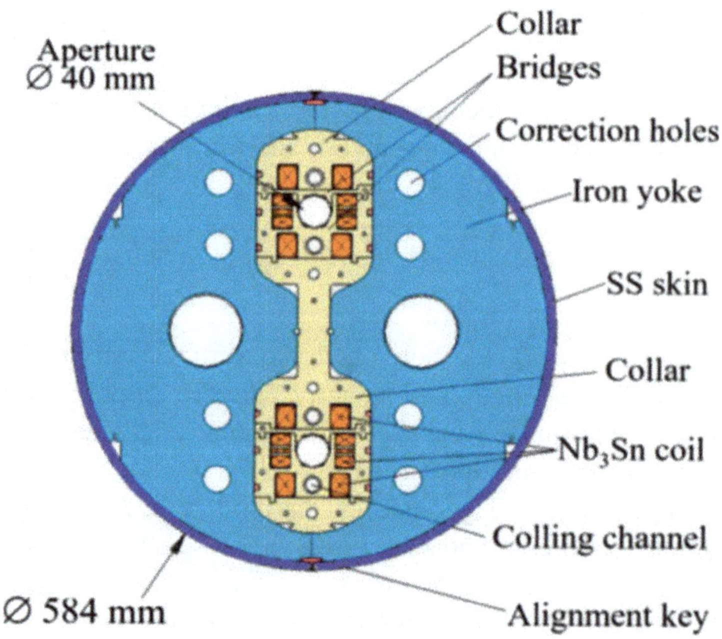

Fig. 15.5 Cross-section of the single-layer common-coil dipole

The cross-section of the optimized single-layer common-coil dipole with cold iron yoke and special collars (HFDC series) is shown in Fig. 15.5. The collar is common for both apertures to simplify coil winding and collared coil handling. The collar structure has rectangular windows for the coil blocks, and round holes for the beam pipes and cooling channels. During magnet yoking and skinning it protects the coil from the horizontal and vertical over-compression. In operation, the structure prevents transferring the vertical Lorentz forces from the pole blocks to the coil central blocks. It also captures a substantial part of the horizontal Lorentz forces in the coils, thereby reducing the force applied directly to the yoke and skin.

The collared coil is located inside the vertically split iron yoke, which is itself enclosed in a stainless-steel skin. The vertical pre-stress of the coils is supplied by the collars. The horizontal pre-compression of the collared coil is produced by the stainless-steel skin through the iron yoke. Thick end plates welded to the skin restrict the axial coil motion under the Lorentz forces.

The described design implies some obvious fabrication steps and conditions. Specifically, both coils have to be wound concurrently and directly into the coil support structure. To do this, the collar is divided into several parts (see Fig. 15.5) that are locked together with keys. After winding, the collared coil assembly is impregnated with epoxy, which produces a very strong mechanical structure. Mold release is used on the coil-collar interfaces for shear stress relief.

15.2.2.2 Mechanical Analysis

Mechanical analysis was performed to validate the design concept of the magnet support structure, optimize the structure dimensions, and choose the materials (Novitski et al. 2001). The main goal of the analysis was to keep the maximum stress in the coil below 150 MPa under all conditions, limit turn displacements by 0.1 mm in a field range up to 11 T, and provide operation of the structure in the elastic regime.

A 2D ANSYS (ANSYS Inc., Canonsburg, PA) model, which included the coil blocks surrounded by 0.5 mm thick polyimide insulation, the stainless-steel coil support structure with 20 mm wide collars, the iron yoke, and the 10 mm thick stainless-steel skin, was created. The thermal and mechanical properties of the materials used in the model are reported by Chichili et al. (2000). Since the coil is impregnated inside the support structure at low compression, the value of the coil modulus of elasticity used in the calculation was 20 GPa.

Calculations were performed for both aluminum and stainless-steel collars. The main benefit of the aluminum collar is its large thermal contraction, which increases the coil pre-stress after cooling down. The analysis revealed, however, that the peak equivalent stress in the aluminum collar exceeds the material yield stress by a factor of two for the chosen collar dimensions. To reduce the stress, a substantial increase of the collar size would have been necessary. The calculated distributions of the equivalent stress in the coil with stainless-steel collars after assembly, after cooling down and at the design field of 11 T, are shown in Fig. 15.6.

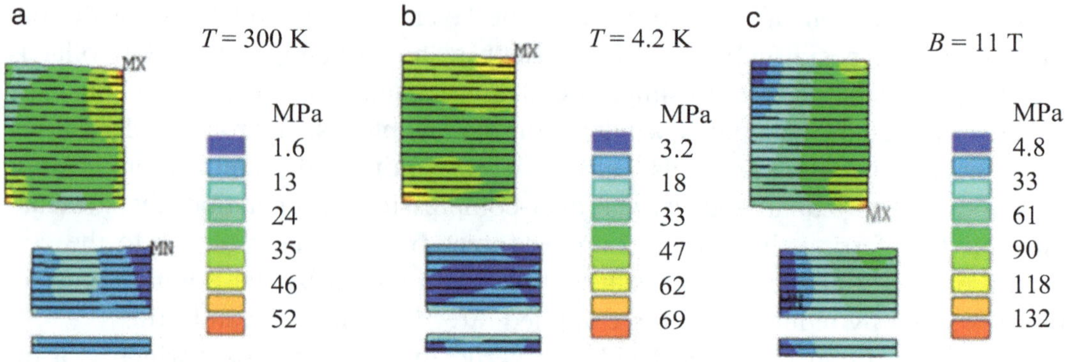

Fig. 15.6 Distributions of the equivalent stress in MPa in the coil quadrant: (**a**) after assembly (300 K); (**b**) after cool-down (4.2 K); and (**c**) at a bore field of 11 T

Table 15.3 Equivalent stress in coil and support structure

Stage	Coil (MPa)	Bridges (MPa)	Collar (MPa)	Yoke (MPa)	Skin (MPa)
300 K	52	118	361	99	125
4.2 K/0 T	69	178	408	194	327
4.2 K/11 T	132	533	455	152	358

At 300 K, the pole blocks are preloaded nearly uniformly in vertical and horizontal directions, whereas the center blocks have a lower, mostly vertical preload. At 4.2 K and maximum Lorentz force, there is no stress on the inner coil–collar interface, while on the outer interface it is large and nearly equally distributed. Analysis shows that the collar structure described intercepts half of the horizontal Lorentz force and prevents transfer of the vertical Lorentz force from the pole to the center blocks.

The stress distribution in the yoke is not uniform. The highest stress is near the collar–yoke interface. The radial force on the yoke–skin interface is uniform, leading to a homogeneous azimuthal stress distribution in the skin. The maximum equivalent stresses in all parts of the magnet are less than the target design values under all conditions.

For the case of glued coil-to-collar interfaces, the coil shape practically follows the shape of the support structure, and the coil displacements are rather small. However, the shear stress on some coil surfaces exceeds 50 MPa, which is critical for epoxy bonding that has a shear strength of 30 MPa (Ohira and Nishijima 2000). Therefore, a stress-relief layer is needed, particularly on the inner and top surfaces of the pole and center blocks. With a sliding coil–collar interface, the horizontal block displacements are less than 0.07 mm, independently from the coil pre-stress. A small vertical coil pre-stress of 30 MPa reduces vertical movements of the pole-block top surface to 0.03–0.04 mm, which is tolerable from the viewpoint of variation of field harmonics in the operating cycle, but may still impact the magnet quench performance.

Table 15.3 presents the calculated equivalent stress in the magnet coil and the support structure at different assembly and operating stages. It can be seen that the target stress and displacement limits are satisfied in this design.

15.3 R&W Common-Coil Dipole R&D

A special R&W R&D program in support of the single-layer common-coil dipole was performed at FNAL in collaboration with LBNL (Ambrosio et al. 2000b, 2002a, b, 2003, 2004a; Bauer et al. 2001). It included the following steps:

1. Wire and cable R&D for the R&W technology;
2. Technology development using sub-scale cable, flat racetrack coils and simple bolted structures;
3. Fabrication and test of a short mechanical model;
4. Fabrication of a full-scale technological model.

15.3.1 *Nb_3Sn Wire and Cable R&D*

The R&W technology requires the use of Nb_3Sn cables with thin strands to reduce the I_c degradation due to the bending of reacted cable. On the other hand, the target design field of 10 T requires a relatively wide cable for the single-layer dipole. A 21 mm wide Rutherford cable with 60 strands of 0.7 mm diameter meets both requirements. Such a cable has an aspect ratio of ~18, which was noticeably larger than most state-of-the-art cables. Fabrication of this cable required equipment that at the time was available only at LBNL.

A first cabling run was performed at LBNL in September 2000 using internal tin (IT) Nb_3Sn wires developed for the International Thermonuclear Experimental Reactor (ITER) project. The wires were drawn down to a nominal diameter of 0.7 mm. The assessment of this cable at FNAL showed that it could be reacted and then wound into racetrack type coils. Moreover, measurements of strands extracted from the cable confirmed that the I_c degradation due to the cabling was relatively small, approximately 10%. Two subsequent runs to produce long cable lengths were then performed. The first run used IT wires produced by Intermagnetics General Corporation (IGC), and the second run used modified jellyroll (MJR) wires produced by Oxford Superconducting Technologies (OST).

To evaluate the expansion of the cable cross-section after reaction, the two different cables were measured in free conditions before and after reaction. It was found that the width of the IT cable increased by 0.7% and the cable thickness increased by 3%, whereas the MJR cable increased by 1.6% in both dimensions.

The conductor R&D program also included studies of I_c degradation of various wire types and diameters, fabrication of cables with and without a stainless-steel core, and measurement of cable I_c degradation due to bending and transverse pressure after bending. The main outcome of the studies was that using synthetic oil during cabling avoided strands sticking during cable reaction, and therefore reduced I_c degradation due to cable bending. The cables with a stainless-steel core, however, which is used to suppress eddy currents in cables, demonstrated larger I_c degradation in cable bending experiments.

15.3.2 Single-Layer Racetracks

To develop procedures (including coil impregnation inside the magnet structure) and select an appropriate cable insulation design and material for the R&W common-coil dipole, simple single-layer racetrack magnets (HFDB series) based on sub-sized pre-reacted cable were built and tested. The racetrack magnet consisted of two flat racetrack coils (Fig. 15.7) connected in the common-coil configuration, and a bolted mechanical structure without an iron yoke (Fig. 15.8).

The cable had 41 Nb_3Sn strands, each 0.7 mm in diameter, and a rectangular cross-section of 15.05 mm width and 1.218 mm thickness. The racetrack mechanical structure included two thick main plates connected by fifty seven 25.4 mm diameter bolts, and two side and two end pushers bolted to the main plates with smaller bolts. The side and end pushers provided the initial coil pre-stress inside the structure. The racetrack magnet parameters are summarized in Table 15.4.

Each racetrack coil was 730 mm long and had a 400 mm long straight section. The minimal cable bending radius in the coil was 90 mm. Each coil end had one spacer, whereas there were no spacers in the coil straight section. All the parts inside the coil were made of G10 and the end shoes were made of bronze. Fiberglass sheets were placed on each side of the coil. Two coils inside the mechanical structure were separated by a 5 mm thick G10 plate. All the structure parts (main plates, side and end pushers) were made of stainless steel.

Three racetrack magnets of HFDB series were fabricated and tested at FNAL from 2001 to 2003.

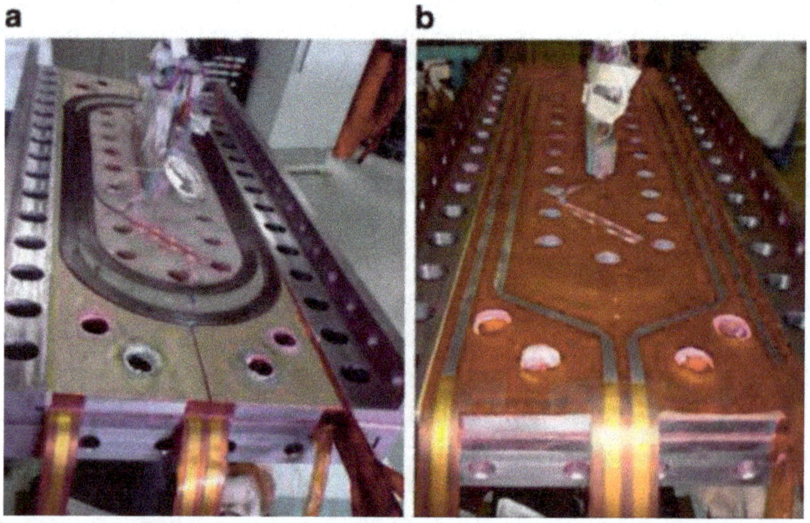

Fig. 15.7 (**a**) A single-layer racetrack coil after winding is being installed into the support structure (HFDB02). Each coil end has one spacer, whereas no spacers in the coil straight section were used. All parts inside the coil are made of G10. The end shoes are split in two pieces and are made from bronze. (**b**) Quench protection heaters are installed on both sides of the coil

Fig. 15.8 Racetrack mechanical structure (lead end). Two thick main plates are connected by 57 bolts, and two side and two end pushers are bolted to the main plates

Table 15.4 Racetrack design parameters

Parameter	Value
Gap between coils (mm)	5
Number of turns per coil	29
Minimum radius in the ends (mm)	90
Transfer function (T/kA)	0.625
Inductance (mH)	0.33
Stored energy at 9 T (kJ)	34.3
Normal horizontal force at 9 T (MN)	1.8

15.3.2.1 HFDB01

This magnet used IT Nb_3Sn wire produced by IGC. The wire had 19 sub-elements surrounded by a tantalum barrier. The Cu/non-Cu ratio was 0.61. The effective filament diameter was large, ~0.176 mm, and together with the low Cu/non-Cu ratio, it was suspected to be the cause of conductor instabilities at fields below 13 T.

The cable was fabricated at LBNL and heat treated at FNAL using the following four-step schedule: 215 °C for 175 h, 340 °C for 120 h, 575 °C for 160 h, and 700 °C for 30 h. Steps 2 and 3 were introduced to prevent tin leaks during cable heat treatment. The duration of step 4 was reduced in an unsuccessful attempt to improve the conductor stability at 10 T. The I_c degradation at 12 T due to cabling was ~16%.

Magnet tests revealed a large reduction of 39% of the magnet quench current with respect to the maximum current estimated from short sample data. Many voltage spikes were observed during the current ramp up.

15.3.2.2 HFDB02

This magnet had several new features, including a different wire, cable insulation, and new techniques of cable insulation and coil winding. The Rutherford cable was made of wires produced by OST using the MJR process with 54 sub-elements and 46.5% copper. Samples of round wire, reacted with the cable, had a critical current of 1.87 kA/mm^2 at 12 T and 4.2 K. The measured I_c degradation due to cabling was only 2%. The calculated maximum field in the coil at 4.5 K was 10.2 T at 16.87 kA. Coils were wound inside the mechanical structure and impregnated in situ. Polyimide films with mold release were used to avoid coils sticking to the mechanical structure.

The magnet reached 7.7 T, which corresponded to 78% of the expected short sample limit. The highest quench current of 12.68 kA was reached at a ramp rate of 75 A/s, showing a reduction of magnet quench currents at both lower and higher current ramp rates. An unusual correlation was also seen between temperature and quench current. Instead of increasing, magnet quench current monotonically decreased when the helium bath temperature decreased from 4.5 to 1.9 K.

15.3.2.3 HFDB03

Based on the HFDB02 results, some of the fabrication steps were further improved. A gap was introduced between the innermost cable turn and the core of the reaction spool. Some voltage taps were made of 0.05 mm thick brass strips instead of copper wires. An appropriate cable support was provided at all stages of the coil winding. Eight instrumented end bullets with resistive strain gauges, as opposed to four in HFDB02, and capacitance gauges in the coil straight sections, were used to monitor coil stresses. The instrumented side bolts were placed according to the locations of peak loads, as identified by the finite element analysis.

The HFDB03 short sample limits at 4.5 K were 10.06 T and 16.59 kA. The maximum quench current was 12.59 kA, or 76% of the expected short sample limit. The ramp rate dependence again had a positive slope at low ramp rates, with a peak at 300 A/s. The dependence of magnet quench current vs. temperature was flat.

Fabrication and tests of HFDB racetrack magnets allowed the development of many aspects of the R&W technology, such as reaction of long cable lengths without strand sintering, cable co-winding with insulation, coil in situ impregnation with epoxy, and coil preload. Magnet quench performance studies did not, however, provide clear answers on the poor conductor performance in these magnets. As understood later (Zlobin et al. 2006), the quench performance of a majority of Nb_3Sn magnets tested at FNAL in the early 2000 was significantly limited by flux jump instabilities in the Nb_3Sn wires.

15.3.3 HFDC Mechanical Model

The mechanical model of the common-coil dipole consisted of a 165 mm long slice of magnet straight section. It was assembled and tested to validate magnet assembly procedures and compare the results of mechanical tests with the predictions of mechanical analyses. The mechanical model typically includes all components of the dipole straight section, i.e., coils, collars, yoke, and skin. Coil blocks were made of 250 mm long hand-insulated cable pieces. A picture of the instrumented mechanical model is shown in Fig. 15.9.

The main goals of the mechanical model were: testing various turn and ground insulation designs; practicing coil–collar assembly and simulating coil winding inside the collar structure using various insulations; and studying the stresses in the skin and in the collars after skin welding and after cool-down.

All cables around one aperture were insulated with a 0.05 mm thick E-glass tape with 45% overlap, all other cables were separated by 0.125 mm polyimide layers as turn insulation. A number of ground insulation materials, such as glass-fiber reinforced epoxy (G10), polyimide layers, and their combinations, were used with both types of cable insulation to select the most appropriate combination. The collared coil model was assembled, vacuum impregnated with CTD-101 epoxy, and the insulation tested at room temperature.

The mechanical model assembly was completed by adding the yoke, which consisted of 36 laminations, and the skin, which was welded under a press. The stresses in the skin, yoke and collars were measured and compared with the predictions of the finite element model. The mechanical model also allowed checking of all the assembly steps.

Fig. 15.9 Instrumented mechanical model before skin welding. The mechanical model includes all components of the dipole straight section, i.e., coils, collars, yoke, and skin

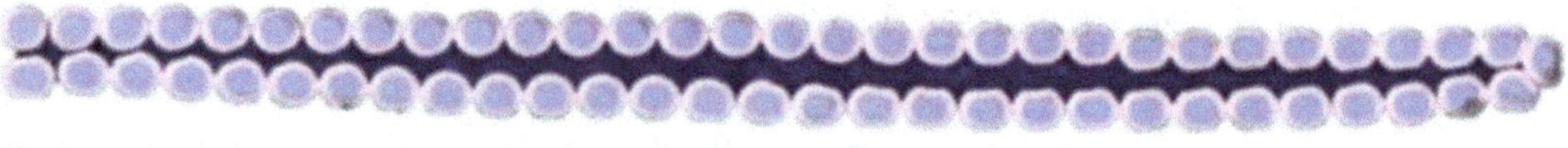

Fig. 15.10 Cross-section of the 60-strand Rutherford cable

It was found that the polyimide strip insulation is acceptable, whereas the glass tape insulation is too soft and may cause loss of coil pre-stress. The collar packs and the skin worked well, as expected. However, some modifications to the collar–yoke shim and instrumentation wires were required.

15.3.4 *HFDC Technological Model*

To test the fabrication tooling and assembly procedures, an 800 mm long technological model was also built using reacted Nb_3Sn cable and actual magnet components.

15.3.4.1 Nb_3Sn Cable

The Rutherford cable for the technological model (Fig. 15.10) had a rectangular cross-section and was made of 60 Nb_3Sn 0.7 mm diameter strands. The strands were produced by IGC using the IT process developed for the ITER conductor. A 300 m long piece of cable for the technological model was produced at LBNL.

Two 120 m long pieces of cable were wound on two single-layer metallic spools together with a mica-glass tape to prevent cable sintering during heat treatment. To minimize cable bending strain during winding, the radius of the reaction spools was 180 mm, which is a factor of two larger than the minimum radius in the coil ends. The cable was reacted in an argon atmosphere using the 4-step heat treatment schedule described above.

15.3.4.2 Insulation

Inter-turn insulation was chosen based on the test results of the mechanical model and the racetrack coils. To minimize the risk of additional cable degradation due to insulation wrapping and additional cable re-spooling during this process, the insulation was co-wound with the cable during coil winding. The insulation consisted of two tapes with the same width as the cable: a 0.163 mm thick pre-impregnated glass tape and a 0.075 mm thick polyimide tape. Both 110 m long glass and polyimide tapes were wound on the same spool.

The ground insulation placed around each current block comprised 0.5 mm thick G10 sheets and a 0.25 mm polyimide film. To fit the slightly wider reacted cable into the nominal collar windows, the nominal thickness of the ground insulation of the coil blocks was reduced by 0.25 mm.

Fig. 15.11 The copper-stabilized pre-formed splice

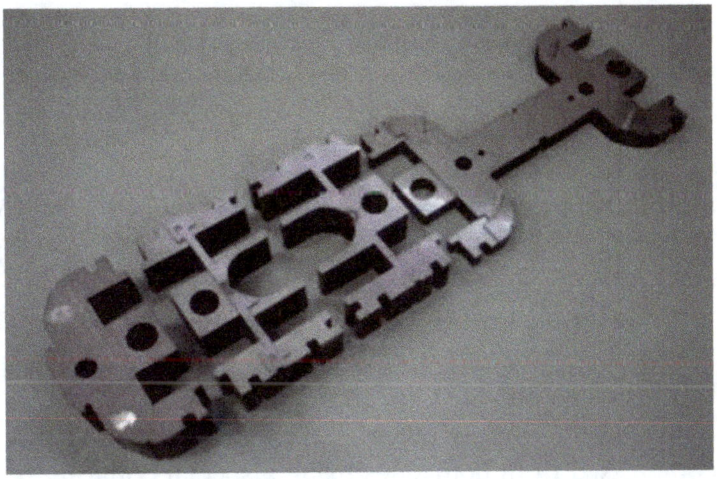

Fig. 15.12 Collar lamination packs are composed of several blocks to confine and lock the coil blocks and provide stress management

15.3.4.3 Splice Joint

The left and right coils were simultaneously wound into the support structure. To achieve the required current direction in the coils the two Nb_3Sn cables were spliced before winding in a special fixture after installation of both spools on separate tensioners. The cables were joined using a U-shaped, pre-reacted multi-strand Nb_3Sn connector and copper stabilizer. The fixture also provided the desired curved shape of the splice. A copper-stabilized pre-formed splice is shown in Fig. 15.11.

15.3.4.4 Coil Winding

Each coil block was wound inside windows formed by the collar laminations (Fig. 15.12) and the stainless-steel ends, and locked in place by keys or screws. Both coils were wound and locked block-by-block (Fig. 15.13). The insulation strips were wound simultaneously with the bare cable. Four independent tensioners were

Fig. 15.13 Coil winding is performed horizontally using a rotating table. Four independent tensioners, shown in the top part of the picture, are used to apply a tension to each cable and to the insulation tapes

used to apply a tension of 90 N to each cable and of 135 N to the insulation. This approach reduced the risk of cable collapse and strand pop-out, and decreased the coil spring-back during winding.

The insulated splice was carefully inserted in the splice slot on the return end of the winding mandrel, thereby avoiding extreme cable bending. The structure formed by the collars provided the proper positioning for the insulation tapes and the cables. Side pushers were used to achieve a dense winding. The entire straight section was compressed from both sides using a special collaring fixture to insert the keys. The fixture was then removed to wind the next coil blocks. Alternated laminations were assembled in ~38 mm long (20 laminations) packs using two pins as a base. Thin stainless-steel washers separated the laminations in the pack, and provided an adequate path for the epoxy during impregnation.

15.3.4.5 Magnet Leads

Nb_3Sn cable leads were spliced with two Nb-Ti cables and two stabilizing copper strips using a special splicing fixture. The splicing procedure was identical to that developed for the technological model. The 150 mm long splices were placed in the lead end block, such that about half of each splice was outside the collared coil and in direct contact with liquid helium.

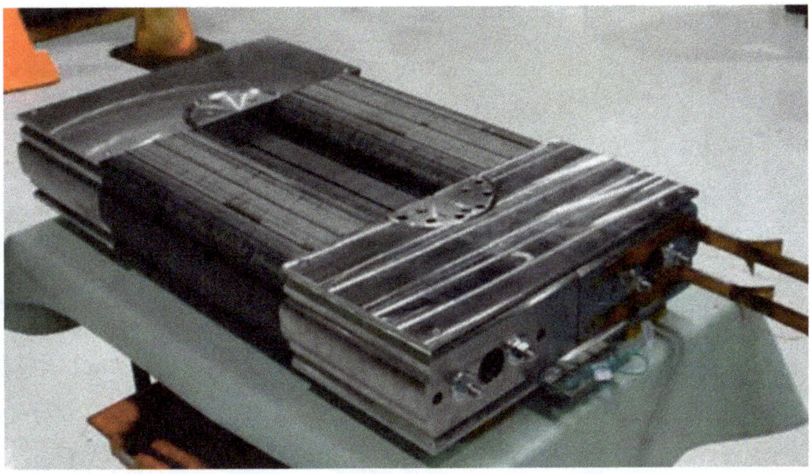

Fig. 15.14 Assembled technological model. The magnet straight section is made of a collar lamination block. The coil ends and the inter-layer splice are supported by solid stainless-steel blocks

15.3.4.6 Instrumentation

The technological model was also used to develop the magnet instrumentation plan and its technology. Instrumentation included voltage taps on each coil block and in the center of the splice box, spot-heaters, temperature sensors on each Nb_3Sn cable coming out of a splice and close to each spot-heater, and quench heaters on the external side of each first block.

The fabrication of the technological model was stopped before epoxy impregnation to reuse the parts in the dipole model. The assembled technological model is shown in Fig. 15.14. The model was disassembled and the parts were used in the common-coil dipole model HFDC01.

15.4 HFDC01 Dipole Model Fabrication

15.4.1 Magnet Design Features

The 3D view of the common-coil dipole model HFDC01 is shown in Fig. 15.15. The magnet parameters are reported in Table 15.5. Based on the results from the mechanical and technological models, the necessary corrections in the design were introduced. The fabrication process for the single-layer common-coil dipole model was based on the experience gained during the technological model fabrication, with specific features as described below.

The Nb_3Sn cable for HFDC01 consisted of 59 MJR strands produced by OST. The number of strands was reduced from 60 to 59 to fit the cable into the gap in the collar packs. To avoid strand sintering during reaction, the cable was impregnated with synthetic Mobil-1® oil. Two 120 m long cable pieces were then co-wound on

Fig. 15.15 Common-coil dipole model

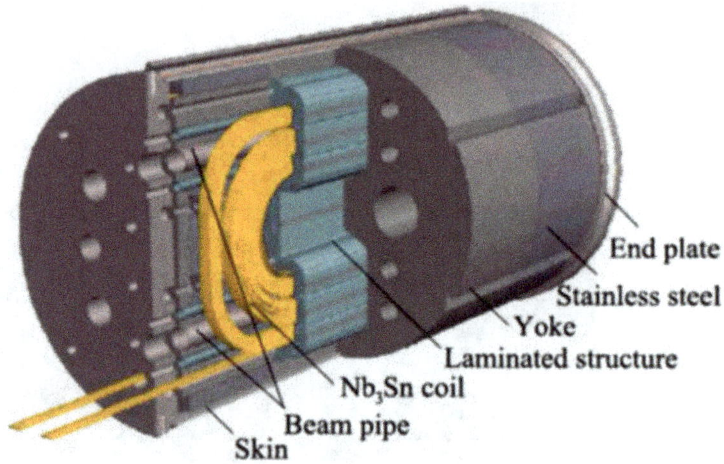

Table 15.5 Magnet design parameters

Parameter	Value
Maximum bore field (T)	10.0
Maximum quench current (kA)	23.6
Aperture diameter (mm)	40
Aperture separation (mm)	290
Iron yoke outer diameter (mm)	564
Iron yoke length (m)	0.4
Straight section length (m)	0.4
Stored energy at 10 T (kJ/m)	820
Inductance at 10 T (mH/m)	2.95

two stainless-steel spools with a 0.1 mm thick, 22 mm wide mica-glass Suritex 0822 tape to prevent sintering of the cable turns during reaction. The cable reaction was performed in an argon atmosphere inside a retort using the heat treatment cycle suggested by OST: 210 °C for 100 h, 340 °C for 48 h, and 650 °C for 180 h.

To monitor cable critical current, several witness samples (both round wires and strands extracted from the cable) were heat-treated on titanium-alloy barrels together with the cable. The calculated magnet short sample limit based on the witness sample data, assuming an additional 6% bending degradation of cable I_c during coil winding, was 23.3 kA, which was very close to the design value presented in Table 15.5.

Before winding, two Nb_3Sn cables were spliced with two U-shaped connectors made of copper-stabilized 41-strand Nb_3Sn cable, similar to the cable used in racetrack coils. The connectors were reacted using the same heat-treatment cycle as the main cable.

The cable insulation comprised two tapes: 0.16 mm thick pre-impregnated fiberglass tape and 0.08 mm thick polyimide tape. Both tapes were co-spooled on a single bobbin. The cable and the insulation were co-wound under tension into the slots of the collar structure. The inner pole blocks were wound first with a cable tension of 140 N and an insulation tension of 110 N. The actual thickness of these first blocks was slightly larger than the design value, due to rather large variations of cable thickness. Therefore, it was decided to reduce the number of turns in each pole block from 18 to 17 and fill the remaining space with 1.27 mm thick G10 spacers.

The wound pair of coil blocks was gently preloaded by sets of laminated packs using a special fixture. The packs were locked together with keys. The coil ends were loaded by stainless-steel end parts with screws, rods, and nuts. The ground insulation was then installed into these new structural slots, and winding was resumed. The second and third coil blocks were wound the same way as the first ones.

The coil instrumentation was installed at the coil ends during winding, since there was no access to the coil after the collaring of each block. Each cable in the transition area between current blocks was equipped with voltage taps and temperature sensors. There were a total of 50 voltage taps, four spot-heaters, and six temperature sensors installed in the coils.

Each Nb_3Sn lead cable was spliced with two Nb-Ti cables and two stabilizing copper strips following the same splicing procedure used for the technological model. The 150 mm long splices were placed in the lead end block, such that about a half of each splice was outside the collared coil to have direct contact with liquid helium.

The collared coil assembly was vacuum-impregnated with epoxy in the closed mold. All internal cooling channels were plugged and sealed with High-Temperature Red RTV Silicone Gasket, and the entire outer surface of the collared coil block was treated with mold release. The block was placed into an impregnation fixture, pumped out, and filled with CTD-101 epoxy. It was then cured in an oven at 125 °C for 21 h.

The impregnated collared coil block was shimmed and placed between the two 0.4 m long yoke halves. Two stainless-steel blocks were installed around each magnet end. The gap between yoke blocks at room temperature was ~1.25 mm at the magnet body, and gradually reduced to zero at the magnet ends with the stainless-steel blocks. The skin halves were partially (50%) hand-welded together inside the press using tack-weld and skip-weld techniques. Four final welding passes were added outside of the press after installation of the end plates.

Thick 50 mm end plates welded to the skin with bullets restricted the axial motion of the coil ends. Magnet ends were preloaded with bullets instrumented with strain gauges. The initial load of 5000 N per bullet was chosen to provide contact between the end plates and magnet ends at all temperatures. A picture of the HFDC01 dipole model is shown in Fig. 15.16.

15.5 Magnet Test

HFDC01 was tested in the Vertical Magnet Test Facility at FNAL in two thermal cycles (Kashikhin et al. 2004a, b).

Fig. 15.16 Common-coil dipole model HFDC01 is ready for cold testing

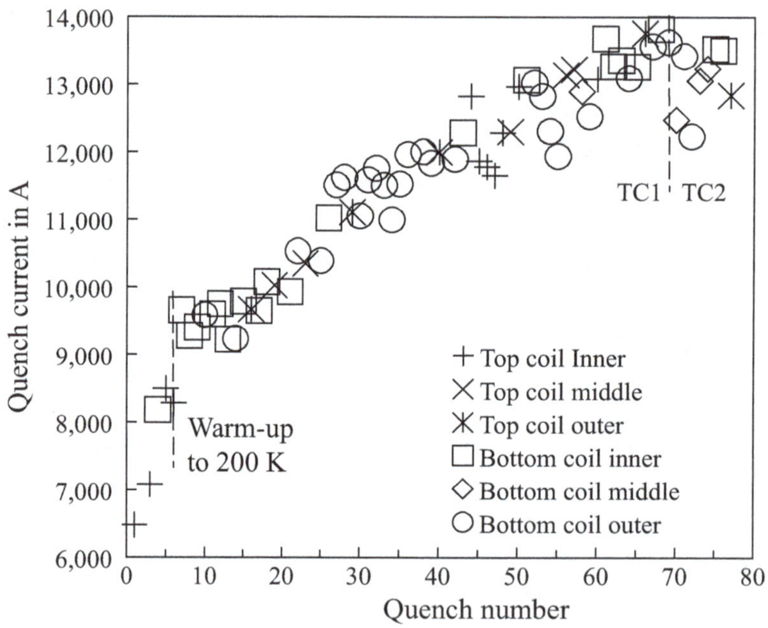

Fig. 15.17 HFDC01 training history

15.5.1 Magnet Training

The HFDC01 training history is plotted in Fig. 15.17. Magnet training at 4.5 K started at a very low current of ~6.5 kA (less than 30% of the magnet short sample limit), and was rather slow with many detraining quenches. Numerous voltage spikes were detected at low currents during the current ramp up between 2 kA and

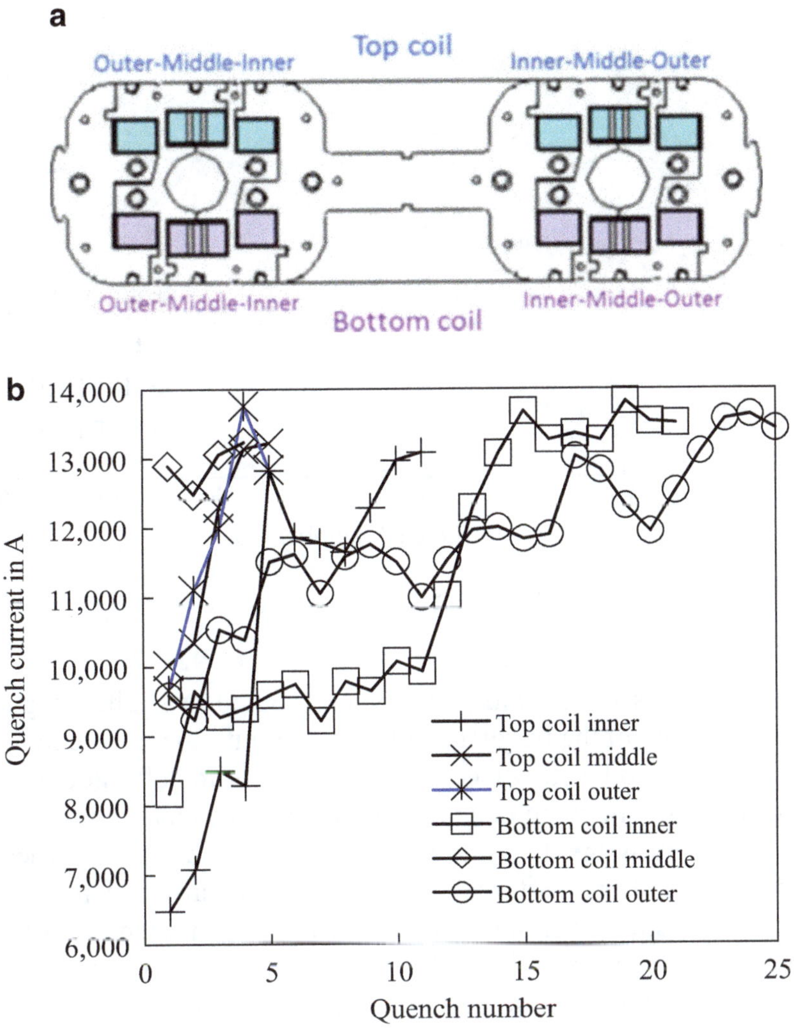

Fig. 15.18 (**a**) Coil block nomenclature; and (**b**) training quench sequence in coil blocks

8 kA. After the first five quenches the magnet was warmed up to 200 K. This intermediate warm-up followed by cooling down to 2.2 K did not impact magnet training. The magnet was warmed up to room temperature after 69 quenches. Some detraining was seen after the second thermal cycle (TC2).

The maximum quench current of 13.67 kA, which corresponds to ~57% of the magnet short sample limit, was reached after 67 quenches. Even after this rather large number of quenches the magnet training was not complete. At this point magnet training was suspended.

The quench sequence in each coil block is shown in Fig. 15.18. A majority (~70%) of quenches started in the so-called "bottom coil" (this coil was at the bottom during collared coil impregnation with epoxy). These quenches were almost equally distributed between the inner and the outer coil pole blocks in this coil, and only a small number of quenches were observed in the middle (central) blocks of both coils.

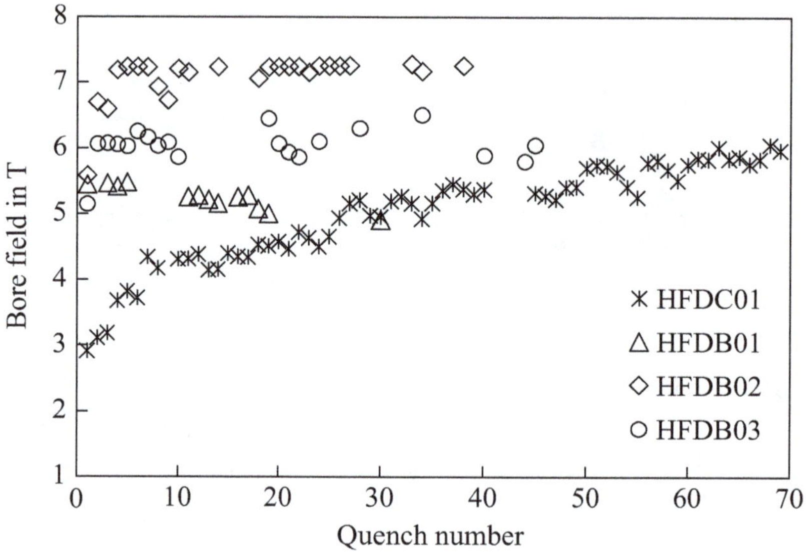

Fig. 15.19 HFDB and HFDC model training data

Although the quenches in the mid-plane blocks started at slightly higher currents (but still at around 50% of the magnet short sample limit), there were indications that their training would also be slow (see the data for Bottom coil middle).

In the pole blocks, where the magnetic field reaches its maximum, large Lorentz force components push the turns towards the coil mid-planes, whereas in the mid-plane blocks the Lorentz force is horizontal. An insufficient coil pre-stress during assembly, or stress lost during cool-down, are consistent with the observed domination of quench origin in the coil pole blocks. The reason why the numerous quenches originated in the bottom coil remained unknown.

Bullet gauge data showed transfer of only around 15% of the axial Lorentz force to the magnet end plates. The rest of the force was transferred to the skin due to friction between the collared coil, iron yoke, and skin, and intercepted by the coil itself.

HFDC01 training data are compared in Fig. 15.19 with the training curves for FNAL R&W single-layer racetrack magnets HFDB01, 02, and 03 described in Sect. 15.3. The maximum field reached in the HFDC01 aperture was 5.97 T, similar to the maximum field in the gap of HFDB01 and HFDB03, and slightly lower than in HFDB02. The dipole training was, however, much longer than the racetrack magnet training. Perhaps, in addition to conductor instabilities, it was due to the more complicated dipole structure and fabrication procedure (especially the impregnation with epoxy), and to insufficient coil preload.

To address all these questions, additional tests were planned using the second dipole model HFDC02. A particularly interesting experiment would have been magnet assembly and test without coil impregnation with epoxy. It might have eliminated epoxy cracking as a source of perturbations in the coil, and improved coil cooling conditions. Nevertheless, the common-coil dipole program was discontinued at this stage due to a change in FNAL high field magnet program priorities.

15.5.2 Field Quality Measurements

Magnetic measurements were performed in both apertures in two thermal cycles using a vertical rotating coil system with a 250 mm long coil, 25 mm in diameter. The aperture closest to the leads is called "aperture I" below, whereas the other one is called "aperture II." Field harmonics are reported at $R_{ref} = 10$ mm in the magnet geometrical center. Since the HFDC01 maximum quench current was limited at 13.7 kA ($B_{max} = 5.9$ T in aperture), magnetic measurements were performed only up to ~5.3 T.

The transfer function B/I and the normalized skew quadrupole a_2, which were measured in a current cycle from 0 to 12 kA and back to 0 in the magnet body and calculated for both 2D and 3D cases, are shown in Figs. 15.20 and 15.21. The values calculated in the 3D case were integrated over a 250 mm region for a correct comparison with the measurements. The iron saturation effect is clearly seen in both plots at fields above 2.5 T. In the 1 m long model this effect reduces the magnet transfer function by 3.5%, and increases the absolute value of the skew quadrupole a_2 by 13 units at 5 T with respect to the long magnet (2D values). The 3D calculations agree very well with the measurements for fields above 1.5 T. The good correlation of measurements with the 3D calculations proved that the iron saturation effect in this magnet type can be predicted and optimized by numerical simulations.

The persistent current effect in the normal sextupole b_3, measured and calculated in the magnet body, is shown in Fig. 15.22 (the geometrical components reported in Table 15.1 were subtracted from the measured data). The ramp-up branches of

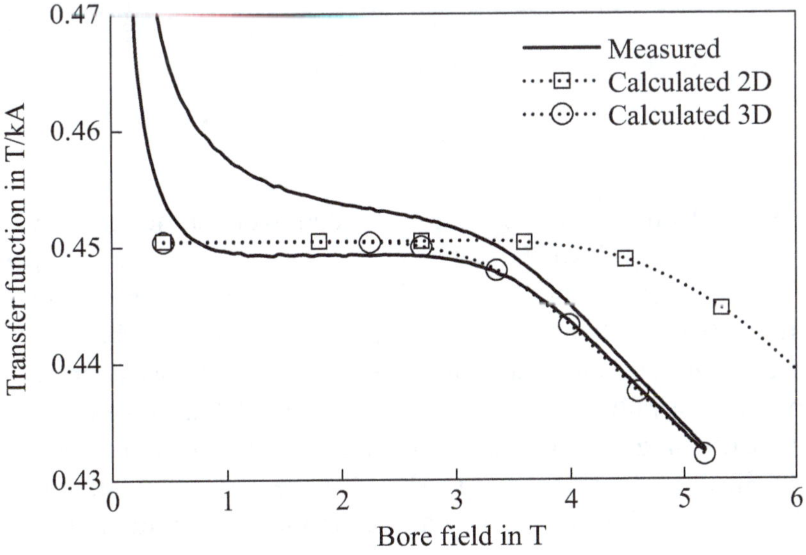

Fig. 15.20 Magnet transfer function vs. bore field

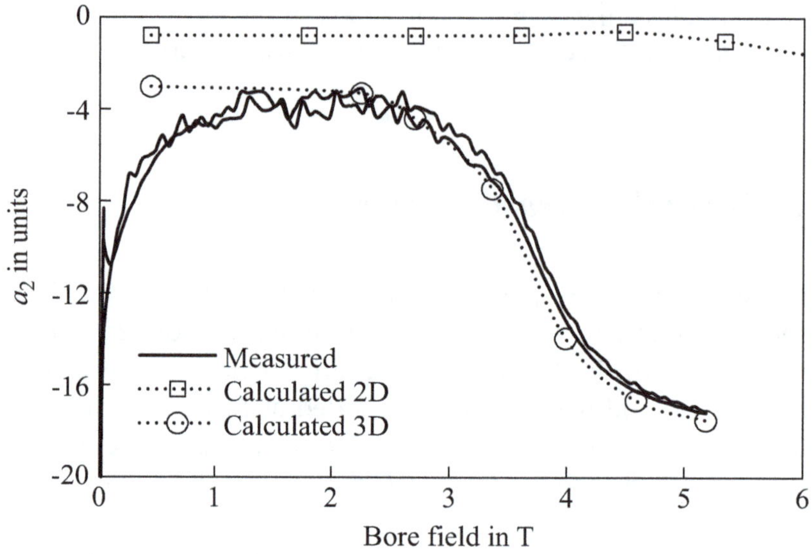

Fig. 15.21 Skew quadrupole a_2 vs. bore field

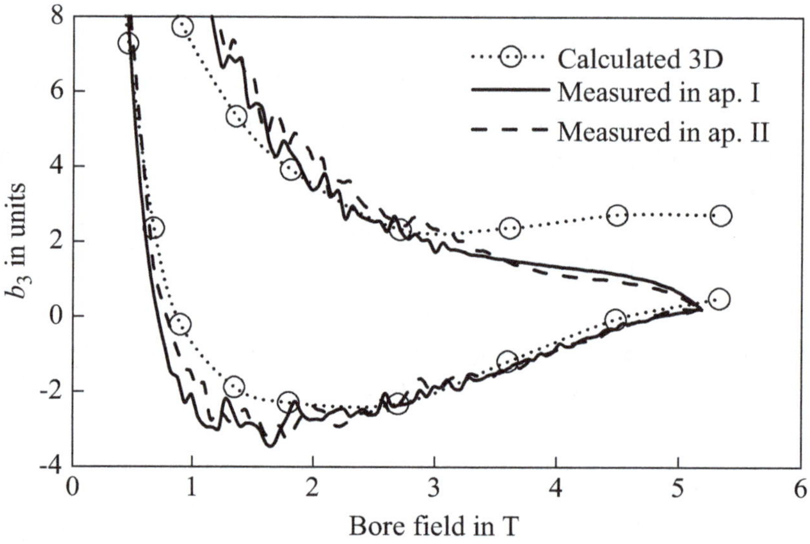

Fig. 15.22 Normal sextupole b_3 vs. bore field

both the measured and calculated curves match well at fields above 2 T. The discrepancy in the ramp-down branches at higher fields is due to superconductor re-magnetization, which was not included in the calculations. One can clearly see the iron saturation effect in both loops, which would otherwise be symmetric with respect to the horizontal axis. The coil geometry of the single-layer magnet design described provides a rather small width for the calculated and measured loops in Fig. 15.22 (Kashikhin and Zlobin 2001b), which is a factor of 5 smaller than in the cos-theta Nb_3Sn dipole models of the HFDA series (Barzi et al. 2002) (see also Chap. 7), despite the larger coil cross-section area in the common-coil magnet.

Table 15.6 Geometrical harmonics in magnet straight section

n	Calculated "as built" values		Measured in aperture I		Measured in aperture II	
	a_n	b_n	a_n	b_n	a_n	b_n
2	−3.30	–	−3.28	0.16	−3.57	−1.26
3	–	10.64	0.23	10.30	−0.20	10.37
4	0.03	–	−0.35	0.02	−0.66	−0.17
5	–	0.35	0.04	0.73	−0.04	0.79
6	−0.04	–	0.00	−0.01	0.01	0.01
7	–	−0.08	0.00	−0.06	−0.00	−0.05
8	−0.00	–	Used for the centering correction			
9	–	−0.01	−0.00	−0.03	−0.00	−0.03

Large fluctuations of b_3 (Fig. 15.22) as well as voltage spikes were seen during HFDC01 tests at low currents. These results are consistent with the large flux jump instabilities observed in the Nb_3Sn MJR54/69 wires used in this magnet.

The harmonic ramp-rate dependence was measured in three consequent current cycles up to 12 kA, with current ramp rates of 20 A/s, 40 A/s, and 80 A/s. The dependence of the field harmonics on the current ramp rate was very small. The absence of eddy currents in the cable was related to the rather large inter-strand resistance, produced during the cable reaction with synthetic oil without transverse pressure on the cable (Ambrosio et al. 2004b).

To evaluate the dynamic effects at injection-like conditions, field measurements were performed at a current plateau of 2.4 kA ($B = 1.125$ T) of 30 min duration, following a pre-cycle up to 12 kA. The measured harmonic decay on the current plateau was very small, less than 0.3 unit, with respect to the b_3 decay in Nb-Ti accelerator magnets. This result agreed well with measurements of dynamic effects in FNAL Nb_3Sn cos-theta dipole models of the HFDA series (Barzi et al. 2002).

The geometrical harmonics were defined as average values between up and down current ramps at 5 kA. A comparison of measured and calculated values in the 3D case for the as-built geometry harmonics is presented in Table 15.6. For a correct comparison, the calculated harmonics were integrated over the probe length of 250 mm. There is an excellent correlation between the calculated and measured geometrical harmonics in the magnet straight section. The integrated harmonics in the magnet return end are also consistent with the calculations. It confirmed the high efficiency of the coil support structure for maintaining the nominal coil geometry, and thereby the field quality during magnet assembly and operation.

15.6 Conclusion

A single-layer common-coil dipole magnet was developed at FNAL for a VLHC. The magnet was designed to provide 10 T nominal field of accelerator quality in two 40 mm diameter apertures at an operating temperature of 4.5 K. To produce the nominal operating field of 10 T with 15% margin, the design described needs Nb_3Sn wires with a high J_c(12 T, 4.2 K) of 3 kA/mm^2.

The magnet has several innovative design and technological features, such as a single-layer coil, a 22 mm wide 60-strand Rutherford-type cable made of 0.7 mm Nb_3Sn wires, and a stainless-steel coil support structure reinforced by horizontal bridges inserted between coil blocks. Both the left and right coils were wound simultaneously into the collar structure and then impregnated in situ with epoxy.

The magnet was designed to use the R&W technique (i.e., the cable is reacted before coil winding). Various aspects of this technique were studied using mechanical and technological models and a series of single-layer racetrack magnets. Three 1 m long racetracks (HFDB01-03) and one common-coil dipole model (HFDC01) based on reacted cables were fabricated and tested from 2001 to 2003. All magnets survived the complicated fabrication process and reached 60–75% of the expected short sample limit. A good, well-understood field quality was achieved in both apertures of the HFDC01 dipole model. However, the dipole model and all the racetrack magnets showed large quench current degradation, and the dipole model also had very slow training. There are several indications that this performance was partially caused by flux jump instabilities in the Nb_3Sn composite wires used at the time. To further explore the potential of the common-coil dipole design and the R&W technology for accelerator magnets, more efforts are required.

References

Ambrosio G, Kashikhin VV, Limon PJ et al (2000a) Conceptual design study of high field magnets for very large hardon collider. IEEE Trans Appl Supercond 10(1):310–313. https://doi.org/10.1109/77.828236

Ambrosio G, Andreev N, Barzi E et al (2000b) Study of the react and wind technique for a Nb_3Sn common coil dipole. IEEE Trans Appl Supercond 10(1):338–341. https://doi.org/10.1109/77.828243

Ambrosio G, Andreev N, Barzi E et al (2001) Development of react and wind common coil dipoles for VLHC. IEEE Trans Appl Supercond 11(1):2172–2175. https://doi.org/10.1109/77.920288

Ambrosio G, Andreev N, Barzi E et al (2002a) R&D for a single-layer Nb_3Sn common coil dipole using the react-and-wind fabrication technique. IEEE Trans Appl Supercond 12(1):39–42. https://doi.org/10.1109/tasc.2002.1018347

Ambrosio G, Andreev N, Barzi E et al (2002b) Development and test of a Nb_3Sn racetrack magnet using react and wind technology. In: Breon S, DiPirro M, Glaister D et al (eds) Advances in cryogenic engineering: proceedings of the cryogenic engineering conference vol 47A. American Institute of Physics, Melville, NY, AIP Conf Proc 613, pp 329–336

Ambrosio G, Andreev N, Barzi E et al (2003) Fabrication and test of a racetrack magnet using pre-reacted Nb_3Sn cable. IEEE Trans Appl Supercond 13(2):1284–1287. https://doi.org/10.1109/tasc.2003.812647

Ambrosio G, Andreev N, Barzi E et al (2004a) Design modifications, fabrication and test of HFDB-03 racetrack magnet wound with pre-reacted Nb_3Sn Rutherford cable. In: Waynert J, Barclay J, Breon S et al (eds) Advances in cryogenic engineering: transactions of the international cryogenic engineering conference—CEC, vol 49A. AIPC, Anchorage, Alaska, 22–26 September 2003. American Institute of Physics, Melville, NY, AIP Conf Proc 710, p 767

Ambrosio G, Barzi E, Chichili D et al (2004b) Measurement of inter-strand contact resistance in epoxy impregnated Nb_3Sn Rutherford cables. In: Balachandran UB, Adams M (eds) Advances in cryogenic engineering: transactions of the international cryogenic engineering conference—CECAIP, Anchorage, Alaska, 22–26 September 2003, vol 711. American Institute of Physics, Melville, New York, pp 828–835

Barzi E, Carcagno R, Chichili D et al (2002) Field quality of the Fermilab Nb_3Sn cos-theta dipole models. In: Terence G (ed) Proceedings of the 8th European particle accelerator conference (EPAC 2002), Paris, France, 3–7 June 2002. European Physical Society, Geneva, pp 2403–2405

Bauer P, Ambrosio G, Andreev N et al (2001) Fabrication and testing of Rutherford-type cables for react and wind accelerator magnets. IEEE Trans Appl Supercond 11(1):2457–2460. https://doi.org/10.1109/77.920360

Chichili DR, Arkan TT, Ozelis JP et al (2000) Investigation of cable insulation and thermo-mechanical properties of epoxy impregnated Nb_3Sn composite. IEEE Trans Appl Supercond 10(1):1317–1320. https://doi.org/10.1109/77.828478

Fermilab (2001) Design study for a staged Very Large Hadron Collider. Fermilab-TM-2149, 4 June. http://lss.fnal.gov/archive/test-tm/2000/fermilab-tm-2149.pdf

Gupta R (1997) A common coil design for high field 2-in-1 accelerator magnets. In: Comyn M, Craddock MK, Reiser M et al (eds) Proceedings of the 1997 particle accelerator conference, Vancouver, 12–16 May 1997. IEEE, Piscataway/New York, pp 3344–3346

Kashikhin VV, Zlobin AV (2001a) Magnetic designs of 2-in-1 Nb_3Sn dipole magnets for VLHC. IEEE Trans Appl Supercond 11(1):2176–2179. https://doi.org/10.1109/77.920289

Kashikhin VV, Zlobin AV (2001b) Correction of the persistent current effect in Nb_3Sn dipole magnets. IEEE Trans Appl Supercond 11(1):2058–2061. https://doi.org/10.1109/77.920260

Kashikhin VS, Ambrosio G, Andreev N et al (2004a) Development and test of single-layer common coil dipole wound with reacted Nb_3Sn cable. IEEE Trans Appl Supercond 14(2):353–356. https://doi.org/10.1109/tasc.2004.829128

Kashikhin VS, Andreev N, DiMarco J et al (2004b) Field quality measurements of Fermilab Nb_3Sn common coil dipole model. IEEE Trans Appl Supercond 14(2):287–290. https://doi.org/10.1109/tasc.2004.829087

Novitski I, Andreev N, Ambrosio G et al (2001) Design and mechanical analysis of a single-layer common coil dipole for VLHC. IEEE Trans Appl Supercond 11(1):2276–2279. https://doi.org/10.1109/77.920314

Ohira S, Nishijima S (2000) Effect of impregnating material failure on stability of superconducting magnet analyzed by wire dynamics simulation. IEEE Trans Appl Supercond 10(1):665–668. https://doi.org/10.1109/77.828321

Russenschuck S (1995) A computer program for the design of superconducting accelerator magnets. In: 11th annual review of progress in applied computational electromagnetics, Monterey, 20–24 March 1995, vol 1, pp 366–377; CERN AT/95-39

Sabbi G, Ambrosio G, Andreev N et al (2000) Conceptual design of a common coil dipole for VLHC. IEEE Trans Appl Supercond 10(1):330–333. https://doi.org/10.1109/77.828241

Zlobin AV, Kashikhin VV, Barzi E et al (2006) Effect of flux jumps in superconductor on Nb_3Sn accelerator magnet performance. IEEE Trans Appl Supercond 16(2):1308–1311. https://doi.org/10.1109/tasc.2006.870557

Part V
Future Needs and Requirements

Chapter 16
Nb_3Sn Accelerator Dipole Magnet Needs for a Future Circular Collider

Davide Tommasini

Abstract The Future Circular Collider (FCC), or the High-Energy Large Hadron Collider (HE-LHC), would require bending magnets operating at 16 T. The large quantity of high-performance conductor required for these projects can only be satisfied by using Nb_3Sn superconductor. This chapter summarizes the main design approaches and parameters for these dipole magnets.

16.1 Introduction

A new proton collider representing a step forward with respect to the Large Hadron Collider (LHC) would provide collisions at a center-of-mass energy of the order of 100 TeV. This could be achieved, as proposed by, the Future Circular Collider (FCC) study (CERN 2013), with bending magnets operating at 16 T in a 100 km long circular machine. Magnets operating in the same field range could also be considered, should interest arise in doubling the energy of the LHC (Todesco and Zimmermann 2011).

With respect to the LHC, this quest for a doubling of the field requires a change of superconducting material, because the upper critical field B_{c2} of Nb-Ti at 1.9 K is limited to, at most, 13.5 T (Bottura 2000), limiting the ultimate field amplitude of Nb-Ti accelerator magnets to about 10 T. A field level of 16 T is also 5 T higher than the field in the Nb_3Sn magnets currently being developed for the High Luminosity LHC (HL-LHC or Hi-Lumi LHC) (Ferracin et al. 2016; Savary et al. 2017). Once installed, they will be the first high-field Nb_3Sn magnets ever operated in a particle collider. It will be shown below that the Nb_3Sn compound is appropriate for 16 T accelerator dipole magnets operating at 1.9 K with a similar margin (14% on the load line) to that used for the LHC magnets. The field amplitude range between 10 T and 16 T is of interest not only for the HL-LHC, the High-Energy LHC (HE-LHC), and

D. Tommasini (✉)
CERN, European Organization for Nuclear Research, Meyrin, Switzerland
e-mail: Davide.Tommasini@cern.ch

D. Schoerling, A. V. Zlobin (eds.), *Nb_3Sn Accelerator Magnets*, Particle Acceleration and Detection, https://doi.org/10.1007/978-3-030-16118-7_16

the FCC, but also for several other initiatives such as, for example, the 100 km long version of the Super Proton–Proton Collider (Gao 2016), a Very Large Hadron Collider (Bhat et al. 2013), or for muon colliders (Kashikhin et al. 2012; Zlobin et al. 2013).

With respect to the LHC dipole magnets, the higher field level yields increases of the forces, size, weight, and stored energy. For magnet cross-sections optimized for using a minimum amount of high-performing conductor, the larger forces also result in greater stress levels in the coil and in the structural components, making the structural design, the magnet's integration into a cryostat, and installation in the accelerator challenging. Finally, the management of the larger stored energy imposes increased challenges in magnet and circuit protection, in particular for the dielectric strength of the coil insulation.

The next sections summarize and discuss the baseline design choices and parameters of these magnets.

16.2 Conductor

16.2.1 Superconductor

As anticipated, the required field amplitude of 16 T cannot be achieved with coils made of Nb-Ti. According to Larbalestier (2017), unless a major discovery is made in the next decade, the candidate conductor materials are to be selected from Nb_3Sn, BSCCO (in particular Bi-2212: $Bi_2Sr_2CaCu_2O_8$), and $REBa_2Cu_3O_7$, where RE stands for rare earth element (REBCO). We exclude MgB_2 due to its low critical field B_{c2}, though the situation may change if the results obtained with thin films (Dai et al. 2011) can be extended to practical conductors. We also have to exclude Fe-based superconductors, mainly because it is not clear whether the limitations of the connectivity between grain boundaries can ever be overcome. We remark, however, that a new development of MgB_2- or Fe-based superconductors may, especially for the latter, open new cost-effective, high-performance opportunities. With regret, however, at this stage there is no basis for considering these materials for high-field accelerator magnets.

Both BSCCO and REBCO can carry engineering current densities of practical use for magnets at field amplitudes well beyond a target field amplitude of 16 T: for that field level, however, their cost cannot yet compete with that of Nb_3Sn. Indeed, a moderate performance increase of the critical current density of Nb_3Sn (in the order of 2300 A/mm^2 at 16 T and 1.9 K, which we will define as the FCC target) with respect to the best state of the art (Larbalestier 2017) would effectively fulfill the requirements for these magnets in terms of conductor current density. To support this statement we consider as an example the 16 T cos-theta magnet operating with a 14% margin on the load line being developed by Istituto Nazionale di Fisica Nucleare (Sorbi 2017). In this design, using the FCC target conductor, the critical current density of the inner coil layers at the short sample peak field (18.7 T) and at an operating temperature of 1.9 K is about 1200 A/mm^2. It will be shown below that to protect the magnet in case of a quench in this configuration, about half of the conductor should contain copper. An additional increase of current density would require a larger fraction of copper, so it would not be linearly exploited in terms of engineering current density. For the outer coil layers, where the short sample peak field is limited to about 14–15 T, the amount of copper needed for magnet protection represents about two times the amount of the non-copper part of the conductor. For example, a 15% further increase of J_c of the outer layer conductor allows a reduction of the amount of superconductor in the wire by 15%, but, as the amount of copper remains the same, it represents a decrease of the wire size by only about 5% (in reality a bit more because a smaller coil is electromagnetically more efficient).

In summary, the FCC target performance for Nb_3Sn seems to be a sufficient conductor for these 16 T magnets: any other option should be considered only if it is more cost-effective. If the target nominal field was higher than 16 T (or if a larger margin would be needed) then the situation would be different: this will be discussed in Sect. 16.3.3.

16.2.2 Nb_3Sn Wire and Cable

An overview on the development of Nb_3Sn wires and cables focused on high-field accelerator magnets can be found in Chap. 3 of this book, and a discussion on targets for research and development (R&D) programs can be found in Ballarino and Bottura (2015).

The wire and cable characteristics and sizes are a compromise between what is required for an optimized electromagnetic design and the achievable electro-mechanical performance of the conductor once submitted to the different manufacturing (including wire cabling and magnet coil winding) and operational phases. For these high-field magnets, there is an interest in using large wires and cables to allow for large currents, and therefore keeping the magnet inductance within reasonably low limits, as required for magnet and circuit powering and protection. Furthermore, the effective filament diameter of the wire should be reasonably small in order to reduce the wire magnetization, which has an impact on the magnet field quality at low fields, on conductor stability against flux jumps, and on the energy losses to be cooled by the cryogenic system. The issue of producing large wires with a small effective filament size has been tackled in two development programs started in the US in 1999 (Scanlan 2001) and in Europe in 2004 for the Next European Dipole program (NED) (Devred et al. 2004). Target values for the effective filament size were < 40 μm for the US program and < 50 μm for NED, and for a strand diameter of up to 1.25 mm in the case of NED. These target values have now been approached, and just recently achieved in critical current density on limited unit lengths (Xu et al. 2019). The 16 T design options explored within the European Circular Collider (EuroCirCol) program (Tommasini et al.

2017) are presently considering a target effective filament size of 20 μm and a maximum strand diameter of 1.20 mm.

The maximum number of strands in a cable is not only dependent on the cabling capacity (presently up to 40 strands in Europe and up to 60 strands in the US), but also on the cable stability during winding and magnet assembly. For example, it is expected that winding a cos-theta geometry with a large cable is more challenging than winding a racetrack coil for a common-coil geometry.

Finally, the maximum allowed reversible and irreversible stress limits for impregnated Rutherford cables at ambient temperatures and cryogenic temperatures represent severe bounds imposed upon the magnet design and may, for example, preclude the use of collared assemblies for dipole magnets in the 16 T field range. For the abovementioned EuroCirCol program, these stress limits have been set at 150 MPa and 200 MPa at ambient and cold temperatures, respectively.

16.2.3 Cost

The target conductor cost set in 1999 for the abovementioned US program was <US $1.50 per kA.m at 12 T, 4.2 K. When scaled to 16 T, 4.2 K, this corresponds to about US$3.5 per kA.m. The target cost currently considered in view of the FCC was set in 2016 at €5 per kA.m at 16 T, 4.2 K (Schoerling et al. 2017).

16.3 Design Parameters

Both the present design of the FCC machine and a possible energy upgrade of the LHC are based on scaling up the LHC machine. The required bending field integral for the FCC is about 1.0 MT.m to achieve a proton–proton center-of-mass energy of 50 + 50 TeV with 66 km of bending length, which, for a 100 km closed orbit length, corresponds to the same filling factor as in the LHC. Considering the same magnetic length of 14.3 m per magnet as in the LHC, the FCC reference lattice needs about 4600 magnets, producing a field of 16.0 T.

16.3.1 Magnet Aperture

The required physical magnet aperture depends on beam dynamics requirements, the beam screen, and the field quality at the reference radius.

For the FCC the physical magnet aperture has been set to 50 mm. A larger aperture would provide more margin for the dynamic aperture as well as for the design and integration of the beam screen: furthermore, it would also make it easier to achieve a given magnetic field quality at a given reference radius (17 mm for the

FCC). A smaller aperture would impose more stringent constraints on the magnets' field quality, the beam screen design and integration, and the magnets' alignment in the accelerator. Besides the fact that the physical aperture has an impact on the magnet stored energy, thus upon the powering and protection scheme, it also determines the amount of conductor needed for the coils. For the FCC it has been estimated (Schoerling et al. 2017) that, for an aperture of around 50 mm, the relative variation of amount of conductor is similar to that of the variation of the aperture: i.e., 10% less aperture corresponds to about 10% less conductor.

Considering the presence of the beam screen and vacuum pipe, such a small reduction of physical aperture would, however, considerably reduce the beam aperture.

16.3.2 Operating Temperature

The choice of the operating temperature has a strong impact on the magnet design and cost as well as on the cryogenic production and distribution system. Operation at 1.9 K, as in the LHC, allows easier and more effective cooling of the magnet coils, and a considerable saving of the amount of conductor needed for producing the 16 T field amplitude when compared to the alternative of working in supercritical helium. Furthermore, operation at 1.9 K greatly simplifies the beam screen design because it enables effective cryo-pumping (Baglin et al. 2013). On the other hand, producing superfluid helium requires additional cold compressors and more energy. It has been estimated (Schoerling et al. 2017) that the additional capital cost and the difference in operation cost over 10 years for cooling the magnets at 1.9 K compared to cooling at 4.5 K is largely compensated for by the savings in the cost of conductor.

16.3.3 Field Level and Margin

Considering the FCC target performance of the conductor discussed above, 16 T seems to be the highest field level that is still economically and technically interesting when the same margin (14%) on the load line of the LHC is considered and the operational temperature is set at 1.9 K. A comparison between optimized designs of different magnet cross-sections, cos-theta, block coils, and common coils, operating at 16 T with 18% and 14% margin on the load line, and using the FCC target performance for the conductor, has been performed in the frame of the EuroCirCol study (Gourlay 2016, 2017). In all cases, the coil cross-sections that were designed with 18% margin were using up to 30% more conductor than the cross-sections designed with 14% margin. At a lower field amplitude the relative difference between margin and field amplitude decreases, due to a combination of the variation of the critical current density with field amplitude and the quantity of copper required to protect the magnet in the case of a quench.

The concept of load line margin is the one most used when comparing magnets performance, because it gives a theoretical performance percentage given by a conductor's electromagnetic characteristics. The load line margin does not, however, have an immediate physical meaning as, for example, in the case where the temperature margin, the enthalpy margin, or the current margin are considered. The use of these latter concepts is, however, not necessarily more representative of magnet performance than using the margin on the load line, it is just different. For example, at a given margin on the load line and a given operating temperature, Nb_3Sn magnets have a much larger temperature margin than Nb-Ti, but not necessarily better performance. At 20% of the load line and 1.9 K the temperature margin for Nb-Ti is 2.1 K and for Nb_3Sn it is 4.5 K; and at 4.2 K it is 1.2 K and 3.0 K for Nb-Ti and Nb_3Sn, respectively. Concerning enthalpy margins, for Nb_3Sn these are typically three times larger than for Nb-Ti for the same conditions of load line margin and operational temperature (Todesco 2017).

16.3.4 Field Quality

The field quality required has an impact on the selection of the conductor, on the manufacturing and assembly tolerances of the magnet and other relevant parts, and on the degrees of freedom in the design of the magnet cross-section and the magnet ends.

We assume that the required control limits for tune and chromaticity are similar to those of the LHC, i.e., in the order of 0.001 of tune and 1 unit of chromaticity (Todesco et al. 2016).

In a hadron collider like the HE-LHC or the FCC, the quadrupole component is mainly due to the two-in-one magnet design, and varies along the powering ramp due to the saturation of the ferromagnetic yoke. To keep these effects small with respect to the force generated by the lattice quadrupoles and to preserve the dynamic aperture at injection, the magnet design should minimize the quadrupole component at the lowest field amplitude. At the highest fields this can be considerably relaxed because the tune of the two beams can be controlled separately as the two quadrupole apertures are individually powered. Indeed, referring to the field error in terms of units (parts of the main field in 10^{-4} at a reference radius two-thirds of the physical aperture), for the FCC the allowed quadrupole component of the dipoles is up to several tens of units at full energy. This limit allows for the design of 16 T dipoles with inter-beam distances in the range 200–250 mm.

Concerning the sextupole component, the main difficulty comes from the fact that controlling 1 unit of chromaticity corresponds to controlling a few percent of units for the sextupole component. For high intensity beams (i.e., during nominal operation), the chromaticity cannot be measured in real time during the magnet ramp, thus requiring an accurate magnetic model to predict the sextupole component during the ramp. Establishing such an accurate model is possible only if the behavior of the sextupole component during the ramp is accurately predictable, i.e., reproducible in time and amplitude. The quality of the prediction depends, on the spread of the conductor characteristics responsible for the sextupole variation during the magnet ramp, and on the shape and amplitude of the variation itself. For the LHC the field model, after several iterations and corrections from ad hoc measurements at low fields, can predict the sextupole component in the dipoles during the ramp by about 0.1 units (Todesco et al. 2016) for a total sextupole variation from injection to a nominal energy of about 10 units.

With respect to the filament diameter of 7 μm and 6 μm in the inner and outer coil layers, respectively, of the LHC dipoles (Rossi 2003), the effective filament size of the best-performing Nb_3Sn conductors is almost one order of magnitude larger. The associated sextupole component due to strand magnetization, decay, and snapback is also therefore much larger, due to both the higher critical current density and the larger filament size. Fortunately, it has been shown that several methods can be effective in mitigating these effects (Izquierdo Bermudez et al. 2016; Kashikhin and Zlobin 2016).

16.3.5 Structural Parameters

Major difficulties for Nb_3Sn with respect to Nb-Ti magnets are related to the lower strain and stress allowed in the brittle Nb_3Sn conductor, and to the higher rigidity of the impregnated Nb_3Sn coils with respect to non-impregnated Nb-Ti coils, which imposes tighter assembly tolerances. These difficulties are enhanced by the higher forces to be handled in a higher field magnet with respect to a lower field magnet, which also yields higher stresses for similar engineering current densities.

The design options explored within the EuroCirCol initiative (Schoerling et al. 2019) show that, by using a conductor with the FCC target performance, the maximum stress in the conductor reaches values in the range of 150 MPa during magnet assembly and of 200 MPa during operation at cold.

As anticipated above, Nb_3Sn coils are in general stiffer than non-impregnated Nb-Ti coils: for example, the average elastic modulus of the coils of the LHC magnets is about 10 GPa (Couturier et al. 2002), and the elastic modulus of the coils built so far for the Hi-Lumi project is in a range between 25–40 GPa. This makes the magnet design and the assembly tolerances of Nb_3Sn magnets more challenging than those of Nb-Ti magnets because, for a given geometric error or deformation, the associated stress variation increases considerably.

Finally, as the performance and integrity of Nb_3Sn is particularly sensitive to strain, the magnet assembly requires particular care in every detail, at the risk of potentially introducing a local discontinuity.

16.3.6 Quench Protection

According to the Hi-Lumi experience (Marinozzi et al. 2016), a conductor hot-spot temperature in the order of 350 K during a quench can be sustained by Nb_3Sn accelerator magnets without degradation. In reality, the epoxy-based impregnation system should allow the reaching of, for short periods, much higher temperatures, in the order of 450 K. In this direction, a study of the effects of high temperatures during quenches performed on a small Nb_3Sn racetrack magnet (Imbasciati et al. 2004) suggested that the temperature threshold could be increased to at least 400 K.

Concerning the protection aspects, with respect to Nb-Ti magnets the higher magnetic field produced by Nb_3Sn magnets engages larger stored energies, which have to be managed in the case of a quench, or a fault, at the magnet and at the circuit level. Considering as a reference that a quench is spread within a time delay of about 40 ms (20 ms for detection and 20 ms for having the coil quenched), the maximum voltage to ground during a quench of a typical EuroCirCol 16 T design option is of the order of 1 kV (Salmi et al. 2017), and the turn-to-turn voltage of the order of up to 100 V. An additional 1 kV should be added to this for the whole circuit, totaling the requirement of an operational dielectric insulation to ground of 2 kV. This is about twice the value of the LHC magnets, and certainly represents an important technological issue to keep in mind.

Both the Hi-Lumi magnets and the EuroCirCol options show that the use of quench heaters and the so-called coupling-loss-induced quench (CLIQ) system (Ravaioli et al. 2015), either in combination or alone, appears appropriate for ensuring reliable protection during a quench.

Concerning the integration of the magnet in a circuit string, a strong limitation comes from the maximum voltage to ground that one is ready to accept in case of a quench. A study performed for the FCC (Prioli et al. 2019) has shown that, if a single magnet can be protected, a string of magnets can also be protected within given constraints (in particular voltage withstand levels and the time constant of the circuit) by a proper subdivision of a powering sector in multiple circuits.

16.3.7 Conductor Grading

In a dipole cross-section the field amplitude is reduced over the coil width as the radius increases. This allows splitting the coil into layers, with the superconductor operating at increasing current densities thanks to the field decrease. This approach, called grading, allows an efficient use of conductor by moving the operating point of each graded layer closer to the conductor critical surface, at the expense of the complication of using and splicing different cables in a coil. Considering an Nb_3Sn conductor with the FCC target performance and at an operational temperature of 1.9 K, the interest in grading arises when the short sample peak field exceeds about 15 T (corresponding to a magnet with an operational field amplitude in the range of

13 T). For lower field levels the critical current density of the superconductor is already in a range (about 3000 A/mm^2) requiring a large fraction of copper for quench protection (two to three times the amount of superconductor), so that the coil size becomes no longer dominated by the conductor performance. For higher field amplitudes we consider the case of a 16 T magnet with 14% of margin on the load line: at the short sample peak field of 18.7 T and at an operating temperature of 1.9 K the FCC target superconductor can carry about 1200 A/mm^2. At this current density, as much copper as superconductor is required to ensure magnet protection in case of a quench. If the coil is graded at a magnetic field of around 3 T lower than the peak field, the critical current density has already doubled, requiring about twice the quantity of copper with respect to the quantity of superconductor. Additional grading levels at a lower field require a larger and larger fraction of copper to non-copper. Therefore, in terms of magnetic efficiency of the magnet cross-section, for a 16 T magnet with 14% margin there is a strong interest in having one or even two grading levels. A further increase of the number of grading levels is of limited interest, because the engineering current density of the conductor does not increase linearly with the number of grading levels. In terms of cost, the raw ingredients of the superconductor material do not currently represent the major component of the conductor cost, but the situation may change in the case of optimized conductor production, requiring that an optimal compromise be found between the technical complexity of additional grading and conductor cost.

Finally, we recall the opportunity of grading with different types of superconductor, towards lower or higher fields. An apparently "cheap" way to grade towards lower fields can be performed using Nb-Ti, which we can imagine being used at 1.9 K and, with some margin, at around 8 T. If this additional grading level is achieved through an internal splice this should, however, be compared to performing the same grading with Nb_3Sn or not performing it at all. At a field amplitude of 8 T plus margin, let us say 10 T, the critical current density of Nb_3Sn at 1.9 K is extremely high, exceeding 5000 A/mm^2: the cost of the superconductor in these conditions may become competitive with that of Nb-Ti.

16.3.8 Quench Performance

A superconducting magnet may not reach its operational field the first time it is powered due to the occurrence of a quench. In most cases, the quench current increases at each subsequent powering until a limiting maximum current (ideally the short sample limit) is reached, going through what is referred to as "training." Provided that the magnet is designed, manufactured, and protected to withstand the occurrence of quenches during its life, depending on the specific situation, we can accept, or not, a certain degree of training. For example, in the case of a single or a few units individually powered, we may decide to accept to perform a few training quenches once the magnet is installed and cooled down before operating the magnet in the facility, if the infrastructure is capable of performing the training campaign.

On the other hand, in the case of a large number of magnets connected in series, the probability that a single magnet unit needs a training quench before reaching its operational field should be minimized. Achieving this objective becomes increasingly expensive as the allowed probability that a magnet needs training quenches becomes smaller. Directions for decreasing the probability of needing training quenches to reach nominal field are: (1) a robust design (typically requiring long and expensive R&D programs); (2) an increased margin (requiring more conductor and also resulting in a larger magnet); and (3) the insertion of training campaigns within the magnet acceptance tests.

In the case where all magnets are individually trained, as should ideally be for a large accelerator, an important element to consider is the so-called "memory," i.e., the ability of a magnet to retain its performance achieved after training in cases where the magnet is submitted to a thermal cycle. A magnet with good memory will not train again once installed and operating in the particle accelerator. Ideally, the training campaign performed during the acceptance tests should include a "thermal cycle" to re-check the performance on a trained magnet after a warm-up and subsequent cool-down.

16.4 Present Development Programs

In these years, the development of the magnets for the Hi-Lumi project will for the first time demonstrate the use of Nb_3Sn magnets in a particle accelerator. To prepare for the next step at higher fields, towards a HE-LHC or a FCC, new R&D programs are being established in the US through the US Magnet Development Program (MDP) (Gourlay et al. 2016), and in Europe through the FCC 16 T development program and the European Circular Collider (EuroCirCol) study (Schoerling et al. 2015; Tommasini et al. 2018). These programs are tackling the main R&D issues in preparation for the large use of high-field Nb_3Sn magnets in a particle accelerator, from conductor development to the training performance and design options. Block type (Felice et al. 2019), common coil (Toral et al. 2018), cos-theta (Valente et al. 2018; Zlobin et al. 2018), and canted cos-theta (Caspi et al. 2017; Montenero et al. 2019) are being studied as design options for twin-aperture dipole magnets accessing the 16 T field range. As a magnet's cost is heavily dependent on the amount of conductor used, for a large number of magnets the cos-theta and block coil options would be preferable because, for the same margin on the load line, they are more efficient than the other two options. The favorable stress management of a canted cos-theta magnet or the regular simple coil geometry of a common-coil magnet may, however, allow these configurations to operate at a lower margin on the load line. This may possibly partially or totally compensate for their lower efficiency in terms of conductor use.

16.5 Conclusions

In the operational field range between 10 T and 16 T the most appropriate, and probably the only, conductor for producing a large series of accelerator magnets is Nb_3Sn. Its present performance is very close to what is ideally required, and its cost should still have a considerable margin of decrease if large-scale production is performed.

At this stage, an operational field range up to about 12 T is becoming reality in the Hi-Lumi project. The operational range between 12 T and 16 T has still to be explored and proven experimentally, showing that the outstanding challenges of stress management and training can be effectively overcome.

References

Baglin V, Lebrun P, Tavian L et al (2013) Cryogenic beam screens for high energy particle accelerators. In: Funaki K, Nishimura A, Kamioka Y et al (eds) Proceedings of the 24th international cryogenic engineering conference and international cryogenic materials conference 2012, Fukuoka, 14–18 May 2012. Cryogenics and Superconductivity Society of Japan, Tokyo p 629; CERN-ATS-2013-006

Ballarino A, Bottura L (2015) Targets for R&D on Nb_3Sn conductor for high energy physics. IEEE Trans Appl Supercond 25(3):1–6. https://doi.org/10.1109/tasc.2015.2390149

Bhat CM, Bhat PC, Chou W et al (2013) Proton–proton and electron–positron collider in a 100 km ring at Fermilab. In: Graf Norman A, Peskin Michael E, Rosner JL et al (eds) Frontier capabilities for hadron colliders, Snowmass, June 2013. https://arxiv.org/abs/1306.2369

Bottura L (2000) A practical fit for the critical surface of NbTi. IEEE Trans Appl Supercond 10 (1):1054–1057. https://doi.org/10.1109/77.828413

Caspi S, Arbelaez D, Brouwer L, Gourlay S, Prestemon S, Auchmann B (2017) Design of a canted-cosine-theta superconducting dipole magnet for future colliders. IEEE Trans Appl Supercond 27 (4):4001505. 2016-12-12. https://doi.org/10.1109/TASC.2016.2638458

CERN (2013) The European strategy for particle physics. CERN-Council-S/106, CERN, Brussels. https://cds.cern.ch/record/1567258/

Couturier K, Ferracin P, Todesco E et al (2002) Elastic modulus measurements of the LHC dipole superconducting coil at 300 K and at 77 K. In: Adams M, DiPirro M, Breon S et al (eds) AIP conference proceedings, Madison, Wisconsin, 16–20 July 2001, 613(1):377–382. https://doi.org/10.1063/1.1472044

Dai W, Ferrando V, Pogrebnyakov AV et al (2011) High-field properties of carbon-doped MgB_2 thin films by hybrid physical–chemical vapor deposition using different carbon sources. Supercond Sci Technol 24(12):125014. https://doi.org/10.1088/0953-2048/24/12/125014

Devred A, Baynham DE, Bottura L et al (2004) High field accelerator magnet R&D in Europe. IEEE Trans Appl Supercond 14(2):339–344. https://doi.org/10.1109/tasc.2004.829121

Felice H et al (2019) F2D2: a block-coil short-model dipole toward FCC. IEEE Trans Appl Supercond 29(5). https://doi.org/10.1109/TASC.2019.2897054

Ferracin P, Ambrosio G, Anerella M et al (2016) Development of MQXF: the Nb_3Sn Low-β quadrupole for the HiLumi LHC. IEEE Trans Appl Supercond 26(4):4000207. https://doi.org/10.1109/tasc.2015.2510508

Gao J (2016) Status of the CEPC project: Physics, accelerator and detector. In: Proceedings of the 38th international conference on high energy physics, vol 282, Chicago, 3–10 Aug 2016, SISSA, Trieste, PoS(ICHEP2016)038, p 5. https://doi.org/10.22323/1.282.0038

Gourlay S (2016) 1st review of the EuroCirCol WP5 Zurich, 11–13 May 2016. CERN, Geneva. https://indico.cern.ch/event/516049

Gourlay S (2017) 2nd review of the EuroCirCol WP5, Zurich, 9–10 Oct 2017. CERN, Geneva https://indico.cern.ch/event/661257

Gourlay SA et al (2016) The US magnet development plan. US Department of Energy, Office of Science, Washington, DC

Imbasciati L, Bauer P, Ambrosio G et al (2004) Study of the effects of high temperatures during quenches on the performance of a small Nb_3Sn racetrack magnet. Supercond Sci Technol 17(5): S389–S393. https://doi.org/10.1088/0953-2048/17/5/060

Izquierdo Bermudez S, Bottura L, Todesco E (2016) Persistent-current magnetization effects in high-field superconducting accelerator magnets. IEEE Trans Appl Supercond 26(4):1–5. https://doi.org/10.1109/tasc.2016.2519006

Kashikhin VV, Mokhov NV, Alexahin Y et al (2012) High-field combined function magnets for a 1.5 × 1.5 TeV Muon Collider Storage Ring. In: Corbett J, Eyberger C, Morris, K et al (eds) 3rd international particle accelerator conference (IPAC 2012), New Orleans, 20–25 May 2012, pp 3587–3589

Kashikhin VV, Zlobin AV (2016) Persistent current effect in 15–16 T Nb_3Sn accelerator dipoles and its correction. In: Power M, Shiltsev V, Schaa VRW et al (eds) Proceedings of NAPAC2016, Chicago, 9–14 Oct 2016, pp 1061–1063. https://doi.org/10.18429/JACoW-NAPAC2016-THA1CO04

Larbalestier D (2017) Superconductors for HEP use in the next few (10?) years (talk). In: Gourlay S, Velev G (chairs) (eds) 1st US magnet development program meeting, Napa, 6–8 Feb 2017. https://conferences.lbl.gov/event/73/session/3/#20170206

Marinozzi V, Ambrosio G, Ferracin P et al (2016) Quench protection study of the updated MQXF for the LHC luminosity upgrade (HiLumi LHC). IEEE Trans Appl Supercond 26(4):1–5. https://doi.org/10.1109/tasc.2016.2523548

Montenero G et al (2019) Coil manufacturing process of the first 1-m-long canted–cosine–theta (CCT) model magnet at PSI. IEEE Trans Appl Supercond. https://doi.org/10.1109/TASC.2019.2897326

Prioli M, Auchmann B, Bortot L et al (2019) Conceptual design of the FCC-hh dipole circuits with integrated CLIQ protection system. IEEE Trans Appl Supercond, accepted for publication

Ravaioli E, Bajas H, Datskov VI et al (2015) Protecting a full-scale Nb_3Sn magnet with the new coupling-loss-induced quench system. IEEE Trans Appl Supercond 25(3):1–5. https://doi.org/10.1109/tasc.2014.2364892

Rossi L (2003) Superconducting cable and magnets for the Large Hadron Collider. In: Andreone A, Pepe GP, Cristiano R et al (eds) Proceedings of the 6th european conference on applied superconductivity, Sorrento, 14–18 Sep 2003, pp 261–268

Salmi T, Prioli M, Stenvall A et al (2017) Suitability of different quench protection methods for a 16 T block-type Nb_3Sn accelerator dipole magnet. IEEE Trans Appl Supercond 27(4):1–5. https://doi.org/10.1109/tasc.2017.2651386

Savary F, Bajko M, Bordini B et al (2017) Progress on the development of the Nb_3Sn 11T dipole for the high luminosity upgrade of LHC. IEEE Trans Appl Supercond 27(4):1–5. https://doi.org/10.1109/tasc.2017.2666142

Scanlan RM (2001) Conductor development for high energy physics-plans and status of the US program. IEEE Trans Appl Supercond 11(1):2150–2155. https://doi.org/10.1109/77.920283

Schoerling D, Bajas H, Bajko M et al (2015) Strategy for superconducting magnet development for a future hadron-hadron circular collider at CERN. In: European Physical Society conference on high energy physics, Vienna, 22–29 July 2015, EPS-HEP2015, p 517. https://doi.org/10.22323/1.234.0517

Schoerling D, Durante M, Lorin C et al (2017) Considerations on a cost model for high-field dipole arc magnets for FCC. IEEE Trans Appl Supercond 27(4):1–5. https://doi.org/10.1109/tasc.2017.2657510

Schoerling D, Areleaez D, Auchmann B et al (2019) The 16 T dipole development program for FCC and HE-LHC. IEEE Trans Appl Supercond. https://doi.org/10.1109/TASC.2019.2900556

Sorbi M (2017) The EuroCirCol 16 T cosine-theta dipole option for the FCC. IEEE Trans Appl Supercond 27(4):1–5. https://doi.org/10.1109/tasc.2016.2642982

Todesco E (2017) Performance of the LHC magnets and margin (talk), In: FCC Week 2017, Berlin, 29 May–2 June 2017. https://indico.cern.ch/event/556692

Todesco E, Zimmermann F (eds) (2011) Proceedings of the EuCARD-AccNet-EuroLumi workshop, Malta, 14–16 Oct 2010, CERN Yellow Report. CERN, Geneva http://cds.cern.ch/record/1344820

Todesco E, Bottura L, Giovannozzi M et al (2016) The magnetic model of the LHC at 6.5 TeV. IEEE Trans Appl Supercond 26(4):4005707. https://doi.org/10.1109/TASC.2016.2549579

Tommasini D, Auchmann B, Bajas H et al (2017) The 16 T dipole development program for FCC. IEEE Trans Appl Supercond 27(4):1–5. https://doi.org/10.1109/tasc.2016.2634600

Tommasini D, Arbelaez D, Auchmann B et al (2018) Status of the 16 T dipole development program for a future hadron collider. IEEE Trans Appl Supercond 28(3):4001305. https://doi.org/10.1109/TASC.2017.2780045

Toral F, Munilla J, Salmi T (2018) Magnetic and mechanical design of a 16 T common coil dipole for an FCC. IEEE Trans Appl Supercond 28(3):1–5 4004305. https://doi.org/10.1109/TASC.2018.2797909

Valente R, Bellomo G, Pasquale F et al (2018) Electromagnetic design of a 16 T cos θ bending dipole for the future circular collider. IEEE Trans Appl Supercond. https://dx.doi.org/10.1109/TASC.2019.2901604

Xu X, Peng X, Rochester J, Sumption M, Tomsic M (2019) Achievement of FCC specification in critical current density for Nb_3Sn superconductors with artificial pinning centers. https://arxiv.org/abs/1903.08121

Zlobin AV, Alexahin YI, Kapin VV et al (2013) Preliminary design of a Higgs factory μ+μ− storage ring. In: 4th international particle accelerator conference (IPAC 2013), Shanghai, 12–17 May 2013, pp 1487–1489

Zlobin AV, Carmichael J, Kashikhin I et al (2018) Conceptual design of a 17 T Nb_3Sn accelerator dipole magnet. In: 9th international particle accelerator conference (IPAC 2018), Vancouver, pp 2742–2744

Index

D. Schoerling, A. V. Zlobin (eds.), *Nb_3Sn Accelerator Magnets*, Particle Acceleration and Detection, https://doi.org/10.1007/978-3-030-16118-7

D

E

F

www.ingramcontent.com/pod-product-compliance
Lightning Source LLC
LaVergne TN
LVHW082108200926
848484LV00008B/1927
9781013271359